TechConnect Briefs 2015

MATERIALS FOR ENERGY, EFFICIENCY AND SUSTAINABILITY

Edited by
Matthew Laudon Bart Romanowicz

TechConnect

CRC Press
Taylor & Francis Group
Boca Raton London New York

CRC Press is an imprint of the
Taylor & Francis Group, an **informa** business

First published 2015 by TechConnect and CRC Press
Taylor & Francis Group
6000 Broken Sound Parkway NW, Suite 300
Boca Raton, FL 33487-2742

ISBN 13: 978-1-4987-4728-8 (pbk)

DOI: 10.1201/9780138749487

Visit the Taylor & Francis Web site at
http://www.taylorandfrancis.com

and the CRC Press Web site at
http://www.crcpress.com

CHAPTER EDITORS - 2015 TechConnect Briefs

Biofuels & Bioproducts

Scott Twary, *Los Alamos National Laboratory, USA*
Srinivas Iyer, *Los Alamos National Laboratory, USA*

ELECTRONICS & MICROSYSTEMS

MEMS & NEMS Devices and Applications

William (Cy) Wilson, *NASA Langley Research Center, USA*

Flexible Electronics

Mandakini Kanungo, *Xerox Innovation Group, USA*

Photonic Materials & Devices

Loucas Tsakalakos, *GE Global Research, USA*

Nano Electronics for 2020 and Beyond

Aaron D. Franklin, *Duke University, USA*
Gernot Pomrenke, *USAF/AFOSR/Optoelectronics and Photonics, USA*
Curt Richter, *NIST, USA*
Charles Ying, *NSF, USA*

Micro, Nano & Bio Fluidics

Edward Furlani, *University at Buffalo SUNY, USA*
Kwang Oh, *University at Buffalo SUNY, USA*

Inkjet Design Materials and Fabrication

Chris Menzel, *Fujifilm Dimatix, Inc., USA*

ENERGY & SUSTAINABILITY

Energy Storage

Imre Gyuk, *U.S. Department of Energy, USA*

Water Technologies

Armin Volkel, *Palo Alto Research Center, USA*

Advanced Materials for Oil & Gas

Willard C. (Bill) Capdevielle, *Oil Patch Engineering, PLLC, USA*
Rushan Sakhibgareev, *Chevron, USA*

Solar Power Technologies

Brian Korgel, *University of Texas at Austin, USA*
Heath Naquin, *University of Texas at Austin, USA*

NanoNuclear Materials, Fuels & Applications

Brent Segal, *Lockheed Martin, USA*
Thomas McKrell, *MIT, USA*
Steven Winston, *Steven Winston & Assoc., USA*

Carbon Capture

Robert Quinn, *Air Products and Chemicals (retired), USA*
Elizabeth Burton, *Lawrence Berkeley National Laboratory, USA*

Materials for Green Building

Elena Enache-Pommer, *Dow Building Solutions R&D, USA*

Key Elements for Successful Commercialization: Safety and Sustainability

Charles Geraci, Jr., *NIOSH, USA*
Phillip Demokritou, *Harvard University, USA*
Sally Tinkle, *Science and Technology Policy Management, USA*

PERSONAL & HOME CARE, COSMETICS & FOOD

Food Materials

Pat Spicer, *University of New South Wales, AU*
Krassimir Velikov, *Unilever R&D, NE*

Materials for Personal and Home Care

Fiona Case, *Nano Science and Technology Institute, USA*

TABLE OF CONTENTS

Carbon Capture & Utilization

Solar Power Technologies

Materials for Green Building

Water Technologies

Materials for Sustainability & Efficiency

PREFACE

TechConnect provides a multi-disciplinary and multi-sector forum focused on accelerating applied research and early-stage innovation towards commercialization. The technologies described in the TechConnect Briefs are provided by innovators working in industry, academic and government laboratories around the world. A review committee comprising more than 300 internationally recognized experts reviewed the research and innovations described and selected them for inclusion in this volume, and for presentation at the 10th annual TechConnect World Innovation Conference & Expo, held jointly with the 18th annual Nanotech Conference & Expo, and the 2015 National SBIR/STTR Conference, June 14-17, 2015 Washington DC.

Through TechConnect prospecting programs, transformational innovations are identified in Advanced Materials, Advanced Manufacturing, Materials for Energy & Sustainability, Electronics & Microsystems, Biotech, Medical & Pharma, Personal & Home Care, Cosmetics, and Food industries. The 2015 TechConnect Briefs are also aligned with the National Nanotechnology Initiative and the National Security Technology Accelerator interest areas.

This volume (one of 4) is focused on research and innovations in the Materials for Energy, Efficiency and Sustainability track, including industrially viable new materials that will transform industry and new platform technologies that enable their discovery and development. The following chapters are included:

- Energy Storage
- Fuel Cells & Hydrogen
- Nanomaterials for Catalysis
- Materials for Oil, Gas & Biofuels
- Carbon Capture & Utilization
- Solar Power Technologies
- Materials for Green Building
- Water Technologies
- Materials for Sustainability & Efficiency

It is our distinct honor to edit the 18th annual TechConnect publication of applied research technologies at the forefront of innovation change for industry and society. The TechConnect Program Committees make every effort to provide a scientifically and commercially relevant publication. The grassroots review and ranking process is a source of pride for these joint committees, and provides a yearly evaluation and technical validation of emerging technologies addressing our world's most pressing needs. We would like to take this opportunity to thank the many individuals who have worked so hard to make this annual publication possible. We hope that you enjoy these TechConnect Briefs, and we hope that the presented technologies are significant steps towards addressing large global challenges.

Matthew Laudon
Austin, Texas

Bart Romanowicz
Austin, Texas

Fiona Case
San Diego, California

Program Review Committee

Wade Adams, *Rice University, USA*
Susan Addo Ntim, *U.S. Food & Drug Administration, USA*
I. Yucel Akkutlu, *Texas A&M University, USA*
Evangelyn Alocilja, *Michigan State University, USA*
Jay Amarasekera, *Sabic, USA*
Mansoor Amiji, *Northeastern University, USA*
Thomas Anchordoquy, *University of Colorado, USA*
Mohamed Ansari, *Universiti Tenaga Nasional Kajang, Malaysia*
Sivaram Arepalli, *Sungkyunkwan University, Korea*
Wolfgang Bacsa, *University of Toulouse, France*
Stephen Bart, *Analog Devices, Inc., USA*
S Rahima Benhabbour, *University of North Carolina at Chapel Hill, USA*
Gary Bernstein, *University of Notre Dame, USA*
Jean Berthier, *CEA Grenoble, France*
Mousumi Mani Biswas, *Western Digital, USA*
Gregory R. Bogart, *Sandia National Laboratories, USA*
Philip Bradford, *North Carolina State University, USA*
Jakob Bredsguard, *Biosyntehtic Technologies, USA*
Elizabeth Burton, *Lawrence Berkeley National Laboratory, USA*
Ahmed Busnaina, *Northeastern University, USA*
Willard C. (Bill) Capdevielle, *Oil Patch Engineering, PLLC, USA*
Fiona Case, *Case Scientific, USA*
Martin Case, *Janssen, Pharmaceutial Companies of Johnson & Johnson, USA*
Mansun Chan, *Hong Kong University of Science & Technology, China*
Alessandro Chiolerio, *Politecnico di Torino, Italy*
Bum-Kyoo Choi, *Sogang University, Korea*
Schuyler Corry, *Life Technologies, USA*
Lance Criscuolo, *Zyvex Technologies, USA*
Jaime Jose Cubillo Capuz, *Acciona, Spain*
Zheng Cui, *Suzhou Institute of Nanotech, Chinese Academy of Sciences, China*
Phil Demokritou, *Harvard University, USA*
Guy DeRose, *California Institute of Technology, USA*
George Elvin, *Green Technology Forum, USA*
Dorothy Farrell, *NCI Alliance for Nanotechnology in Cancer, USA*
Aaron Franklin, *Duke University, USA*
Nicolas Feltin, *Laboratoire national de métrologie et d'essais, France*
Thomas Fengyuan Li, *Freescale Semiconductor, USA*
Lisa Friedersdorf, *National Nanotechnology Coordination Office, USA*
Ian Fuller, *Nanotek Instruments, USA*
Edward Furlani, *University at Buffalo SUNY, USA*
Charles Geraci, *Jr., NIOSH, USA*
Harris Goldberg, *InMat, Inc., USA*
Imre Gyuk, *U.S. Department of Energy, USA*
Greg Haugstad, *University of Minnesota, USA*
Tim Hebrink, *3M Corporate Research Process Laboratory, USA*
Michael Helmus, *Consultant / Synogen, Inc., USA*
Andreas Hieke, *Stanford University, USA*
Bin Hu, *University of Tennessee, USA*
Zhaohua Huang, *University of California, San Francisco, USA*
Kiki Ikossi, *Defense Threat Reduction Agency, USA*
Srinivas Iyer, *Los Alamos National Laboratory, USA*
Mark Jack, *Florida A&M University, USA*
Oana Jurchescu, *Wake Forest University, USA*
Mandakini Kanungo, *Xerox, USA*
Barbara Karn, *National Science Foundation, USA*
Brian Korgel, *University of Texas at Austin, USA*

Jana Kukutschova, *Technical University Ostrava, Czech Republic*
James Lamb III, *Brewer Science, USA*
William A Lambert,i *ExxonMobil, USA*
David Lashmore, *Nanocomp Technologies, Inc., USA*
Matthew Laudon, *TechConnect, USA*
Kock-Yee Law, *Research and Innovative Solutions, USA*
Stephen Lehrman, *National Nanotechnology Coordination Office, USA*
James Alexander Liddle, *National Institute of Standards & Technology, USA*
Igor Linkov, *U.S. Army Engineer Research and Development Center, USA*
Yuri Lyubchenko, *University of Nebraska, USA*
Katrin Mackenzie, *Helmholtz Centre for Environmental Research - UFZ, Germany*
Michael Makowski, *PPG Industries R&D, USA*
Stuart Maloy, *Los Alamos National Laboratory, USA*
Anil Mane, *Argonne National Laboratory, USA*
Larry Mann, *Shoei Electronic Materials, Inc., USA*
Thomas McKrell, *Massachusetts Institute of Technology, USA*
Scott McNeil, *National Cancer Institute, USA*
Joey Mead, *University of Massachusetts Lowell, USA*
Michael Meador, *NASA Glenn Research, USA*
Chris Menzel, *Dimatix, Inc., USA*
Meyya Meyyappan, *NASA Ames Research, USA*
Stephen Miller, *Northwestern University Medical School, USA*
Moein Moghimi, *University of Copenhagen, Denmark*
Ron Munson, *Global CCS Institute, Australia*
Chang-Yong Nam, *Brookhaven National Laboratory, USA*
Andrzej Napieralski, *Technical University of Lodz, Poland*
Heath Naquin, *University of Texas at Austin, USA*
Tinh Nguyen, *National Institute of Standards & Technology, USA*
Patricia Nieva, *University of Waterloo, CA*
Greg Nordin, *Brigham Young University, USA*
Kwang Oh, *University at Buffalo SUNY, USA*
Pierre Panine, *Xenocs SA, France*
Hyung Gyu Park, *ETH Zurich, Switzerland*
Anil Patri, *National Institutes of Health, USA*
Joachim Piprek, *NUSOD Institute, USA*
Elena Polyakova, *Graphene 3D Lab, USA*
Gernot Pomrenke, *U.S. Air Force, USA*
Sotiris Pratsinis, *Swiss Federal Institute of Technology, Switzerland*
Serge Prudhomme, *University of Texas at Austin, USA*
Robert Quinn, *Air Products and Chemicals, USA*
YuanQiao Rao, *The Dow Chemical Company, USA*
Robin Ras, *Aalto University School of Science, Finland*
Somilkumar Rathi, *Intel Corporation, USA*
Mark Reed, *Yale University, USA*
Erik Reimhult, *Universität für Bodenkultur Wien, Austria*
Curt Richter, *National Institute of Standards & Technology, USA*
Bart Romanowicz, *TechConnect, USA*
Robert Rudd, *Lawrence Livermore National Laboratory, USA*
Richard Sadus, *Swinburne University of Technology, Australia*
Rushan Sakhibgareev, *Chevron, USA*
Robert Savinell, *Case Western Reserve University, USA*
Derek Schorzman *Liquidia Technologies, Inc., USA*
David Schut, *Shoei Electronic Materials, Inc,. USA*
Keana Scott, *National Institute of Standards and Technology, USA*
Brent Segal, *Lockheed Martin, USA*
Philippe Serp, *Institute National Polytechnique de Toulouse, France*
Neeraj Sharma, *3M, USA*
Emilie Siochi, *NASA Langley Research Center, USA*

CONFERENCE OPERATIONS MANAGER

TechConnect Briefs 2015 Volumes

ADVANCED MATERIALS:

TechConnect Briefs 2015, Vol. 1, ISBN: 978-1-4987-4727-1:

1. Nanoscale Materials Characterization
2. Informatics, Modeling & Simulation
3. Carbon Nano Structures & Devices
4. Graphene
5. Nanoparticle Synthesis & Applications
6. Nanocomposite Materials
7. Advanced Materials for Engineering Applications
8. Coatings, Surfaces & Membranes
9. Polymers
10. Soft Matter & Colloids
11. Nanotech Commercialization: Safety & Sustainability

MATERIALS FOR ENERGY, EFFICIENCY AND SUSTAINABILITY:

TechConnect Briefs 2015, Vol. 2, ISBN: 978-1-4987-4728-8:

1. Energy Storage
2. Fuel Cells & Hydrogen
3. Nanomaterials for Catalysis
4. Materials for Oil, Gas & Biofuels
5. Carbon Capture & Utilization
6. Solar Power Technologies
7. Materials for Green Building
8. Water Technologies
9. Materials for Sustainability & Efficiency

BIOTECH, BIOMATERIALS AND BIOMEDICAL:

TechConnect Briefs 2015, Vol. 3, ISBN: 978-1-4987-4729-5:

1. Materials for Drug & Gene Delivery
2. Cancer Nanotechnology
3. Sensors, Diagnostics & Imaging
4. Micro, Nano & Bio Fluidics
5. Biomaterials
6. Materials for Personal/Home Care & Cosmetics
7. Soft Matter & Colloids
8. Food Materials

ADVANCED MANUFACTURING, ELECTRONICS AND MICROSYSTEMS:

TechConnect Briefs 2015, Vol. 3, ISBN: 978-1-4987-4730-1:

1. Advanced Manufacturing
2. Sustainable Nanomanufacturing
3. NanoFab: Manufacturing & Instrumentation
4. Flexible Electronics
5. Photonic Materials & Devices
6. Nanoelectronics, Materials & Devices
7. Modeling & Simulation of Microsystems
8. MEMS & NEMS Devices & Applications
9. Micro, Nano & Bio Fluidics
10. Inkjet Design, Materials & Fabrication

TECHCONNECT BRIEFS 2015: FOUR VOLUME SET

ISBN: 978-1-4987-4731-8 (hardcopy)

TECHCONNECT BRIEFS 2015: FOUR VOLUME SET

ISBN: 978-1-4987-4733-2 (DVD)

Inclusion of Electromagnetic Stress to explain Lithiated Silicon Nanowires

Donald C Boone*

***-Nanoscience Research Institute - College Park, Maryland**

Abstract: This study considers the electromagnetic stresses and simulates the lithium insertion into a silicon nanowire. The resulting model uses a multi-physics approach to explain the two detrimental effects that could result during the lithiated silicon process: (1) The partial lithiation effects that are observed in some silicon nanowires under negligible volume expansion; (2) The excessive volume expansion that is observed after full lithium insertion with a resulting Cassini oval shaped silicon nanowire. Magnetic and electric fields are included in this computational model in order to introduce additional tensile and compressive stresses and as a result alters the band structure of the silicon nanowire as it transition from crystalline silicon (c-Si) structure to an amorphous lithiated silicon (a-LiSi) heterogeneous material.

1. The Continuum Model

The computational model that will be presented is based on the elastic-diffusion model of a silicon spherical nanoparticle that was presented in Golmon et-al. [1]. This model described the lithiation process from an electrochemical point of view. Lithium ions diffused through the crystalline silicon (c-Si) lattice, breaking silicon-silicon bonds and reforming lithium-silicon bonds, creating tensile stress throughout the nanoparticle while simultaneously Li-Si alloy material concentration builds up a compressive stress uniformly within the nanoparticle shell. Although the elastic-diffusion model does explain the volume expansion, it fails to explain for the slowing of the interfacial reaction front which is the interface between the c-Si and a-LiSi alloy that causes partial lithiation within the Si nanowire as was reported by Liu et-al. [2]. Therefore, an extension of this model is performed on a cylindrical silicon nanowire with the addition of electromagnetic theory. In the computational model or simulation that was performed, a series of partial differential equations (PDE) and trail functional (TF) were utilized to determine the effects of four solutions that comprised the varies stresses within our silicon nanowire model. The solutions of PDE with the use of TF are: 1) Elastic Displacement **u**, 2) Concentration $\overline{c}$, 3) Magnetic Field **B**, and 4) Electric Field **E** which culminates in radial stress equation

$$\boldsymbol{\sigma}_{\mathbf{r}}=\bar{\mathbf{E}}\left[\frac{v}{2}\left(\frac{\mathbf{u}}{\mathbf{R_c}}\right)-\frac{\Omega}{6}\overline{c}\,(1+v)+\frac{\mathbf{aB}}{\rho\mathbf{P}}(1-2v)\right] \quad (1)$$

where $\mathbf{P}=\epsilon_0\chi\mathbf{E}$ **(2)**

where $\bar{\mathbf{E}}$ is the modulus of elasticity, (v) is the Poisson constant, (a) is the lattice constant and $\mathbf{R_c}$ is the radius of curvature.

2. The Quantum Mechanical Model

Unlike continuum mechanics to describe the lithiation process in terms of varies stresses where a computer generated model was created from a large group of differential equations, the quantum mechanics approach that will be presented will use a heuristic method in examining the lithiated silicon process. Prior to the beginning of the lithiation process, there is no electron flow through the silicon nanowire. We will start the description of our model from the quantum mechanical perspective by developing the band structure of our pre-lithiated silicon nanowire. Silicon, which is a semiconductor material, is made up of band structures composed of valence and conduction bands which are separated by indirect band gap energy $\mathbf{E_g}$. Silicon is a diamond cubic crystalline structure made up of tetrahedral molecules with its hybridized $\mathbf{sp^3}$ orbitals within their valence shells filled with covalent bonding electrons from neighboring silicon atoms. Because the valence orbitals are completely filled, electron are not able to move freely within the crystal until a voltage is applied to the band structure. For silicon material, electrons within the valence band needs a minimum band gap energy of 1.1 electron volts in order to transition by way of trans-versing the band gap from the valence band to conduction band to create current flow. The Schrödinger equation is used to calculate the band structure by solving this eigenvalue equation in order to derive the band energies and wavefunctions of the individual electrons. For our model, we will use the Time-Dependent Schrödinger equation:

$$\mathbf{ih}\frac{\partial\Psi}{\partial t}=\mathrm{H}\Psi \quad \mathbf{(3)}$$

The Hamiltonian is defined as

$$H = H_o + H_p \quad \textbf{(4)}$$

The total Hamilton (H) is composed of an initial or unperturbed Hamiltonian H_o that represents the state of the silicon nanowire before the start of lithiation and a perturbed Hamiltonian H_p that define the additional energy the nanowire will be subjected to by electromagnetic field. The unperturbed Hamiltonian H_o is defined as:

$$H_o = \frac{\hbar^2}{2m}\nabla^2 + V(r) + \frac{\hbar}{4m^2c^2}\frac{1}{\hat{\mathbf{R}}}\cdot \boldsymbol{\nabla} V\, \hat{\mathbf{S}}\cdot\hat{\mathbf{L}} \quad \textbf{(5)}$$

where $\hat{\mathbf{S}}$ **and** $\hat{\mathbf{L}}$ are defined as the spin and angular momentum operators respectively. The spin operator $\hat{\mathbf{S}}$ are the Pauli matrices that are commonly used in quantum mechanics. The mass (m) is defined as the rest mass of the electron, h-bar $(\bar{h})$ is the Planck's constant, **c** the velocity of light and $\hat{\mathbf{R}}$ is the position operator. The potential V(r) is a sum of several terms that are the interactions of the constitutive particles of this nanowire model namely silicon, lithium and electrons. The overall electric potential is defined as:

$$V(r) = V_{e-e} + V_{e-Si} + V_{e-Li} + V_{Si-Si} + V_{Si-Li} + V_{Li-Li} \quad \textbf{(6)}$$

The unperturbed Hamiltonian H_o is used in the **k·p** theory which is a mathematical method used to calculate the band structure in semiconductor materials such as silicon. In equation 5, the first term is the kinetic energy term (the **k·p** method formulates this term as being positive instead of negative as is the case in most Hamiltonians in Schrödinger equation), the second term as was mention is V(r) and the third term is the spin-orbital interaction term responsible for the energy that is generated from the interaction between the particles spin and orbital angular momentum. Figure 1 illustrates the conduction band structure due to H_o prior to the application of the electromagnetic fields at the moment lithium insertion begins. The band structure is crystalline silicon (c-Si) with an indirect band gap energy which are represented by the dispersion relation equations that are the result of the solution of the Schrödinger equation using the unperturbed Hamiltonian H_o.

Minimum conduction band

$$E_c = E_{k0} + \frac{\bar{h}^2k^2}{2m} + \frac{\bar{h}^2}{E_{Si}m^2}\sum_n |\langle u_c|\mathbf{k}\cdot\mathbf{p}|u_v\rangle|^2 \quad \textbf{(7)}$$

and the maximum valence band

$$E_v = -\frac{\bar{h}^2k^2}{2m} \quad \textbf{(8)}$$

where $E_{k0} = E_{k0}(0,t)$ is an energy coefficient, **k** is the wavevector, $\hat{\mathbf{p}}$ the momentum operator, $\mathbf{u_c}$ and $\mathbf{u_v}$ are gaussian functions for the conduction and valence bands respectively, and E_{Si} is the band gap energy of silicon. This as was stated previously is a constant at 1.1 eV. From equations 7 the effective mass $\mathbf{m_{eff}}$ can be derived as

$$\frac{1}{m_{eff}} = \frac{1}{m} + \frac{2}{E_g m^2}\sum_n |\langle u_c|\mathbf{k}\cdot\mathbf{p}|u_v\rangle|^2 \quad \textbf{(9)}$$

and $\mathbf{E_g}$ is in general the band gap energy.

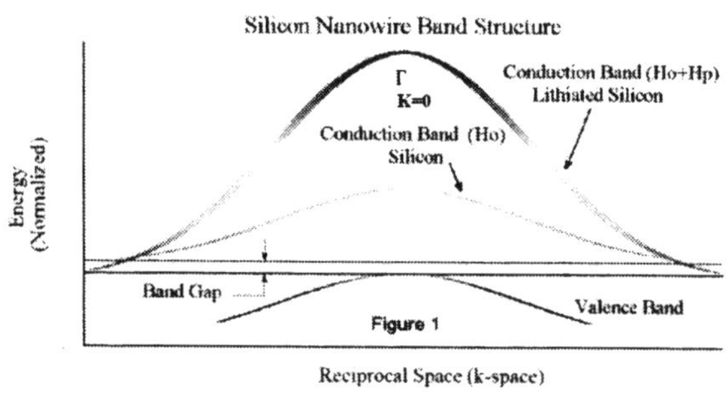

Figure 1

The perturbed Hamiltonian $\mathbf{H_p}$ is the energy added to the silicon nanowire the moment lithiation process begins. $\mathbf{H_p}$ is defined as:

$$H_p = -\frac{e}{2\,m_{eff}c}\mathbf{B}\cdot\hat{\mathbf{L}} + \frac{e^2}{8\,m_{eff}c}[\mathbf{B}^2\mathbf{r}^2 - (\mathbf{B}\cdot\mathbf{r})^2] + \frac{e^2E^2}{2\,m_{eff}\omega^2} \quad \textbf{(10)}$$

where (e) is the electron charge and (ω) is the angular frequency of the electric field. In our computational continuum model, the applied electric field **E** and magnetic field **B** magnitudes are of a small scale to a degree that the electric field **E** is just above the threshold to allow electrons from the silicon to transition from valence band to the conduction band. The magnitude of the applied magnetic field **B** is several orders below the electric field **E**. When the lithiation process begins, this initiates a process of transforming from crystalline silicon (c-Si) to an

TechConnect Briefs 2015, TechConnect.org, ISBN 978-1-4987-4728-8

amorphous lithiated silicon (a-LiSi) material. Lithium in its natural metallic crystalline form has electrons in its conduction band and therefore an energy band gap of zero. As lithium atoms diffuses into the silicon structure, the silicon band gap transition to a band gap that is representative of lithiated silicon. As lithium becomes the predominated material within the structure, the newly formed a-LiSi material band gap approaches zero. This transforms the effective mass of the electron from that of c-Si prior to lithiation to a value that of a-LiSi which also approaches zero. As a result, the conduction band energy at the gamma point Γ at k=0 of lithiated silicon (a-LiSi) increases while at the same time the slope on the curve increases as illustrated in figure 1.

3. Continuum Model vs Quantum Model

The perturb Hamiltonian in equation 10 is analogous to the third term of the radial stress equation in equation (1).

$$\boldsymbol{\sigma}_{\mathbf{em}} = \bar{\mathbf{E}}\left[\frac{\mathbf{aB}}{\rho\epsilon_0\chi\mathbf{E}}(1-2\mathrm{v})\right] \quad \mathbf{(11)}$$

$$\mathrm{H_p} = -\frac{\mathrm{e}}{2\,\mathrm{m_{eff}c}}\mathbf{B}\cdot\hat{\mathbf{L}} + \frac{\mathrm{e}^2}{8\,\mathrm{m_{eff}c}}[\mathbf{B}^2\mathbf{r}^2 - (\mathbf{B}\cdot\mathbf{r})^2] + \frac{\mathrm{e}^2\mathrm{E}^2}{2\,\mathrm{m_{eff}}\omega^2} \quad \mathbf{(12)}$$

In equation 11 the resistivity parameter ρ approaches zero as lithium atoms diffuses into the c-Si nanowire which is analogous to the effective mass approaching zero as the band gap narrows due to lithium diffusion.

Continuum Model:

Resistivity $\rho \longrightarrow 0$ (approaches zero)

Quantum Model:

Band Gap Energy $E_g \longrightarrow 0$

Effective Mass $m_{eff} \longrightarrow 0$

The partially lithiated silicon nanowire from the viewpoint of the continuum model, as the resistivity ρ decreases due to lithium insertion, a compressive electromagnetic stress is generated. The current is minute and constant at 1 nA which results in the magnetic **B** and electric **E** fields close to zero [2]. In addition **E** and **B** < ρ. As the resistivity approaches zero due to lithium diffusion, the resulting electromagnetic stress is small in comparison to a much larger stress observed when the silicon nanowire is fully lithiated and excessive volume expansion is the outcome. Although the resistivity for the lithiated silicon is approaching zero, the electron flow or current density is extremely small which results in an electromagnetic stress in the order of $\mathbf{10^2}$ smaller in magnitude when the nanowire is fully lithiated with a volume 300% larger in size then the undeformed nanowire (figure 2). Similarly, from the quantum mechanics point of view, as the band gap between the conduction and valence bands become narrower due to lithium diffusion, this decreases the effective mass, however as with the continuum model, the magnetic field and electric field are extremely small and **E** and **B** < $\mathbf{m_{eff}}$. Therefore the perturbed Hamiltonian $\mathrm{H_p}$ approaches zero.

In the case of the fully lithiated silicon nanowire, when analyzing the lithiation process from a continuum perspective, an applied constant voltage of 2-V causes an increase in the current density as the resistivity ρ decreases from the result of lithium diffusion [3]. This results in ρ < **E** and **B**. The resulting electron flow that is several orders of magnitudes larger than that of the partially lithiated nanowire, results in an increase in the electromagnetic stress which is the cause of the excessive volume expansion of the nanowire. From the quantum mechanics perspective, the magnetic **B** and electric **E** fields grows in magnitude due to an increase in electron flow from lithium diffusing into the silicon nanowire. As a result the band gap decreases and the perturbed Hamiltonian increases from the fact that $\mathbf{m_{eff}}$ < **E** and **B**. The lithiated silicon conduction band structure shift upwards along the energy axis as noted in figure 1. This results in greater momentum and kinetic energy for the individual electrons as they transition from the valence band of the lithiated silicon molecules to the conduction band.

4. Electron Scattering

Up until now our discussion on the quantum mechanical model has focus solely on the Hamiltonian. We now state the other part of the Schrödinger equation, the wavefunctionΨ. Each individual electron wavefunction will be defined prior to collision with an atom as

$$\boldsymbol{\Psi}(\mathbf{r},\boldsymbol{\theta},\boldsymbol{\phi},\mathbf{t}) = \sum_{\mathbf{k}}\mathbf{c_k}\left[\mathbf{u_{inc}}(\mathbf{r})\mathbf{e}^{\mathbf{ikr}} + \mathbf{u_{sc}}(\boldsymbol{\theta},\boldsymbol{\phi})\frac{\mathbf{e}^{\mathbf{ikr}}}{\mathbf{r}}\right]\mathbf{e}^{-\frac{\mathbf{iE_k t}}{\hbar}} \quad (13)$$

The wavefunction consist of two terms, the first term is known as the incident wavefunction and the second term is called the scattering wavefunction. When an electron is traveling in the conduction band the first term of the wavefunction describes the electron from a point before impact with the oncoming atom (incident wavefunction) and the second term of the wavefunction describes the electron after impact with the colliding atom (scattering wavefuction). The second term of the wavefunction dissipates with the inverse of the radial distance r^{-1} which begins from the point of the scattering electron with the atom. The current density is calculated with the use of the wavefunction

$$\mathbf{J} = -\frac{i\hbar}{2m_r}(\Psi^*\nabla\Psi - \Psi\nabla\Psi^*) \quad \mathbf{(14)}$$

where $\mathbf{m_r}$ is defined as the reduced mass between the electron and the colliding atom. The current densities of the incident and scattering parts of the wavefunction are determined separately. If the scattering current density is solved by their individual spherical components - $\mathbf{J_r}, \mathbf{J_\theta}, \mathbf{J_\phi}$ - it is discovered that the radial current density component $\mathbf{J_r}$ disperses at a rate of $\mathbf{r^{-2}}$ while both of the angular current density components, $\mathbf{J_\theta}, \mathbf{J_\phi}$, dissipates at a rate of $\mathbf{r^{-3}}$

Incident current density:

$$\mathbf{J_{inc}} = \frac{k\hbar}{m_r}|u_{inc}(\theta, \phi)|^2 \quad \mathbf{(15)}$$

Scattering current density:

$$\mathbf{J_r} = \frac{k\hbar}{r^2 m_r}|u_{sc}(\theta, \phi)|^2 \quad \mathbf{(16)}$$

$$\mathbf{J_\theta} = \mathbf{J_\phi} = \frac{k\hbar}{r^3 m_r}|u_{sc}(\theta, \phi)|^2 \quad \mathbf{(17)}$$

Since the current density leads to the determination of other observables or measureable quantities i.e. angular momentum, energy, stress etc. – than these observables that are described in the wavefunction should also disperse in a similar way. As displayed in figure 2, the anisotropic cross sectional shape of the silicon nanowire (Cassini oval shape) is the result of excessive volume expansion in the radial direction while the expansion in the other two orthogonal directions is of a lesser extent. The scattering of electrons off of lithium and silicon atoms has the resulting electromagnetic stress that is larger and disperse at a slower rate in the radial direction as opposed to the zenith angle θ and azimuth angle ϕ directions which dissipates at a much faster rate and therefore the electromagnetic stress is weaker in these orthogonal directions.

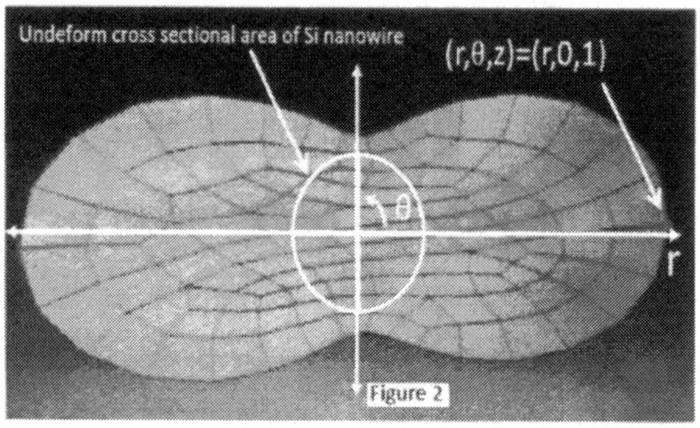

5. Summary

This report presented the possible explanation of two adverse effects that researchers has encountered separately, partially lithiated and excessive volume expansion in silicon nanowires. This paper offers three root causes to these effects that operates in conjunction with each other: 1) The change in band gap energy as c-Si transform to a-LiSi, 2) The current density differences within the nanowire and 3) The collisions of electrons with lithium and silicon atoms. Future research is needed in the way of molecular dynamics and/or density function theory in order to confirm these assertions.

Acknowledgement

I would like to thank Dr. Reza Shahbazian-Yassar of Michigan Technological University - Mechanical Engineering Department for introducing the author to the research of lithiated silicon nanowires.

References

[1] Stephanie Golmon, et.al, Stress generation in silicon particles during lithium insertion, Applied Physics Letters 97 033111, 2010

[2] Xiao Hua Liu, et.al, Self-Limiting Lithiation in Silicon Nanowires, ACS NANO 7 (2) p1495–1503 2013

[3] Xiao Hua Liu et.al, Anisotropic Swelling and Fracture of Silicon Nanowires during Lithiation, Nano Letter 11 p3312-3318 2011

High energy density and high temperature multilayered polymer film capacitors

Deepak Langhe and Michael Ponting

PolymerPlus LLC, 7700 Hub Pkwy,
Valley View, OH 44125, dlanghe@polymerplus.net

ABSTRACT

Pulsed power DOD applications like railguns utilize metalized biaxially oriented polypropylene (BOPP) based capacitors for energy storage. Although, BOPP films offer extremely low loss, stable capacitance and long life at ambient temperature; the energy density in these films is limited to 2.4 J/cc at film level, which leads to large capacitor volume. Many other disadvantages include DC conduction losses at high temperature leading to capacitor failure. Increased energy density decreases the life expectancy exponentially. Increased energy density capability, high temperature stability while maintaining low losses is critical in pulsed power capacitors. PolymerPlus developed multilayered film technology demonstrated capacitors with higher energy storage capacity, low losses, while eliminating the high temperature limitations of present railgun capacitors. Using PolymerPlus coextrusion process, fabrication of high energy density films and first generation capacitor prototypes was demonstrated.

Keywords: high temperature capacitors, multilayered films, dielectrics, energy storage

1 INTRODUCTION

Capacitors are used in nearly all electronics and electrical applications such as computers, power electronics, hybrid vehicles, grid converters, and pulsed power. [1-3] Development of compact devices necessitated the size reduction of electronic components including capacitors. Development of compact capacitors while maintaining the desired energy storage capacity requires increased energy density capacity. Energy density of the insulating material is directly proportional to the dielectric constant and the square of the dielectric strength of the material. Therefore, high dielectric constant, high breakdown strength, and low losses are desired in a capacitor material to achieve high energy density.

Polymer film capacitors with graceful failure are more advantageous than ceramic and electrolytic capacitors for high-voltage/high current pulsed power, power conditioning applications. However, they also suffer from few disadvantages. Existing pulsed power applications utilize metalized, biaxially oriented polypropylene (BOPP) film capacitors due to their extremely low loss, stable capacitance, and long life at ambient temperature. However, the energy density of BOPP films is limited to approximately 2.4 J/cc in these applications and is accompanied by a significant DC conduction loss at 100 °C. It is known that continuous shooting (6-12 shots/min at 6.5-11 kV) tends to increase the capacitor temperature creating hot spots inside a railgun capacitor. Life expectancy of the BOPP capacitor is limited by the energy density. BOPP capacitors survive longer if operated at lower energy density. Also, thermal issues become significant in BOPP-based capacitors in railgun applications. For a commercial railgun capacitor, the life expectancy decreases exponentially as the required energy density increases. Meanwhile, the low energy density of BOPP films results in a large volume, as compared to the other types of capacitors. Therefore, it is highly desired to develop new dielectric films with higher energy density (at least twice that of BOPP), low loss (dissipation factor, tanδ < 0.01), and high operating temperature (> 125 °C) capabilities.

In another example, electric vehicles (EVs) and hybrid electric vehicles (HEVs) use DC-link capacitor, which supplies periodic high currents enabling quick energy transfer to the insulated gate bipolar transistor (IGBT) circuit and typically runs at higher temperature (> 100 °C). Although commercially available electrolytic capacitors offer higher energy density, they have short life at high temperature. Other capacitors such as ceramic and glass-based capacitors are still in research stage. Alternately, BOPP and polyethylene terephthalate (PET) based polymer film capacitor suffer from low energy density and limited high temperature performance for EV and HEV applications. Additionally, BOPP-based capacitors occupy more than 1/3rd of the total volume in power electronics components.

Polymer film capacitors are widely favored due to their high dielectric strength, low dielectric losses, high energy density and ease of film processing. Additionally, polymers are lightweight and inexpensive as compared to other capacitors. The most commonly used polymer films for capacitor applications are BOPP and PET. BOPP dielectric films have low loss (tan δ~0.0002) and low energy density (4.88 J/cm^3 at breakdown of 730 MV/m), while biaxially oriented PET film has a slightly higher energy density (6.0 J/cm^3 at breakdown of 600 MV/m) and moderate dielectric loss (tan δ~0.003). [2] Both BOPP and PET capacitor films show low energy density, unsatisfactory high temperature and high voltage performance. Both polymers exhibit excellent room temperature performance, however, significant deterioration of dielectric properties above 80 °C. Improving the high temperature and high voltage capability of polymer dielectric films can enable fabrication of compact and more efficient systems.

To overcome these challenges, the development of high energy density and high temperature capacitors, while maintaining the low losses, is desired for current and future applications. Multiple approaches combining high dielectric constant inorganic particles such as barium titanate, titanium dioxide with high breakdown strength polymers negatively impacted the breakdown properties. [4] Other approaches included development of new copolymers and modification of BOPP. These efforts had limited commercial viability and scale up challenges [5]

PolymerPlus has demonstrated, using licensed coextrusion technology, multilayering a high dielectric constant polymer (e.g., PVDF and its copolymers) and a high electric breakdown polymer (PC) results in the achievement of high breakdown strength, high discharged energy density, and relatively low loss. Combining poly(vinylidene fluoride) (PVDF) and its copolymers with a high breakdown strength linear polymer, such as polycarbonate (PC) or PET, into a layered structure using the multilayer film coextrusion technology led to the development of novel dielectric films with improved dielectric properties. [6-10] The PC/P(VDF-HFP) 70/30 32-layer film shows a high discharged energy density at breakdown (~13 J/cm^3), low dissipation factor (tanδ~0.005) and low hysteresis loss (<15% at 600 MV/m). Different from the ferroelectric P(VDF-*co*-hexafluoropropylene) [P(VDF-HFP)], which exhibits significant hysteresis in the D-E loop, the PC/P(VDF-HFP) 50/50 32-layer film showed a much higher energy density than that of the state-of-the-art BOPP, but a dramatically reduced hysteresis with linear D-E loops. It is known that PVDF and its copolymers are ferroelectric thereby causing significant hysteresis loss during dynamic poling processes. However, the ferroelectric hysteresis is completely suppressed in PC/P(VDF-HFP) multilayered films.

Since the T_g of the PC (145 °C) and the melting temperature (T_m) of PVDF-HFP (125 °C) in these films limits the high temperature use, newer material grades were investigated in this study. PolymerPlus has investigated multilayered film extrusion of PVDF homopolymer (melting temperature, T_m = 165 °C) with a high glass transition (T_g) polycarbonate (PC, T_g = 175 °C). Dielectric properties of 33 layered PC/PVDF films were investigated at room temperature (RT) and at 120 °C. The results provide insights into a unique opportunity to use multilayer coextrusion to meet the need for high energy density and higher operating temperature capacitors in many applications.

2 EXPERIMENTAL

PVDF homopolymer (Solef 6010 from Solvay), and high temperature PC (APEC 1745 from Bayer MaterialScience) were acquired and used for coextrusion trials. 33 layered films of PC/PVDF were fabricated using forced assembly layer multiplying melt coextrusion process as described previously.[10] PC was dried overnight at 80 °C under vacuum prior to the coextrusion processing. The film structure constituted alternating layers of PC and PVDF as shown in **Figure 1**. A sacrificial surface layer of LDPE was also added before the melt exited from the film die. The LDPE surface layers were used to achieve excellent surface finish and improved uniformity in the multilayered films (MLFs). The film thickness of 8 µm was ensured during the film processing. MLFs were tested for various dielectric properties including breakdown strength (BDS), energy density, and dielectric hysteresis. The film take-off unit was equipped with in-line removal of LDPE layers.

The breakdown strength (BDS) was measured with a Quadtech Guardian 20 kV HiPot tester using plane-plane electrodes and a voltage ramp of 500 V/s. The voltage was increased until breakdown occurred. Test samples were sandwiched between flexible electrodes consisting of metalized polypropylene films. The sample area was 2 cm^2 and controlled using a thick sample mask. Each measurement was repeated at least ten times.

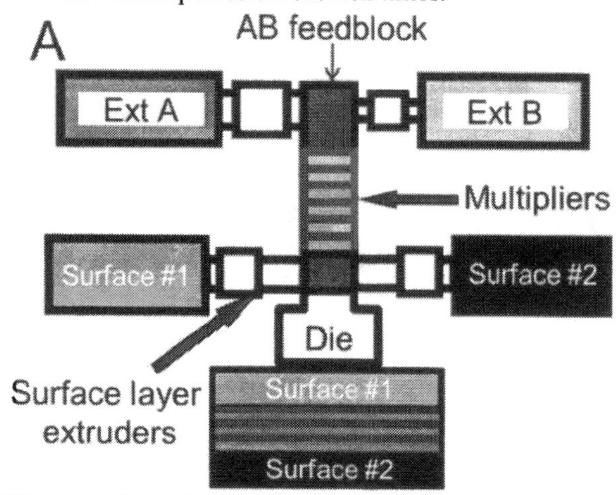

Figure 1: Multilayer coextrusion process at PolymerPlus

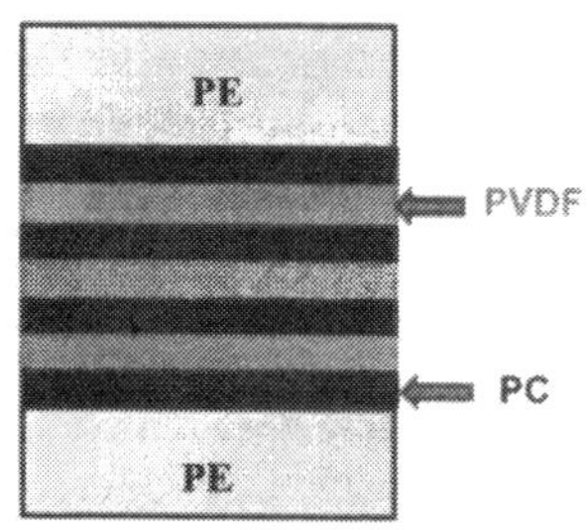

Figure 2: 33 layered PVDF/PC film structure with protective PE layers (removed in-line during processing)

The energy density of multilayered films was measured with a charge/discharge circuit. 1 cm gold electrodes were coated on both sides of the film samples using a sputter coater. A pulsed voltage with duration of 500 ms was applied using a Quadtech Guardian 20 kV HiPot tester in increments of 0.5 kV. The electrodes and sample were immersed in mineral oil to minimize surface and corona discharge. The resulting charging and discharging currents were measured by placing two resistors (150MΩ and 100 kΩ) in parallel with the sample and recording the voltage across the precision 100 kΩ resistor with a data acquisition card (NI AT-MIO-16E-1). The energy density was determined by integrating the discharge current with respect to time.

Electric displacement–electric field (D–E) hysteresis measurements were carried out using a Premiere II ferroelectric tester from Radiant Technologies, Inc. The applied voltage was a bipolar triangular waveform at 1 Hz. Gold coated film samples with a 1 cm diameter were used for testing. For each sample, electric fields from 100 to 700 MV/m were applied to the sample in 50 MV/m increments.

All the measurements were carried out in a silicone oil (DPDM-400 Diphenyl Dimethyl Silicone Fluid) bath. High temperature measurements were performed at 120 °C in a heated oil bath.

The film rolls were metallized and wound into capacitors. Film capacitors in 1.1 to 6 μF range were fabricated using multilayer films. The capacitors were further tested for capacitance, dissipation factor, equivalent series resitance (ESR), and insulation resistance (IR). These properties were tested over 25 to 160 °C temperature range.

3 RESULTS

3.1 Multilayer Film Production

Continous coextrusion process was successfully used to fabricate multilayer film rolls up to 3000 ft length demonstrating process scalability. Typically 4-6 splices were required to fabricate these rolls. The multilayer films were then characterized for dielectric properties and then used for metallization and capacitor winding.

3.2 High temperature performance of PC/PVDF films

High temperature dielectric characterization of PC/PVDF films was carried out at 120 °C. The hysteresis increased with increasing PVDF content and poling electric field, which was attributed primarily to the enhanced DC conduction and increased ion migration in multilayered films. **Figure 3** shows the percentage energy loss as a function of electric field for different compositions at 120 °C. Below 300 MV/m, the hysteresis loop loss is below 20%, which is significantly lower than that for BOPP at 125 °C (i.e., over 50%). At 125 °C, ion migration loss is seen at 10-100 Hz and crystalline PVDF α_c loss is seen at 70 kHz A reasonably low tanδ of 1.5% is seen for the high T_g PC/PVDF 50/50 33-layer MLF at 1 kHz.

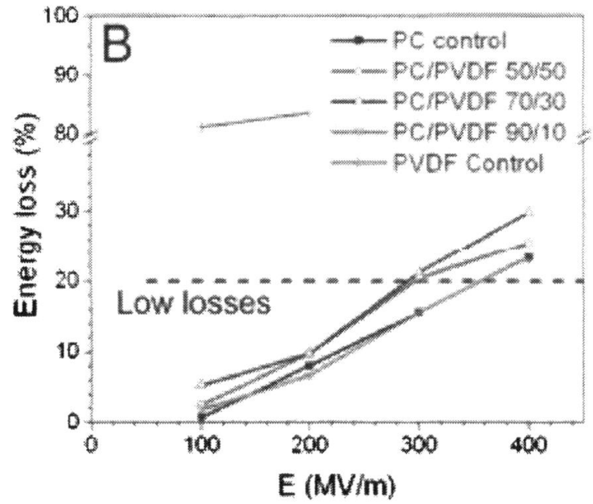

Figure 3: Energy loss as a function of electric field in PC/PVDF multilayer films

3.3 Capacitor Fabrication and Testing

PolymerPlus manufactured a first generation MLF capacitor prototype using high temperature PC/PVDF films. 2000 ft. film rolls are successfully metallized and slit using conventional techniques used for BOPP films **(Figure 4a)**. The metallized films were further wound into capacitor prototypes. Two examples of capacitor prototypes – 1.4 and 3.2 μF – are shown in **Figure 4b**. The analysis of the capacitor was performed at 25-160 °C. Capacitance change and dissipation factor change over this temperature range are shown in **Figures 5a** and **5b**. The multilayer film capacitors show a very stable capacitance at 25-160 °C. The measured capacitance change is less than 0.1% between 25 and 160 °C. The DF (tanδ) is 0.6% at room temperature, which increases to 1% at 80 °C and further decreases to 0.8% at 140 °C. In addition, the breakdown voltage properties of the capacitors also show improvements over comparable BOPP and PET film capacitors. The properties are compared with BOPP and PET film capacitors in **Table 1**

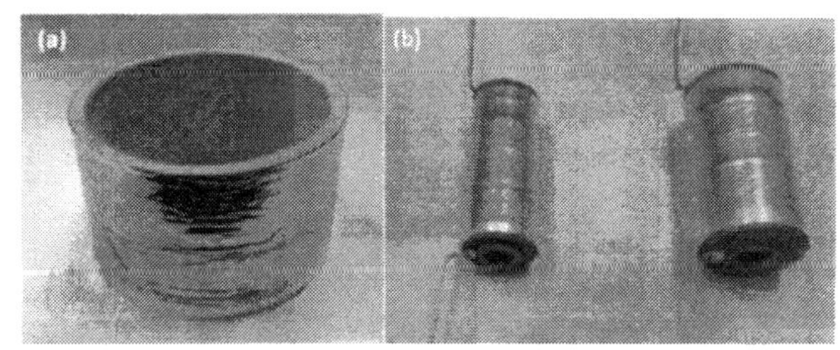

Figure 4: (a) 38.1 mm metallized MLF film roll produced by slitting 9" wide x 2000 ft long film, (b) examples of two prototoypes – 1.4 μF and 3.2 μF

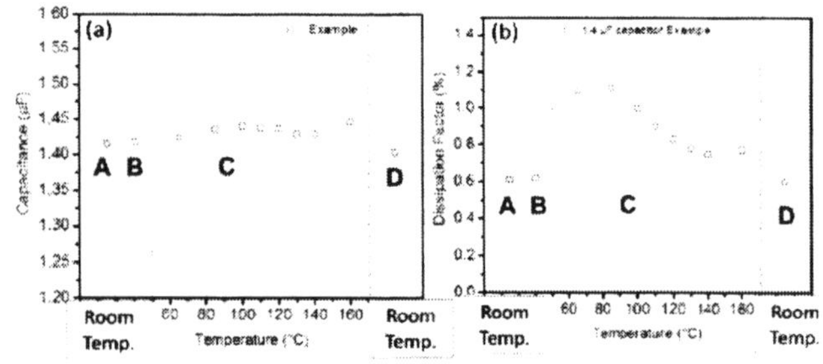

Figure 5: (a) capacitance change over 25 to 160 °C temperature range, (b) dissipation factor change over 25 to 160 °C temperature range. [A= tested after drying for 12 h at 40 °C, B= tested after clearing at 100- VDC for 2 min, C= 25 to 160 degree temperature range, D = tested after temperature sweep]

Capacitor Prototype	Dissipation Factor (%)	Capacitance change	Breakdown strength (MV/m)
Multilayer Films	0.6%	<0.1 % (25-160 °C)	200-350
BOPP	0.02- 0.1%	3 % (25 -100 °C)	200
PET	0.5%	10-15% (-25-125 °C)	170
PPS	<1%	<2 % (25 -125 °C)	--

Table 1: Comparison of multilayer film capacitor with commerical BOPP and PET capacitors

3.4 Conclusions

Multilayered film based capacitors have demonstrated signficantly improved dielectric properties and improved high tempearature performances. Significantly improved energy desity as compared to BOPP and PET, with reasonably low losses. The first generation capacitor prototype demonstrated less than 1% change in the capacitance over a wide temeprature range of 25 to 160 °C, which was significantly better than many commercial capacitors. The improved properties allow use of multilayer film capacitors in many applications including railguns and power electronics in electric vehicles or hybrid electric vehicles.

REFERENCES

[1] W.J. Sarjeant, J. Zirnheld, F.W. MacDougall, IEEE Transactions on Plasma Sciences, 26, 1368 (1998)

[2] I. Husain, Electric and Hybrid Vehicles: Design Fundamentals, CRC Press: Boca Raton, FL, (2010)

[3] R. Teodorescu, M. Liserre, P. Rodriguez, Grid Converters for Photovoltaic and Wind Power Systems, Wiley: Chichester, U.K., (2011)

[4] J. Wang, F. Guan, Q. Wang, J. Huang, W. Li, L. Zhu, PMSE Preprints, 100, 437 (2009)

[5] B. Chu, X. Zhou, K. Ren, B. Neese, M.R. Lin, Q. Wang, F. Bauer, Q. M. Zhang, Science, 313, 334 (2006)

[6] M. Mackey, A. Hiltner, E. Baer, L. Flandin, M.A. Wolak, J. S. Shirk, Journal of Physics D: Applied Physics, 42, 175304 (2009)

[7] M.A. Wolak, A.S. Wan, J.S. Shirk, M. Makey, A. Hiltner, E. Baer, Journal of Applied Polymer Science, 123, 2548 (2012)

[8] M. Mackey, D.E. Schuele, L. Zhu, L. Fandin, M.A. Wolak, J.S. Shirk, A. Hiltner, E. Baer, Macromolecules, 45, 1954 (2012)

[9] Z. Zhou, M. Mackey, J. Carr, L. Zhu, L. Flandin, E. Baer, Journal of Polymer Science Polymer Physics, 50, 993 (2012)

[10] M. A. Wolak, M-J Pan, A. Wan, J.S. Shirk, M. Mackey, A. Hiltner, E. Baer, L. Flandin, Applied Physics Letters, 92, 113301 (2008)

Ionic Liquid Flow Battery Materials and Prototyping

T. M. Anderson* and H. D. Pratt III*

*Power Sources R&D, Sandia National Laboratories,
Albuquerque, NM, USA, tmander@sandia.gov

ABSTRACT

SNL has developed a series of ionic-liquid electrolytes with accompanying non-aqueous compatible membranes and flow cell designs for improved energy density redox flow batteries targeted to support increasing demands for stationary energy storage. The new electrolytes yield a higher energy density by chemically incorporating an electro-active transition metal element into the solvent's molecular formula. Although ionic liquids have higher viscosities than conventional non-aqueous electrolytes, they are promising for higher energy densities due to higher metal concentrations and wider voltage windows. We have addressed high viscosity by developing new materials through careful ligand and anion selection. We have also developed tunable membranes for non-aqueous compatibility and rapid laboratory-scale prototyping to quickly screen materials and cell designs. We are projecting a four-fold improvement in energy density over the next two years.

Keywords: ionic liquids, redox flow batteries, energy density

1 INTRODUCTION

Worldwide energy demand is projected to significantly increase by 2050, and this increased need will be met, at least in part, through the use of renewable energy sources such as solar and wind [1]. However, the intermittent output of these sources requires the development of large-scale stationary energy storage systems. Among the most promising technologies are redox flow batteries (RFBs) that utilize the oxidation and reduction reactions of redox couples to perform reversible charging and discharging [2, 3]. RFBs can have long lifetimes due to the fact that their electrode reactions do not involve the mechanical stresses associated with the insertion or removal of the active species. In addition, because the energy-bearing complexes are stored externally rather than in the electrode compartment, the energy capacity can be tailored independently from the power output.

Despite many compelling attributes, work is needed to increase energy density and decrease costs [4]. As part of our own efforts to address energy density, we previously reported proof-of-concept work on the synthesis of a family of low-cost metal ionic liquids (MetILs) that contain a first-row transition metal coordination cation and weakly pairing monovalent anion(s) [5–8]. The advantage is the higher concentrations of active metal that can be achieved relative to dissolution of a salt into a solution.

This paper highlights the cell testing capabilities developed at SNL for flow battery R&D. It also discusses the synthetic processes required to produce MetILs and the variables that can be exploited to fine tune their properties with emphasis on lower viscosity and higher electrochemical reversibility. Finally, we end with a brief discussion of the performance of MetILs in laboratory-scale test fixtures.

2 EXPERIMENTAL

All starting materials were purchased from either Alfa Aesar or Sigma-Aldrich. The stabilities of the compounds were monitored using a Thermo Nicolet iS5 FT-IR equipped with a Smart Orbit (Diamond) ATR accessory. Electrochemical experiments on half-cells were performed at room temperature under Ar using a BASi Epsilon potentiostat. All solutions were freshly prepared and degassed prior to electrochemical testing. Cyclic voltammograms (CVs) were collected using a three-electrode setup with a freshly polished 3 mm diameter glassy carbon working electrode, Ag/AgCl (in 3 M KCl) reference electrode, and a platinum wire counter electrode.

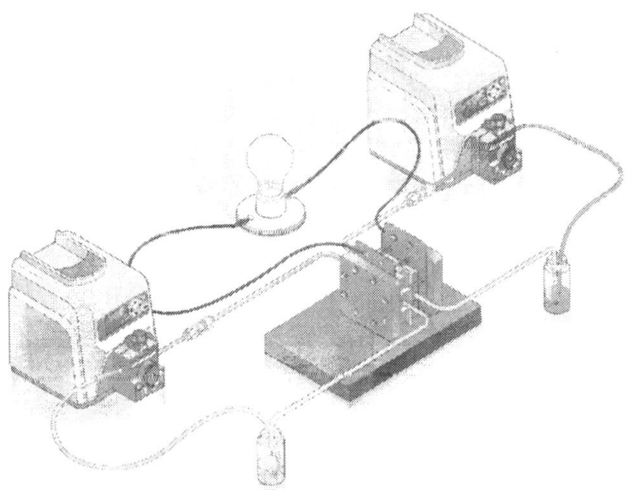

Figure 1: Schematic of a laboratory-scale flow battery test setup illustrating the peristaltic pumps, reservoirs, and electrochemical cell.

3 CELL TESTING CAPABILITIES

In a typical flow cell experiment (Figure 1), the cells consist of carbon-felt electrodes, current collectors, flow fields, gaskets, and a membrane. The flow cell testing system consists of a single cell, peristaltic pumps, polypropylene reservoirs, and chemically compatible tubing. A Solartron 1287 potentiostat/galvanostat is used to control the charging and discharging of the cell at variable temperatures and flow rates.

3.1 Pump Heads and Tubing

The test apparatus consisted of drive assemblies, head assemblies, tubing, and flow cell geometries. For all testing, we utilized Masterflex L/S series drives and various drive compatible pump heads were employed, depending on viscosities and desired flow rates. As an initial starting point, the 60 series Easy Load II was chosen because it features an adjustable occlusion to fine tune for the compression to allow higher pressure and suction lift for more viscous fluids and lower pressure for less viscous fluids. In addition, the four-roller rotor reduces the flow pulsation, which delivers fluids at a more consistent rate and provides more uniform wetting of the cell. For the Easy Load II pump head, Cole-Parmer provides an array of materials in Masterflex tubing that are compatible with the Easy Load II; the tubing was chosen based on compatibility charts provided by Masterflex/Cole-Parmer. It is important to note that solvent compatibility alone is inadequate in determining long-term performance, since friction created by potential nanoparticulates can degrade the material over time. We have made this assessment based on the fact that we notice perforations at the tubing roller interface. Hose size inner diameter and outer diameter (ID/OD) is generally based upon viscosity for the Easy Load II with increasing viscosity necessitating greater tubing size. In addition to the Easy Load II, we have utilized the PTFE tubing head. The PTFE tubing head utilizes six rollers and can pump fluids at even higher pressures than the Easy Load II (up to 100 psi). The PTFE tubing head utilizes Masterflex tubing sets; these sets come in either 2 mm ID by 4 mm OD or 4 mm ID by 6 mm OD. We utilize Swagelok stainless steel reducing unions to transition between metric size tubing sets and SAE size PTFE tubing, generally feeding in flow with eighth-inch OD tubing with quarter-inch OD outflow. This variation in ID/OD is employed to reduce bottlenecking at the exit pathway and reduce overall back pressure in the system/membrane.

3.2 Flow Cell Design

The basis of our redox flow cell was based on the classic methanol fuel cells that are commercially available. In particular, we utilized several designs from Fuel Cell Technologies. This design provides for separate positive and negative half-cells and it provides a compression mechanism for the membrane, flow fields, and current collectors. It is common for these materials to be made out of graphite to resist harsh chemical conditions. However, the relatively benign nature of our MetILs compounds allowed for other materials to be used including stainless steel (304). This can potentially allow some cost savings over machining graphite materials. The stainless steel designs allow the viscous MetIL solutions to flow with more fluid turns, in contrast to sharp angles generally present in methanol fuel cells. Our stainless steel assemblies could be put together using their own bolt circles, and they were also designed to fit into the Fuel Cell Technologies hardware assemblies. The latter design also provided a cost effective means of making iterations to the flow path. Sample geometries that were tested include a straight trough (eighth-inch in width and depth with radial ends), serpentine patterns, and circular flow fields. In addition, all of these geometries could be oriented in either the vertical or horizontal feed orientation. Fluids with viscosity values between 500 and 700 cP (at 25 °C) generally require larger flow paths than those provided by the classic methanol fuel cells. When assembling a cell, a compression of twenty percent to the carbon felt was generally achieved. In addition, plasma treating the membranes, carbon felt, or both prior to cell assembly was beneficial to performance. Finally, it is important to note that machined graphite is available as either Poco or pyrosealed Poco. However in most of our experiments more favorable results were obtained with pyrosealed Poco. In summary, these are the tools that we have on hand to construct a MetIL based redox flow battery.

Table 1: Summary of cupric ionic liquids as a function of chemical constitution.

Ligand[1]	Anion 1[2]	Anion 2[2]	State at 25 °C	ΔE[3] [mV]
EA	EHN	EHN	Liquid	244
EA	OTf	OTf	Solid	158
EA	EHN	OTf	Solid	158
EA	BF_4	BF_4	Liquid	102
EA	BF_4	OTf	Solid	256
EA	BF_4	EHN	Liquid	187
EN	BF_4	BF_4	Liquid	112
DEA	EHN	EHN	Liquid	522
DEA	OTf	OTf	Liquid	566
DEA	EHN	OTf	Solid	507
DEA	BF_4	BF_4	Liquid	150
DEA	BF_4	OTf	Liquid	159
DEA	BF_4	EHN	Liquid	201

[1]EA = ethanolamine, EN = ethylenediamine DEA = diethanolamine; [2]BF_4 = tetrafluoroborate, EHN = 2-ethylhexanoate, OTf = CF_3SO_3; [3]0.1 M solutions in 1-butyl-3-methyl-imidazolium hexafluoro-phosphate with a glassy carbon (working) electrode (scan rate 50 mV s^{-1}).

4 SYNTHESIS AND CHARACTERIZATION

A series of copper-based MetILs were prepared with varying ligands and anions in order to elucidate trends for lower viscosity (<1000 cP) and higher electrochemical reversibility (i.e. peak separations less than 200 mV), and the results are summarized in Table 1. The results showed the most promising compounds are those containing either EA or EN ligands and BF_4^- anions. In addition, several of the compounds have electrochemical reversibility values that were comparable or even better than ferrocene (~150 mV) under identical conditions (i.e. in an ionic liquid).The cyclic voltammograms (CVs) of $Cu(DEA)_6(CF_3SO_3)_2$, $Cu(EA)_6(BF_4)_2$, and $Cu(DEA)_6(BF_4)_2$, performed at a 100 mV s^{-1} scan rate with a glassy carbon working electrode, are shown in Figure 2. The open circuit potential was around 3.0 V in all three compounds (i.e. the fully oxidized species was present at the start of the experiment). The CV for $Cu(EA)_6(BF_4)_2$ indicated that electrodeposition and stripping of the bulk metal on the working electrode occurred at 0.5 V and 2.4 V, respectively, with the Cu(II)/Cu(I) redox pair centered at 2.5 V. The redox processes at 1.1 V, 1.4 V, and 2.0 V were attributed to the ethanolamine ligand, and this assignment was verified by comparing the CV with that of a pure sample. Compounds $Cu(DEA)_6(BF_4)_2$ and $Cu(DEA)_6(CF_3SO_3)_2$ underwent Cu(II)/Cu(I) reductions at 2.3 V and 2.1 V, respectively, which was consistent with the differences in the ion pairing

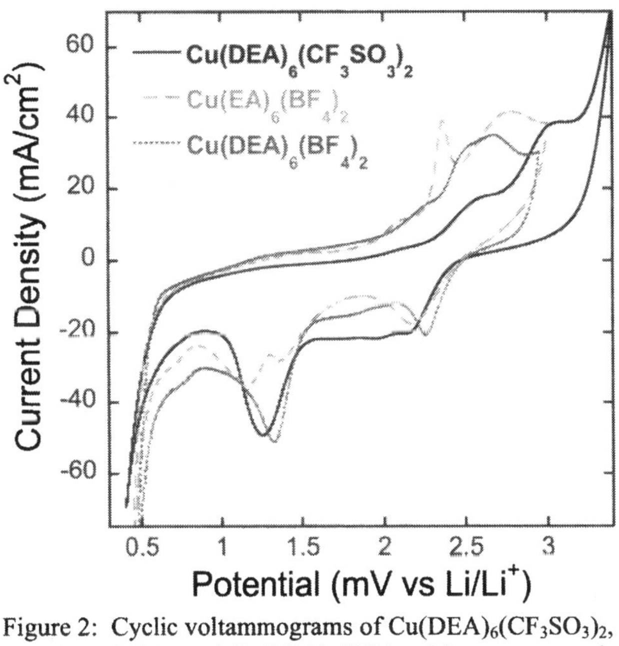

Figure 2: Cyclic voltammograms of $Cu(DEA)_6(CF_3SO_3)_2$, $Cu(EA)_6(BF_4)_2$, and $Cu(DEA)_6(BF_4)_2$ with a scan rate of 100 mV s^{-1} and a glassy carbon working electrode.

of BF_4^- and $CF_3SO_3^-$. In contrast to $Cu(EA)_6(BF_4)_2$, neither $Cu(DEA)_6(BF_4)_2$ nor $Cu(DEA)_6(CF_3SO_3)_2$ had any evidence of stripping although copper deposition was slightly more facile (i.e. a slightly higher potential) in $Cu(DEA)_6(BF_4)_2$. Given that DEA is a more strongly coordinating ligand, the results suggested that DEA hindered stripping. Finally, static cell testing of a cell containing $Cu(EA)_6(BF_4)_2$ and $Fe(C_{20}H_{27}N_2)(PF_6)$ in separated compartments showed significantly improved discharge profiles that were consistent with more reversible plating and stripping of copper. In addition, the coulombic efficiency increased from 50 to 65%. Additional efforts to further improve copper plating and stripping reversibility are currently underway, and flow cell testing of these compounds has commenced.

5 FLOW CELL TESTING

Flow cell testing in a serpentine flow field was performed using copper- and iron-based MetILs with a 1.5 V separation. A number of commercial membranes were examined but Neosepta AHA gave the best results. The Coulombic efficiency as a function of cycle number is shown in Figure 3 for two different active materials concentrations. The increased concentration gave more scatter in the data, but a higher energy density was achieved. We are currently looking at tunable SNL-made membranes to increase performance.

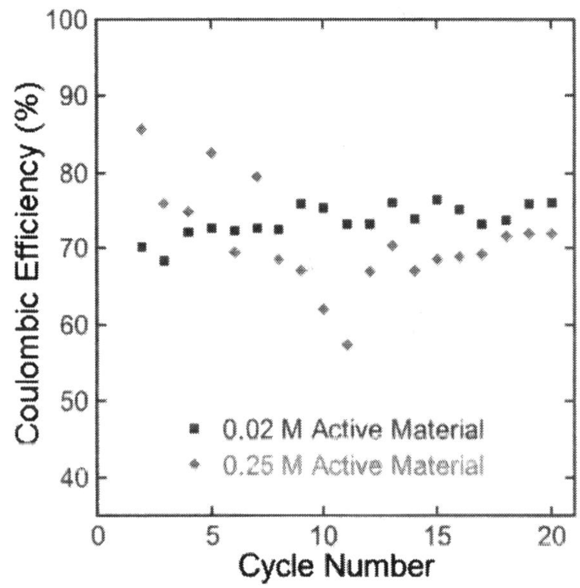

Figure 3: Coulombic efficiency of a copper- and iron-based MetIL redox flow battery as a function of cycle.

6 ACKNOWLEDGEMENT

We thank the U. S. Department of Energy, Office of Electricity Delivery and Energy Reliability (Dr. Imre Gyuk, Energy Storage Program), for funding. Sandia National Laboratories is a multi-program laboratory operated by Sandia Corporation, a wholly owned subsidiary of Lockheed Martin Company, for the U. S. Department of Energy's National Nuclear Security Administration under contract DE-AC04-94AL85000.

REFERENCES

[1] C. Wadia, P. Albertus, V. Srinivasan, "Resource constraints on the battery energy potential grid and transportation applications," J. Power Sources, 2011, 1593.

[2] M. Skyllas-Kazacos, M. H. Chakrabarti, S. A. Hajimolana, F. S. Mjalli, and M. Saleem, "Progress in flow battery research and development," J. Electrochem. Soc., 2011, R55.

[3] A. Z. Weber, M. M. Mench, J. P. Meyers, P. N. Ross, J. T. Gostick, Q. Liu, "Redox flow batteries: a review," J. Appl. Electrochem., 2011, 1137.

[4] R. M. Darling, K. G. Gallagher, J. A. Kowalski, S. Ha, F. R. Brushett, " Pathways to low-cost electrochemical energy storage: a comparison of aqueous and nonaqueous flow batteries," Energy Environ. Sci., 2014, 3459.

[5] T. M. Anderson, D. Ingersoll, A. J. Rose, C. L. Staiger, and J. C. Leonard, "Synthesis of an ionic liquid with an iron coordination cation," Dalton Trans., 2010, 8609.

[6] H. D. Pratt III, A. J. Rose, C. L. Staiger, D. Ingersoll, and T. M. Anderson, "Synthesis and characterization of ionic liquids containing copper, manganese, or zinc coordination cations" Dalton Trans., 2011, 11329.

[7] H. D. Pratt III, J. C. Leonard, L. A. M. Steele, C. L. Staiger, T. M. Anderson, "Copper ionic liquids: examining the role of the anion in determining physical and electrochemical properties" Inorg. Chim. Acta, 2013, 78.

[8] H. D. Pratt III, D. Ingersoll, N. S. Hudak, B. B. McKenzie, T. M. Anderson, "Copper ionic liquids: Tunable ligand and anion chemistries to control electrochemistry and deposition morphology" J. Electroanal. Chem. 2013, 153.

Low-Cost, High-Performance Redox Flow Battery for Renewable Energy Storage

A. Colclasure[1], D. Eldridge[2], and B. Berland[3]

[1]ITN Energy Systems, Inc., Littleton, CO, USA, acolclasure@itnes.com
[2]ITN Energy Systems, Inc., Littleton, CO, USA, deldridge@itnes.com
[3]ITN Energy Systems, Inc., Littleton, CO, USA, bberland@itnes.com

ABSTRACT

ITN's innovative Redox Flow Battery (RFB) combines advanced materials engineering with stack design breakthroughs to provide the high-power, low-cost energy storage solution that grid-based and off-grid power suppliers have been looking for. ITN's RFB provides higher power capability in a smaller space while using more economical materials to bring the marketplace an affordable and scalable battery to offset the fluctuations found in renewable energy generation (i.e., wind and solar). Combining this low-cost battery with renewable sources, grids, microgrids, and standalone power generators can deliver consistent and reliable power by using the battery to compensate for the intermittent nature of renewable energy production.

Keywords: redox flow battery, energy storage, grid storage, renewable energy, microgrid

1 REDOX FLOW BATTERIES

Redox flow batteries, due to their inherent safety, scalability, and long term durability, are the only technology that has the potential to satisfy all of the requirements for large-scale storage in applications where long discharge times are required. The power is governed by the stack area, while the total stored energy can be independently designed by simply adjusting the size of the tanks, i.e. the amount of electrolyte available to the system.

Figure 1 Redox Flow Battery Components

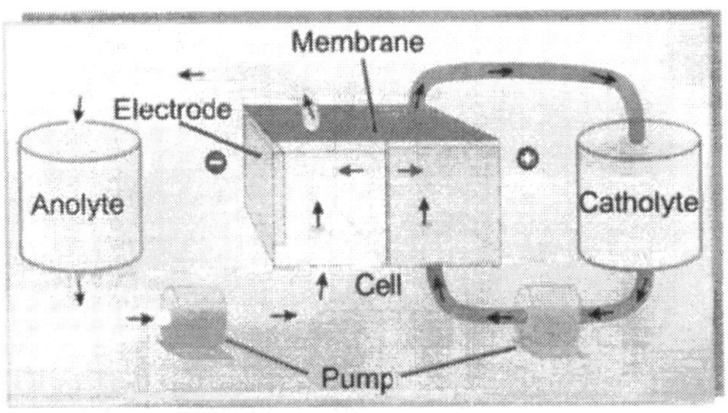

Cathode: $VO_2^{+} + 2H^{+} + e^{-} \xrightarrow{\text{Discharge}} VO^{2+} + H_2O$

Anode: $V^{2+} - e^{-} \xrightarrow{\text{Discharge}} V^{3+}$

Cell: $VO_2^{+} + V^{2+} + 2H^{+} \xrightarrow{\text{Discharge}} VO^{2+} + V^{3+} + H_2O \quad E_o=1.26V$

The primary advantages of RFB over other battery technologies include:

- **Safety:** No fire hazard
- **Long Life:** Greater than 10,000 charge/discharge cycles = low life cycle cost of energy
- **Low Maintenance:** Automated, sensor based battery maintenance
- **Scalability:** Supports systems from kWh to MWh and avoids complex cell level battery management systems required for li-ion, etc.

The inherent safety, durability, and scaling of RFB are derived from the simple mechanism that RFBs store energy via simple conversion of ion states in a liquid.[1] In contrast, lithium ion batteries store energy via intercalation of lithium ions into solid state crystalline matrix leading to mechanical strain and cracking of the intercalation materials that ultimately limit the lifetime of the batteries. Further, lithium-ion's use of liquid electrolytes introduces a flammability hazard, and leads to formation of interface materials that further limit battery life.

There are three main components of an RFB: (1) The Stack Power Assembly (SPA), with catalytically active electrodes to promote conversion of ion states and an ion exchange membrane to separate the anolyte and catholyte chemicals, (2) The Tank Electrolyte Assembly (TEA), and (3) The Power Conditioning Unit (PCU), which interfaces the battery to a grid/microgrid and monitors the health and state of charge. While RFBs have long been recognized as having superior potential to meet large scale storage challenges, their historical cost has been prohibitive. ITN has focused on a low cost SPA and TEA, introducing low cost materials and efficient designs to make an appealingly affordable RFB solution.

1.1 The Opportunity

Renewable energy generation is expanding rapidly throughout the world. Unfortunately, renewable sources (wind and solar) are intermittent and unpredictable.

- Wind turbines only produce electricity when the wind is blowing.
- Solar panels only produce electricity when the sun is shining.
- Electricity demand is based on human activity – which may not correlate to the schedule of the energy that is being generated.

Without storage, electricity must be used the moment it is generated. Thus, when a power source generates more energy than can be consumed:

- Utilities throttle down their coal, nuclear or hydro generation
- Solar systems offload excess energy to the grid putting more pressure on utilities' energy management
- Wind Turbines are turned off

Affordable RFBs solve the problems caused when energy generation doesn't precisely match user demand. Attaching batteries to power generation sources enables power providers to:

- Extend the life of conventional power generation assets
- Stabilize inconsistent power supplied by solar and wind energy sources
- Provide power for peak usage periods without needing to invest in excess capacity just to satisfy temporary demand, while having assets sit idle when demand is low

The integration of renewable energy generation and storage capability provides benefits of increased energy security, emergency preparedness, reliable energy pricing, etc. across commercial and military applications. Figure 2 shows a typical deployment where renewable solar and wind are the primary generators, the RFB provides storage, and a diesel generator and/or grid connection provide backup to meet localized energy demands with the lowest carbon footprint and cost. [2], [3]

Figure 2—ITN's RFB enables smart and economical networks for grid and microgrid applications.

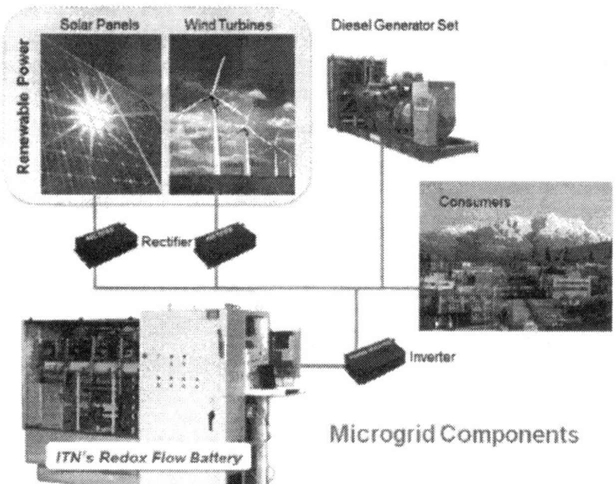

Figure 3 shows the impact of energy storage on the system operation. The RFB effectively buffers periods of excess demand and generation capacity to minimize the use of grid/genset assets to meet local energy supply. Periods of battery charge are shown in green, discharge in violet.

Figure 3—ITN's RFB helps match renewable power generation supply with customer electricity demand.

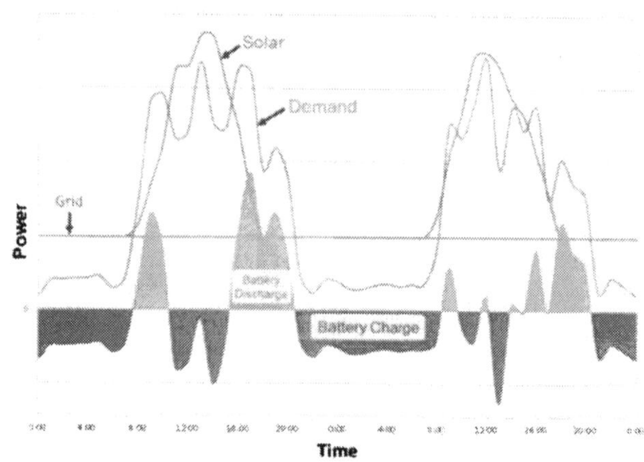

1.2 ITN's RFB Innovations

Building on its extensive membrane and fuel cell experience, ITN began to pursue RFB technology in 2010 leading to a Gen 1 RFB employing all vanadium chemistry, which incorporates a low-cost, high power density stack. Because the RFB's membrane and electrolyte constitute a large fraction of cost, ITN also pursued new designs to further reduce cost. In 2012, ITN won a $2.1M grant from the Advanced Research Projects Agency (ARPA-e) to investigate innovative membrane materials and electrolyte chemistries. With this funding, ITN is developing:

- Low Cost Membranes
- Low Cost Mn Electrolytes
- Lowest Cost Organic (Nonmetallic) Electrolytes

As a result of these efforts, ITN's team has developed materials and design breakthroughs that reduce the overall RFB costs by 50%, resulting in a Gen 2 battery that is significantly cheaper than current state of the art. Integrating these key differentiators, ITN has the lowest cost, highest performing redox flow battery for load shifting applications. These developments come from the company's long history of technical and process innovation.

Figure 4—ITN's Mn/V Electrolyte Offers Higher Voltage as Well as Lower Cost

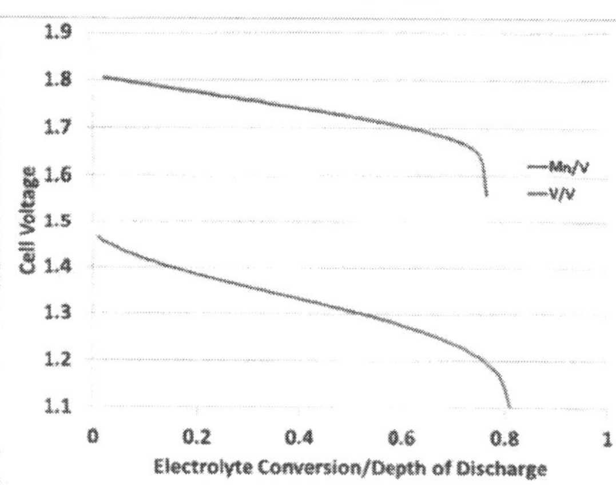

Figure 4 shows that, in addition to lower cost materials, ITN's next generation Mn/V electrolyte provides a higher cell/stack voltage to further improve the performance. In addition, ITN's membrane has shown strong cycling data and performance similar to Nafion®.

2 CONFIGURATION AND PERFORMANCE

2.1 ITN's Battery Configuration

ITN's advanced RFB consists of three main assemblies, the Stack Power Assembly, the Power Conditioning Unit, and the Tank Electrolyte Assembly (Figure 5).

Figure 5—ITN's RFB Design Concept

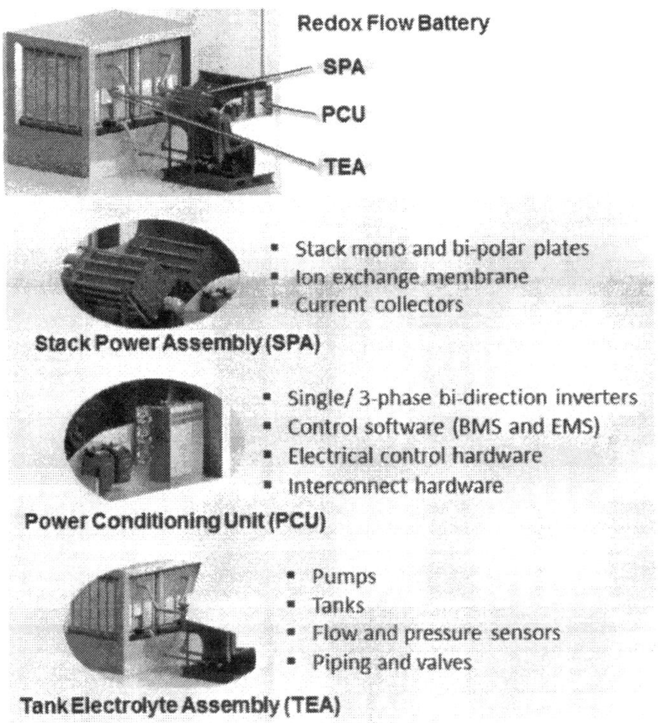

2.2 ITN's Battery Performance

Figure 6 shows ITN's modular stack design that enables the SPA to be constructed from modules that can be connected in series/parallel combinations to provide the desired power and energy. Each stack shown can produce up to 5 kW of power. Figure 7 shows a typical discharge curve for ITN's subassembly stack design with coulombic efficiency ~98% and energy efficiency approaching 85%, even at high operating power density.

3 COMMERCIALIZATION

The world's industrial nations are expecting electricity storage needs to grow rapidly. To capture this emerging market, ITN has established strategic partnerships with two European partners to demonstrate and commercialize its RFB technology. With these partners, ITN has fielded two units (Figure 9). Figure 10 shows operation of one of these demonstration systems. Similar relationships are now being developed in the US and other regions of the world.

Figure 6—ITN's improved stack design provides high power and efficiency at low cost.

Figure 7—ITN's RFB achieves high power density and performs with industry leading efficiency.

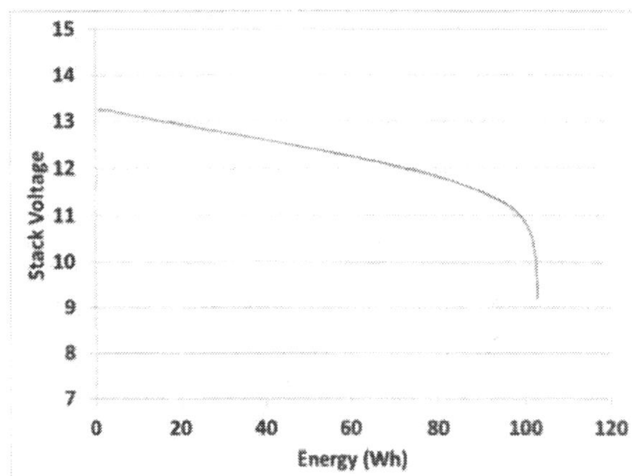

The world's industrial nations are expecting electricity storage needs to grow rapidly and urgently need efficient and cost-effective large-scale energy storage systems (Figure 8). While the 2012 grid storage market of $200 million was substantial, market forecasters predict the market will grow exponentially to reach $19 billion by 2017. Advanced flow batteries are predicted to account for approximately 17% of this market.

To capture this emerging market, ITN has initially established strategic partnerships with two European partners to demonstrate and commercialize its RFB technology. With these partners, ITN has fielded two RFB units in 2014/2015 (Figure 9). Figure 10 shows operation of one of these demonstration systems during field testing. Similar relationships are now being developed in the US and other regions of the world.

Figure 8—Worldwide installed grid storage is predicted to exceed 14 GW in the next ten years. [4]

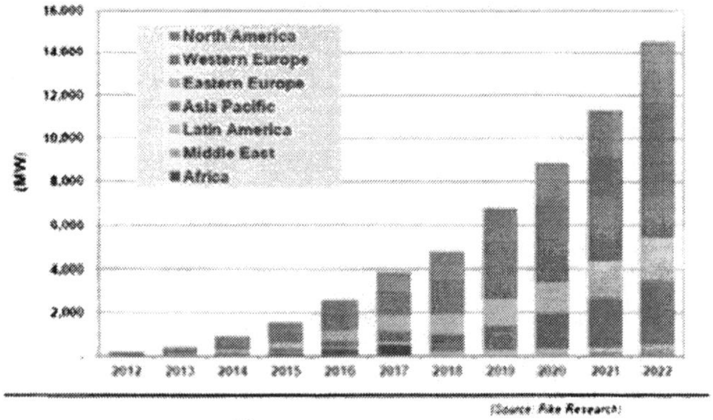

Figure 9—ITN's RFB Demonstration with a European Partner

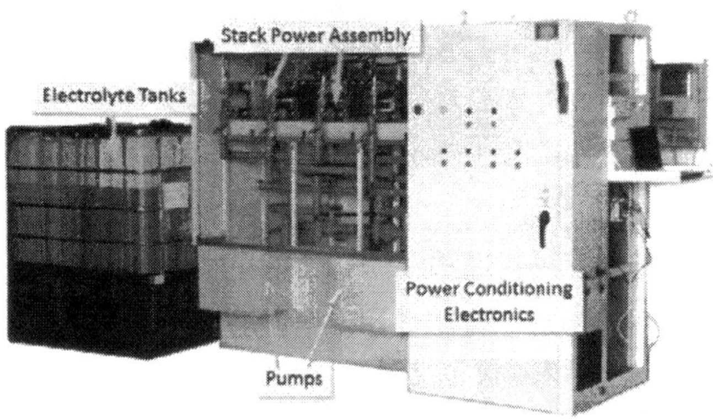

Figure 10—Field Testing of ITN RFB.

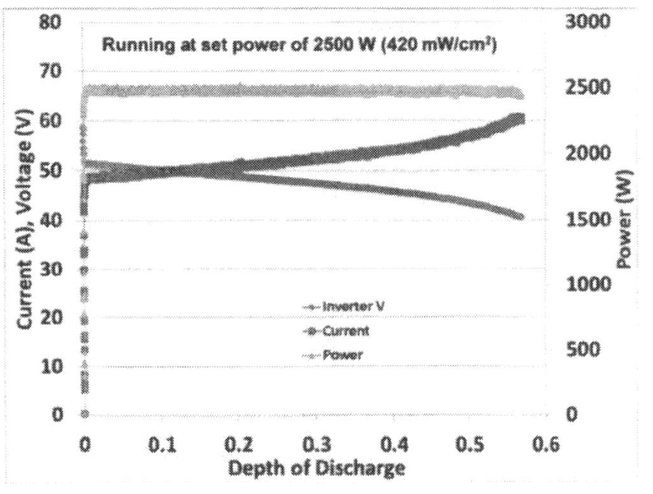

4 PRODUCTION TIMELINE

ITN will form additional strategic partnerships in different parts of the world via licensing agreements or joint ventures to leverage local expertise and established distribution channels and established business presence. The goal is to initially focus on 10 - 100 kWh systems (e.g., residential backup, telecom towers, military missions, etc.).

ITN will initially manufacture and supply the stack power assembly (SPA) and will provide specifications to our partners for the rest of the system.

Figure 11—ITN's Commercialization Timeline

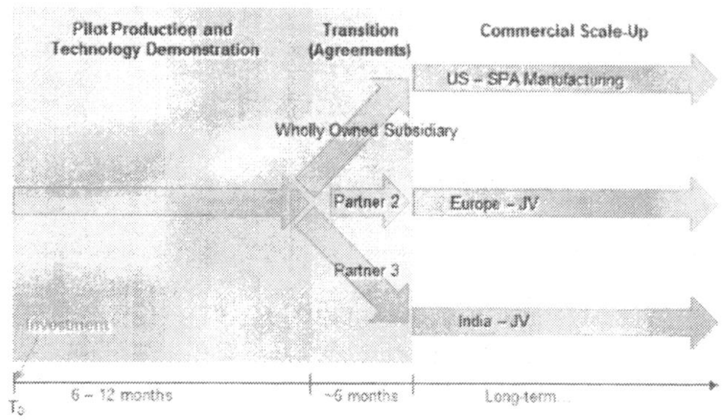

Based on successful testing in Europe, ITN expects to secure a large purchase order and begin full commercial stack production in mid-2015. ITN is currently planning and procuring financing for the volume expansion to handle these orders as well as to handle expanded market demand based on these initial successes.

5 SUMMARY

With advances in stack design, membrane and electrolyte chemistry, and overall system efficiencies, ITN has advanced the state of low-cost RFBs for Distributed Power Systems, making redox flow batteries affordable for grid and off-grid storage solutions for the first time. ITN is beginning commercial production, based on existing relationships with initial customers and market efforts to establish new customers in commercial and defense markets.

REFERENCES

1. Ding, C. *et al.* (2013). Vanadium Flow Battery for Energy Storage: Prospects. *Journal of Physical Chemistry*, 1281-1283.
2. U.S. Department of Energy. (2013). *Grid Energy Storage.* Washington, DC: U.S. Dept. of Energy.
3. Weber, A. Z., *et al.* (2011). Redox flow batteries: a review. *J Appl Electrochem*, 1-5.
4. Dehama, A. & Adamson, K.-A. (2012). *Energy Storage on the Grid.* Boulder, CO: Pike Research.

Sodium-Ion Batteries: "Beyond Lithium-Ion"

M.Shirpour*, X. Zhan** and M. Doeff***

*University of Kentucky, Lexington, KY, USA, mona.shirpour@uky.edu
** University of Kentucky, Lexington, KY, USA, xiaowen.zhan@uky.edu
*** Lawrence Berkeley National Laboratory, Berkeley, CA, USA, mmdoeff@lbl.gov

ABSTRACT

Aqueous sodium-ion batteries may solve the cost and safety issues associated with the energy storage systems for the fluctuating supply of electricity based on solar and wind power. Aqueous, or water-based, sodium-ion batteries offer multiple cost savings using less expensive electrode materials and much cheaper electrolyte solutions compared to the lithium ion cells. In the present work, we for the first time report the electrochemical performance of $NaTi_2(PO_4)_3$ anode material paired with Na_2FePO_4F cathode material in an aqueous system. This aqueous sodium-ion battery exhibits a higher voltage and a greater specific energy (50 Wh/Kg) than its organic analog (45 Wh/Kg).

Keywords: sodium-ion battery, grid energy storage, abundant, low-cost, aqueous

1 INTRODUCTION

The drastic increase in demand for lithium-ion batteries in the electronics and automotive industries is currently putting enormous pressure on lithium supplies, and has resulted in a steep price rise for the lithium precursors used in the manufacture of electrode and electrolyte materials.[1] Sodium-ion batteries are an appealing lower cost alternative to lithium-ion batteries because of vast resources of sodium and much lower materials costs ($5000/ton for lithium in comparison with $160/ton for sodium).[2,3] Lithium is also a strategic element for energy because more than half of the world's lithium reserves are located in only three countries: Argentina, Bolivia, and Chile[4], while the United States has the world's largest deposits of sodium and is the lowest cost producer of sodium precursors. One-third of the world's sodium precursor comes from Wyoming's Green River Basin.[5]

Low cost energy storage systems are the key technologies for the fluctuating supply of electricity based on solar and wind power. Aqueous, or water-based, sodium-ion batteries are promising candidates for large-scale grid storage applications because they offer multiple cost savings using less expensive electrode materials, much cheaper electrolyte solutions compared to the lithium-ion cells, and less costly manufacturing conditions. They also have the extra advantage of being safer for grid applications,[6] and higher cycling efficiency due to the higher ionic conductivity of the aqueous electrolyte.

In the water-based cells, both the anode and cathode should undergo redox processes approximately within the voltage stability region of water (Figure 1a).

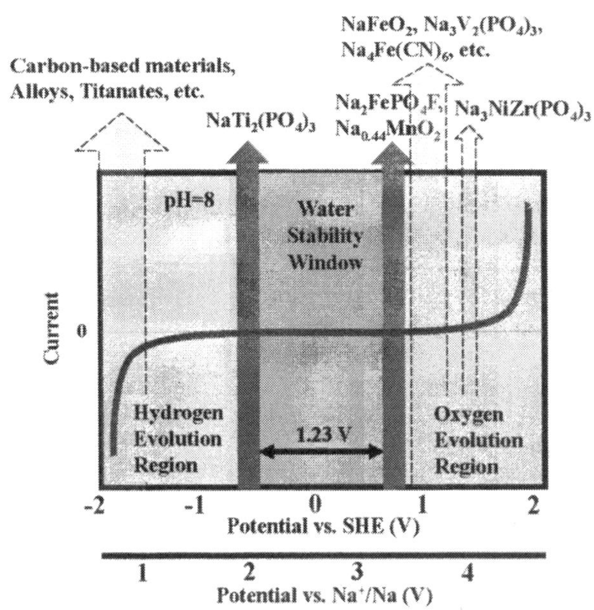

Figure 1: Voltage stability window of water; the highlighted regions show the operating voltage of most of the reported cathode and anode materials for sodium-ion cells.

Most of the anode materials [3,7,8] for sodium-ion batteries, including carbon-based compounds and titanates[9,10] have a very low operating voltage vs. sodium, and fall in the region of hydrogen evolution. NASICON ($NaTi_2(PO_3)_4$) as negative electrode and Na_2FePO_4F as positive electrode operate within the voltage stability region of water, 2.1 V and 3.0-3.2 V versus Na/Na^+, respectively. $NaTi_2(PO_3)_4$ accommodates 2 Na and exhibits a theoretical capacity of 133 mAh/g, and Na_2FePO_4F has a theoretical capacity of 124 mAh/g based on intercalation of one sodium.[3,6,8,11-13] These values correspond to a theoretical specific energy of 64 Wh/Kg based on the total weight of active electrode materials. In this work, we for the first time discuss aqueous sodium-ion batteries based on the combination of these two electrode materials. The comparison between organic and aqueous electrolytes in this work highlights the potentials and limitations of water-based sodium-ion batteries.

2 EXPERIMENTAL

Synthesis and Electrode Fabrication: $NaTi_2(PO_4)_3$ and Na_2FePO_4F were prepared by solid state methods. For the $NaTi_2(PO_4)_3$ NASICON phase, a mixture of Na_2CO_3, TiO_2, and $6NH_4H_2PO_4$ was calcined at 800°C for 24 and at 1000°C for 12h, using intermediate milling and grinding (300rpm/2h). For the Na_2FePO_4F, a mixture of $Fe(C_2O_4).2H_2O$, $NaHCO_3$, NaF and $NH_4H_2PO_4$ was heated in an Alumina boat crucible at 350°C for 4h under Ar. Then dry milled for 4h at 100 rpm, and calcined at 600°C for 6h under Ar flow. Prior to electrode fabrication, the obtained active materials were milled with acetylene black (300rpm for 2h) in order to enhance the electronic conductivity of the electrodes. The electrodes were prepared by casting a mixture of the active materials, acetylene black and Polyvinylidene fluoride (PVDF) (in a weight ratio of 70:20:10 for positive electrode, and 70:25:5 for negative electrode) on aluminum foil for organic cells and on stainless steel foil for aqueous cells (Multipurpose 304 Stainless Steel Foil, Bright Finish, 0.001 inch thick).

Cells with Organic Electrolyte: Full sodium-ion cells with organic electrolyte were assembled using 2032-type coin cells inside in a helium-filled glovebox. Celgard 2400 was used as separators and 1M $NaPF_6$ (Sigma Aldrich) in EC:DMC (3:7 mol, from Novolyte Technologies) as the electrolytic solution.

Cells with Aqueous Electrolyte: Each electrode material was tested in a three electrode cell (Plate material evaluating cell from Bio-logic) with Pt foil counter electrode and Ag/AgCl (3 M NaCl) reference electrode. For full sodium-ion cells with aqueous electrolyte (1 M Na_2SO_4), the electrodes together with filter paper as separator were sealed in pouche cells. Electrochemical evaluation was carried out using a Bio-logic VMP3 potentiostat/galvanostat. Electrochemical impedance spectra were recorded from 100 kHz to 50 mHz.

3 RESULTS AND DISCUSSIONS

Our sodium-ion battery (full cell) consisting of $NaTi_2(PO_4)_3$ anode, Na_2FePO_4F cathode and organic electrolyte (1M $NaPF_6$ in EC/DMC) operates at about 0.9V with a reversible charge capacity of about 100 mAh/g cathode (Fig. 2), corresponding to a specific energy of about 45 Wh/Kg based on the total weight of active electrode materials. In the charge and discharge curves, there are two evident plateaus, indicating the redox reactions of Na_2FePO_4F.

When these two electrode materials (coated on stainless steel current collector instead of aluminum foil) are assembled in a pouch cell with aqueous electrolyte, similar features in the voltage profile are observed (Fig. 3a), which refer to similar intercalation mechanism of the sodium ions in both organic and aqueous systems. The specific capacity of the aqueous cell during the first and second cycles is about 90 mAh/g cathode, slightly lower than the reversible capacity of the organic cell (100 mAh/g). The voltage of the cell is slightly higher than the organic cell, most likely due to the smaller internal resistance in the aqueous system. The cell capacity of 90 mAh/g and potential of 1.1 V correspond to a specific energy of about 50 Wh/Kg based on the total weight of active materials.

Despite these similarities for the first few cycles, the capacity of the aqueous cell drops significantly, and goes down to 50% of the initial capacity on the 15th cycle (Fig. 3b). The fast capacity fade in aqueous lithium- and sodium-ion batteries has often been reported.[7,14-17] The mechanism responsible for this severe capacity fading upon cycling has not been clarified.

In order to learn about the mechanism of capacity loss, we studied the positive and negative electrode materials separately, using a three electrode cell with Ag/AgCl reference electrode.

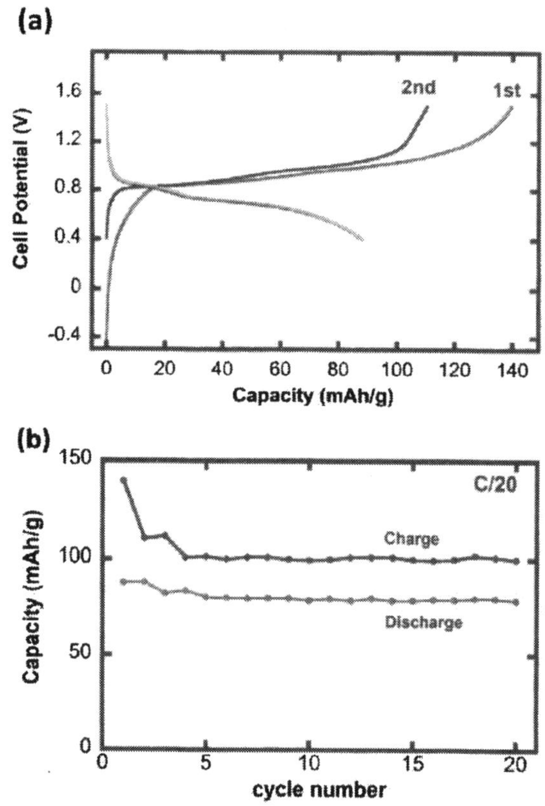

Figure 2: Electrochemical performance of a sodium-ion battery with $NaTi_2(PO_4)_3$ anode, Na_2FePO_4F cathode, and organic electrolyte cycled between 0.4 V and 1.5 V at 0.027 mA/cm^2.

Electrochemical performance of $NaTi_2(PO_4)_3$ at different C-rates in aqueous three-electrode cells is shown in Fig. 4a. At 5C, the plateau at -0.5V vs. Ag/AgCl suggests that electrochemical side reactions happen during the cell discharge. At higher C-rates i.e. 20C and 50C, the extra voltage plateau does not exist. Therefore, at slower charge/discharge rates the side reactions are eliminated or happen more slowly. The electrochemical impedance spectra of the assembled cells before cycling is shown in Fig. 4b. We did not observe any high frequency semicircle in the initial spectra of the as-made cells, but the impedance spectra of the cycled cells consist of one semicircle at high frequency range (Fig. 4c).

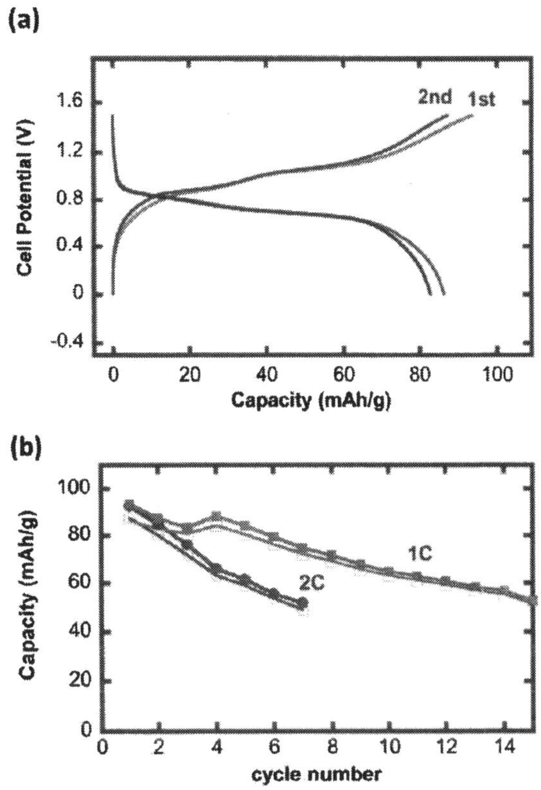

Figure 3: Electrochemical performance of sodium-ion batteries with $NaTi_2(PO_4)_3$ anode, Na_2FePO_4F cathode, and aqueous electrolyte (a) Voltage profile for the cell cycled between 0 V and 1.5 V at 0.21 mA/cm^2, (b) Capacity vs. cycle number for cells cycled at 1C (0.21 mA/cm^2) and 2C (0.42 mA/cm^2).

Formation of surface electrolyte interface (SEI) layer in non-aqueous lithium-ion batteries is a common observation and the high frequency semicircle is usually assigned to the diffusion of lithium ions through this SEI layer. In aqueous electrolytes, the formation of SEI is not extensively proved. It is reported that SEI layer is not formed in the presence of aqueous electrolytes, or even if it is formed the resistance of such an ultra-thin layer is negligible. Our results show that the size of the high frequency semicircle decreases at higher charge/discharge rates, and according to our electrochemical measurements side reactions are less pronounced at higher charge/discharge rates. Therefore, the high frequency semicircle originates from the products of the side reaction, and its size changes with the thickness of the surface layer. We see a better capacity retention for the 1C compared to the 2C (Fig. 3b), therefore the surface layer, similar to the SEI layer, could act as a protective barrier between the anode and the electrolyte.

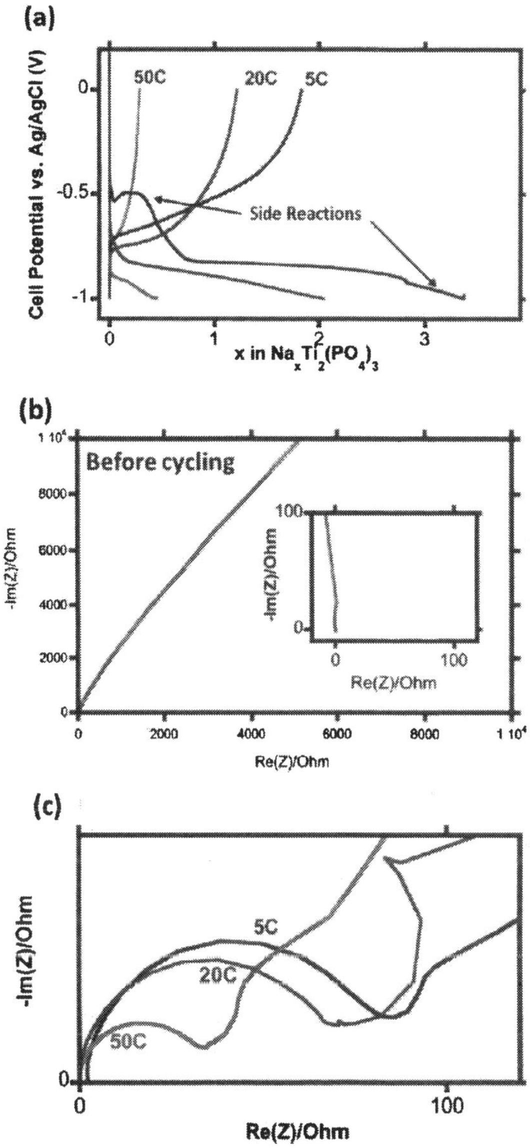

Figure 4: (a) Voltage profile of the $NaTi_2(PO_4)_3$ cycled in aqueous electrolyte between -1V and 0 V vs. Ag/AgCl at different C-rates. (b) Nyquist plots of the cells before cycling. (c) Nyquist plots of the cells after 20 cycles.

Similar measurements on Na_2FePO_4F positive electrodes in aqueous three-electrode cells were performed (Fig. 5a). At C/20 there was an additional, relatively large, plateau at around 0V vs. Ag/AgCl. This plateau became smaller at slower rates. Our conclusion is that side reactions

are eliminated or postponed in faster charge/discharge rates. Similarly, a high frequency semicircle formed after the cycling, and the size of this semicircle chnages with the charge/discharge rate and side reactions.
Based on our high resolution synchrotron XRD measurements (not shown here) the electrode compounds do not decompose during the cycling, and remain crystalline. We are currently studying the composition of the electrode surface using XPS which is a highly surface sensitive technique.

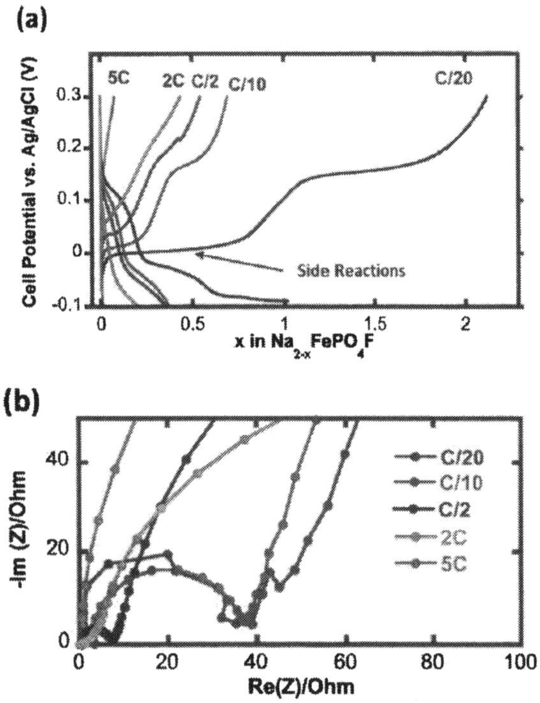

Figure 5: (a) Voltage profile of the Na_2FePO_4F cycled in aqueous electrolyte between -0.1V and 0.3 V vs. Ag/AgCl at diffrenet C-rates. (b) Nyquist plots of the cells after 20 cycles.

Strategies for improvement of cyclability in aqueous sodium-ion batteries are essential for implementation of these batteries in grid-scale storage. Our most recent experiments show enhancement of cycling behavior in $Na_{0.44}MnO_2$ cathode material by partial substitution of Mn with Ti.

4 CONCLUSIONS

We have developed an aqueous sodium-ion battery based on a NASICON-type anode material, and a fluorophosphate cathode material. The specific energy of the cell is 50Wh/Kg, which is larger than the specific energy of similar cell with organic electrolyte. The higher specific energy is mainly due to the lower electrolyte resistance, and therefore higher cell voltage. The similarities between the organic and aqueous systems suggest that the electrochemical reactions of electrodes in organic cells are valid in the aqueous systems. The electrochemical impedance data show that irreversible side reactions in slow cycling rates result in formation of a protective layer on both negative and positive electrodes.

5 ACKNOWLEDGMENTS

This work was supported by the Laboratory Directed Research and Development Program of Lawrence Berkeley National Laboratory under U.S. Department of Energy Contract DE-AC02- 05CH11231, and Department of Chemical and Materials Engineering at the University of Kentucky.

REFERENCES

[1] Wu, C. MRS Bulletin 2010, 35, 650.
[2] Ellis, B. L.; Nazar, L. F. Current Opinion in Solid State and Materials Science 2012, 16, 168.
[3] Palomares, V.; Serras, P.; Villaluenga, I.; Hueso, K. B.; Carretero-Gonzalez, J.; Rojo, T. Energy & Environmental Science 2012, 5, 5884.
[4] USGS (United States Geological Survey), http://minerals.usgs.gov/minerals/pubs/commodity.
[5] WSGS (Wyoming State Geological Survey), http://www.wsgs.uwyo.edu.
[6] Li, Z.; Young, D.; Xiang, K.; Carter, W. C.; Chiang, Y.-M. Advanced Energy Materials 2013, 3, 290.
[7] Kim, H.; Hong, J.; Park, K. Y.; Kim, H.; Kim, S. W.; Kang, K. Chemical Reviews 2014, 114, 11788.
[8] Slater, M. D.; Kim, D.; Lee, E.; Johnson, C. S. Advanced Functional Materials 2013, 23, 947.
[9] Shirpour, M.; Cabana, J.; Doeff, M. Energy & Environmental Science 2013, 6, 2538.
[10] Shirpour, M.; Cabana, J.; Doeff, M. Chemistry of Materials 2014, 26, 2502.
[11] Park, S. I.; Gocheva, I.; Okada, S.; Yamaki, J.-i. Journal of The Electrochemical Society 2011, 158, A1067.
[12] Kawabe, Y.; Yabuuchi, N.; Kajiyama, M.; Fukuhara, N.; Inamasu, T.; Okuyama, R.; Nakai, I.; Komaba, S. Electrochemistry Communications 2011, 13, 1225.
[13] Il Park, S.; Gocheva, I.; Okada, S.; Yamaki, J. Journal of the Electrochemical Society 2011, 158, A1067.
[14] Wu, W.; Mohamed, A.; Whitacre, J. F. Journal of the Electrochemical Society 2013, 160, A497.
[15] Wang, G. J.; Fu, L. J.; Zhao, N. H.; Yang, L. C.; Wu, Y. P.; Wu, H. Q. Angewandte Chemie-International Edition 2007, 46, 295.
[16] Li, Z.; Young, D.; Xiang, K.; Carter, W. C.; Chiang, Y. M. Advanced Energy Materials 2013, 3, 290.
[17] Luo, J. Y.; Cui, W. J.; He, P.; Xia, Y. Y. Nature Chemistry 2010, 2, 760.

Development of an alternative binder for lithium ion batteries based on epoxy resin

F. Bunzel*, M. Wagner*, E. Stammen* and K. Dilger*

*Technische Universität Braunschweig, Institute of Joining and Welding, Braunschweig, Germany, f.bunzel@tu-braunschweig.de

ABSTRACT

The development of batteries for the e-mobility continuously creates new challenges for the researchers. To improve the performance of lithium ion batteries, different approaches might be considered. Therefore the Institute of Joining and Welding (ifs) focuses its research on the development of alternative binders for active materials.

Hence, different reactive adhesive systems were characterized on the basis of acrylates, epoxy resins and silicones with different curing mechanisms and were examined for the use as a binder. However, epoxy resins in particular showed a good performance. On account of the possibility of being able to adjust the properties of epoxy resins to a certain extent, this system was investigated in more detail. The results of the present study indicated that the use of epoxy resins is beneficial in terms of their properties and the simplified process technology via using PVdF. It is possible to adjust the properties of the epoxy resin to use it as a binder in lithium ion batteries.

Keywords: lithium ion battery, binder, adhesive, epoxy resin

1 INTRODUCTION

The development of batteries for the e-mobility continuously creates new challenges for the researchers. To improve the performance of lithium ion batteries, such as higher energy/power density and longer shelf/cyle life, intense research has been focused on the development of different cell materials, such as positive and negative electrode active material, separators and electrolytes. So far, much less attention has been given to the binders, the electrochemically inactive material. The function of the binder is to bond the electrode particles, second to bond the active material to its substrate. Furthermore, the binder is responsible for the life cycle. The life cycle of a lithium ion battery is determined by different degradation mechnanisms in the cell. The electrochemical properties of the cell, as for example the available capacity, energy and power or the mechanical integrity of the cell deteriorate due to these degradation reactions. The causes of ageing of the cell are the dynamic stresses with charge and discharge and the structural changes caused thereby. Furthermore, the interaction between electrode and electrolyte and the chemical decomposition reactions originating from it and the modification of surface films contributes to the ageing. For example, the intercalation of lithium ions results in volume expansion of the electrode material from which macroscopic shear forces result by which the contact of the particle in the active material between them, the contact between active material and binder and further the contact of the binder with the substrate deterioates and thus the electron conductivity strongly decreases [1-2].

To raise the life cycle of a lithium ion battery, the degradation mechanisms described above must be prevented. One way of doing this is to substitute the generally used binder. However, this binder must show several requirements. On the one hand the binder in a lithium ion battery must provide the mechanical stability of the electrode, on the other hand it must guarantee the bonding of the particles of electrochemically active components to each other as well as to the current collector. Generally used binders are based on fluoridated polymers and copolymers, above all polyvinylidene fluorid (PVdF) and polyvinylidene fluorid-hexafluoropropylene (PVdF-HFP) [3-5]. From an ecological and economic point of view the application of the so-called fluorinated polymers is questionable, because their processing requires the use of organic solvents like N-methylpyrrolidon-2-on (NMP). NMP is identified as a substance of very high concern. Hence, there are numerous efforts to use healthwise safe binder systems and thus to improve the aspect of the environmental compatibility. In the literature numerous alternative binder systems are described for the anode as well for the cathode.

On the one hand different water-soluble binder systems were examined, mostly based on renewable raw materials. These include, for example, sodium carboxymethyl cellulose (Na-CMC) [4], [6-13]. To improve the mechanical stability a styrene butadien rubber (SBR) is added. In the anode as well as in the cathode, the Na-CMC shows an adhesive strength similar to that of the PVdF, however, the interaction to the active material is much stronger due to the high number of hydroxyl groups, and moreover the ion conductivity is higher than with the PVdF. In addition Na-CMC is significantly more elastic than PVdF, so that a higher stability and a longer life cycle are achieved. Another advantage of the Na-CMC is the discontinuation of calendering which otherwise would be necessary, due to the drying shrinkage of the binder. Other water-soluble binders based on renewable raw materials are alginate [14], chitosan [15], catechol [16] which show similar properties to an Na-CMC. Besides the water-soluble binders, only a few alternative polymers soluble in other solvents were used as binders in lithium ion batteries. A usable binder is a polyvinyl carbonate [17]. It was used as a binder in the graphite anode and led to a higher capacity as well as a much better performance at low temperatures in comparison to PVdF. Polysiloxane cross-linked with epoxide could be successfully used as a binder [18]. Toluene was used as solvent. It shows that, due to the flexibility of polysiloxane and due to the good chemical resistance of both components, a binder system with very good adhesion properties can be developed.

As regards operating completely without solvent, this work represents a very new approach. Reactive adhesive systems were investigated due to their low viscosity, high adhesion strength to aluminum and copper, high mechanical strength and chemical resistance. Furthermore, it is possible to modify the reactive adhesive systems very easily in view of curing time, elasticity and porosity. Different reactive adhesives are available, such as silicone, acrylate, PU and epoxide. In this work an epoxy resin was investigated in more detail.

2 EXPERIMENTAL

Active anode material – SMG (graphite powder) - was purchased from Hitachi Chemical Co. The active cathode material – LMO - was purchased from Toda Kogyo Corp. PVDF binder was supplied by Arkema, two-component epoxy resin, flexibilizer and foamer were purchased from R&G. Isopropyl alcohol was supplied by Merck.

For swelling experiments small plates of cured adhesives were stored in dimethylcarbonate (DMC) for 72 h. DMC was acquired from Merck. The thickness of the small plates was measured before and after swelling, and ATR spectra (Bruker Tensor 27) were also recorded.

Uniform slurries with different LMO/epoxy resin or graphite/epoxy resin ratios were made by adding dried LMO or graphite powder into the epoxy resin and then mixing them with a Speedmixer for 3 min. The slurries were then cast onto aluminum or copper foil using the doctor-blade method with a thickness of 200 µm. The adhesive was cured at 65°C in a oven for 12 h. The porosity was determined by a Laser Scanning Microscop (Keyence VK/9710).

For T-peel tests the slurries were cast between two foils with a thickness of 200 µm. The adhesion strength was determined using an Instron 5567.

For Dynamic Mechanical Analysis (DMA) measurements circular samples with a thickness of 2 mm were prepared. DMA measurements were carried out in shear mode with a fixed frequency of 1 Hz and a heating rate of 2 K/min.

3 RESULTS AND DISCUSSION

Various adhesives were investigated as regards the use of a binder in lithium-ion batteries, e.g. silicone, acrylates and epoxides. First of all the chemical resistance of the adhesives was validated. Swelling experiments shows that acrylates are not stable in the electrolyte. The ATR spectrum (Figure 1) shows degradation of the acrylate chain. The acrylate sample decomposed completely in the electrolyte. In contrast, an epoxide adhesive is stable in the electrolyte. The ATR spectrum shows no differences before and after the swelling test (Figure 2). Hence, only the epoxide adhesive is applicable as a binder in lithium-ion batteries.

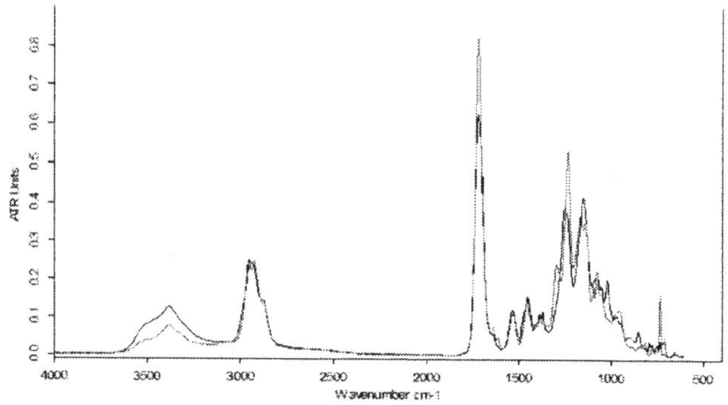

Figure 1: ATR spectrum of an acrylate adhesive; blue line before swelling, red line after 72 h swelling.

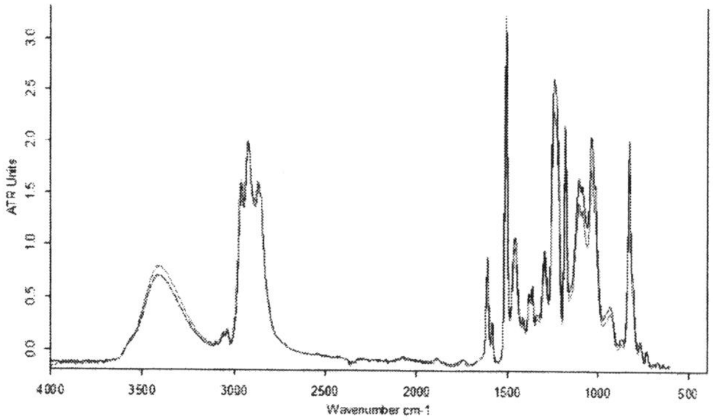

Figure 2: ATR spectrum of an epoxide adhesive; blue line before swelling, red line after 72 h swelling.

Mixtures with different contents of LMO and graphite were prepared. Up to 60 % graphite and 70 % LMO no solvent is neccesary. For the preparation of a mixure with a content of 80 % graphite or LMO, isopropyl alcohol was used as a solvent. Homogeneous mixtures were obtained with suitable viscosity for casting onto aluminum or copper foil using the doctor-blade method.

The adhesion strength was measured with a T-peel test (Figure 3). The results show that the adhesion strength improved with the use of a reactive epoxy resin as a binder. The adhesion strength is higher to aluminum foil than to copper foil and with increasing the content of the active electrode material a decrease of the adhesion strength is obtained. Different surface pretreatments, such as flame treatment and plasma, improve the adhesion strength. The best results were obtained with flame treatment. Hence, a delamination could be reduced with the use of a reactive epoxy resin.

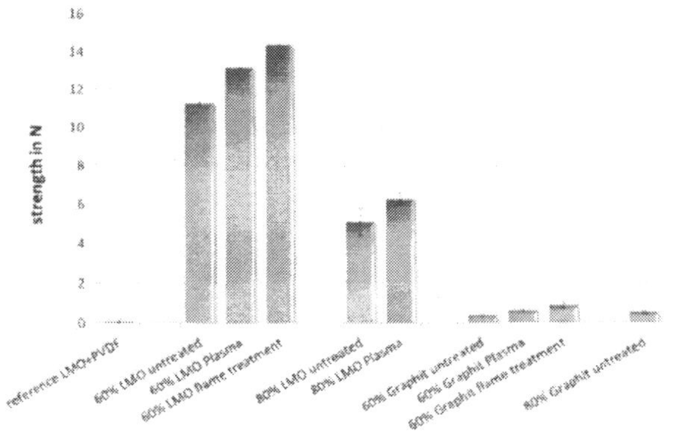

Figure 3: Adhesion strength determined by T-peel test

To get a sufficient porosity without a solvent, a foamer was used. The LSM images show that without a foamer low porosity of the coating can be seen (Figure 4b). By adding a foamer the porosity increased (Figure 4c) and is comparable to the reference (Figure 4a).

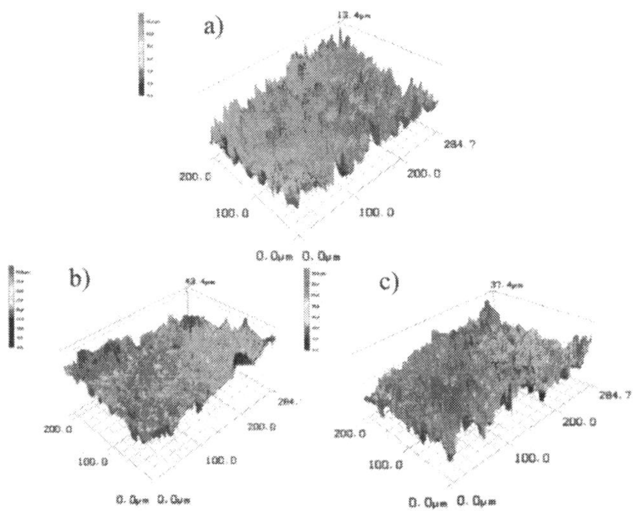

Figure 4: LSM images of a) PVdF graphite, b) epoxy graphite and c) epoxy graphite with foamer

An epoxy resin is normally very strong but brittle. To get a flexible binder in order to avoid crack formation by charging and discharging additives were used to modify the epoxy resin. By means of DMA measurements the elasticity and the T_g can be determined. Mixtures with LMO and graphite were investigated.

In Figure 5 the DMA curves of epoxy-LMO mixtures are shown. With increasing content of LMO the T_g decreases significantly, but the storage module E′ increases. This indicates an interaction between LMO and epoxy resin. Hence, the chemical structure and thus the properties of the epoxy resin are influenced by the LMO. Due to a high content of LMO the mechanical strength of the cured epoxy resin deteriorates, because at 0 °C the storage module E′ decreases significantly. With the addition of a flexibilizer the T_g decreases, too. But the change is not so significant. Furthermore, the storage module increases by adding a flexibilizer.

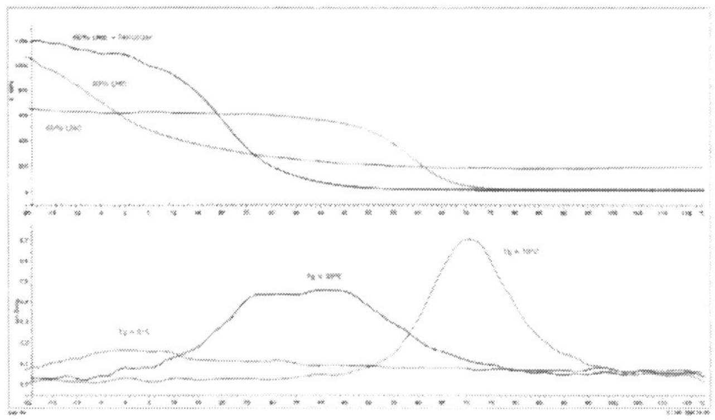

Figure 5: DMA curves of different epoxy-LMO mixtures, red: 60 % LMO, green: 80 % LMO and blue: 60 % LMO with flexibilizer

In Figure 6 the DMA curves of different epoxy-graphite mixtures are shown. In this case the T_g and the storage module show no shifting with increasing content of graphite. The chemical structure and thus the properties of the epoxy resin are not influenced by the graphite. By addition of a flexibilizer the T_g decreases and the storage module increases slightly.

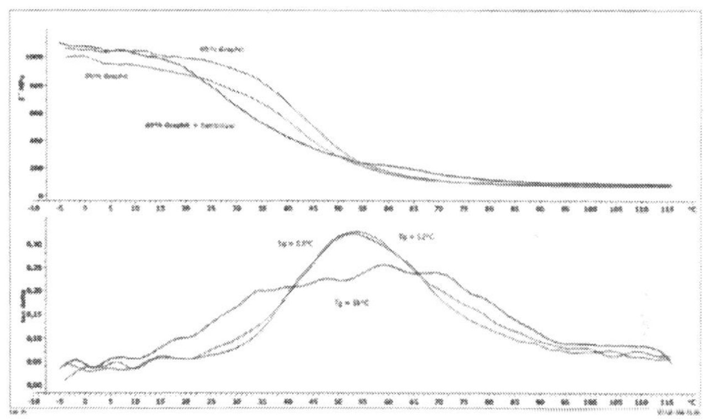

Figure 6: DMA curves of different epoxy-graphite mixtures, red: 60 % graphite, green: 80 % graphite and blue: 60 % graphite with flexibilizer

The DMA measurements show that it is possible to get a flexible binder by the addition of a flexibilizer in order to avoid crack formation.

4 CONCLUSIONS

Reactive adhesion systems such as epoxy resins are suitable for being used as a binder in lithium-ion batteries. Epoxy resins are chemically resistant and miscible with active electrode materials. They show a very high adhesive strength to aluminum and copper foil. The porosity of the coating can be adjusted by different foaming agents. The elasticity of the epoxy resin can be set by different flexibilizers. Advantages of an epoxy resin as binder in lithium-ion batteries are the chemical resistance, high adhesion strength, elasticity and no solvent is neccesary. Hence, the life cycle of a lithium-ion battery with an epoxy resin as binder should be improved.

5 ACKNOWLEDGEMENTS

The authors would like to thank the MWK Niedersachsen for the financial support of the project GEENI.

REFERENCES

[1] Chen, J., Liu, J., Qi, Y., Sun, T., Li, X., „Unveiling the roles of Binder in the Mechanical Integrity of Electrodes for Lithium-Ion Batteries“, Journal of The Electrochemical Society, 160, A1502-A1509, 2013.

[2] Rahani, E.K., Shenoy, V.B., "Role of Plastic Deformation of Binder on Stress Evolution during Charging and Discharging in Lithium-Ion Battery Negative Electrodes", Journal of The Electrochemical Society, 160, A1152-A1162, 2013.

[3] Barsykov, V., Khomenko, V.,"The Influence of Polymer Binders on the Performance of Cathodes for Lithium-Ion Batteries"; Scientific Journal of Riga Technical University, 67–71, 2010.

[4] Eom, J.-Y, Cao, L., Wang, C.-Y., "Effect of anode binders on low-temperature performance of Li-ion batteries"; The Electrochemical Society, 2012.

[5] Zheng, H., Yang, R., Liu, G., Song, X., Battaglia, V. S., "Cooperation between Active Material, Polymeric Binder and Conductive Carbon Additive in Lithium Ion Battery Cathode"; The Journal of Physical Chemistry, 116, 4875–4882, 2012.

[6] Hochgatterer, N., "Elektroden und Lithium-Ionen-Zellen mit neuartigen Elektrobinder", EP2 179 464 B1, 2008.

[7] Buqa, H., Holzapfel, M., Krumeich, F., Veit, C., Novák, P., „Study of styrene butadiene rubber and sodium methyl cellulose as binder for negative electrodes in lithium-ion batteries“, Journal of Power Sources, 116, 617-622, 2006.

[8] Moretti, A., Kim, G.-T., Bresser, D., Renger, K., Paillard, E., Marassi, R., Winter, M., Passerini, S., "Investigation of different binding agents for nanocrystalline anatase TiO2 anodes and its application in a novel, green lithium-ion battery"; Journal of Power Sources, 221, 419–426, 2013.

[9] Li, J., Klöpsch, R., Nowak, S., Kunze, M., Winter, M., Passerini, S., "Investigations on cellulose-based high voltage composite cathodes for lithium ion batteries"; Journal of Power Sources. 18, 7687–7691, 2011.

[10] Kim, G.T., Jeong, S., Joost, M., Rocca, E., Winter, M., Passerini, S., Balducci, A., "Use of natural binders and ionic liquid electrolytes for greener and safer lithium-ion batteries", Journal of Power Sources, 4, 2187–2194, 2011.

[11] Xu, J., Chou, S.-L., Gu, Q.-f., Liu, H.-K., Dou, S.-X., "The effect of different binders on electrochemical properties of LiNi1/3Mn1/3Co1/3O2 cathode material in lithium ion batteries", Journal of Power Sources, 225, 172–178, 2013.

[12] Li, C.-C., Wang, Y.-W., "Importance of binder compositions to the dispersion and electrochemical properties of water-based LiCoO2 cathodes"; Journal of Power Sources, 227, 204–210, 2013.

[13] Jeong, S.S., Böckenfeld, N., Balducci, A., Winter, M., Passerini, S., "Natural cellulose as binder for lithium battery electrodes", Journal of Power Sources, 331–335, 2012.

[14] Xu, J., Chou, S.-L., Gu, Q.-f., Liu, H.-K., Dou, S.-X., "The effect of different binders on electrochemical properties of LiNi1/3Mn1/3Co1/3O2 cathode material in lithium ion batteries", Journal of Power Sources, 225, 172–178, 2013.

[15] Yue, L., Zhang, L., Zhong, H., "Carboxymethyl chitosan: A new water soluble binder for Si anode of Li-ion batteries", Journal of Power Sources 247, 327–331, 2014.

[16] Ryou, M.-H., Kim, J., Lee, I., Kim, S., Jeong, Y.K., Hong, S. et al., "Mussel-Inspired Adhesive Binders for High-Performance Silicon Nanoparticle Anodes in Lithium-Ion Batteries", Advanced Materials, 25, 1571–1576, 2013.

[17] Zhao, H.; Zhou, X., Park, S.-J., Shi, F., Fu, Y., Ling, M. et al., "A polymerized vinylene carbonate anode binder enhances performance of lithium-ion batteries", Journal of Power Sources 263, 288–295, 2014.

[18] Xue, Z.; Zhang, Z.; Hu, L.; Amine, K., "Polysiloxane-Epoxide as Cross-Linkable Binders for Lithium-Ion Batteries", Journal of the Electrochemical Society, 160, A1819–A1823, 2013.

Li salt rich separator for Li-ion batteries

D. Mourzagh[1], S. Jouanneau[1], F. Rouillon[1], H. Rouault[1], M. Chapuis[1], S. Patoux[1]

M-D. Braida[2] O. Buisine[3] C. Mercier[3], C. Hamon[4]

1- CEA – Grenoble
LITEN/DEHT
17, rue des Martyrs 38054 Grenoble cedex 9- France

2- Solvay R&I
52 rue de la Haie Coq, 93300 Aubervilliers – France

3- Solvay Aroma performance
190 Avenue Thiers, 69457 Lyon Cedex 06 – France

4- Solvay Specialty Polymers
20, Viale Lombardia, 20021 Bollate - Italy

Abstract

In order to reduce the duration of the electrolyte filling step, a new concept of separator embedding itself Li salt thanks to a Li salt rich film coated on. The objective of study was to quantify the impact (benefits and drawbacks) of a Li salt containing PVdF based film deposited by coating onto a conventional polyolefin based separator. A protocol for elaborating the salt based polymeric solution to be coated onto a commercial polyolefin separator was developed at CEA/LITEN. Separators were coated with solutions at a defined coating thickness. The samples were characterized in terms of salt loading, morphology of the layers, adhesion of the film on the separator, ionic conductivity and ability to be soaked by the electrolyte. Some cycling tests were undertaken in full coin cells Cgr / LiFePO4 and soft pouch cells with samples of coated separators in order to evaluate the impact of the manufacturing conditions on the electrical performances. The impact of the storage conditions of the coated separator was also studied

Keywords: coated separator, Li-ion batteries, soft pouch cells, salt loading, SEM

I - Introduction

Solvay proposes a new concept of separator embedding itself Li salt thanks to a Li salt rich film coated on (Patent Solvay).
The objective of the present study was to quantify the impact (benefits and drawbacks) of a Li salt containing PVdF based film deposited by coating onto a conventional polyolefin based separator. The thin coated layer is composed of Solef® PVdF containing LiTFSI salt in appropriate amount. The separator selected for the study was the Celgard® 2320.

After having developed the method for manufacturing such a coated separator, electrical evaluations were conducted in representative intermediate capacity-level Li-ion cells, trials with pilot coating machines were also performed to demonstrate the feasibility of coating a separator at industrial level and the impact of the storage were also studied.

II- Development of the formulation of the ink

The aim was to design an ink composed of PVdF dissolved in solvent and containing LiTFSI salt with appropriate rheological characteristics, especially in terms of viscosity, to be coated on a polyolefin separator.

A protocol for elaborating the LiTFSI based polymeric solution to be coated onto a commercial polyolefin separator was developed at CEA/LITEN. The investigations lead on its preparation and some various characterizations as the dilution of the materials and the viscosity.

The lab coatings conditions of the solution onto the separator to reach the loading target were investigated: in anhydrous air (typically in a dry room at dew point -40°C).

As mentioned previously, Celgard® 2320 separators were coated with solutions at a defined coating thickness and dried in temperature.

III- Characterization of the LiTFSI coated separator

The different samples were so characterized in terms of shrinkage, salt loading, morphology of the layers, adhesion of the LiTFSI film on the separator, ionic conductivity and ability to be soaked by the electrolyte.

- *Skrinkage of separator*

To evaluate the shrinkage of the neat separator, several samples with the same dimensions (43mm x 43mm) were prepared and stored in an oven at 80°C. Their length and width were so measured at different time.

- *Salt loading*

The objective of the study was so to determine the coating thickness required to reach a target loading. Coating trials were done at different wet thickness.

- *Morphology of the coated separator*

Samples of separators coated with solutions were prepared for observations by Scanning Electronic Microscopy (SEM). They were cryo-fractured and gold metallized prior to the observations.(Figure 1 and 2).

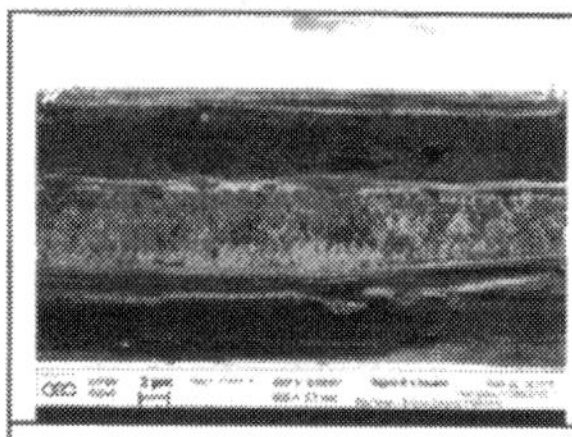

Figure 1: Separator Celgard®

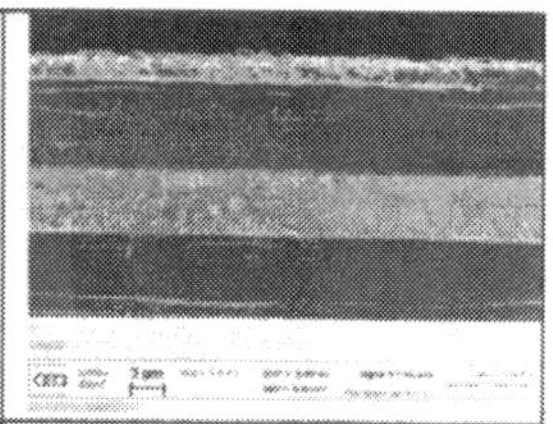

Figure 2: Separator Celgard® with thin coated layer.

- Wettability of the coated separator by the electrolyte

In order to evaluate the wettability of the coated separator by the electrolyte, measurements of contact angle were carried out in dry room. The measurements were performed by deposition of a drop of electrolyte on the coated separator. (Figure 3 and 4).

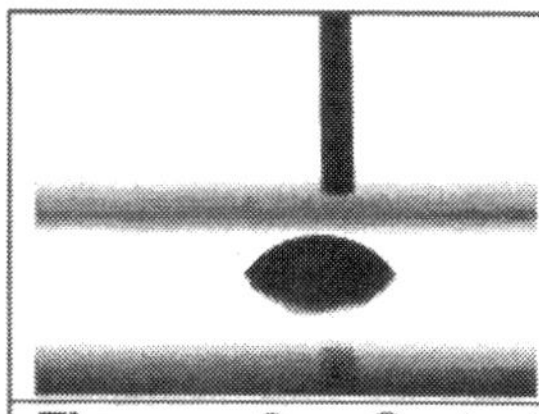

Figure 3: Contact angle measurement on separator Celgard® 2320

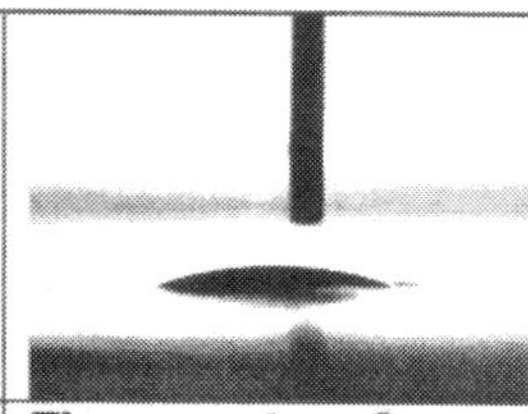

Figure 4: Contact angle measurement on Separator Celgard® 2320 with thin coated layer

- *Adhesion of the coated film on Celgard 2320 separator*

To estimate the adhesion of the film on the separator, peeling tests were performed on the coated separators in dry room.

- Ionic conductivity of the separator

In order to evaluate the ionic conductivity of the coated separator, coin cells constituted of two stainless steel disks separated by a pellet of coated separator (thickness Ep) and soaked with electrolyte were assembled.
The measurement of their resistance (R) by Electrochemical Impedance Spectrometry (EIS) thanks to Bio-Logic VMP multipotentiostat allows determining the ionic conductivity of the coated

separator.

The values of the ionic conductivity calculated from the resistance measured on the coated separators and their thickness are close to $0.3mS.cm^{-1}$, comparable with the ionic conductivity measured on the neat separator (i.e. ~$0.4\ mS.cm^{-1}$).

- *Evaluation of LiTFSI salt dissolution into the electrolyte by atomic absorption spectroscopy*

Celgard® 2320 separator was coated and dried. Samples of the coated separators (10cm x 10cm) were then immersed into 5 mL of neat carbonate during 2 weeks. The concentrations of LiTFSI were analyzed by Atomic Absorption in order to determine accurately the Li content (see Figure 5).

Figure 5 : Perkin Elmer AAnalyst 400 Atomic Absorption Spectrometry

III – Coating trials on large length

The objective of the coating trials on the industrial coating machines was:

- To demonstrate the feasibility of coating a LiTFSI rich polymeric film on large length of separator;
- Then to manufacture enough amount of coated separator for the assembly of some 53437 pouch cells in order to validate the coated separator on large wound prismatic configuration.

Two coating technologies were tested:

- One based on the reverse roll principle (Figure 6 : INGECAL equipment)
- The second on the use of slot die (Figure 7 :COATEMA equipment)

Figure 6: Reverse roll coating machine

Figure 7: Slot die coating machine

The tests revealed promising, demonstrating the feasibility of the separator coating at industrial level.

IV- Electrochemical characterization

Some cycling tests were undertaken in full coin cells Cgr / $LiFePO_4$ with samples of (coated) separators in order to evaluate the impact of the manufacturing conditions on the electrical performances. Thus the effect of the drying temperature on the neat separator and the effect of the composition of LiTFSI based polymeric solution used for the coated films was investigated through some discharge/charge cycles.

To reach the battery format dedicated to the consumer electronics market, CEA also tested in cycling the coated Celgard® 2320 separator in CEA standard wound prismatic cells, i.e. in soft pouch (500mAh – Figure 9).

Figure 8: Coin cell

Figure 9: CEA soft pouch cells (500mAh)

V-Impact of the storage conditions of coated separator on the performances in cycling

The impact of the storage conditions was also studied. Bands of Celgard®2320 separator were coated by LiTFSI rich. A half of the samples was stored in normal conditions during 1 week without any particular precautions when the second part was stored in dry conditions (dry room, dew point of -20°C).

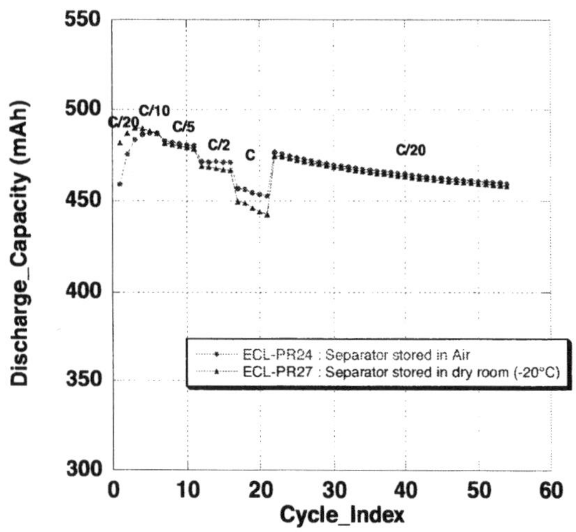

Figure 9: Electrical performances in cycling of Cgr/$LiFePO_4$ based 53437 pouch cells with different storage conditions of Celgard® 2320 separator.

It seems that the storage of such coated separators in air without any specific conditions (during one week) does not induce any consequences on the electrical performances.

VI – Conclusion and perspectives

In a preliminary study, a protocol for elaborating the LiTFSI based polymeric solution to be coated onto a commercial Celgard®2320 polyolefin separator was developed.

To coat a commercial polyolefin separator with a LiTFSI rich PVdF based film seems to improve clearly its wettability towards the electrolyte, without modifying its performances in cycling.

some coating trials were performed on pilot equipment using two technologies: one based on the reverse roll principle (INGECAL equipment) and the second on the use of slot die technique (COATEMA equipment). The tests revealed promising, demonstrating the feasibility of the separator coating at industrial level.

Moreover, it seems that the storage of such coated separators in air without any specific conditions (during one week) does not induce any consequences on the electrical performances.

However it would be interesting to evaluate the electrical behavior of bigger cells with coated separator.

Cluster-assembled Carbon Thin Films and Nanocomposites for Electrochemical Energy Storage

L.G. Bettini[*,§], P. Milani[*] and P. Piseri[*]

[*]CIMaINa and Physics Department, University of Milano, Via Celoria 16, 20133 Milano, Italy
[§]lucagiacomo.bettini@unimi.it

ABSTRACT

Supercapacitors are extremely versatile energy storage devices that store energy by the electrostatic separation of charged species at the interface between a solid electrode and an electrolyte forming the so called electric double layer (EDL). Although the supercapacitor market is rapidly growing, the continuous claim to improved energy and power storage pushes a lot of efforts to the synthesis of carbon based materials with tailored properties. Moreover, powering mobile/personal small-scale devices with on-chip integrated micro-supercapacitors is still a challenge that can only be addressed by the development of a scalable technique for the deposition and patterning of highly porous carbons in the form of thin films. Here the deposition of nanostructured carbon (ns-C) thin films and nanocomposites by Supersonic Cluster Beam Deposition (SCBD) is reported to be a viable route toward the fabrication of miniaturized and/or integrated electrochemical energy storage devices.

Keywords: carbon, supercapacitors, nanostructured thin films, gas-phase deposition, supersonic cluster beam deposition

1 INTRODUCTION

The rapid growth of mobile/personal electronics for applications in sensors, communication, health care, and environmental monitoring is pushing the demand for the sustainable and independent operation of small-scale systems. Currently, the powering of these devices still relies on batteries. The amount of batteries required increases in proportion with the increase in the number of mobile electronic devices used resulting in challenges for recycling and replacement of the batteries as well as concerns about potential environmental pollution [1]. Scientific efforts to develop sustainable, efficient, miniaturized, planar and integrable on-chip energy-storage systems are dramatically needed [2]. Power generation and storage for microelectromechanical systems (MEMS), miniaturized biomedical devices, sensors and integrated on-chip components, can significantly benefit from the fabrication and integration of planar thin film supercapacitors [2].

However, porous carbon fabrication protocols commonly employed in the manufacturing of supercapacitors are not easily integrable with thin film deposition processes and/or micro-patterning techniques. Although the amount of works in literature reporting on novel techniques for the synthesis of planar thin film carbon supercapacitors is rapidly increasing [2-5] for an effective miniaturization and planarization of supercapacitors it is essential to develop the ability of synthetize and deposit nanostructured porous materials in an efficient and reliable way that allow a fine control over the shape, thickness, morphology and structure of the carbon electrodes, possibly avoiding expensive and/or complicated fabrication steps such as photolithographic techniques, high-temperature processing and aggressive chemical treatments.

The deposition of carbon nanoparticles with low kinetic energy by Supersonic Cluster Beam Deposition (SCBD) is an effective bottom-up approach to the room temperature production and integration of nanostructured films on any substrate which is vacuum-compatible, allowing semi-continuous manufacturing in clean high-vacuum conditions.

2 SUPERSONIC CLUSTER BEAM DEPOSITION

The fabrication of complex devices by integrating of gas-phase nanoparticles with planar microtechnologies requires a very high degree of control on the lateral resolution of the nanoparticle deposition process, mild processing conditions (avoiding high temperature and aggressive chemicals), low contamination. In order to be industrially competitive the process must be compatible with standard top-down planar microfabrication technologies, highly scalable and low-cost.

Supersonic Cluster Beam Deposition (SCBD) fulfills these requirements: it consists in the deposition under standard high vacuum conditions (used for PVD coatings) of neutral nanoparticles (few nanometers in diameter) accelerated in a free jet expansion to form a supersonic seeded beam [6, 7]. The kinetic energy acquired by the nanoparticles is of the order of fraction of ev/atom thus preventing fragmentation of the nanoparticles upon landing

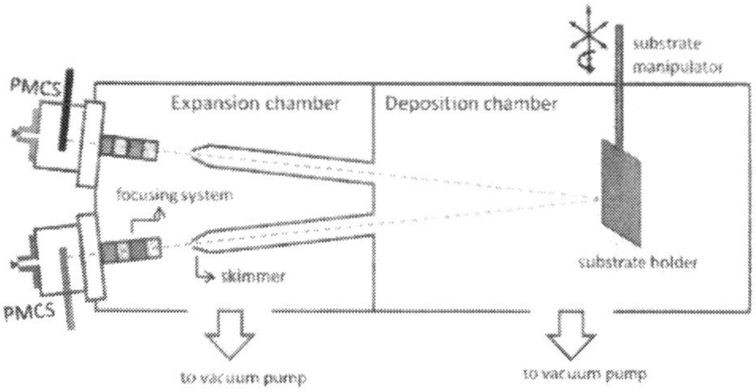

Figure 1: Sketch of a Supersonic Cluster Beam (SCBD) apparatus equipped with two Pulsed Microplasma Cluster Sources (PMCS)

on the substrate, while assuring good cohesion and adhesion. This results in a random stacking of nanoparticles leading to nanostructured films with very low-density and high porosity [6]. Due to the relatively low kinetic energy no substantial substrate heating is produced thus allowing the use of any kind of substrate, including the polymeric ones [8]. The use of supersonic expansions for cluster deposition allows to obtain very high deposition rates and highly collimated beam (divergence <20 mrad) which are suitable for the deposition of patterned nanostructured films by using stencil masks on substrates kept at room temperature. The resulting material is characterized by a low density compared to that of films assembled by depositing particles at high energies and it shows different degrees of order, depending on the scale of observation. The characteristic length scales are determined by cluster dimensions and by their degree of fragmentation and coalescence after deposition. The low degree of coalescence of clusters does not favor the formation of a hard material, however, it causes a large porosity and surface corrugation which can be beneficial for electrochemical applications [9, 10].

The coupling of the SCBD apparatus with a Pulsed Microplasma Cluster Source (PMCS) [11] and the application of aerodynamic separation concepts and techniques [12] to supersonic cluster beams for the purpose of nano-particle filtering, and for the production of high intensity particle beams allows to produce stable and intense nano-particle beams from most metallic elements and assures lateral resolution, deposition conditions, cleanliness and high throughput required by standard microfabrication technologies [6]. Furthermore, by rotating and rastering the substrate in the plane perpendicular to the nanoparticle beam it is virtually possible to coat samples of any large size area and geometry (from flat to complex curved surfaces).

Recently the feasibility of the Supersonic Cluster Beam Deposition (SCBD) technique for the deposition of nanostructured carbon (ns-C) thin films with promising electrochemical capacitive properties has been reported [10, 13].

Moreover, SCBD technique easily allow the synthesis of porous nanocomposite thin films by the mixing of clusters of different materials separately produced in distinct cluster sources (see Figure 1). In this way the two different materials do not interact during the cluster formation process and are mixed after the supersonic expansion, at the deposition stage. The low kinetic energy of the cluster in the beam avoids substantial fragmentation and coalescence upon clusters landing on the substrate and the result is a uniform mixing of the clusters. This strategy represents a powerful way to synthesize a virtually infinite range of porous nanocomposites with controllable and reproducible tuning of properties and composition. The embedding of nanoparticles in the carbon matrix can significantly affect ns-C properties such as porosity, density, surface area and material structure leading to improved physico-chemical properties.

3 PLANAR THIN FILM CARBON-BASED SUPERCAPACITOR FABRICATED BY SCBD

The supersonic cluster beam deposition (SCBD) of carbon clusters produced in a Pulsed Microplasma Cluster Source (PMCS) is a versatile technique for the fabrication of nanostructured carbon (ns-C) thin films with high porosity.

Ns-C deposited by SCBD consists in a high surface area disordered carbon characterized by substantial dominance of sp2 hybridization [9]. This material presents a typical density of ca. 0.5 g cm^{-3} [10] and a specific surface area measured by BET analysis of ca. 700 m^2 g^{-1} [9]. When ns-C is employed as active material in carbon-based capacitive electrodes this structure has been shown to originate an electric double layer gravimetric capacitance above 75 F g^{-1} both in aqueous [10] and ionic liquid electrolyte [13] .

These electrochemical performances, together with the relatively simple ns-C fabrication method, its compatibility with planar microfabrication technologies and the high control over the deposited material properties (e.g. thickness, surface roughness and porosity) obtained by the tuning of the deposition parameters empower the nanostructured carbon deposited by SCBD as an interesting material in electrochemical energy storage applications especially where extremely thin films of material are requested or where fragile substrates (e.g. polymers) are needed. The synthesis of porous carbon via SCBD presents several advantages, such as the compatibility with temperature sensitive substrates and with standard planar microtechnology processes, that may be pivotal toward the development of miniaturized thin film supercapacitors. The avoidance of binders to hold the ns-C to the current collector is another advantage of technological relevance toward the integration of nanostructured carbon electrodes on almost any kind of substrate (e.g. glass, silicon, paper

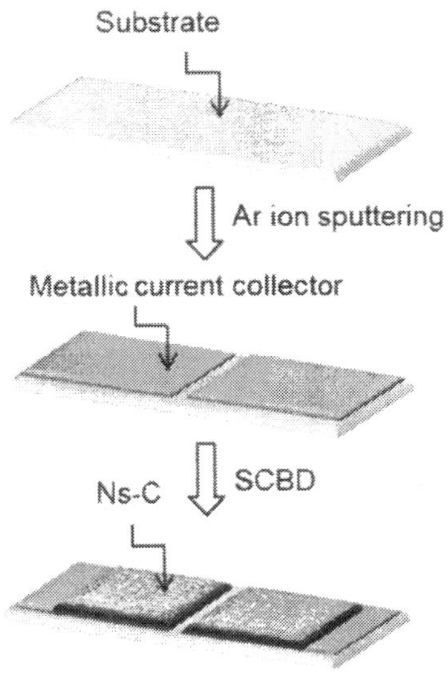

Figure 2: Sketch of the SCBD fabrication process of the planar ns-C based supercapacitors as reported in [13, 14].

and polymers). Moreover, the SCBD fabrication method allows to go beyond the traditional prismatic design of supercapacitors by replacing it with a 2D in-plane architecture that is pivotal toward their on-chip integration.

Supersonic Cluster Beam Deposition technique has been succesfully used for the room temperature fabrication of the proof of concept of planar thin film supercapacitors based on cluster assembled carbon (ns-C) and ionic liquids on both glass and flexible mylar substrates [13, 14]. The supercapacitors consisted of two in-plane ns-C electrodes with an area of ca. 1 cm^2 and thickness in the range between 200 and 500 nm separated by a non conductive gap of ca. 0.5 mm (Figure 2). A stencil mask was used during the SCBD process to obtain the desired electrode geometry. These devices showed an operating potential of 3 V with a capacitance density approaching 10 F cm^{-3} and delivering maximum specific energy and power densities of 10 mWh cm^{-3} and 8-10 W cm^{-3} with long cycling stability over more than 10^4 cycles even at temperatures up to 350 K [14].

Although the measured performances in terms of energy and power density, together with the intrinsic versatility of SCBD, are appealing for the fabrication of thin electrodes for microscale energy storage devices, the use of ns-C produced via SCBD in practical applications requires a further development of material physico-chemical properties (e.g. nanostructure and electric conductivity).

The codeposition of nickel and carbon clusters by a SCBD apparatus equipped with two separate PMCSs (see Figure 1) has been reported to be a feasible route to deposit porous nanocomposites with improved electrochemical properties [15]. The presence of embedded Ni nanoparticles in the carbon film preserves the high specific surface area of ns-C and drastically improves its electrical conductivity. The electric double layer capacitance of Ni:C films featuring the same thickness (200 nm) and different nickel volumetric concentrations (0÷35%) has been investigated by electrochemical impedance spectroscopy employing an aqueous solution of potassium hydroxide (KOH 1M) as electrolyte solution [15]. Evidence of increased electric conductivity, facilitated EDL formation and negligible porous structure modification was found as consequence of Ni embedding. This results in the ability to synthesize carbon-based electrodes with tailored specific power and energy density by the accurate control of the amount of deposited Ni and C clusters. Embedded nickel has been also shown to promote the formation of ordered graphitic nanostructures in heated nanocomposites which critically increase the rate capability of the electrodes without reducing their volumetric energy density [15].

The embedding of nanoparticles in the carbon matrix via SCBD can be thus an effective method to significantly improve the ns-C electrochemical performances not only in terms of electrode conductivity but also electrolyte reversible absorption and pseudocapacitive behavior, opening the way for the synthesis of advanced capacitive materials with enhanced energy storage properties.

4 CONCLUSIONS

Supersonic cluster beam deposition (SCBD) is a high throughput technique for the integration of nanoparticles into almost any kind of substrate. The microfabrication of planar and flexible thin film supercapacitors by the SCBD method has been demonstrated. The synthesis of nanostructured porous carbon thin films via SCBD presents several advantages, such as the compatibility with temperature sensitive substrates and with standard planar microtechnology processes, that may be pivotal toward the development of miniaturized, planar and flexible supercapacitors. The relatively simple ns-C fabrication method and the high control over the deposited material properties (e.g. thickness, surface roughness , porosity and composition) obtained by the tuning of the deposition parameters empower the nanostructured carbon deposited by SCBD as an interesting material in electrochemical energy storage application especially where extremely thin films of material are requested or where fragile substrates (e.g. polymers) are needed.

REFERENCES

[1] Z.L. Wang, W. Wu. Angewandte Chemie 51, 11700–11721, 2012.

[2] M. Beidaghi, Y. Gogotsi. Energy Environ. Science, 7, 867-884, 2014

[3] J. Chmiola, C. Largeot, P. Taberna, P. Simon, Y. Gogotsi. Science 328, 480-483, 2010
[4] D. Pech, M. Brunet, H. Durou, P. Huang, V. Mochalin, Y. Gogotsi, P. Taberna, P. Simon. Nature Nanotechnology 5, 651–654, 2010.
[5] M.F. El-Kady, R. B. Kaner. Nature Communications 4, 1475, 2013.
[6] K. Wegner, P. Piseri, H.V. Tafreshi, P. Milani. Journal of Physics D: Applied Physics 39, R439-R459, 2006.
[7] K Wegner, S Vinati, P Piseri, A Antonini, A Zelioli, E Barborini, C Ducati, P Milani. Nanotechnology, 23, 185603, 2012
[8] G. Corbelli, C. Ghisleri, M. Marelli, P. Milani, L. Ravagnan. Advanced Materials 23, 4504–4508, 2011.
[9] G. Bongiorno, A. Podestà, L. Ravagnan, P. Piseri, P. Milani, C. Lenardi, S. Miglio, M. Bruzzi, C. Ducati. Journal of Materials Science: Materials in Electronics 17, 427-441, 2006.
[10] L.G. Bettini, G. Bardizza, A. Podestà, P. Milani, P. Piseri. Journal of Nanoparticle Research 15,1429-38, 2013.
[11] E. Barborini, P. Piseri P. Milani. Journal of Physics D: Applied Physics 32, L105-L109, 1999.
[12] P. Piseri, A. Podestà, E. Barborini, P. Milani. Review of Scientific Instruments 72, 2261-2267, 2001
[13] L.G. Bettini, M. Galluzzi, A. Podestà, P. Milani, P. Piseri. Carbon 59, 212-220, 2013.
[14] L.G. Bettini, P.Piseri, F. De Giorgio, C. Arbizzani, P. Milani, F. Soavi. Electrochimica Acta 2015 (accepted).
[15] L.G. Bettini, G. Divitini, C. Ducati, P. Milani, P. Piseri. Nanotechnology 25, 435401, 2014.

Degradation Mechanisms in Electric Double Layer Capacitors Subjected to Voltage and Temperature Stress Testing

A. Peter*, M. H. Azarian** and M. Pecht***

Center for Advanced Life Cycle Engineering, University of Maryland, College Park, MD 20740
*anto@calce.umd.edu, **mazarian@calce.umd.edu, ***pecht@calce.umd.edu

ABSTRACT

To ensure that EDLCs are capable of satisfying reliability requirements, original equipment manufacturers (OEMs) need to be able to test and screen samples from several suppliers. In order to accomplish this, the screening test conditions would need to accelerate EDLC degradation without over-stressing them. In this paper commercially available EDLC samples from different manufacturers were subjected to voltage and temperature step stress testing. Each sample's capacitance and DC (internal) resistance were monitored. The variation of these electrical parameters with the applied stress levels is used to identify margins outside the rated temperatures and voltages, within which accelerated testing can be performed. We also identify the failure modes associated with increasing and voltage and temperature stresses. The underlying failure mechanisms associated with the observed failure modes are also determined through post-test failure analyses.

Keywords: Supercapacitors, screening, reliability exfoliation, SEI.

1 INTRODUCTION

Unlike other commonly used capacitor technologies, electric double layer capacitors (EDLCs) do not derive their capacitive energy storage properties from a conventional dielectric, as shown in Figure 1. Commercially used EDLCs use graphitic carbon deposited on aluminum or copper current collectors as electrodes. These electrodes are wound together with a separator soaked in an electrolyte containing ionic salts. Upon application of an external potential (during charging), the salt dissociates into its constituent anions and cations, and each of these species forms a double layer at the oppositely charged electrode. The charged species are held in place by electrostatic forces of attraction. The work done in separating these charges can be recovered when the externally applied potential is removed and replaced with a load. Unlike Li-ion batteries, the charge carriers do not diffuse into the electrodes in EDLCs. Hence these carriers have greater mobilities. Consequently, EDLCs have higher power densities compared to Li ion batteries.

Due to their high power densities, EDLCs are used to supplement high energy density power sources (such as lithium ion batteries and fuel cells) in applications such as automotive kinetic energy recovery systems. At the other end of the spectrum, EDLCs are also used in data storage technologies such as redundant array of independent disks (RAID). Such applications demand high reliability from the EDLCs used. Original equipment manufacturers (OEMs) can choose from several different manufacturers. However, in order to choose a supplier, OEMs must be able to screen samples from several different suppliers in a short period of time by using accelerated tests.

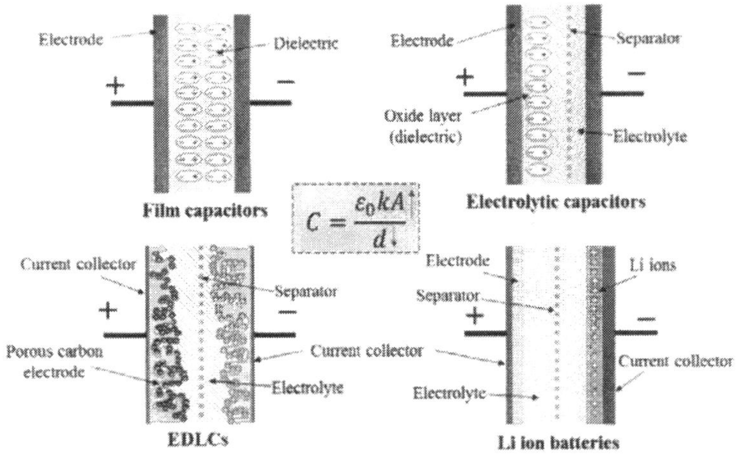

Figure 1: Charge storage in various energy storage technologies

EDLCs have usable lives of well over 100,000 cycles [1]. They are also known to be fairly robust under temperature ageing and thermal shock conditions [2]. Hence in order to accelerate test times, in this paper, we use progressively increasing voltage and temperature stresses. We then identify the associated failure modes seen in EDLCs. Based on results and trends from the stress tests and, we recommend voltage and temperature conditions within which accelerated testing can be performed. Such guidelines are currently unavailable.

We validate the suggestions by identifying the underlying failure mechanisms. This is achieved by performing material level analyses on deconstructed, tested and untested samples. Fourier transformed infrared spectroscopy (FTIR) was used to analyze the electrolytes, and scanning electron microscopy was (SEM) was used to analyze the electrode material. We describe how the reaction of the electrode and electrolyte could lead to formation of resistive films and degrade the structural integrity of the electrodes. These mechanisms irreversibly affect the electrical performance of EDLCs.

2 TEST DESIGN

Seven samples from six different manufacturers were subjected to voltage and temperature step stress testing. The samples along with their electrical ratings are shown in Figure 2. Temperature ratings are shown in Table 1.

The two parameters which were periodically measured for each sample over the course of the test were capacitance (C) and DC resistance (DCR). The equipment that was used in our study included the Arbin BT-2000 (battery tester, for electrical characterization) the Agilent N6705B (DC power analyzer, for voltage stressing) and a Yamato DVS402 (constant temperature oven for temperature stress testing).

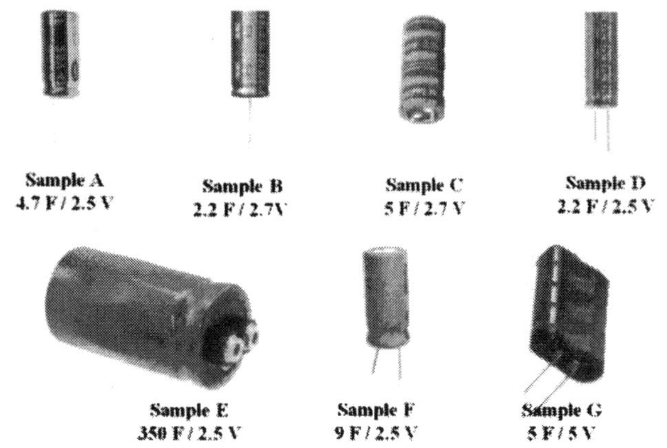

Figure 2: Samples used in the study

2.1 Measurement of Capacitance and DCR

Capacitance (C) and DCR measurements were made in accordance with the standards IEC 62391-1 and JIS 5160-1. For each measurement, the samples were first charged up to their rated voltages, and held there for 30 minutes. They were then discharged with currents I_D, which were calculated based on the capacitance and voltage ratings of the sample as described in the IEC standard. The currents used to discharge the samples for DCR measurements (DC resistance method) were 10 times larger than the currents used for capacitance measurements as shown in Table 1.

Sample	Discharge Currents, I_D (A)		Rated Temp (°C)
	Constant Current	DC Resistance	
A	0.004	0.04	70
B	0.002	0.02	70
C	0.005	0.05	65
D	0.002	0.02	70
E	3.500	35.00	60
F	0.009	0.09	60
G	0.010	0.10	70

Table 1: Discharge currents used during measurement of electrical parameters of test samples

A sample charge-discharge curve for capacitance and DCR measurements is shown in Figure 3. Based on the constant current (capacitance) discharge profile, the time t_2-t_1 taken for the capacitor to discharge from 80% of its rated voltage (U_1) to 20% of its rated voltage (U_2) can be calculated, and then used to measure the capacitance C as shown in equation 1. Similarly, the ohmic potential drop (ΔU) at the beginning of discharge can be used to calculate the DC Resistance (R_{DC}) as shown in equation 2.

$$C = \frac{I_D \times (t_2 - t_1)}{U_1 - U_2} \tag{1}$$

$$R_{DC} = \frac{\Delta U}{I_D} \tag{2}$$

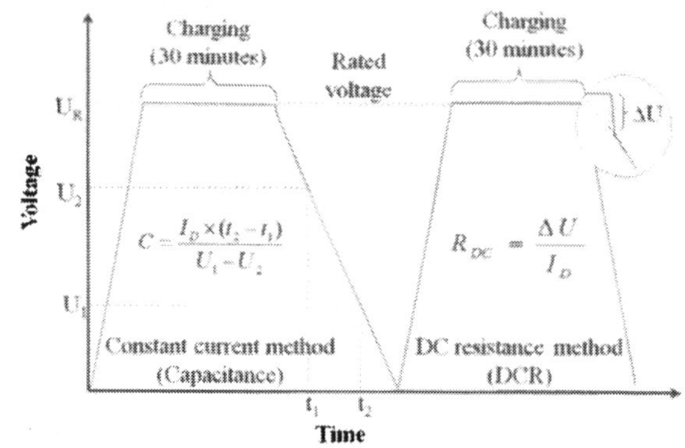

Figure 3: Sample charge discharge curve used for C and DCR measurements

2.2 Step Stress Test Methodology

All samples were first characterized at room temperature, before being stressed. Two sets of step stress tests were performed; one each subjected to temperature and voltage stresses. One sample of each type was included in each test. For the temperature step stress test, samples were first heated to their maximum rated temperatures. They were then held there for an hour, before being cooled down to room temperature. Following the cool-down, capacitance and DCR were measured at room temperature. The process was repeated with the temperature set point being raised by 10°C each time, until the capacitors were stressed at a maximum temperature of 150°C. The capacitors were not biased during temperature stress test.

Similarly, for voltage step stress testing, the samples were charged up to their rated voltages, and held there for 30 minutes. They were then discharged completely. Capacitance and DCR were then measured at room temperature. The process was repeated by charging the samples to a voltage level that was 0.5 V higher than the previous set point. The voltage stress tests were performed at room temperature. Samples E and G could not be included in the voltage test, as their test requirements exceeded the instrumentation capability.

3 RESULTS AND DISCUSSION

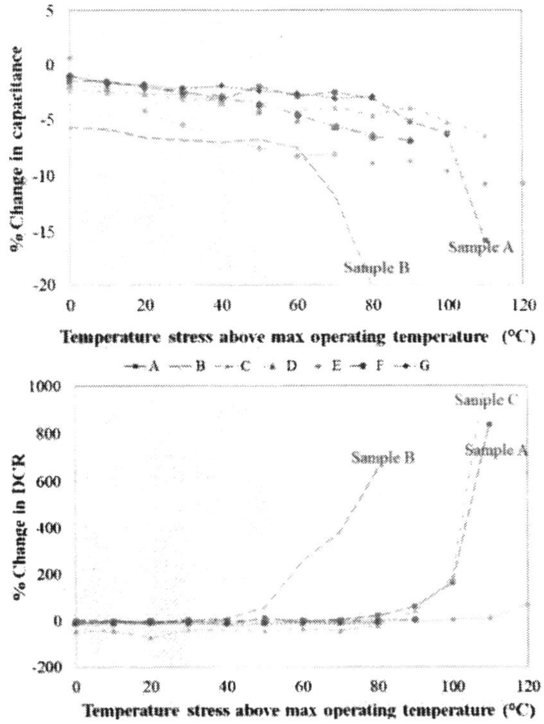

Figure 4: Capacitance variation (above) and DCR variation (below) for temperature step stress tests

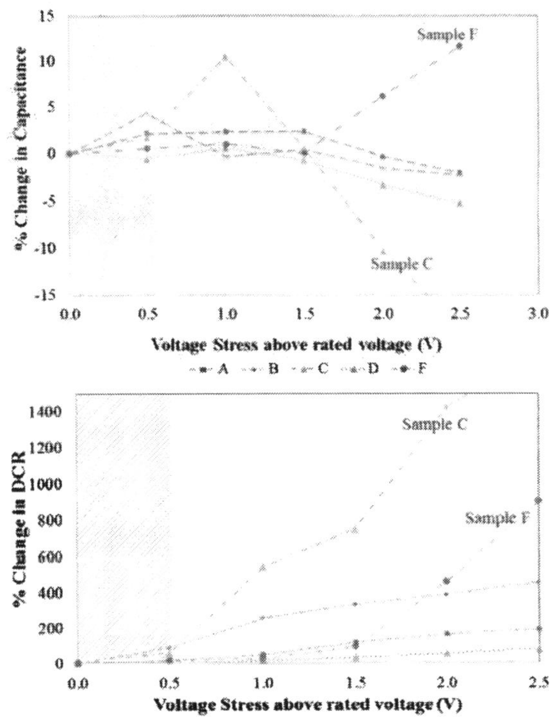

Figure 5: Capacitance variation (above) and DCR variation (below) for voltage step stress tests

Figure 4 depicts the results from the temperature step stress tests. It is seen that both capacitance and DCR either remain relatively constant, or degrade very gradually as the samples are stressed at temperature levels within 50°C of their rated temperatures. However, as the samples continued to be stressed at higher temperatures, a marked decrease in capacitance and increase in DCR could be seen in some samples.

In the voltage stress tests, the capacitance of samples initially increased with voltage stressing (Figure 5). However, with further increase in voltage stress levels, the capacitance of most samples decreased, with the only exception being sample F. The DCR trends for voltage stress tests resembled those seen in the temperature step stress test.

With regard to capacitance degradation, it is clearly evident from our results that voltage and temperature stresses induce different failure mechanisms within the EDLCs. A decrease in capacitance often seems to be accompanied by an increase in DCR. This correlation could imply that the underlying mechanisms behind this behavior be coupled. However, since there are cases where a significant drop in capacitance is not accompanied by a large increase in DCR (sample E under temperature stressing, and sample D under voltage stressing), the two degradation mechanisms could be distinct.

The failure modes in both sets of stress tests were increase in DCR and decrease in capacitance. A change in the degradation trends is seen when samples are subjected to temperatures more than 50°C above their rated temperature. Similarly, subjecting samples to voltage stresses more than 0.5V above their rated voltages causes the degradation to accelerate rapidly. Hence, while designing accelerated tests, OEMs should try to stay within these limits (indicated by the shaded regions in Figure 4 and Figure 5).

4 FAILURE ANALYSIS

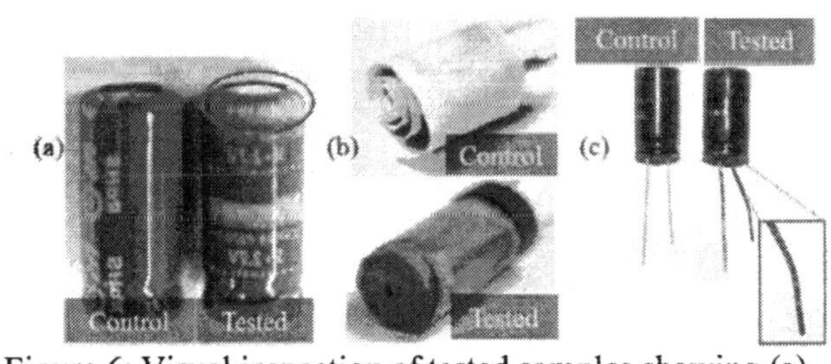

Figure 6: Visual inspection of tested samples showing (a) bulging of can (b) separator degradation (c) lead corrosion

Figure 6 shows the physical changes observed in tested samples. Bulging and corrosion of leads were seen on most samples; more prominently so on samples subjected to thermal stressing. Both these changes indicate that samples experienced a loss of electrolyte, either due to side reactions occurring within the EDLC, or leakage through the seal. This implies that there was a reduction in the number of active charge carriers, which could lead to a decrease in capacitance. Removal of the external metallic can (packaging) off some samples revealed degraded separators, showing signs of oxidation. This degradation could affect ion mobilities, as the porosity of the separator may have been altered. This would be reflected as an increase in DCR, as seen in our results.

Strips of the electrode assembly from samples subjected to temperature and voltage step stress testing were removed, and washed in distilled deionized water to remove the salts from the electrolyte. They were then examined under the scanning electron microscope. It was seen that in the temperature stress tested samples' electrodes, the electrode material agglomerated in some regions (Figure 7). In the voltage tested samples, the electrodes showed signs of exfoliation and cracking. These were visible at fairly low magnifications as shown in Figure 8. FTIR analysis was also performed on EDLC electrolytes, and it was seen that all samples except B used propylene carbonate (PC) as a primary solvent. Sample B used Acetonitrile (AN).

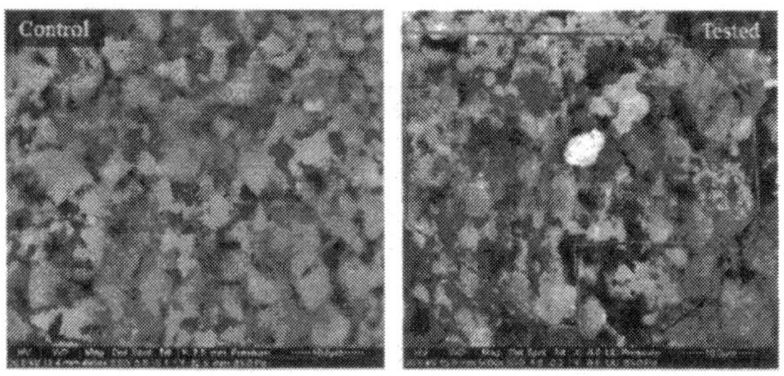

Figure 7: SEM micrographs showing active material agglomeration in electrodes of temperature tested sample

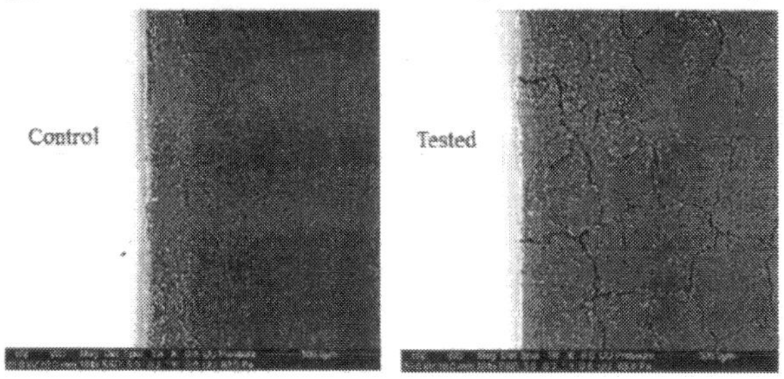

Figure 8: Electrode exfoliation in voltage tested sample

5 FAILURE MECHANISMS

During solvation of the EDLC salts in PC, the dissociated ions from the salt are surrounded by PC molecules to form a sphere known as solvation sheath (more information in [3]). These solvation sheaths are significantly bulkier than the original ions, which are represented as X^+ in Figure 9. When an external potential is applied to the EDLC, such as the voltage stresses in our test, the solvation sheaths, as a whole, are driven through the porous carbon electrode. The electrodes would experience greater stresses with increasing electric field strengths that result from greater voltage stresses. As a result, the electrode is prone to cracking and exfoliation, as seen in Figure 8. This would also be reflected as an increase in capacitance due to the increased surface area available for double layer formation. In designing accelerated tests, excessive exfoliation would represent an overstress.

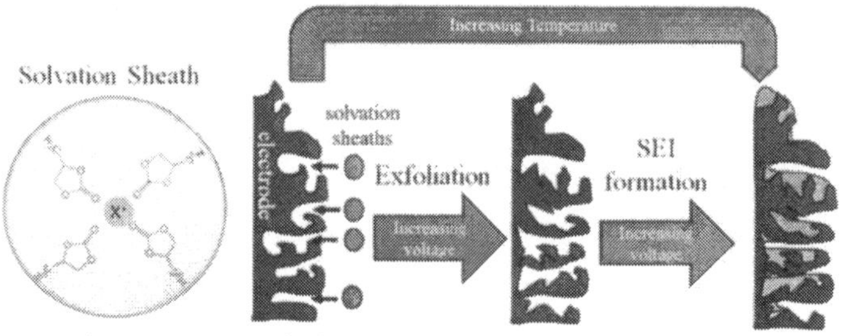

Figure 9: Degradation mechanisms in EDLC electrodes

Continuous increase in the voltage stress applied would lead to decomposition and reduction of the electrolyte with the electrode material. This would lead to formation of a new resistive layer on the electrode referred to as a solid electrolyte interphase (SEI). Thermodynamically, SEI formation would also be favorable at elevated temperatures, and hence the similarity in results from both our tests. The agglomeration seen in Figure 7 could be an SEI layer. Since SEI formation by decomposition of PC is known to be accompanied by gas generation [4], we would see bulging in tested samples. The increased pressure within the EDLCs could cause the electrolyte leakage seen in our tests. SEI formation also consumes active charge carriers and causes reduction of the surface area of the electrode, both of which would lead to a loss of capacitance. The resistive nature of SEI would also cause increased DCR of EDLCs.

6 CONCLUSIONS

When designing accelerated tests for EDLCs, temperature stresses should be limited to a maximum of 50°C over the rated temperature. Similarly, for voltage stress tests, the stress levels should be limited to 500mV over the rated voltage. Commonly observed failure modes under these stress conditions is an increase in DC resistance and a loss of capacitance. These failure modes can be explained by formation of resistive SEI layers which leads to a loss of both active charge carriers as well as surface area available for double layer formation. A temporary increase in capacitance may be obtained due to the exfoliation of carbon electrodes by solvation sheaths under voltage stresses.

REFERENCES

[1] A. Burke, Journal of Power Sources, vol. 91, no. 1, pp. 37-50, 2000.

[2] H. Gualous, R. Gallay, G. Alcicek, B. Tala-Ighil, A. Oukaour, B. Boudart and P. Makany, Microelectronics Reliability, vol. 50, no. 9, pp. 1783-1788, 2010.

[3] K. Xu, Y. Lam, S. S. Zhang, T. R. Jow and T. B. Curtis, The Journal of Physical Chemistry C, 111(20), 7411-7421, 2007.

[4] P. Verma, P. Maire, and P. Novák, Electrochimica Acta 55.22, 6332-6341, 2010.

Renewables-source Energy Storage for < $1.00 / kWh CAPEX as Hydrogen and Ammonia C-free Fuel Systems

Hydrogen and Ammonia Fuels via Underground Pipelines with Storage in Caverns and Tanks

W. Leighty, The Leighty Foundation, Box 20993, Juneau, AK 99802 wleighty@earthlink.net
www.leightyfoundation.org/earth.php

ABSTRACT Humanity's grand challenge now is converting the world's largest industry from ~ 85% fossil to 100% renewable energy (RE) sources as quickly as we prudently and profitably can, to "Run the World on Renewables" -- perhaps including some nuclear, now very hard to predict. Nothing less will allow us to escape these synergistic global consequences of large-scale fossil fuel combustion:

- Rapid climate change, generally warming
- Accelerating sea level rise
- Ocean acidification
- Species extinctions
- Violent human conflicts

Jacobson and Delucchi have demonstrated that we can run the world on wind, water, and solar (WWS). The wind energy of the twelve Great Plains states, if fully harvested on about 50% of these states' aggregate land area, transmitted to distant markets, and "firmed" at annual scale with storage, could supply the entire annual energy demand of the USA: about 10,000 terawatt-hours (TWh = billion kWh), or about 100 quads (quadrillion btu). However, existing Great Plains electric transmission export capacity is insignificant relative to this resource. Any large, new electric transmission systems, or fractions thereof dedicated to wind energy, will:

- Be very costly to build;
- Be difficult to site: FERC has no authority for permitting interstate electric lines;
- Be difficult to route and permit, because of public objection, as in NIMBY;
- Suffer the same low capacity factor (CF) (typically 40%) as the windplants and other RE plants they serve, unless RE gen curtailed;
- Provide no affordable "firming" (weekly-to-annual scale) energy storage, taxing "system balancing" ability of the electricity grid;
- Be vulnerable to damage by acts of God and man.

Relieving RE generation and conversion equipment of the requirement to deliver "grid quality" AC (Voltage, frequency, power factor, and harmonics) as well as "low voltage ride through" (LVRT), because RE electricity is instead entirely delivered to electrolyzers and NH_3 synthesis systems, may significantly reduce the capital and O&M costs of RE -- at the plant gate and at end-user.

Therefore, the RE industry needs to look beyond the electricity grid, which now supplies only about one-third of a modern country's total energy, to supply all energy, from all sources, for all purposes, without CO_2 emissions, as firm and dispatchable energy. Electricity systems are not well suited for delivering on-demand energy services from the time-varying output -- at scales of seconds to seasons -- of most renewables, while simultaneously and affordably providing:

1. Gathering and transmission, from both large, stranded, "centralized" RE plants and smaller "distributed" generation (DG)
2. Annual-scale firming storage, rendering RE dispatchable
3. Distribution, integration, CHP, and delivering energy services to end-users

A technically and economically superior strategy may be to:

1. Convert the wind or solar plant's entire energy output to the only two C-free fuels, hydrogen, H_2, and / or anhydrous ammonia, NH_3
2. Transport the fuels in underground pipelines as gaseous hydrogen (GH2) and / or liquid NH_3
3. Completely avoid the high capital and O&M costs of:

a. Electricity grid connection: gathering, field transformers, substation, and HV line links
b. Sophisticated generating systems required to deliver grid-quality AC with LVRT

4. Acquire, store, and deliver surplus energy from PV and other distributed RE generation.

Having left electricity at the RE sources, we would now deliver C-free fuel(s) to distant markets for stationary combined-heat-and-power (CHP), transportation fuel, and the complete other spectrum of residential, commercial, and industrial uses. We are thus emulating the natural gas industry's ubiquitous pipeline network and freedom from large, looming "Smart Grid" investments, enabled by their large and diverse array of low-cost energy storage and inherent long time constants.

The electricity grid, as we've known it, is rapidly transforming, perhaps atrophying, under pressure from DG, nascent demand for BEV charging, and new Smart Grid complexity, which together may make the grid more susceptible to cyberattack. While the Grid may always be with us, we need to look beyond it in our compulsion to run the world on renewables.

Figure 1 shows that GH2 and NH_3 underground pipeline transmission capital costs compare favorably with electricity, oil, and gas per MW-km, which should be the common measure of any transmission system's service. Underground pipelines are better protected; O&M costs per MWh-km are lower. New pipelines are generally easier to site and permit than new electric transmission lines. For example, these pipelines have 8 GW capacity:

- 1 m diameter, 800 km long, GH2 pipeline, at 100 bar input, 30 bar output, with no midline compression
- 50 cm diameter NH_3 pipeline, with midline pumping, of any length

Figure 2 shows the system simplicity and inherent low-cost energy transmission and storage when high-pressure-output electrolyzers input the GH2 pipeline; midline compression for < 1,000 km length is not needed. "Packing" the GH2 pipeline, as natural gas pipelines routinely are, provides "free" storage. If a national or continental GH2 pipeline "grid" can access a region of domal salt geology, very large and low-cost RE storage is available in large, deep, solution-mined, man-made, salt caverns, in Figure 3. Multiple caverns may be manifolded together, operating at the same pressure, sharing a surface facility, to reduce capital cost to well below \$ 1.00 / kWh at multi - 100 GWh scale. Two GH2 storage salt caverns, near Moss Bluff, TX, have operated well, with low O&M: Chevron-Phillips cavern for > 30 years; Praxair cavern for > 7 years.

Figure 3. USA has enough domal salt geology along the Gulf of Mexico coast to install ~ 15,000 large GH2 storage caverns, providing enough storage to firm the entire Great Plains wind resource, to provide all ~ 100 Quads of USA's total energy demand as annually firm and dispatchable. Northern Germany is also endowed with good salt geology for large storage caverns.

The large, nascent H_2 fuel market for fuel cell buses and light duty vehicles (LDV's) should motivate interest in integrated, optimized, RE-source systems. Toyota North America chief Jim Lentz recently told *Automotive News* that Toyota sees battery-electric vehicles (BEV's) only viable in "... a select way, in short range vehicles ..." Transportation fuel may be a larger market than the Grid, and may motivate turbine suppliers to offer equipment optimized for off-grid H_2 and NH_3 fuel production, with simpler, more robust, and less-costly electric generating systems, using variable-speed induction motors, as in Figure 5. Polymer-metal tubing and linepipe developed at ORNL may replace steel for high-pressure transmission pipelines, solving the hydrogen embrittlement and other corrosion problems.

NH_3, the "other hydrogen", is a higher-hydrogen-density energy carrier and storage medium than liquid hydrogen, LH2. A liter of NH_3 contains more H atoms than a liter of LH2. The modified internal combustion engine, combustion turbine, and fuel cell run well on NH_3, while RE transmission and storage costs are even lower than for GH2.

Figure 4 shows the Agrium terminal where twin mild steel tanks together hold 80,000 tons of refrigerated, liquid anhydrous ammonia, NH_3, which stores ~ 500,000 MWh as the energy content in NH_3, which may be directly used as fuel or may be easily reformed to H_2 fuel and harmless, vented N_2. At this scale, energy storage costs ~ $ 0.10 / kWh capex. Over 4,000 km of NH_3 transmission pipeline is already in place in USA, where NH_3 is used primarily as N-fertilizer. These pipelines are low-cost, mild steel, operating at ~ 15 bar.

Figure 5: By any measure, the low cost and high capacity of GH2 and NH3 storage is "off the charts" vis-a-vis the several technologies considered for "electricity" storage, as adjuncts to the Grid.

Great Plains wind is very seasonally-variable: about 20% above average in Winter and Spring, 30% below average in Summer. Firming 1,000 MW (nameplate), to allow firm energy delivery every day of the year, would require ~ 300,000 MWh of storage: ~ 4 GH2 salt caverns or ~ 2 NH_3 tanks. In either case, the storage incremental capex is ~ 3 % of total windplant cost. Compressed air or reflow battery storage is many times higher. If we consider investing in the energy conversion equipment necessary to avail us of this low-cost storage, we should consider solving the transmission and integration problems as well, via complete, integrated, optimized, RE systems -- from sources to delivered energy services -- employing GH2 and NH_3 as the transmission and storage media. We should not allow ourselves to be limited to defining "storage" as an appurtenance to the electricity Grid.

Japan is researching three strategies for importing C-emissions-free H_2 from distant sources, as high-hydrogen-density liquids in ocean tankers, to build a significant "hydrogen sector" in their national energy system:

- Liquid hydrogen, LH2: Kawasaki Heavy Industry
- Anhydrous ammonia, NH_3: Sumitomo Chemic
- Toluene - Methylcyclohexane cycle: Chiyoda Chemical and Heavy Industry

Alaska's diverse, abundant, stranded RE resources are strategically close and convenient for export to Japan via any of these H_2 transmission schemes, and have been approximately surveyed.

For over a decade, researchers have been bravely and brilliantly engineering wind turbines and solar PV and CSP systems, and the Grid, to accommodate an increasing share of RE, attempting to deliver firm and dispatchable electricity with system stability while displacing fossil generation. This is ever more costly and technically difficult, at the margin, for both generation and transmission. This work is well reported by the IEEE Power and Energy Society in their journal, *Power and Energy*. But we have only begun to ask how RE may serve energy markets beyond the extant electricity sector, whether via an expanded electricity sector and Grid, or perhaps via alternatives to electricity. Do we imagine that we may supply all humanity's energy, for all purposes, from diverse RE sources, entirely, or even primarily, via electricity systems? The Grid, assaulted by PV systems large and small, connected to distribution circuits, may so evolve and atrophy as to be inadequate and unsuitable to "Running the World on Renewables"

Therefore, we now need to assemble a collaborative of private enterprise, academia, national labs, funders, and others to thoroughly and expertly model, technically and economically, complete RE systems -- from photons and moving air and water molecules to delivered energy services -- in the context of humanity's responsibility to expeditiously "Run the World on Renewables". Before we invest billions in a larger and smarter electricity Grid, we need to be sure we've considered alternatives, that we're not trying to stuff a square peg into a round hole. This serious consideration has never been done; our co-authors attempted analysis have been provocative, but inadequate and unpersuasive.

Please contact the author if you wish to consider participating. Conference presentations and papers on this and related RE topics are at: www.leightyfoundation.org/earth.php

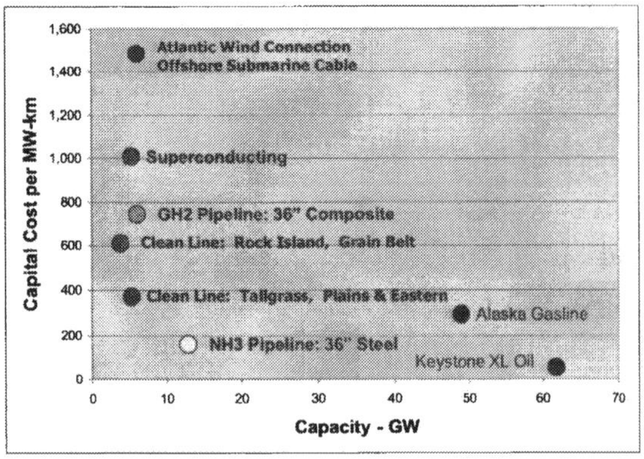

Figure 1. Approximate transmission capital costs per MW-km of service, $ US

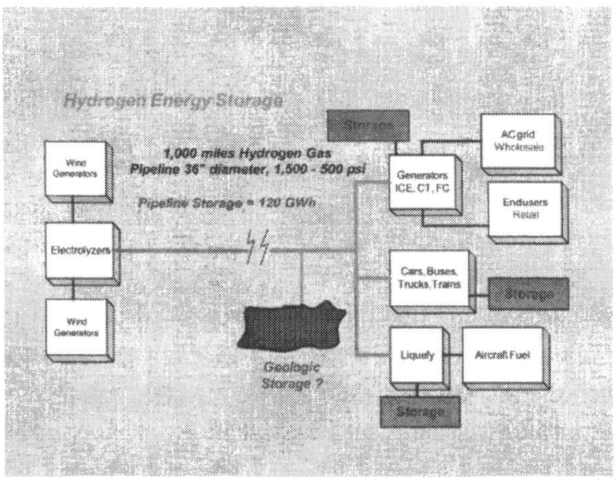

Figure 2. Complete Renewable Energy GH2 system: annual-scale firming, dispatchable storage

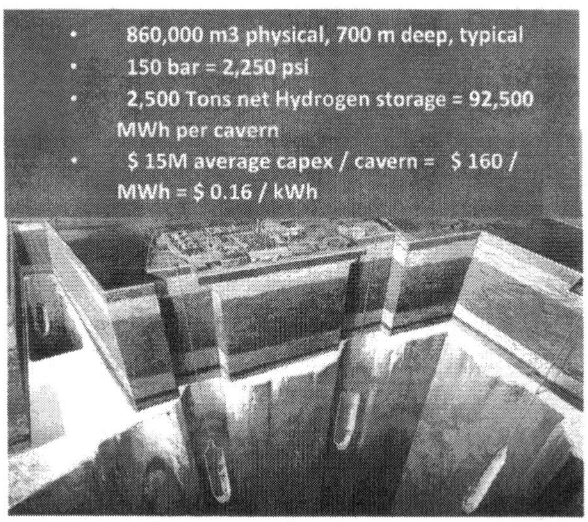

Figure 3. Solution-mined compressed Hydrogen storage caverns in deep domal salt. Multiple caverns manifolded together, share a common surface facility, for very large energy storage.

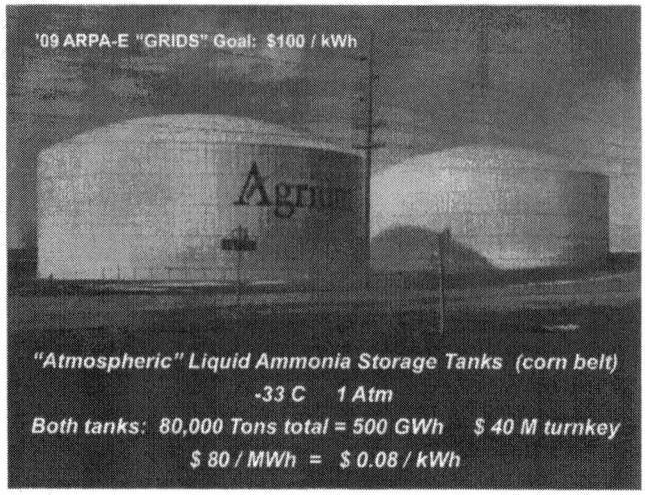

Figure 4. Agrium US, ammonia storage terminal at Early, Iowa, supplying N-fertilizer for corn ag.

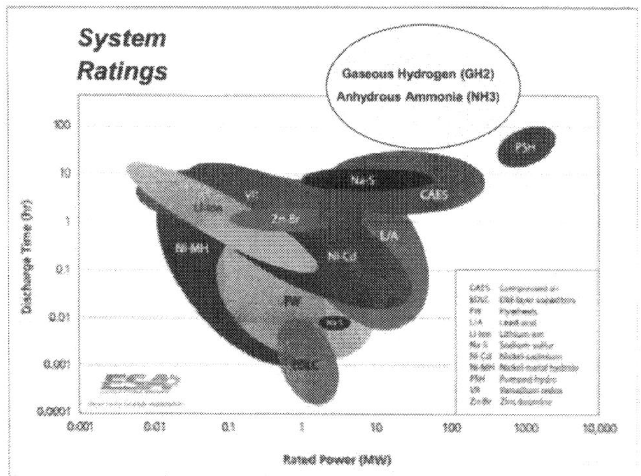

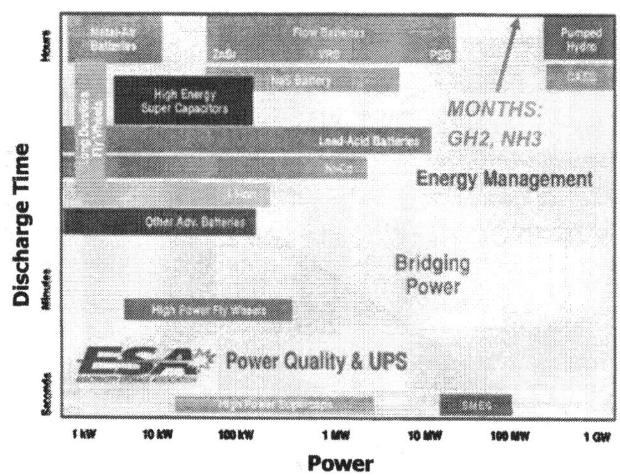

Figure 5. By any measure, the low cost and high capacity of GH2 and NH_3 storage is "off the charts" vis-a-vis the several technologies considered for bulk "electricity" storage as adjuncts to the Grid.

Hydrothermal Synthesis of Uniform Porous Carbon Nanospheres and their gas storage capacity

Mun Keun Lee*, Sang Woo Nam*, Ji Sun Han* and Shinhoo Kang*

*Department of Materials Science and Engineering, Seoul National University, 1 Gwanak-ro, Gwanak-gu, Seoul, 151-742, Republic of Korea, shinkang@snu.ac.kr

ABSTRACT

We investigated the synthesis of porous carbon nanosphere via two-steps process, using phenolic resol as a carbon source and block copolymer F127 as a soft template via hydrothermal process followed by carbonization. The porous nanospheres have high surface area of 1619 m^2/g after KOH activation measured by nitrogen sorption isotherms. The uniformity in the size of the nanospheres was verified by SEM and TEM analyses. The KOH activated porous carbon nanospheres show H_2 uptake of 2.42 wt.% (1 bar at 77K) and CO_2 adsorption capacity of 6.38 mmol/g (1 bar at 273K). The results indicate that this carbon material should have a high potential as gas adsorption carbonaceous materials.

Keywords: carbon spheres, nano materials, gas adsorption

1 INTRODUCTION

Due to depletion of fossil fuels and other environmental issues, developments of green energy materials are strongly demanded. Abundant gases such as carbon dioxides (CO_2) and hydrogen (H_2) are gaining considerable attention as an alternative potential energy source. With increasing environmental policy issue, CO_2 capturing can not only control the greenhouse gas emission from fossil fuel power plants, but also these captured CO_2 can be recycled by converting into useful products. Moreover, exhaust gases in hydrogen-powered applications will not contain any harmful-gases.

For aforementioned reasons, porous nanomaterials have been actively researched over the past decades due to its number of benefit from advantageous potential applicability in energy storage, adsorption, catalysis, supercapacitors, and biomedical applications [1-3]. Among the various candidates, carbonaceous materials including amorphous carbon, CNT and graphene have been considered due to their good thermal and chemical stabilities, mass production, and physical characteristic of light weight. However, the adsorption properties are greatly influenced by the porous network, pore size distribution, shapes and volume of void space within the material [4].

For these reasons, particular attention are being focused on controlling and manipulating the size, shape and uniformity of the pores. Also, number of studies have reported that the morphology is also a crucial factor for some applications [1].

With growing interest in designing porous nanomaterials, various synthetic approach such as hydrothermal [5], template [5, 6], sonication [7] methods have been developed to synthesize spherical nanomaterial, as it has revealed superior performance in the range of applications [8]. Above all, hydrothermal polymerization process using carbon source such as glucose or phenolic resins with soft or hard templates followed by carbonization are now became one of the most general method. From the recent study, Fang et al. [9] have reported a simple hydrothermal approach based on the organic-organic assembly between phenolic resol as a carbon precursor and controlled low concentration of triblock copolymer (Pluronic F127) as a surfactant to produce ordered mesoporous carbon nanospheres, during carbonization step. The open mesopore size was estimated around 3.0 nm with BET specific surface area of 1131 m^2/g.

Furthermore, in terms of gas adsorption properties, many study has revealed that nitrogen doping of carbons can enhance the surface properties due to polarity, electric conductivity, and electron-donor affinity [10], where recent study from Wickramaratne et al. [10] has revealed that nitrogen doped porous carbon spheres synthesized using polymer spheres via hydrothermal reaction can uptake up to 6.2 mmol/g of CO_2. However, nitrogen source generally involves toxic precursors such as ammonia based organic solutions, which may limit some applications. Therefore, other approaches such as activation and doping with metal nanoparticles are also being considered for altering the physical and chemical properties of carbons.

Herein, we report the synthesis of porous carbon nanosphere via two-steps process using phenolic resol as a carbon source and block copolymer (Pluronic F127) as a soft template via hydrothermal polymerization followed by carbonization. The manipulation of porosity was further conducted using KOH activation at various ratios. It is shown that increasing KOH activation ratio between carbon nanospheres and KOH resulted increase of higher specific surface area up to 1619 m^2/g with high H_2 and CO_2 uptake of 2.42 wt.% at 77K and 6.38 mmol/g at 273K, respectively, showing these carbon nanospheres with a high potential as gas adsorption materials.

2 EXPERIMENTAL

2.1 Materials

Phenol (bioXtra, ≥ 99.5%), formaldehyde solution (37 wt.%), sodium hydroxide solution (~0.1M), and Pluronic F127 were all purchased from Sigma-Aldrich. All chemicals were used as received.

2.2 Synthesis of porous carbon nanospheres

Porous carbon nanospheres (CNS) were prepared through aqueous hydrothermal synthesis route described by Fang et al. [9] In general, 0.6 g of phenol, 2.1 ml of formaldehyde solution (37 wt.%) and 15 ml of (0.1 M) of NaOH aqueous solution were mixed and stirred at 70 °C for 30 minutes. Then, 0.96 g of F127 dissolved in 15 ml of distilled water solutions are added into the mixture and temperature was lowered to 66 °C for 2 hours followed by further addition of 50 ml of distilled water. During this reaction, stirring was adjusted to 340 rpm and colorless solution has transformed into pink and crimson. After 18 hours, reaction was stopped.

17.7 ml of mixed solution was transferred into autoclave and diluted with 56 ml of distilled water and heated at 130 °C for 24 hours. The obtained product was collected and washed with distilled water and freeze-dried for 12 hours. The carbonization was proceeded at 700 °C in N2 atmosphere for 3 hours at rate of 5°C/min and CNSs were collected for further characterizations.

2.3 KOH activation of porous CNS

For KOH activation, prepared CNSs and KOH at ratios of 1:1, 1:2, 1:3, and 1:4 (wt.%) are hand mixed together and heated at 700 °C in N_2 atmosphere for 1 hour at rate of 3 °C/min. Activated CNSs were washed with distilled water then freeze-dried for overnight for further characterizations.

2.4 Characterization

Transmission electron microscopy (high-resolution TEM) measurements were conducted on a JEM-3000F microscope (JEOL, Japan) operated at 300 KV. Field emission scanning Electron microscopy (FE-SEM) images were recorded on Merlin Compact electron microscopy operating at 1KV with Pt coating. The nitrogen (N_2) and hydrogen (H_2) (at 77K), and carbon dioxide (CO_2) (at 273K and 298K) adsorption measurements were conducted on Belsorp-mini (BEL, Japan). Surface area was calculated using the Brunauer-Emmett-Teller (BET) method from nitrogen adsorption-desorption isotherm.

3 RESULTS AND DISCUSSION

3.1 Structure of porous CNSs

Figure 1 shows the SEM images of hydrothermally synthesized resins (Figure 1a), CNSs carbonized at 700°C for 3 hours in nitrogen atmosphere (Figure 1b), and its KOH activated CNSs with different ratio (Figure 1c to 1f). The obtained particles all possessed spherical morphology with almost uniform diameters. Moreover, the ordered mesostructures are also observed in the SEM images. It is interesting that even with increase of KOH activation ratio, the particles still remains its spherical morphology with slight of dent possibly due to surface activation by KOH treatment.

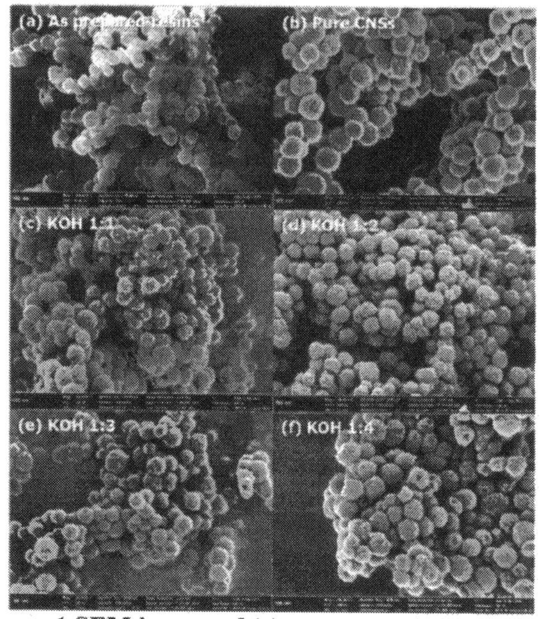

Figure 1 SEM images of (a) as-prepared, (b) pure porous carbon nanospheres (CNS), (c) 1:1 KOH activated CNS, (d) 1:2 KOH activated CNS, (e) 1:3 KOH activated CNS, and (f) 1:4 KOH activated CNS.

Figure 2 illustrates TEM images of hydrothermally synthesized resins (Figure 2a), CNSs carbonized at 700°C for 3 hours in nitrogen atmosphere (Figure 2b), and its KOH activated CNSs with different ratio (Figure 2c to 2f). Again, spherical morphology has been confirmed throughout all samples, but ordered mesoporous sites cannot clearly be observed in as-prepared sample (Figure 2a). This is because during the carbonization process, the template (F127) gets removed and opening the mesopores structured obtained during hydrothermal process [9].

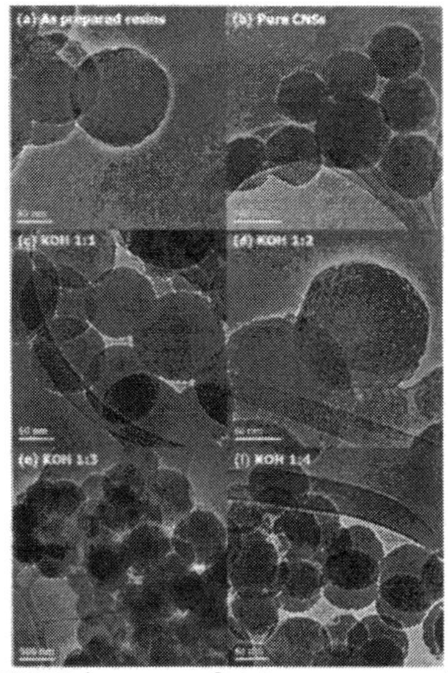

Figure 2 TEM images of (a) as-prepared, (b) pure CNS, and KOH activated CNS at ratio of (c) 1:1, (d) 1:2, (e) 1:3, and (f) 1:4.

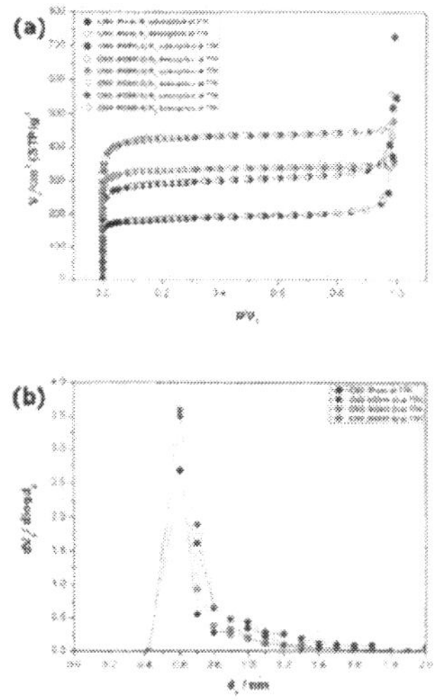

Figure 3 Nitrogen adsorption-desorption isotherms and pore size distribution measured at 77K for KOH activated CNSs at different ratios at 700 °C for 1 h in N_2 atmosphere.

3.2 Gas adsorption-desorption studies

We evaluated the porosity of pure CNSs and KOH activated CNSs using nitrogen adsorption-desorption isotherms at 77K and the corresponding pore size distribution curves, illustrated in Figure 3a. All adsorption isotherms shows type I of adsorption isotherm with slight H1 hysteresis loop at high relative pressure. This behavior is typically associated with micropores and mesopores. Moreover, with increase in KOH ratios, increment of N_2 adsorption properties can be observed indicating that KOH activation has successfully increased the surface areas CNSs.

The summarized adsorption parameters obtained by analysis of the experimental isotherms are provided in Table 1. The BET surface areas of pure CNSs is 691 m^2/g, whereas, KOH activation at 1:4 ratio shows 1619 m^2/g almost higher than twice of the pure CNSs. Although, the pore size distribution (Figure 3b) shows uniform micropore size of 0.6 nm throughout all pure CNSs and KOH activated CNSs samples, the area under the graph also confirms the increase in surface areas with higher ratios of KOH activation.

The H_2 and CO_2 adsorption performance of all samples was also examined and illustrated in Figure 4.

The summarized H_2 and CO_2 adsorption parameters obtained by analysis of experimental isotherms are also provided in Table 1.

The H_2 adsoprtion-desorption is shown in Figure 4a, where, pure CNSs only shows 1.60 wt.% of hydrogen uptake, whereas 1:4 KOH activated CNSs shows 2.42 wt.% of hydrogen measured at 77K. the higher surface area associated with KOH activation of CNSs lead to increased H_2 adsorption. Micropore size also contributes to increases in hydrogen uptake since KOH activation has etched the surface of CNSs producing smaller pore size.

The CO_2 adsorption-desorption at 273K (Figure 4b) and 298K (Figure 4c) are illustrated. Again, higher ratios of KOH activation shows higher adsorption of CO_2, where 6.38 mmol/g and 3.84 mmol/g of CO_2 adsorptions were measured for 273K and 298K, respectively. This is very interesting as, in general, N-doped porous carbon is known to be one of the most promising candidates for CO_2 adsorption [10]. However, in this study, there is no presence of nitrogen but just carbon. As, CO_2 adsorption is not limited by the total pore volume but only by pores smaller than a hydrothermal synthesis followed by carbonization and investigated their H_2 and CO_2 adsorption properties.

Table 1 Adsorption parameters for pure CNS and KOH activated CNSs obtained by analysis of the N_2, H_2 and CO_2 adsorption isotherms shown in Figure 4 and 5.

Samples*	S_{BET}/m^2g^{-1}	V_{total}/cm^3g^{-1}	V_{micro}/cm^3g^{-1}	V_{meso}/cm^3g^{-1}	H_2 uptake at 77K/wt.%	CO_2 uptake at 273K/mmol g^{-1}	CO_2 uptake at 298K/mmol g^{-1}
CNS700	691	0.9312	0.2918	0.0957	1.60	4.16	2.93
CNS700-2	1099	0.9404	0.4370	0.1147	1.87	5.40	3.61
CNS700-3	1253	0.5502	0.5024	0.0352	2.21	6.12	3.63
CNS700-4	1619	0.7376	0.6446	0.0560	2.42	6.38	3.84

* CNS700 denotes carbon nanospheres carbonized at 700 °C for 3 hours, CNS700-number gives the KOH ratios.

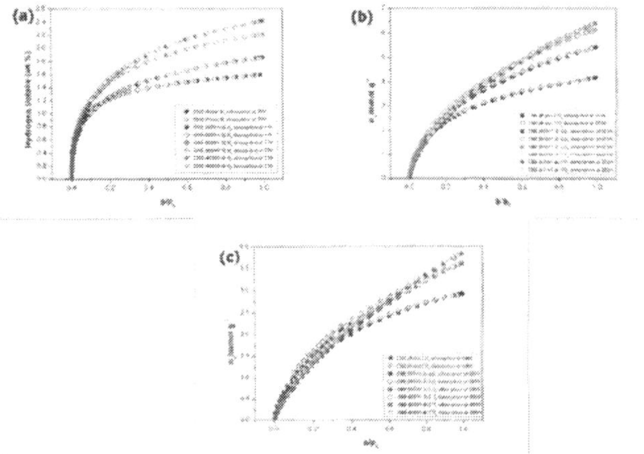

Figure 4 (a) H_2 uptake at 77K and CO_2 adsorption isotherms at (b) 273K and (c) 298 K for pure CNS and KOH activated CNSs with different ratios.

certain diameters [4], CNSs already having both micro-mesopores structures, KOH activation has not only increased the existing micropores volume but also has created more narrow micropores by reducing the mesopore volume, therefore increase in CO_2 adsorption.

4 CONCLUSIONS

In this study, we have produced KOH activated uniform porous carbon nanospheres using phenolic resol-F127 under It is interesting that even after KOH activation up to 1:4 ratios, the CNSs still remained its spherical morphology and increased its BET specific surface area up to 1619 m2/g but also increased the micropore volume up to 0.6446 cm^3/g. The H_2 adsorption capacity can reach 2.42 wt.% at 77K, which is high value for spherical morpherical carbon nanoparticles. Also, CO_2 adsorption capacity can reach up to 6.38 mmol/g and 3.84 mmol/g at 273K and 298K, respectively. This is very high value for carbon only substance as, in general, nitrogen doped porous carbon usually exhibits high uptake of CO_2, whereas CNSs produced in this study has no presence of nitrogen source. The study indicate that this porous carbon nanospheres should have a high potential as gas adsorption carbonaceous materials.

REFERENCES

[1] P. K. Tripathi, L. H. Gan, M. X. Liu, and N. N. Rao, "Mesoporous Carbon Nanomaterials as Environmental Adsorbents," Journal of Nanoscience and Nanotechnology, vol. 14, pp. 1823-1837, Feb 2014.

[2] K. P. Gong, F. Du, Z. H. Xia, M. Durstock, and L. M. Dai, "Nitrogen-Doped Carbon Nanotube Arrays with High Electrocatalytic Activity for Oxygen Reduction," Science, vol. 323, pp. 760-764, Feb 6 2009.

[3] J. Zhu, L. Liao, X. J. Bian, J. L. Kong, P. Y. Yang, and B. H. Liu, "pH-Controlled Delivery of Doxorubicin to Cancer Cells, Based on Small Mesoporous Carbon Nanospheres," Small, vol. 8, pp. 2715-2720, Sep 10 2012.

[4] S. Wang, W. C. Li, L. Zhang, Z. Y. Jin, and A. H. Lu, "Polybenzoxazine-based monodisperse carbon spheres with low-thermal shrinkage and their CO2 adsorption properties," Journal of Materials Chemistry A, vol. 2, pp. 4406-4412, 2014.

[5] J. Choma, D. Jamiola, K. Augustynek, M. Marszewski, M. Gao, and M. Jaroniec, "New opportunities in Stober synthesis: preparation of microporous and mesoporous carbon spheres," Journal of Materials Chemistry, vol. 22, pp. 12636-12642, 2012.

[6] Y. H. Deng, C. Liu, T. Yu, F. Liu, F. Q. Zhang, Y. Wan, et al., "Facile synthesis of hierarchically porous carbons from dual colloidal crystal/block copolymer template approach," Chemistry of Materials, vol. 19, pp. 3271-3277, Jun 26 2007.

[7] V. G. Pol, L. K. Shrestha, and K. Ariga, "Tunable, Functional Carbon Spheres Derived from Rapid Synthesis of Resorcinol-Formaldehyde Resins," Acs Applied Materials & Interfaces, vol. 6, pp. 10649-10655, Jul 9 2014.

[8] Z. C. Yang, Y. Zhang, J. H. Kong, S. Y. Wong, X. Li, and J. Wang, "Hollow Carbon Nanoparticles of Tunable Size and Wall Thickness by Hydrothermal Treatment of alpha-Cyclodextrin Templated by F127 Block Copolymers," Chemistry of Materials, vol. 25, pp. 704-710, Mar 12 2013.

[9] Y. Fang, D. Gu, Y. Zou, Z. X. Wu, F. Y. Li, R. C. Che, et al., "A Low-Concentration Hydrothermal Synthesis of Biocompatible Ordered Mesoporous Carbon Nanospheres with Tunable and Uniform Size," Angewandte Chemie-International Edition, vol. 49, pp. 7987-7991, 2010.

[10] N. P. Wickramaratne, J. T. Xu, M. Wang, L. Zhu, L. M. Dai, and M. Jaroniec, "Nitrogen Enriched Porous Carbon Spheres: Attractive Materials for Supercapacitor Electrodes and CO2 Adsorption," Chemistry of Materials, vol. 26, pp. 2820-2828, May 13 2014.

Electrocatalytic Activity of Core-shell Carbon-metal Nanocomposites Derived From Carbon Dioxide

Jae W. Lee*, Junshe Zhang, Yeeun Kim, and Ayeong Byeon

Department of Chemical and Biomolecular Engineering, Korea Advanced Institute of Science and Technology (KAIST), 291 Daehak-ro, Yuseong-gu, Daejeon, 305-701, South Korea, *jaewlee@kaist.ac.kr

ABSTRACT

The polymer electrolyte fuel cell (PEFC) is one of promising devices to convert chemical energy of fuels to electrical energy with high energy efficiency, high power density, and low environment impact. Despite wide applications in many areas, the large-scale commercialization of PEFCs is difficult because of the high cost of platinum-based electrocatalysts for oxygen reduction reaction (ORR) at cathode. There have been numerous efforts to developing alternative materials to platinum-based electrocatalysts and here we introduce boron-doped carbon/iron nanocomposites (Fe/B/C) as oxygen reduction electrocatalysts. They were synthesized through the reduction of CO_2 by $NaBH_4$ with a Fe precursor at 500 ℃ and atmospheric pressure. Furthermore we performed heat treatments of the resulting Fe/B/C at 850℃, 1050 ℃ (sample named as FeBC050, FeBC850 and FeBC105). Fe/B/C composites have a core@shell structure, in which the iron-containing nanoparticles are confined within onion-like graphitic carbon shells. Electrochemical analyses in cyclic voltammetry (CV) and rotating disk electrode (RDE) showed Fe/B/C composites enhanced ORR activity and especially FeBC105 presented excellent performance. Through XRD analyses, XPS, SEM and HRTEM observations, the thermal annealing is proved to be the reason for this better performance that changes the surface state and more active sites are generated by both the reduction of γ-Fe_2O_3 and the decomposition of B_4C species.

Keywords: oxygen reduction reaction (ORR), B-doped, electrocatalyst, CO_2 conversion, core-shell nanocomposites

1 INTRODUCTION

A fuel cell is a device to convert chemical energy of fuels to electrical energy. Among many types of fuel cells, polymer electrolyte fuel cells (PEFCs) have demonstrated one of promising devices because of its many advantages such as high energy efficiency, high power density, low environment impact [1]. However it is difficult to commercialize because of the expensive platinum-based eletrocatalyst for oxygen reduction reaction (ORR) at cathode [2-5]. Therefore, developing cost-effective ORR catalysts is required for the sustainable development of fuel cells. There have been many attempts to replace platinum-based electrocatalyst with other efficient, low-cost electrodes. Iron (Fe), a non-precious metal catalyst, attested and showed the high electrocatalytic performance originating from the metal cations coordinated by nitrogen atoms of aromatic rings. These are incorporated into graphitized carbon matrices.

Heteroatom-doped such as nitrogen (N) [3,6,7-10] and boron (B) [11-14] can enhance the electrocatalytic activity of carbon-based materials toward ORR. Here, we introduce boron(B)-doped carbon/iron nanocomposites (Fe/B/C) as efficient ORR electrocatalysts because the electrocatalytic evaluation by introducing the Fe/B/C composite have never been explored.

2 EXPERIMENTAL SECTION

2.1 Synthesis of B-doped carbon/iron composites and charaterization

The synthesis method of boron-doped carbon/iron composites are composed of three steps. The first step is Fe precursor synthesis. The second step is preparation of boron-doped carbon/iron composites. The final step is a thermal annealing step. By adding 50 cm^3 sodium carbonate solution slowly to 50 cm^3 $FeCl_2$ solution in a glass beaker, Fe precursor is synthesized. When the particles precipitate, the supernatant was decanted, and then 150 cm^3 D.I. water was added. These steps were repeated ten times. Finally, the suspension was filtered and dried in oven at ca. 120 °C. The main crystalline phase of Fe precursor is magnetite (Fe_3O_4, PDF #19-0629) and goethite (FeOOH, PDF #29-0713) which is identified from the XRD spectrum. The reduction of CO_2 by $NaBH_4$ with the Fe precursor at 500 °C and atmospheric pressure forms the B-doped carbon/iron composites. DI water washing can leach salts from the solid product. Sequentially, the resulted Fe/B/C composite is annealed at 850 °C and 1050 °C in argon atmosphere.

To identify physical and chemical characteristics of Fe/B/C composites, we carried out powder X-ray diffraction (XRD) measurements, Raman spectra, X-ray photoelectron spectroscopy (XPS), high resolution transmission electron microscope (TEM) images and scanning electron microscope (SEM) images.

2.2 Electrochemical measurements.

Cyclic voltammetry (CV) and linear sweeping voltammetry (LSV) were performed at room temperature on a RRDE-3A rotating ring disk electrode rotator. The Ag/AgCl is a reference electrode and the platinum wire is served as a counter electrode. The glassy carbon (GC) rotating disk electrode was used as a working electrode. The inks deposited on the disk were made by mixing 5.0 mg catalyst and 1.0 mL solutions of water, methanol, and Nafion® with a volume ratio of 3:1:0.1 by ultrasonication for 10 min. About 7.5 μL of the catalyst ink was loaded onto the polished carbon disk electrode for CV measurements.

3 RESULTS AND DISCUSSION

Cyclic voltammetry (CV) was employed to evaluate the electrocatalytic performance of the Fe/B/C composites and Pt (20 wt.%)/carbon. In the case of Pt/carbon in an alkaline solution (1 M NaOH) saturated with 1 atm argon (Ar), oxide formation/reduction peaks and H_2 adsorption /desorption peaks exist [15, 16]. FeBC050 is featureless at 0 to -1.2 V, whereas redox peaks of FeBC085 and FeBC105 in CV curve reveal well-defined at -1.2 to -0.4 V and become sharp along with the annealing temperature increase. In the O_2-saturated solution, cathodic peak and onset potentials decrease in the order of Pt/carbon, FeBC105, FeBC085, and FeBC050. For the comparison purpose, the peak potential of boron-doped carbon nanotubes is more negative than -0.31 V [6]. These show that thermal annealing enhances drastically the ORR activity of the Fe/B/C composites. We conducted the linear sweep voltammetric (LSV) measurements on a rotating disk electrode (RDE) in order to further analyze the ORR activity of the Fe/B/C composites. The LSV curves of FeBC105 is the same as those of Pt/carbon in terms of the limiting current.

Through the X-ray diffraction (XRD), the crystalline form of irons restricted within the B-doped carbons can be identified. The analysis showed that the main crystalline phase of FeBC050 is maghemite (γ-Fe_2O_3). As the treatment temperature increases to 850 °C, maghemite is reduced to iron. After the heat treatment at 1050 °C, the carbide phase (Fe_xC) occurs. Thus, the thermal treatment affects both the property of iron compounds and the nature of solid carbon. Also, Raman spectra support these changes. The D band shifts from 1370 to 1350 cm^{-1} and becomes sharp after annealing at 850 and 1050 °C. This result is due to incorporating heteroatoms into the carbon network. It breaks its hexagonal symmetry and induces an increased D band in the Raman spectrum. After the heat treatment, 2D band becomes a pronounced peak in the spectra because of occurrence of metastable crystalline carbon.

From X-ray photoelectron spectroscopy (XPS), the chemical composition and bonding state of elements of a solid surface can be detected. For the FeBC050, the B 1s spectrum exhibits two envelops. The first envelop is fitted by two peaks and these are respectively corresponding to B atoms bound to B atoms and B atoms connected to C carbons in B_4C species [17]. The other envelop is only fitted by one peak, corresponding to BCO_2 species. The C 1s envelop consists of three peaks which are respectively matched with sp^2-hydridized carbon, defective carbon, and >C=O or C-O [18,19]. The O1s spectral envelop can be deconvoluted into three peaks which are assigned to Fe-O, C=O, and C-O species, respectively [18-20]. The characteristic signal of BCO_2 species was not identified, possibly because of its low surface concentration. The Fe 2p spectrum has two envelops, one for Fe $2p_{3/2}$ and the other for Fe $2p_{1/2}$ and a satellite peak for Fe $2p_{3/2}$. Fe $2p_{3/2}$ which has four states is fitted by multiplets and a high binding energy (BE) surface peak. But Fe $2p_{1/2}$ has only two, so it is deconvoluted into two peaks.

After heat treatment at 1050 °C (FeBC105), the most interest thing is that the B 1s spectrum has only one high BE envelope; these two peaks are assigned to BC_xO_y species. Also, which has high BE envelope occurs only for FeBC085. The C 1s spectral envelop consists of three peaks and these are respectively assigned to sp^2-hydridized carbon, defective carbon, and –COO. After annealing at 1050 °C, the relative concentration of defective carbon decreases which is corresponding to the appearance of metastable crystalline carbon in the XRD spectra. The O 1s spectral envelop is fitted by three peaks matched with Fe-O, C=O, and C-O species, respectively. For FeBC105, two new peaks appear; the former is ascribed to the metallic iron and the latter is possibly related to iron carbides. All of XPS and XRD observations demonstrate that B_4C species disappear, a decrease of defects in the carbon structure and a reduction of iron oxide to iron after the thermal annealing.

By scanning electron microscopy (SEM) and high-resolution transmission electron microscopy (HRTEM), its effect on the microstructure of the composite is discovered. As the thermal annealing increases, the porosity of boron-doped carbon/iron composites also amplifies. Sodium borohydride reacts with CO_2 after melting to produce carbon which deposits on the solid iron precursor, resulting in a core-shell structure. HRTEM images show this specific structure. It looks like a onion because iron-containing nanoparticles are encapsulated into well-defined graphitic carbon shells. Its interplanar distance is ca. 0.34 nm, corresponding to the d-spacing of the (002) basal planes in graphite. In case of FeBC050, the interplanar distance of cores is about 0.24 nm which close to the d-spacing of the (311) planes in γ-Fe_2O_3. For FeBC105, the interplanar distance of cores is around 0.19 nm, approximating to the d-spacing of the (110) planes in α-Fe.

The average electron transfer number is 1.96 at the range from -0.2 to -0.4 V but it increases up to 4.7 between -0.5 and -0.8 V. It indicates that oxygen reduction goes on by a two-electron pathway at negative potentials and a four-electron ORR reaction at more negative potential than -0.7 V. In proportion to the increase of the potential, the oxygen reduction mechanism changes accordingly.

In addition, FeBC105 shows an excellent tolerance of methanol crossover, which is another main challenge faced by metal-based cathode catalysts in fuel cells. In the absence of methanol, the current maintains of FeBC105 is around 0.17 and that of Pt/carbon is about 0.29 mA. When 1 cm^3 methanol added to 80 cm^3 electrolyte solution, the current of FeBC105 remains around 0.17 mA, but the current of Pt/carbon changes to -0.37 mA, and then only increases back to 0.03 mA. These results indicate that the annealed B-doped carbon/iron composites have excellent activity as electrocatalyst for ORR, high tolerance to the crossover effect, and relatively a small generation of peroxide.

4 CONCLUSION

We have introduced a synthesis method of boron-doped carbon/iron composites from CO_2 at 500 °C and 1 atm. The synthesized composite has a core-shell structure composed of the iron-containing nanoparticles and the surrounding graphitic carbon shells. When the higher temperature is applied for the heat treatment, the electrocatalytic activity toward ORR in NaOH solutions enhanced. In addition to the improved catalytic activity, the high tolerance of methanol crossover and long-term stability of the heat treated composite were identified. The surface state change affects on the ORR performance. Tuning the microstructure of graphitic carbon shells increases the surface area to contact oxygen and thus, a direct 4-electron pathway could become favorable.

REFERENCES

[1] R. Borup, J. Meyers, B. Pivovar, Y. S. Kim, R. Mukundan, N. Garland, D. Myers, M. Wilson, F. Garzon, D. Wood, P. Zelenay, K. More, K. Stroh, T. Zawodzinski, J. Boncella, J. E. McGrath, M. Inaba, K. Miyatake, M. Hori, K. Ota, Z. Ogumi, S. Miyata, A. Nishikata, Z. Siroma, Y. Uchimoto, K. Yasuda, K. I. Kimijima and N. Iwashita, "Scientific Aspects of Polymer Electrolyte Fuel Cell Durability and Degradation ", Chem. Rev, 107, 3904, 2007.

[2] B. Bashyam and P. Zelenay, "High-Performance Electrocatalysts for Oxygen Reduction Derived from Polyaniline, Iron, and Cobalt", Nature, 443, 63, 2006.

[3] K. Gong, F. Du, Z. Xia, M. Durstock and L. Dai, "A class of non-precious metal composite catalysts for fuel cells", Science, 323, 760, 2009.

[4] R. Liu, D. Wu, X. Feng and K. Müllen, "Nitrogen-Doped Ordered Mesoporous Graphitic Arrays with High Electrocatalytic Activity for Oxygen Reduction", Angew. Chem. Int. Ed., 49, 2565, 2010.

[5] S. Yang, X. Feng, X. Wang and K. Müllen. "Graphene-Based Carbon Nitride Nanosheets as Efficient Metal-Free Electrocatalysts for Oxygen Reduction Reactions", Angew. Chem. Int. Ed., 50, 5339, 2011.

[6] L. Yang, S. Jiang, Y. Zhao, L. Zhu, S. Chen, X. Wang, Q. Wu, J. Ma, Y. Ma and Z. Hu, "Boron-Doped Carbon Nanotubes as Metal-Free Electrocatalysts for the Oxygen Reduction Reaction", Angew. Chem. Int. Ed., 50, 7132, 2011.

[7] A. Byeon and J. W. Lee, "Electrocatalytic Activity of BN Codoped Graphene Oxide Derived from Carbon Dioxide", J. Phys. Chem. C, 117, 24167–24173, 2013.

[8] G, Wu, K. L. More, C, M. Johnston and P. Zelenay, "High-Performance Electrocatalysts for Oxygen Reduction Derived from Polyaniline, Iron, and Cobalt", Science, 332, 443-447, 2011.

[9] C. Zhang, R. Huo, H. Yin, F. Liu and Y. Hou, "Fabrication of cone-shaped CNF/SiC-coated Si-nanocone composite structures and their excellent field emission performance", Nanoscale, 4, 7362-7368, 2012.

[10] Z. S. Wu, S. Yang, Y. Sun, K. Parvez, X. Feng and K. Müllen, J. Am., "3D Nitrogen-Doped Graphene Aerogel-Supported Fe3O4 Nanoparticles as Efficient Electrocatalysts for the Oxygen Reduction Reaction", Chem. Soc., 134, 9082–9085, 2012.

[11] J. Zhang, J. W. Lee, "Production of boron-doped porous carbon by the reaction of carbon dioxide with sodium borohydride at atmospheric pressure", Carbon, 53, 216-221, 2013.

[12] J. Zhang, A. Byeon and J. W. Lee, "Boron-doped electrocatalysts derived from carbon dioxide", J. Mater. Chem. A, 1, 8665-8671, 2013.

[13] A. Byeon, J. Park, S. Baik, Y. Jung and J. W. Lee, "Effects of boron oxidation state on electrocatalytic activity of carbons synthesized from CO2", J. Mater. Chem. A, 3, 5843-5849, 2015.

[14] S. Baik and J. W. Lee, "Effect of boron-nitrogen bonding on oxygen reduction reaction activity of BN Co-doped activated porous carbons", RSC Adv., 5, 24661-24669, 2015.

[15] C. Zhang, F. F. Fan and A. J. Bard, J. Am., "Electrochemistry of Oxygen in Concentrated NaOH Solutions: Solubility, Diffusion Coefficients, and Superoxide Formation", Chem. Soc., 131,177–181, 2009.

[16] W. Jin, H. Du, S. Zheng, H. Xu and Y. Zhang, "Comparison of the Oxygen Reduction Reaction between NaOH and KOH Solutions on a Pt Electrode: The Electrolyte-Dependent Effect", J. Phy. Chem. B, 114, 6542–6548, 2010.

[17] T. Shirasaki, A. Derré, M. Ménétrier, A. Tressaud and S. Flanrois, "Synthesis and characterization of boron-substituted carbons", Carbon, 38, 1461-1467, 2000.
[18] H. Estrade-Szwarchopf, "XPS photoemission in carbonaceous materials: A "defect" peak beside the graphitic asymmetric peak", Carbon, 42, 1713-1721, 2004.
[19] T. I. T. Okpalugo, P. Papakonstantinu, H. Murphy, J. McLaughlin and N. M. D. Brown, "High resolution XPS characterization of chemical functionalised MWCNTs and SWCNTs", Carbon, 43, 153-161, 2005.
[20] A. P. Grosvenor, B. A. Kobe, M. C. Biesinger and N. S. Mclntyre, "Investigation of multiplet splitting of Fe 2p XPS spectra and bonding in iron compounds", Surf. Interface Anal., 36, 1564-15774, 2004.

Catalytic Performance Enhancement by Low Temperature Plasma Treatment

George H. Miley*, Shailendra N. Srivastava**, George Chen**

*NPL Associates, Inc.
912 W Armory Ave, Champaign, Illinois 61821
**Illinois Applied Research Institute
2100 S Oak St Suite 206, Champaign, Illinois 61820

ABSTRACT

Increasing the performance of fuel cell electrocatalysts can help make fuel cell technology economically viable. One way to improve the catalytic activity of fuel cell electrocatalysts is through plasma treatment of the surfaces of such catalyst to create defects which can serve as possible catalytic reaction sites. The focus of this plasma modification process is on fuel cell electrocatalysts. The purpose of the plasma treatment process is to increase the catalytic performance of the fuel cell electrocatalysts possibly making fuel cells a commercially viable product.

Keywords: plasma, heterogeneous catalysts, fuel cell electrocatalysts, plasma treatment / modification / irradiation

1 INTRODUCTION

In this paper we emphasize the application of the plasma treatment method to improve PEM fuel cell catalysts, however in a closely related paper at this meeting we stress broader use for a variety of commercial processes involving catalysts [1]. Heterogeneous catalysts which produce good reaction rates are sometimes composed of noble metals (e.g. platinum, palladium, etc.) which are scarce and expensive [2]. Such noble metals are used as electrocatalysts in PEM fuel cell technology. There are a variety ways to try to circumvent the use of these scarce and costly catalysts [3] [4], but none have been fully successful to date. The process under study here uses a low temperature plasma to enhance catalytic performance of fuel cell electrocatalysts.

2 BACKGROUND

In surface "structure-sensitive" catalysts, certain sites on the surface of the catalysts enhance the rate of the chemical reactions [4] [5] [6] [7]. The reaction rate of these structure-sensitive catalyst greatly increases with the number of surface defects (e.g. steps, kinks, etc.) [6] [7] [8]. This observation is based on experiments comparing reaction rates of catalyst with single-crystal surfaces and catalyst containing many surface defects [7]. Other factors that enhance catalytic activity include increasing the surface area of the heterogeneous catalyst [9]. Furthermore catalytic activity of materials with equal surface areas may also differ depending on their actual form [9]. For example, materials that are normally inert in macroscopic solid form (e.g. gold), demonstrate high catalytic activity in submicron solid form [10].

In the semiconductor manufacturing industry low temperature plasmas are used to manipulate the surfaces of solid materials. Plasma ions can damage the surfaces of materials through ion bombardment which can eject surface atoms; this process is called sputtering. Plasmas can also induce the formation of submicron structures on the surfaces of solid materials. The effects caused by plasma-surface interaction (notably surface damage and defects, as well as the formation of submicron structures) are utilized to possibly alter PEM fuel cell electrocatalysts surface so the catalytic activity can be enhanced. The plasma-catalysts treatment process has been studied for some time [11].

Polymer electrolyte membrane (PEM) fuel cells have been proposed as one of the technologies that can displace the use of fossil fuels in the transportation sector. The issues that hold back PEM fuel cells from being commonplace are: (1) the membrane, (2) the catalyst, (3) the bipolar plate, and (4) water management [12] [13] [14] [15] [16] [17] [18] [19] [20] [21]. The membrane is a thin layer of electrolyte which has high ionic conductivity, but prevents the transport of electrons. The membrane accounts for ~6% of the PEM fuel cell cost; additionally the membrane suffers from degradation due to mechanical stress and chemical attack. Thus the membrane suffers from low lifetimes [13] [18]. The electrocatalysts are where the electrochemical reactions in the fuel cell take place. In PEM fuel cells the electrocatalyst is ~77% of the fuel cell cost, making it the key factor causing the high cost for fuel cell. The catalysts also plays a key role in the durability of a fuel cell; catalyst degradation is common and causes gradual loss of fuel cell performance [13] [18]. The bipolar plate separates fuel and oxidants in a fuel cell. The bipolar plates also conducts electrons and diffuses heat [13] [18]. There are two types of bipolar plate materials, they are metal or carbon. Carbon (bipolar) plates are thicker, more massive, and are structurally weaker than metal plates, however metal plates especially less expensive metal plates suffer from corrosion in the fuel cell environment [13] [16]. Water management involves damage to PEM fuel cell due to freezing and thawing of water. The freezing and thawing of water in fuel cells significantly lower fuel cell component lifetimes [13].

3 EXPERIMENT

The focus of the present work is on the electrocatalysts, because the electrocatalyst represents the majority of the cost in PEM fuel cells. Electrocatalysts are a subcategory of catalysts and incorporate similar characteristics as regular heterogeneous catalysts. Characteristics that affect electrocatalysts performance include crystal orientation and structural defects at the surface [22] analogous to regular heterogeneous catalyst. PEM fuel cell electrocatalysts are shown to be structure-sensitive; this is because experiments have shown that the catalytic activity of platinum and platinum-nickel alloys change with crystal structure and surface form [13]. This indicates that certain sites on the surface of PEM fuel cell electrocatalysts allow for more reactions to take place. The goal is to demonstrate that plasma-surface treatment of electrocatalysts in PEM fuel cells can enhance catalytic performance. Figure 1 shows a schematic of a DC plasma discharge system with the cathode (exposed to the plasma) made out of an electrocatalysts. Electrocatalysts as the name suggest are electrically conducting.

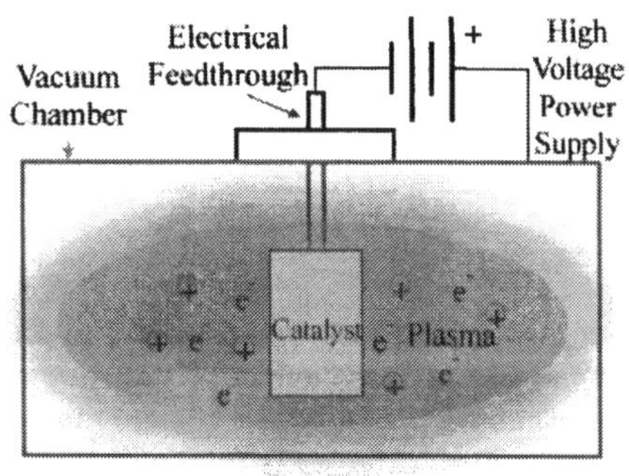

Figure 1: Schematic of DC plasma source modifying the surface of a conducting heterogeneous catalysts.

A simple plasma-surface modification method is being studied to treat PEM fuel cell catalytic material (e.g. platinum, etc.). The plasma-surface treatment modifies the surface of the catalytic material creating very pack submicron structures and defects [23] [24] [25] [26] [27] [28]. These submicron formations and defects on the catalytic material can serve as reactant adsorption and reaction sites [4] [5] [6] [7] [29] [30] [31]. The plasma modification can potentially increase the number reaction sites this can greatly improve the performance of the fuel cell catalytic material which would therefore improve the economic viability of fuel cells technology. Furthermore plasma irradiating some of the more expensive catalysts can also increase their catalytic activity, thereby increasing reaction rates and ultimately improving the economics without changing the catalysts employed in the process.

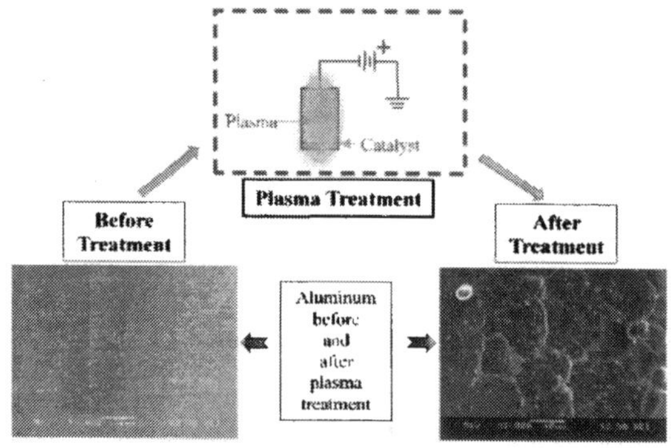

Figure 2: Plasma treatment of aluminum foil showing the effects of plasma on material surfaces.

Figure 2 shows a plasma treatment system currently set up to study and optimize the PEM fuel cell electrocatalysts treatment process. As shown, before and after measurements using SEM of the surfaces of conducting materials. The case shown is for an aluminum foil, but this work will eventually cover catalysts for interest for fuel cells. In addition to SEM scans, changes in the catalysts activity will be measured. To do that for fuel cell electrocatalysts an electrochemical test setup is being constructed.

4 CONCLUSION

A successful outcome can lead to economic fuel cell applications in automobile transportation or distributed stationary power generation. This plasma treatment technique can also be applied to other electrochemical systems that use electrocatalysts. Catalysts are ubiquitous in chemistry and enhancing the activity of fuel cell electrocatalyst through simple plasma treatment can ripple through the automotive and distributed stationary power sector. This should

TechConnect Briefs 2015, TechConnect.org, ISBN 978-1-4987-4728-8

accelerate the use of less polluting fuel cell vehicles in the transportation industry.

REFERENCES

[1] G. H. Miley, S. Srivastava and K.-J. C. G. Kim, "Low Temperature Plasma Surface Treatment of Catalytic Material," in *TechConnect World 2015*, Washington, DC, 2015.

[2] M. Armbruster, K. Kovnir, M. Friedrich, D. Teschner, G. Wowsnick, M. Hahne, P. Gille, L. Szentmiklosi, M. Feuerbacher, M. Heggen, F. Girgsdies, D. Rosenthal, R. Schlogl and Y. Grin, "Al13Fe4 as a low-cost alternative for palladium in heterogeneous hydrogenation," *Nature Materials,* vol. 11, pp. 690-693, 2012.

[3] M. P. de Almeida and S. A. C. Carabineiro, "The Best of Two Worlds from the Gold Catalysis Universe Making Homogeneous Heterogeneous," *CHEMCATCHEM,* vol. 4, pp. 18-29, 2012.

[4] S. He, C. Li, H. Chen, D. Su, B. Zhang, X. Cao, B. Wang, M. Wei, D. G. Evans and X. Duan, "A Surface Defect-Promoted Ni Nanocatalyst with Simultaneously Enhanced Activity and Stability," *Chemistry of Materials,* vol. 25, pp. 1040-1046, 2013.

[5] M. Boudart, "Heterogeneous Catalysis By Metals," *Journal of Molecular Catalysis,* vol. 30, pp. 27-38, 1985.

[6] T. Zambelli, J. Wintterlin, J. Trost and G. Ertl, "Identification of the "Active Sites" of a Surface-Catalyzed Reaction," *Science,* vol. 273, pp. 1688-1690, 1996.

[7] Z.-P. Liu and P. Hu, "General Rules for Predicting Where a Catalytic Reaction Should Occur on Metal Surfaces: A Density Functional Theory Study of C-H and C-O Bond Breaking/Making on Flat, Stepped, And Kinked Metal Surfaces," *Journal of the American Chemical Society,* vol. 125, pp. 1958-1967, 2003.

[8] R. T. Vang, K. Honkala, S. Dahl, E. K. Vestergaard, J. Schnadt, E. Laegsgaard, B. S. Clasusen, J. K. Norskov and F. Besenbacher, "Controlling the catalytic bond-breaking selectivity of Ni surfaces by step blocking," *Nature Materials,* vol. 4, pp. 160-162, 2005.

[9] U. Heiz, F. Vanolli, A. Sanchez and W. D. Schneider, "Size Dependent Molecular Dissociation on Mass-Selected, Supported Metal Clusters," *Journal of the American Chemical Society,* vol. 120, pp. 9668-9671, 1998.

[10] A. Sanchez, S. Abbet, U. Heiz, W. D. Schneider, H. Hakkinen, R. N. Barnett and U. Landman, "When Gold Is Not Noble: Nanoscale Gold Catalysts," *Journal of Physical Chemistry A,* vol. 103, pp. 9573-9578, 1999.

[11] C.-j. Liu, G. P. Vissokov and B. W.-L. Jang, "Catalyst preparation using plasma technologies," *Catalysis Today,* vol. 72, pp. 173-184, 2002.

[12] B. R. Limoges, R. J. Stanis, J. A. Turner and A. M. Herring, "Electrocatalyst materials for fuel cells based on the polyoxometalates," *Electrochimica Acta,* vol. 50, pp. 1169-1179, 2005.

[13] J. Macinkoski, J. P. Kopasz and T. G. Benjamin, "Progress in the US DOE fuel cell subprogram efforts in polymer electrolyte fuel cells," *International Journal of Hydrogen Energy,* vol. 33, pp. 3894-3902, 2008.

[14] J. J. Zhang, PEM Fuel Cell Electrocatalysts and Catalyst Layers, London: Springer-Verlag, 2008.

[15] S. Zhang, X.-Z. Yuan, J. N. C. Hin, H. Wang, K. A. Friedrich and M. Schulze, "A review of platinum-based catalyst layer degradation in proton exchange membrane fuel cells," *Journal of Power Sources,* vol. 194, pp. 588-600, 2009.

[16] R. A. Antunes, M. C. L. Oliveira, G. Ett and V. Ett, "Corrosion of metal bipolar plates for PEM fuel cells: A review," *International Journal of Hydrogen Energy,* vol. 35, pp. 3632-3647, 2010.

[17] Z. Chen, D. Higgins, A. Yu, L. Zhang and J. Zhang, "A review on non-precious metal electrocatalyst for PEM fuel cells," *Energy & Environmental Science,* vol. 4, pp. 3167-3192, 2011.

[18] Y. Wang, K. S. Chen, J. Mishler, S. C. Cho and X. C. Adroher, "A review of polymer electrolyte membrane fuel cells: Technology, applications, and needs on fundamental research," *Applied Energy,* vol. 88, pp. 981-1007, 2011.

[19] R. Othman, A. L. Dicks and Z. Zhu, "Non precious metal catalysts for the PEM fuel cell cathode," *International Journal of Hydrogen Energy,* vol. 37, pp. 357-372, 2012.

[20] D. C. Higgins and Z. Chen, "Recent Progress In Non-Precious Metal Catalysts For PEM Fuel Cell Applications," *The Canadian Journal of*

Chemical Engineering, vol. 91, pp. 1881-1895, 2013.

[21] P. Pei and H. Chen, "Main factors affecting the lifetime of Proton Exchange Membrane fuel cells in vehicle applications: A review," *Applied Energy,* vol. 125, pp. 60-75, 2014.

[22] V. S. Bagotsky, "Fundamentals of Electrochemistry," in *Fundamentals of Electrochemistry*, Hoboken, John Wiley & Sons, Inc., 2006, pp. 530-531.

[23] Y. Zayachuk, A. Manhard, M. H. J. 't Hoen, W. Jacob, P. A. Zeijlmans can Emmichoven and G. van Oost, "Depth profiling of the modification induced by high-flux deuterium plasma in tungsten and tungsten-tantalum alloys," *Nuclear Fusion,* vol. 54, no. 123013, 2014.

[24] H. Y. Xu, G. N. Luo, H. Schut, Y. Yuan, B. Q. Fu, A. Godfrey, W. Liu and G. D. Temmerman, "Enhanced modification of tungsten surface by nanostructure formation during high flux deuterium plasma exposure," *Journal of Nuclear Materials,* vol. 447, pp. 22-27, 2014.

[25] M. Laurent-Brocq, N. Job, D. Eskenazi and J.-J. Pireaux, "Pt/C catalyst for PEM fuel cells: Control of Pt nanoparticles characteristics through a novel plasma deposition method," *Applied Catalysis B: Environmental,* vol. 147, pp. 453-463, 2014.

[26] S. M. M. Coelho, F. D. Auret, P. J. Janse van Rensburg and J. M. Nel, "Unexpected properties of the inductively coupled plasma induced defect in germanium," *Physica B,* vol. 439, pp. 97-100, 2014.

[27] O. V. Ogorodnikova, J. Roth and M. Mayer, "Deuterium retention in tungsten in dependence of the surface conditions," *Journal of Nuclear Materials,* Vols. 313-316, pp. 469-477, 2003.

[28] T. Takimoto, K. Takeshita, S. Nakamura and T. Okumura, "Effects of plasma-induced defects on electrical characteristics of AlGaN/BaN heterostructure before and after low-temperature annealing," *Thin Solid Films,* vol. 557, pp. 212-215, 2014.

[29] R. C. Egeberg, S. Ullmann, I. Alstrup, C. B. Mullins and I. Chorkendorff, "Dissociation of CH4 on Ni(1 1 1) and Ru(0 0 0 1)," *Surface Science,* vol. 497, pp. 183-193, 2002.

[30] H. Uetsuka, K. Watanabe, H. Kimpara and K. Kunimori, "Structure Sensitivity in the Kinetics and the Dynamics of CO Oxidation over Stepped Pd(335) Studied by the Molecular Beam Infrared Chemiluminescence Technique: Determination of Working Sites during the Steady-State Reaction," *Langmuir,* vol. 15, pp. 5795-5799, 1999.

[31] B. Hammer, "Bond Activation at Monatomic Steps: NO Dissociation at Corrugated Ru(0001)," *Physical Review Letters,* vol. 83, no. 18, pp. 3681-3684, 1999.

Tuning of Nafion® by HKUST-1 as Coordination Network to Enhance Proton Conductivity for Fuel Cell Applications

Hee Jin Kim*†, Krishan Talukdar**† and Sang-June Choi**

*Research Institute of Advanced Energy Technology, Kyungpook National University, Daegu 702-701, Korea, zammanbo814@knu.ac.kr

**Department of Environmental Engineering, Kyungpook National University, Daegu 702-701, Korea, krishantu@yahoo.com, sjchoi@knu.ac.kr

ABSTRACT

Metal-organic frameworks (MOFs) can be intentionally coordinated to achieve improved proton conductivity, making MOFs candidate materials for fuel cell application. Using the concept of a coordination network, the conductivity of Nafion® was tuned by the incorporation of HKUST-1 that provides an improved protonic channel depending on the water content. Increased ion exchange capacity, water uptake, proton conductivity, and thermal stability (as confirmed by thermogravimetric analysis) were achieved. The structure of HKUST-1 was confirmed via field emission scanning electron microscopy and X-ray diffraction, while the porosity and adsorption desorption capacity were characterized by porosity analysis.

Keywords: PEMFC, coordination network, MOF, HKUST-1.

† These authors contributed equally.

1 INTRODUCTION

Metal-organic frameworks (MOFs) are a new class of exciting materials that have been explored for a wide variety of popular applications [1]. Apart from their conventional applications, MOFs may potentially be applied as proton conductors in fuel cells [2]. The factors that may lead to enhanced proton conductivity in a MOF include an increase in the number of charge carriers, greater motional entropy, lower activation energy, and a 3-dimensional conduction pathway.

HKUST-1 is a nanoporous crystalline material constructed from paddlewheel-coordinated (CuII)2 nodes and 1,3,5-benzenetricarboxylate (BTC) linkers [3]. Although HKUST-1 is neutral, coordination of water molecules in the structure to CuII renders these molecules sufficiently acidic to contribute protons to acidic molecules in the pores, thereby enhancing the alternating current conductivity. Three methods (types *a, b, c*) have been suggested for the fundamental design of highly proton-conductive MOFs that contain proton carriers and conducting media inside the pores. In the type *a* approach, the proton carriers are introduced directly into the pores as countercations; in the type *b* approach, acid groups are positioned on the frameworks; and in the type *c* methodology, acidic molecules are incorporated into the voids [4].

Polymer-based electrolyte membranes utilize acidic groups as proton carriers and hydrogen-bonding networks as proton-conducting pathways to facilitate proton transport or transfer. Nafion® has been widely used as a solid electrolyte in proton exchange membrane fuel cells (PEMFCs). Nafion® consists of a perfluorinated polyethylene backbone grafted with side chains possessing terminal sulfonic acid groups. The acidic groups function as proton carriers, and this membrane has been the popular material of choice. However, Nafion® suffers from several limitations. Active humidification is necessary for adequate performance, which adds a layer of complexity, while the potential for physical degradation due to hydration/dehydration results in a loss of efficiency. In addition, there is an energy requirement for powering the humidifiers. To date, three methods have been used to fine-tune the proton conductivity of MOFs. The first involves modification of the acidity of the functional group. The second involves injecting proton conducting chemicals such as imidazole molecules and H_3PO_4 into the empty pores of insulating MOFs. The third is to control the quantity of the conducting media, which assists proton transport through the hydrogen bonding networks [4].

In this study, HKUST-1 is incorporated into the Nafion® polymer matrix as a filler and imbibed with H_3PO_4 to form composite membranes to achieve enhanced proton conductivity; the developed composite is very promising for fuel cell application.

2 EXPERIMENTAL

HKUST-1 was prepared by using a novel solvothermal method based on the reported procedures [5]. Nafion®/HKUST-1 composite membranes were prepared by the solution casting method [6]. Each of the prepared membranes was carefully and individually soaked in 100 mL of 50 wt% H_3PO_4 (85%; Duksan Pure Chemicals Co.) in a Teflon dish for 24 h at room temperature. At the end of the process, the membranes were placed on other Teflon dishes and dried in a vacuum oven for 24 h at 60 °C. The dried membranes were ready for further characterization. All other chemicals were purchased from Sigma-Aldrich and used without further purification.

The morphology and molecular structure of the membranes were determined by field-emission scanning electron microscopy (FE-SEM) (Hitachi, JSM-6701F, JEOL) and powder X-ray diffraction (XRD; Bruker D2 PHASER, Germany). The water uptake and ion exchange capacity of the composite membranes was evaluated to explore their applicability for PEMFC. These characterizations were performed according to the method reported by Choi et al. [6]. The Fourier transform infrared (FTIR) spectra of the sulfonated membrane and MOF were acquired using a Perkin-Elmer (Spectrum GX & Autoimage) instrument. The proton conductivities of the acid doped HKUST-1/Nafion® composite membranes were measured between 0.1 kHz and 1 MHz using AC impedance spectroscopy (Versastat 3, Ametek).

3 RESULTS AND DISCUSSION

Evaluation of the crystallinity and morphology of HKUST-1 and the composite via FESEM (Figure 1(a)) demonstrated that the MOF had an octahedral crystalline structure with average diameter of 15–20 μm. The rough surface and nano-pores are evident in Figure 1(b). As shown in Figure 1(c) and (d) the MOF crystals were split into sub-micron rods and were well dispersed over the Nafion® membrane.

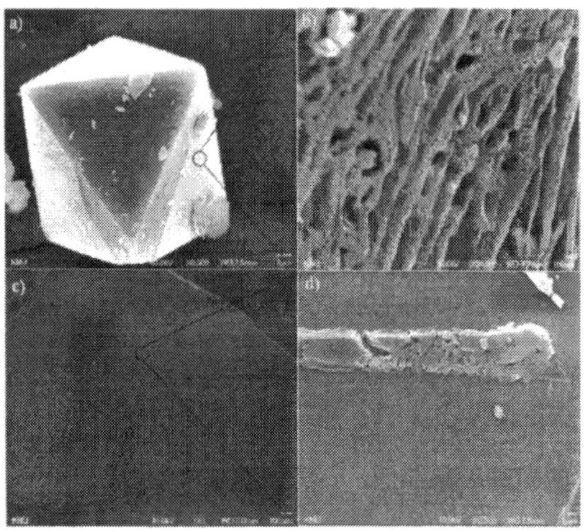

Figure 1. FESEM images of a) HKUST-1 crystal, b) nano-pore of HKUST-1, c) composite membrane, and d) sub-micron rods in the composite membrane.

These sub-micron rods have a length about 5–8 μm and an average diameter of 1–2 μm. In this study, the HKUST-1 MOF sub-micron rods are the dispersed phase and the Nafion® polymer is the continuous phase. The HKUST-1 sub-micron rods were well-mixed and homogeneously distributed in the Nafion® membrane, which may be highly beneficial for enhancing the proton conductivity of the composite membrane [7].

The phase purity of pristine HKUST-1 was confirmed via XRD measurement (Figure 2(b)) [8]. The XRD patterns of Nafion® and the HKUST-1/Nafion® composite membrane [9] were also acquired. Because Nafion® is not crystalline, no peaks were observed in the XRD profile. The non-crystallinity of the HKUST-1/Nafion® composite membrane is also demonstrated by the absence of peaks in the XRD profile, indicative of segregation of the octahedral MOF crystals into sub-micron-rods in the Nafion® membrane.

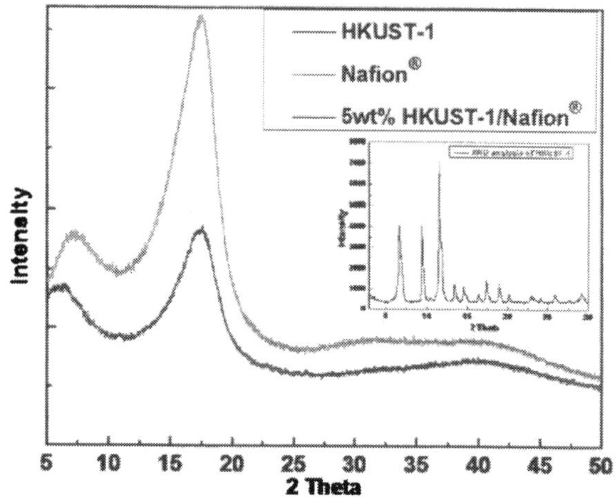

Figure 2. Powder XRD patterns of HKUST-1, Nafion®, and composite membrane.

The thermogravimetric (TG) profiles of the MOF and the composite membranes are shown in Figure 3. TG analyses revealed an initial weight loss for all samples, which was attributed to water elimination, in a process that occurred in a single step and ended at temperatures ≥ 300 °C. Since the samples were dried before the analyses, the water that was eliminated was likely to have been rapidly adsorbed during sample handling because of the hydroscopic nature of the MOF and membranes. Based on the TG curves, the HKUST-1/ Nafion® composite membranes had greater thermal stability than Nafion® at the intended operation temperature, i.e., around 300 °C, which suggests the prospective suitability of the composite for PEMFC application.

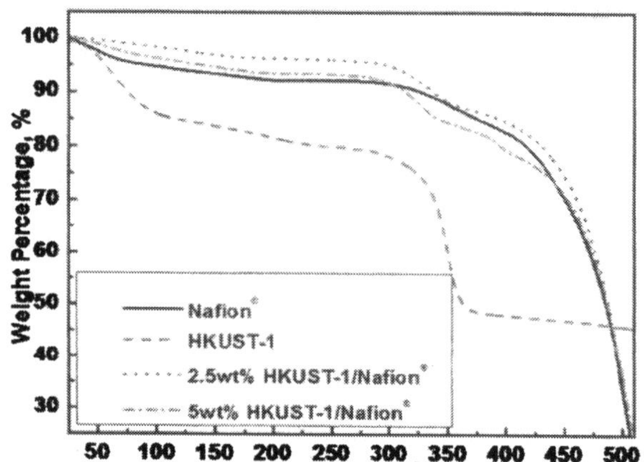

Figure 3. Thermogravimetric curves of Nafion®, HKUST-1, and composite membranes.

FTIR analysis shows the characteristic peaks of HKUST-1 at 1645, 1613, 1551, and 1445 cm^{-1}, corresponding to the C=O symmetric and asymmetric and Ar-C=C absorptions, and a peak at 1369 assigned to C-O [10]. The FTIR spectrum of Nafion® shows –COC– symmetric stretching bands at 974, $–SO_3–$ symmetric stretching vibrations bands at 1056 and asymmetric stretching of the $–SO_3–$ groups at 1205 cm^{-1}. The double broadening, observed in the bending vibrations of the O–H deformation peak at 1728 cm^{-1} and the bands at 3455 cm^{-1} are indicative of intermolecular hydrogen bonding [11]. The spectrum of the composite was dominated by peaks associated with HKUST-1 in the region of 1100–1200 cm^{-1}, obscuring the peaks attributable to the –CO– and OH functional groups of Nafion®; this indicates interactions between the dispersed graphite oxide and the hydrate groups of the Nafion® polymer matrix.

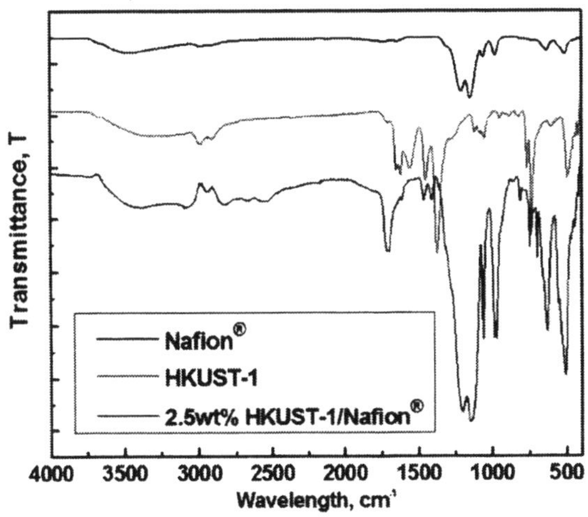

Figure 4. FTIR spectra of Nafion®, HKUST-1, and composite membrane.

The water uptake capacity of polymeric membranes is an important factor that influences the effectiveness of a membrane under high temperature operation. However, excess water uptake can lead to dimensional changes in the membrane, which in turn result in impaired mechanical properties [12]. The presence of water molecules in Nafion® polymer membranes has a significant effect on the transport of protons and the hydrated structures formed around the negatively charged fixed ions in the acids such as sulfonic acid, phosphoric acid, or carboxylic acid groups. The water uptake of the HKUST-1/Nafion® composite membranes was lower than that of Nafion®; this occurs because of modification of Nafion® by the HKUST-1 sub-micron rods distributed in the membrane, which decreases the size of the pores that can generally accommodate water molecules. Based on the data in Figure 5, increasing the weight percentage of HKUST-1 in the Nafion® matrix decreases the water uptake value.

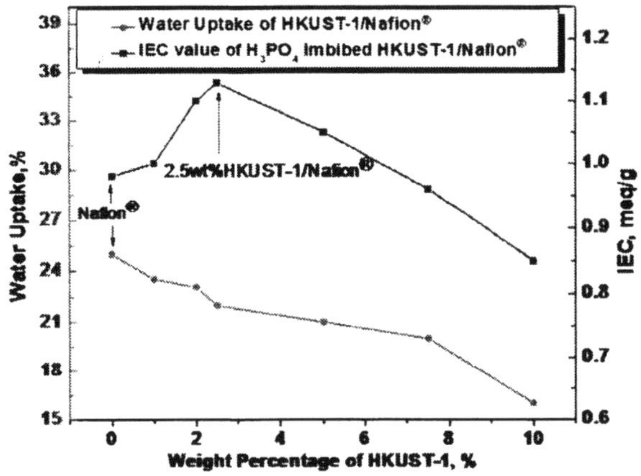

Figure 5. Water uptake and ion exchange capacity of Nafion®, and H_3PO_4-doped HKUST-1/Nafion® composite membranes wrt weight percentage of MOF.

The IEC provides an indication of the amount of acid groups present in a polymer matrix; these groups are responsible for the conduction of protons, and thus the IEC provides an indirect and reliable approximation of the proton conductivity. The variation of the IEC value of the acid-doped composite membranes with the weight percentage of the MOF is also depicted in Figure 5. The observed increase in the IEC of the doped composite membranes may be because PO_4^{-3} in the composite membrane can provide Brønsted acid sites that can be exchanged. The membrane, by virtue of its IEC and water uptake, regulates the water content at the membrane/electrode interface, which in turn influences the electrochemical characteristics [13]. An increase in the IEC values as the weight percentage of HKUST-1 incorporated into the membrane increased was apparent. The membrane containing 2.5 wt% HKUST-1 gave rise to the maximum IEC; it is likely that H_3PO_4 formed a better acid-MOF-polymer complex in the composite membrane.

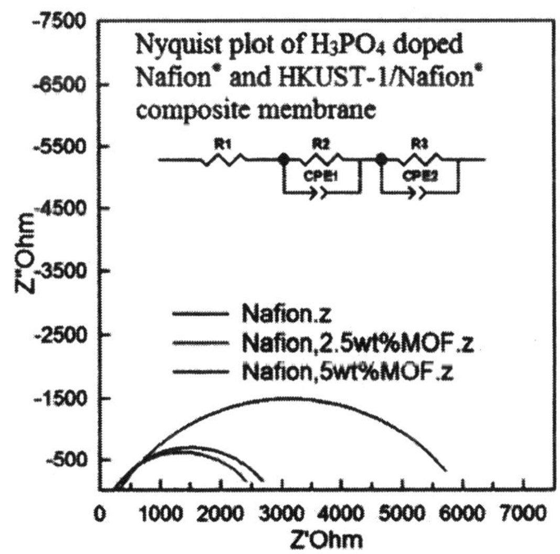

Figure 6. Nyquist impedance plot for Nafion® and H_3PO_4 doped HKUST-1/Nafion® composite membranes along their equivalent circuit.

Two-probe impedance analysis of the H_3PO_4 Nafion® and composite membranes was performed at room temperature to acquire the Nyquist plot according to the described method [13,14]. From Figure 6, it is apparent that the impedance of the composite membrane is less than that of Nafion®. The circuit used to determine the impedance analysis is also specified in the figure. The proton conductivities of the membranes were also determined and are as follows: (I) cast Nafion®: 1.1×10^{-2} Scm^{-1}, (II) acid doped 2.5 wt% HKUST-1/Nafion® composite membrane: 1.8×10^{-2} Scm^{-1}, and (III) acid doped 5 wt% HKUST-1/Nafion® composite membrane: 1.5×10^{-2} Scm^{-1} at 100% relative humidity (RH) and 298 K. Based on this data it can also be concluded that the increased proton conductivity of the composite membranes is due to the inclusion of PO_4^{-3} and as well as H^+ ions as a result of doping the membranes with 50 wt% H_3PO_4 solution. Herein, 2.5 wt% HKUST-1/Nafion® was found to exhibit the best proton conductivity among the evaluated membranes and is also favorable due to the higher IEC values.

4 CONCLUSION

The H_3PO_4-doped HKUST-1/Nafion® composite membrane is demonstrated to be a promising material based on its proton conductivity. HKUST-1 has an average particle diameter of around 15–20 μm. The proton conductivity, IEC values, and the thermal stability of the 2.5 wt% HKUST-1/Nafion® composite membrane suggest that HKUST-1 may be a promising candidate as a proton conductive material in the polymer electrolyte fuel cell membrane due to its reasonable proton passageway, favorable surface area, and facile synthesis.

Acknowledgments: This work was supported by the Priority Research Centers Program through the National Research Foundation of Korea (NRF), funded by the Ministry of Education, Science, and Technology (2009-0093819).

REFERENCES

[1] S. Goswami, A. Adhikary, H. S. Jena, S. Biswas and S. Konar, Inorg. Chem. 52, 12064, 2013.

[2]G. Xu, T. Yamada, K. Otsubo, S. Sakaida and H. J. Kitagawa, J. Am. Chem. Soc. 134, 16524, 2012.

[3] S. S. Y Chui, S. M. F. Lo, J. P. H. Charmant, A. G. Orpen and I. D. Williams, Science 283, 1148, 1999.

[4] M. Sadakiyo, T. Yamada, K. Honda, H. Matsui and H. Kitagawa, J. Am. Chem. Soc. 136, 7701, 2014.

[5] I. Ahmed, Z. Hasan, N. A. Khan and S. H. Jhung, Appl. Catal. B. Environ. 129, 123, 2013.

[6] M. D. A. Sayeed, K. Talukdar, H. J. Kim, A. I. Gopalan, Y. Park, Y. H. Kim, K. P. Lee and S. J. Choi, J. Nanosci. Nanotechnol. 14-12, 9329, 2014.

[7] S. Zhang, H. Liu, C. Sun, P. Liu, L. Li, Z. Yang, X. Feng, F. Huo and X. Lu, J. Mater. Chem. A 3, 5294, 2015.

[8] N. C. Jeong, B. Samantha, C. Y. Lee, O. K. Farha and J. T. Hupp, J. Am. Chem. Soc. 134, 51, 2012.

[9] H.-C. Kuan, C.-S. Wu, C.-Y. Chen, Z.-Z Yu, A. Dasari and Y.-W. Mai, Electrochem. Solid-State Lett. 9 (2), 76, 2006.

[10] D. Laura, O. Neil, H. Zhang and Bradshaw, J. Mater. Chem. 20, 5720, 2010.

[11] R. Kumar, C. Xu and K. Scott, RSC Adv. 2, 8777, 2012.

[12] S. Bose, T. Kuila, T. X. H. Nguyen, N.-H. Kim, K. T. Lau and J.-H. Lee, Prog. Polym. Sci.36, 813,2011.

[13] K. Talukdar, H. J. Kim, Y. H. Kim, H. C. Lee and S. J. Choi, Sci. Adv. Mater. 6, 1, 2014.

[14] S. Ma, A. Kuse, Z. Siroma and K. Yasuda, Measuring conductivity of proton conductive membranes in the direction of thickness, Espec Technology Report No20, 12, 2005

Gasification Technology Developed for the Production of Renewable Hydrogen

D. Dodd and J. Hazard

Sierra Energy
221 First St, Davis, CA, USA, info@sierraenergycorp.com

ABSTRACT

Sierra Energy's FastOx technology is a fixed-bed gasification system that breaks down waste at the molecular level by injecting oxygen and steam at highly-concentrated and rapid rates. The main product is a clean and high-quality syngas, which is an intermediate fuel used to produce valuable energy products, such as renewable hydrogen. The FastOx gasifier can produce hydrogen at a cost of \$1–\$5/kg(H2) with systems ranging from 12.5 MTPD to 100 MTPD. For vehicular transportation, a 100 MTPD system can produce hydrogen at an industry-leading \$0.43/GGE. Coupled with the benefits of eliminating waste, reducing greenhouse gas emissions, and fueling economic growth, the production of renewable hydrogen with FastOx gasification is a paradigm shift from conventional operations. This paper will introduce the FastOx gasification process, review the performance of the pilot system in comparison with conventional technologies, examine the process and economics of hydrogen production using FastOx and explain the environmental impact with a developed hydrogen case study using California waste.

Keywords: gasification, hydrogen, renewable hydrogen, greenhouse gases

1 INTRODUCTION

Hydrogen fuels from renewable sources are considered some of the most promising alternative fuels for reducing oil consumption, greenhouse gas emissions, and air pollution. Municipal solid waste (MSW) and other waste resources represent a distributed and renewable source of hydrogen capable of addressing multiple environmental concerns associated with energy generation while remaining cost competitive with conventional sources of hydrogen.

Sierra Energy's FastOx gasifier represents a new hybrid of waste gasification developed to directly address the limitations of conventional and plasma-arc systems. The FastOx technology is a fixed bed gasification system that breaks down feedstock at the molecular level by injecting oxygen and steam at highly-concentrated and rapid rates. The injections result in high temperatures, around 2,200°C, without capital-intensive plasma-arcs or difficult-to-operate molten baths. The FastOx gasifier has a cold gas efficiency of 66–79% and a parasitic load of 16–20%, which maximizes the output and minimizes natural resource inputs. FastOx gasification drives off 90% of the organic components in waste, directly reforming it to clean and high-quality syngas, typically in the region of 60% CO and 30% H2. The syngas is an intermediate fuel used to produce valuable energy products, such as electricity, renewable diesel, ethanol, methane, hydrogen and more. The high temperature of FastOx gasifiers allow the remaining 10% inorganic materials and toxic compounds in waste to be melted and recovered as recycled metal and inert stone (slag) that can be sold for reuse.

Sierra Energy's pilot gasification system was located at U.S. Department of Defense's Renewable Energy Testing Center (RETC) at the former McClellan Air Force Base, Sacramento, California. The gasifier, named Mark 4 (Mk4), is a functional pilot that was designed to test the conversion of multiple waste streams and the composition of the resulting syngas, to troubleshoot operational logistics, and to collect data for further system automation.

The production of hydrogen from waste materials via the FastOx process is very compelling, given that the FastOx technology can be deployed in small-scale, distributed production plants that could be located to solve multiple local municipal problems. This paper will introduce the FastOx gasification process, review the performance of the pilot system in comparison with conventional technologies, examine the process and economics of hydrogen production using FastOx and explain the environmental impact with a developed hydrogen case study using California MSW.

2 FASTOX GASIFICATION TECHNOLOGY

The FastOx process begins by feeding a waste material, such as municipal solid waste, medical waste, biomass, industrial waste or hazardous waste, into the top of the gasifier. The waste descends in the FastOx gasifier (by gravity) through four reaction zones inside the chamber (Figure 1):

Drying occurs when the hot, rising gases from the tuyere zone heat the waste in the top zone of the unit. The first reaction removes moisture (free water) from the waste material.

Devolatization is where the majority of the organic matter is driven off into syngas. The waste undergoes pyrolysis (heating in the absence of oxygen between 150°C and 500°C), during which the volatile matter is removed from the waste. The volatile matter is typically where a large amount of the chemical energy in the waste is released, as a mixture of light gases, hydrocarbons and condensable tars. Afterwards, the only remnants are char (or fixed carbon) and ash.

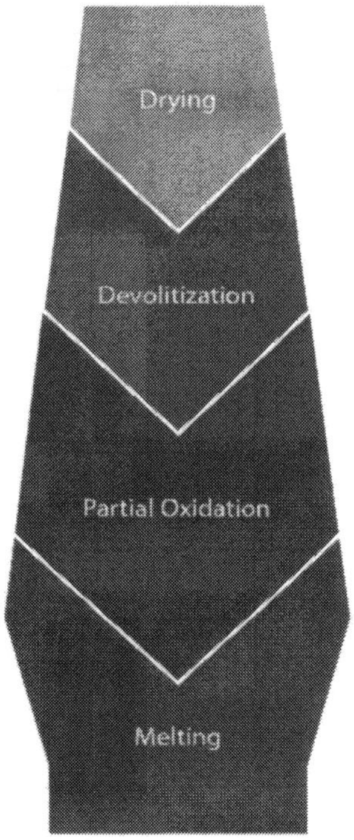

Figure 1. Reaction zones in a FastOx gasifier

Partial oxidation occurs when carbon-containing materials in the waste react with the injectants. In the lower regions of the FastOx gasifier, oxygen and steam are injected through the tuyeres to consume all of the remaining char material and convert it into syngas. This exothermic oxidation reaction raises the local temperatures to over 2,200°C and provides all of the energy for the self-sustaining FastOx process to occur.

Melting of inorganic compounds result from the high temperatures occurring in the partial oxidation zone. These compounds collect at the bottom of the unit and are continuously removed as inert stone (slag) and recycled metals.

Unlike conventional gasification which typically injects atmospheric air (composed of 80%vol. N2 that dilutes and impedes the reactions), the FastOx process injects solely oxygen and steam. In theory, this has several favorable effects on the gasification process:

- With the exclusion of nitrogen, the total gas volume inside the gasifier is significantly reduced, which increases the partial pressure of reacting species inside the vessel and shifts the reaction equilibrium to more products;
- Lower gas volumes also allow for smaller solid material sizing, which increases the total surface area and the number of reaction sites of the gasifiable material, causing an increase in reaction rates;
- Oxygen increases local temperatures, which are then moderated by the addition of steam, subsequently producing hydrogen which is a higher-diffusing molecule, increasing internal reaction rates;
- The presence of nitrogen interferes with the mass-transfer mechanisms of reactants and products. Therefore by excluding nitrogen, a significant increase in reaction rates is observed; and
- Sensible heat energy which is generally absorbed by inert nitrogen during conventional gasification, may instead be used for other endothermic reactions in the upper zones, allowing for more solid materials to be converted with FastOx, increasing the productivity.

3 SIERRA ENERGY'S VALIDATION TEST PLAN

A total of sixty-seven test runs were conducted on the Mk4 unit to establish an understanding of the FastOx gasifier's operability. For each test run, the syngas composition and the key operational measurements (i.e. temperature, pressures, and mass flows) were recorded during conventional compressed air gasification and then for oxygen-steam injection gasification. Additionally, slag samples collected from baseline test runs were sent to a third party laboratory to confirm non-leaching characteristics.

3.1 Feedstock Flexibility

During the Feedstock Variability Testing Campaign, a total of sixty-seven test runs were conducted to establish an understanding of the FastOx gasifier's operability with various feedstocks. Most of these runs generated baseline data from charcoal briquettes, lump charcoal, and biomass woodchips. In addition to these baseline feedstocks, the gasifier was also able to convert a variety of feedstock (e.g., plastics, glass, tires, ecopellets, walnut shells, metallurgical coke and oil shale) and synthetic municipal solid waste (sMSW) into a high-quality syngas. These feedstocks were tested because of their varying ash, moisture, and volatiles contents to help determine process requirements (Table 1).

Table 1. Proximate analysis for tested feedstock [%wt.]

Feedstock	Ash	Moisture	Volatiles
Briquettes	13.8	5.2	26.8
Lump Charcoal	9.1	6.1	30.1
Plastics	2.0	0.2	95.8
Glass	98.0	2.0	--
Tires	23.2	0.8	54.2
Woodchips	0.1-2.0	25.0-60.0	77.0-87.0
Ecopellets	1.5	7.1	71.2
Walnut shells	2.6	10.9	--
Met coke	5.0-9.0	--	--
Oil shale	59.8	0.05	41.0

TechConnect Briefs 2015, TechConnect.org, ISBN 978-1-4987-4728-8

Because it is difficult to obtain a consistent and pre-sized baseline MSW stream for direct run comparisons, Sierra Energy used a synthetic post Materials Recycling Facility's (MRF) MSW recipe developed by the University of California, Davis [1]. Sierra Energy adjusted the UC Davis recipe to replicate actual post-MRF MSW by including the inorganic fraction that the university removed from the recipe. The academic personnel were testing in a non-slagging gasifier and therefore wanted to limit the ash content of the waste material. Items that were difficult to obtain or unsafe to handle on the non-commercial pilot unit were replaced with chemically-compatible and easy to obtain alternatives. For example, 'Food scraps' were replaced with dog food pellets and grass or 'yard wastes' were replaced with alfalfa pellets. The final feed material composition used by Sierra Energy is shown in Table 2.

Table 2. Synthetic municipal solid waste recipe [%wt. Wet basis]

Ingredient Name	Fraction of sMSW
Paper	
Recycled paper	33%
Plastics	
PETE (polyethylene terephthalate)	7%
LDPE (low density polyethylene)	7%
PS (polystyrene)	4%
Glass	
Glass (recycling scrap from MRF)	2%
Organics	
Food waste (dog food)	13%
Grass (alfalfa pellets)	8%
Trimmings and lumber	15%
Metals	
Various metals (scraps from MRF)	4.5%
Aluminum pellets	0.5%
Other	
Construction and demolition concrete, soil and dirt (limestone)	6%

Results: The Mk4 processed feedstocks with high moisture content, such as fresh woodchips whose mass is approximately 50% moisture. Generally, feedstocks with high moisture content reduce the thermal efficiency of the gasifier and the syngas energy density (MJ/m3). The FastOx gasfier, however, is able to handle these moisture fluctuations due to the "updraft" configuration of the gasifier allowing volatiles and moisture to be driven off the material at the top of the gasifier before it descends into the zones where the main gasification and melting reactions take place. While there was a decrease in gasifier efficiency, the system was able to maintain steady-state operations with high moisture feedstocks, with no detrimental operational side effects.

Another feature of the FastOx gasifier is its ability to process feedstock with large ash contents (e.g., oxides and inorganics), which is known to affect slagging capabilities. The FastOx gasifier operates at a high enough temperature to melt and vitrify the ash. Consequently, the toxic ash is rendered chemically inert and exits the system as an inert slag.

3.2 Syngas Analysis

In order to compare and validate the syngas composition and heating value with conventional compressed air gasification, six test runs were employed for the Real Time Syngas Analysis Campaign. Each run lasted for 6 to 12 hours and the feedstock varied between lump charcoal and woodchips.

Syngas composition and quality were evaluated by installing a Wuhan Cubic Optoelectronic Model 3100P portable coal gas analyzer downstream of the FastOx gasifier. This analyzer uses multiple detectors to determine the composition of the major components found in syngas: A Non-Dispersive Infrared (NDIR) detector to measure CH4, CnHm, CO, and CO2, Electrochemical Detector (ECD) to measure O2, and a Thermal Conductivity Detector (TCD) to measure H2. However, since the TCD is not selective for hydrogen, the analyzer uses a software algorithm to correct the TCD reading for the other gases detected by the NDIR and ECD.

Results: Compared to the conventional compressed air gasification, FastOx gasification increased the quality and heating value of the syngas produced (Table 3). Testing indicated that there was a 215% increase in H2 and 108% increase in CO, respectively, which leads to a 118% increase in heating value. During compressed air mode operation, excess nitrogen acts as a diluent and lowers the heating value of the syngas. Because the FastOx process injects oxygen and not high-N2 air, a higher BTU value syngas is produced (12.23 MJ/Nm3), which allows for the syngas to be used in higher-efficiency thermodynamic power cycles.

Table 3. Averaged real time gas analysis results [%vol.]

Gasification Type	H_2	CO	CO_2	O_2	N_2
FastOx Gasification	27.70%	68.29%	2.90%	0.43%	0.04%
Compressed Air Gasification	8.79%	32.87%	2.63%	0.89%	55.37%
Percent Increase	215.01%	107.77%	--	--	--

3.3 Slag Analysis

To confirm the non leaching characteristics of the slag produced from the FastOx process, six test runs were conducted to produce sufficient slag samples. Each run was scheduled for 6 to 12 hours and the feedstock varied between lump charcoal and woodchips. Two slag samples were tested by third party laboratories and confirmed non-toxic by the EPA's Total Threshold Limit Concentration (TTLC) analysis and Toxicity Characteristic Leaching Procedure (TCLP). The TTLC analysis ensured that the materials were in compliance with Title 22 of the California Code of Regulations, while the TCLP analysis made sure the materials met federal guidelines.

Results: The TTLC results (Table 4) verify that the slag samples contain a variety of heavy metals; however, the concentrations of these components are significantly less than the regulatory levels. Furthermore, the TCLP results confirm that the leachability of these components are significantly less than the regulatory levels, and therefore, the slag samples are considered non-toxic to the environment and can be used for making cement clinkers, road base, bricks, and other construction material.

Table 4. TTLC results

Component	Concentration [mg/kg]	Regulatory Level [mg/kg]
Beryllium	2.0	75
Vanadium	590	2,400
Chromium	44	2,500
Cobalt	20	8,000
Nickel	420	2,000
Copper	55	2,500
Zinc	Not Detected	5,000
Arsenic	3.2	500
Selenium	Not Detected	100
Molybdenum	5.6	3,500
Silver	Not Detected	500
Cadmium	Not Detected	100
Antimony	Not Detected	500
Barium	1,500	10,000
Mercury	Not Detected	20
Thallium	Not Detected	700
Lead	Not Detected	1,000

3.4 Gasifier Performance Metrics

Various metrics were used to evaluate the gasifier's performance in both compressed air mode and FastOx mode by obtaining key process parameters (e.g., pressures, temperatures, mass flow rates and material throughput rates).

The 'specific productivity' of a blast furnace or gasifier is defined as the mass of material consumed per unit-time, per unit-internal volume. The 'oxidant productivity rate' is defined as the burden consumption per unit mass of oxidant.

Results: A selection of the testing data for the Mk4 operation in compressed air mode shows that the gasifier consumes the feedstock at a specific productivity of 176.8 kg/hr/m3. In comparison to the compressed air mode, the FastOx mode consumed the feedstock at a specific productivity of 381.6 kg/hr/m3, a 116% increase in specific productivity.

The FastOx process has an oxidant productivity rate of 0.52 kg (O_2)/kg, which is significantly lower than the conventional air-blown process of 1.50 kg(O_2)/kg. Fewer oxidant injectants are needed for the FastOx process because excess heat energy is used for additional endothermic reactions, rather than being absorbed by the large volume of nitrogen present in air injection.

3.5 Overall Findings

At this phase of testing and validation, the Mk4 has satisfied the following technical objectives:

- Verified the Mk4's operability with numerous feedstocks of varying ash and moisture content;
- Validated an increase in syngas quality through real time gas analysis, which demonstrated a 118% increase in the dry syngas energy density when compared to compressed air mode;
- Confirmed the non-leaching characteristics of the slag byproduct in accordance with the EPA's standard TCLP and TTLC test methods; and
- Analyzed the Mk4 gasifier's performance metrics, which indicated a 116% increase in specific productivity of operations and a 185% increase in oxidant productivity.

4 HYDROGEN PRODUCTION USING FASTOX GASIFICATION

The FastOx gasifier is able to produce low-cost hydrogen using the following systems in series:

- The FastOx gasifier
- Gas Cleaning, Water Gas Shift and H_2 Purification
- Hydrogen Fuel Cell

4.1 Gas Cleaning, Water Gas Shift and H_2 Purification

Additional gas conditioning and preparation is required for the production of hydrogen. This additional equipment is not required for the production of electricity using gas turbines, or reciprocating internal combustion engines. After additional cleanup, the syngas moves on to the gas conditioning and preparation unit, where it is compressed and mixed with high-pressure saturated steam and fed into the Water-Gas Shift (WGS) reactor where the following reaction takes place to maximize H_2 production:

$$CO + H_2O \rightarrow H_2 + CO_2 \quad (1)$$

The WGS reaction is mildly exothermic, and excess heat is removed from the shifted syngas to makes high-pressure steam.

A hydrogen membrane module is used to remove the hydrogen from the post-WGS gas stream, before sending that hydrogen stream to a pressure swing adsorption (PSA) module to purify the hydrogen product stream to the very high purity requirements.

4.2 Hydrogen Fuel Cell

The purified hydrogen produced by the system is fed into a hydrogen fuel cell. A proton exchange membrane (PEM) fuel cell is an established, low temperature hydrogen-fueled technology with proven operational abilities converting conventional hydrogen sources into electricity. PEMs have a variety of end use applications (stationary, transportation), high electrical efficiency ranging from 40-60%, and no emissions other than heat and water.

4.3 CO_2 Recovery

After the syngas goes through the water gas shift and hydrogen membrane, a system upgrade can be added to include enhanced CO_2 recovery. Gas treatment to the off-gas at this point in the process (gases other than the hydrogen) can be installed to generate a purified stream of CO_2. At this stage, one of the several methods of converting CO_2 to transportation fuels becomes viable. The addition of this capability to a FastOx system will maximize the energy output and leave no wasted emissions.

5 HYDROGEN ECONOMIC ANALYSIS USING FASTOX GASIFICATION

The US Department of Energy (DOE) developed the Hydrogen Analysis Project (H2A) to model the unit cost of producing hydrogen from various industrial processes [2]. Sierra Energy used the H2A model and worked with experts from the DOE's National Renewable Energy Laboratory (NREL) to calculate key metrics for the FastOx process. It was calculated that the FastOx gasifier can produce hydrogen at a cost of $1–$5/kg(H2) with systems ranging from 12.5 MTPD to 100 MTPD.

For vehicular transportation, proton exchange membrane (PEM) fuel cells that utilize hydrogen are significantly more efficient compared to internal combustion (IC) engines. A fuel cell can power a vehicle approximately 2.5 times as far as a gallon of gasoline in an IC engine. Therefore, 1 kg(H2) has the potential to replace 2.5 gallons of gasoline, which allows the hydrogen production cost of $5.00/kg(H2) to be equivalent to $2.00/gal(gasoline) or $2.00/Gallon of Gasoline Equivalent (GGE).

A small-scale FastOx gasification system (12.5 MTPD) can produce hydrogen at $1.94/GGE, while a 100 MTPD system can produce hydrogen at an industry-leading $0.43/GGE. For comparison, see the Table 5 below:

Table 5. Hydrogen production cost comparison

Production Method	Production Cost	
FastOx Gasification (systems ranging from 12.5 to 100 MTPD)	\$1-\$5/kg(H_2)	\$0.40-\$2/GGE
Natural Gas (produced via steam reforming at fueling station)	\$4-\$5/kg(H_2)	\$1.60-\$2/GGE
Natural Gas (produced via steam reforming off-site)	\$6-\$8/kg(H_2)	\$2.40-\$3/GGE
Wind (via electrolysis)	\$8-\$10/kg(H_2)	\$3.20-\$4/GGE
Solar (via electrolysis)	\$10-\$12/kg(H_2)	\$4-\$4.80/GGE

6 CALIFORNIA CASE STUDY AND IMPACT

Sierra Energy has developed a ChemCAD model of the FastOx gasification process and has validated it against actual empirical data from Sierra Energy's pilot demonstration unit.

Using the waste composition of average California post-recycled MSW residual composition and the validated ChemCAD model, Sierra Energy created a case study examining the potential impacts associated with the production of hydrogen via FastOx gasification [3].

Table 6. Summary of california post-recycled msw residual composition

Item	[%wt. Dry basis]
Organics	30.3
Plastics	19.9
Electronics	1.3
Metals	6.6
Glass	2.7
Paper	37.8
Mixed Residue	0.7
Special Waste	0.7
TOTAL	100

In general, one metric ton of as-received California post-recycled MSW feedstock generates 0.73 metric tons of syngas with approximately 8.8 MJ of chemical energy. This syngas can be further converted into 58.0 kg of H_2. Table 7 shows the renewable energy potential and the climate warming impact using FastOx gasification [4].

Table 7. Climate Warming Impact of Califonia's Trash and Renewable Hydrogen Potential Using FastOx Gasification

Emission Data of California	
Waste Disposed of in California Landfills	30.4 MMT/yr
Climate Warming Impact	116.55 MMTCO2e/yr
Renewable Energy Potential with FastOx	
Hydrogen	1,562,749,200 kg(H_2)/yr
Gallons of Gasoline Equivalent	3,307,322,780 GGE/yr
Emissions Avoided	29.39 MMTCO_2e/yr

REFERENCES

[1] Cattolica, R. and Jenkins, B., "Research and Demonstartion Project for Bioenergy and Biofuels Production from Post-MRF Residuals," California Integrated Waste Management Board, 2009.

[2] DOE H2A Production Analysis, http://www.hydrogen.energy.gov/h2a_production.html.

[3] Beck, R., "Targeted Statewide Waste Characterization Study: Characterization and Quantification of Residuals from Materials Recovery Facilitie," California Integrated Waste Management Board, 2006

[4] Sierra Energy, "Waste to Hydrogen in California Using Sierra Energy's FastOx Gasification System," 2014. https://gallery.mailchimp.com/73ba9fcff85ec9b3768d2848c /files/California_Hydrogen_Report_rev2.pdf

Renewables-source Hydrogen Systems: Alternatives to Electricity for Transmission, Storage, and Integration

Hydrogen and Ammonia Fuels via Underground Pipelines with Storage at < $1.00 kWh CAPEX

W. Leighty, The Leighty Foundation, Box 20993, Juneau, AK 99802 wleighty@earthlink.net
www.leightyfoundation.org/earth.php

ABSTRACT Humanity's grand challenge now is converting the world's largest industry from ~ 85% fossil to 100% renewable energy (RE) sources as quickly as we prudently and profitably can, to "Run the World on Renewables" -- perhaps including some nuclear, now very hard to predict. Nothing less will allow us to escape these synergistic global consequences of large-scale fossil fuel combustion:

- Rapid climate change, generally warming
- Accelerating sea level rise
- Ocean acidification
- Species extinctions
- Violent human conflicts

Jacobson and Delucchi have demonstrated that we can run the world on wind, water, and solar (WWS). The wind energy of the twelve Great Plains states, if fully harvested on about 50% of these states' aggregate land area, transmitted to distant markets, and "firmed" at annual scale with storage, could supply the entire annual energy demand of the USA: about 10,000 terawatt-hours (TWh = billion kWh), or about 100 quads (quadrillion btu). However, existing Great Plains electric transmission export capacity is insignificant relative to this resource. Any large, new electric transmission systems, or fractions thereof dedicated to wind energy, will:

- Be very costly to build;
- Be difficult to site: FERC has no authority for permitting interstate electric lines;
- Be difficult to route and permit, because of public objection, as in NIMBY;
- Suffer the same low capacity factor (CF) (typically 40%) as the windplants and other RE plants they serve, unless RE gen curtailed;
- Provide no affordable "firming" (weekly-to-annual scale) energy storage, taxing "system balancing" ability of the electricity grid;
- Be vulnerable to damage by acts of God and man.

Relieving RE generation and conversion equipment of the requirement to deliver "grid quality" AC (Voltage, frequency, power factor, and harmonics) as well as "low voltage ride through" (LVRT), because RE electricity is instead entirely delivered to electrolyzers and NH_3 synthesis systems, may significantly reduce the capital and O&M costs of RE -- at the plant gate and at end-user.

Therefore, the RE industry needs to look beyond the electricity grid, which now supplies only about one-third of a modern country's total energy, to supply all energy, from all sources, for all purposes, without CO_2 emissions, as firm and dispatchable energy. Electricity systems are not well suited for delivering on-demand energy services from the time-varying output -- at scales of seconds to seasons -- of most renewables, while simultaneously and affordably providing:

1. Gathering and transmission, from both large, stranded, "centralized" RE plants and smaller "distributed" generation (DG)
2. Annual-scale firming storage, rendering RE dispatchable
3. Distribution, integration, CHP, and delivering energy services to end-users

A technically and economically superior strategy may be to:

1. Convert the wind or solar plant's entire energy output to the only two C-free fuels, hydrogen, H_2, and / or anhydrous ammonia, NH_3
2. Transport the fuels in underground pipelines as gaseous hydrogen (GH2) and / or liquid NH_3
3. Completely avoid the high capital and O&M costs of:

a. Electricity grid connection: gathering, field transformers, substation, and HV line links
b. Sophisticated generating systems required to deliver grid-quality AC with LVRT

4. Acquire, store, and deliver surplus energy from PV and other distributed RE generation.

Having left electricity at the RE sources, we would now deliver C-free fuel(s) to distant markets for stationary combined-heat-and-power (CHP), transportation fuel, and the complete other spectrum of residential, commercial, and industrial uses. We are thus emulating the natural gas industry's ubiquitous pipeline network and freedom from large, looming "Smart Grid" investments, enabled by their large and diverse array of low-cost energy storage and inherent long time constants.

The electricity grid, as we've known it, is rapidly transforming, perhaps atrophying, under pressure from DG, nascent demand for BEV charging, and new Smart Grid complexity, which together may make the grid more susceptible to cyberattack. While the Grid may always be with us, we need to look beyond it in our compulsion to run the world on renewables.

Figure 1 shows that GH2 and NH_3 underground pipeline transmission capital costs compare favorably with electricity, oil, and gas per MW-km, which should be the common measure of any transmission system's service. Underground pipelines are better protected; O&M costs per MWh-km are lower. New pipelines are generally easier to site and permit than new electric transmission lines. For example, these pipelines have 8 GW capacity:

- 1 m diameter, 800 km long, GH2 pipeline, at 100 bar input, 30 bar output, with no midline compression
- 50 cm diameter NH_3 pipeline, with midline pumping, of any length

Figure 2 shows the system simplicity and inherent low-cost energy transmission and storage when high-pressure-output electrolyzers input the GH2 pipeline; midline compression for < 1,000 km length is not needed. "Packing" the GH2 pipeline, as natural gas pipelines routinely are, provides "free" storage. If a national or continental GH2 pipeline "grid" can access a region of domal salt geology, very large and low-cost RE storage is available in large, deep, solution-mined, man-made, salt caverns, in Figure 3. Multiple caverns may be manifolded together, operating at the same pressure, sharing a surface facility, to reduce capital cost to well below \$ 1.00 / kWh at multi - 100 GWh scale. Two GH2 storage salt caverns, near Moss Bluff, TX, have operated well, with low O&M: Chevron-Phillips cavern for > 30 years; Praxair cavern for > 7 years.

Figure 3. USA has enough domal salt geology along the Gulf of Mexico coast to install ~ 15,000 large GH2 storage caverns, providing enough storage to firm the entire Great Plains wind resource, to provide all ~ 100 Quads of USA's total energy demand as annually firm and dispatchable. Northern Germany is also endowed with good salt geology for large storage caverns.

The large, nascent H_2 fuel market for fuel cell buses and light duty vehicles (LDV's) should motivate interest in integrated, optimized, RE-source systems. Toyota North America chief Jim Lentz recently told *Automotive News* that Toyota sees battery-electric vehicles (BEV's) only viable in "... a select way, in short range vehicles ..." Transportation fuel may be a larger market than the Grid, and may motivate turbine suppliers to offer equipment optimized for off-grid H_2 and NH_3 fuel production, with simpler, more robust, and less-costly electric generating systems, using variable-speed induction motors, as in Figure 5. Polymer-metal tubing and linepipe developed at ORNL may replace steel for high-pressure transmission pipelines, solving the hydrogen embrittlement and other corrosion problems.

NH_3, the "other hydrogen", is a higher-hydrogen-density energy carrier and storage medium than liquid hydrogen, LH2. A liter of NH_3 contains more H atoms than a liter of LH2. The modified internal combustion engine, combustion turbine, and fuel cell run well on NH_3, while RE transmission and storage costs are even lower than for GH2.

Figure 4 shows the Agrium terminal where twin mild steel tanks together hold 80,000 tons of refrigerated, liquid anhydrous ammonia, NH_3, which stores ~ 500,000 MWh as the energy content in NH_3, which may be directly used as fuel or may be easily reformed to H_2 fuel and harmless, vented N_2. At this scale, energy storage costs ~ $ 0.10 / kWh capex. Over 4,000 km of NH_3 transmission pipeline is already in place in USA, where NH_3 is used primarily as N-fertilizer. These pipelines are low-cost, mild steel, operating at ~ 15 bar.

Figure 5: By any measure, the low cost and high capacity of GH2 and NH3 storage is "off the charts" vis-a-vis the several technologies considered for "electricity" storage, as adjuncts to the Grid.

Great Plains wind is very seasonally-variable: about 20% above average in Winter and Spring, 30% below average in Summer. Firming 1,000 MW (nameplate), to allow firm energy delivery every day of the year, would require ~ 300,000 MWh of storage: ~ 4 GH2 salt caverns or ~ 2 NH_3 tanks. In either case, the storage incremental capex is ~ 3 % of total windplant cost. Compressed air or reflow battery storage is many times higher. If we consider investing in the energy conversion equipment necessary to avail us of this low-cost storage, we should consider solving the transmission and integration problems as well, via complete, integrated, optimized, RE systems -- from sources to delivered energy services -- employing GH2 and NH_3 as the transmission and storage media. We should not allow ourselves to be limited to defining "storage" as an appurtenance to the electricity Grid.

Japan is researching three strategies for importing C-emissions-free H_2 from distant sources, as high-hydrogen-density liquids in ocean tankers, to build a significant "hydrogen sector" in their national energy system:

- Liquid hydrogen, LH2: Kawasaki Heavy Industry
- Anhydrous ammonia, NH_3: Sumitomo Chemic
- Toluene - Methylcyclohexane cycle: Chiyoda Chemical and Heavy Industry

Alaska's diverse, abundant, stranded RE resources are strategically close and convenient for export to Japan via any of these H_2 transmission schemes, and have been approximately surveyed.

For over a decade, researchers have been bravely and brilliantly engineering wind turbines and solar PV and CSP systems, and the Grid, to accommodate an increasing share of RE, attempting to deliver firm and dispatchable electricity with system stability while displacing fossil generation. This is ever more costly and technically difficult, at the margin, for both generation and transmission. This work is well reported by the IEEE Power and Energy Society in their journal, *Power and Energy*. But we have only begun to ask how RE may serve energy markets beyond the extant electricity sector, whether via an expanded electricity sector and Grid, or perhaps via alternatives to electricity. Do we imagine that we may supply all humanity's energy, for all purposes, from diverse RE sources, entirely, or even primarily, via electricity systems? The Grid, assaulted by PV systems large and small, connected to distribution circuits, may so evolve and atrophy as to be inadequate and unsuitable to "Running the World on Renewables"

Therefore, we now need to assemble a collaborative of private enterprise, academia, national labs, funders, and others to thoroughly and expertly model, technically and economically, complete RE systems -- from photons and moving air and water molecules to delivered energy services -- in the context of humanity's responsibility to expeditiously "Run the World on Renewables". Before we invest billions in a larger and smarter electricity Grid, we need to be sure we've considered alternatives, that we're not trying to stuff a square peg into a round hole. This serious consideration has never been done; our co-authors attempted analysis have been provocative, but inadequate and unpersuasive.

Please contact the author if you wish to consider participating. Conference presentations and papers on this and related RE topics are at: www.leightyfoundation.org/earth.php

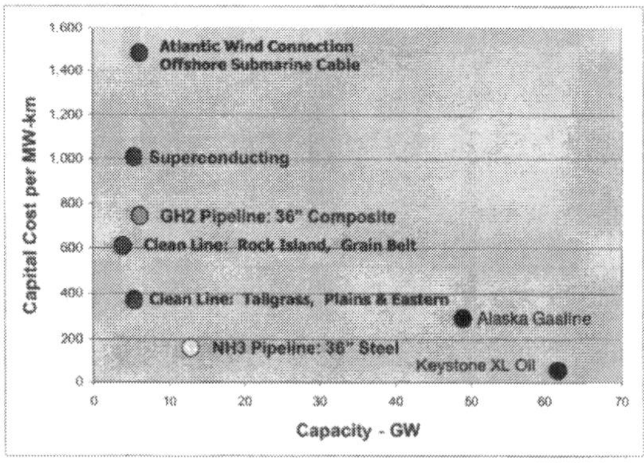

Figure 1. Approximate transmission capital costs per MW-km of service, $ US

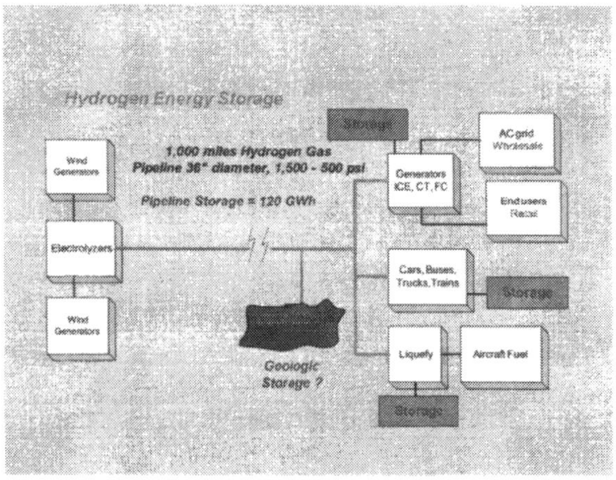

Figure 2. Complete Renewable Energy GH2 system: annual-scale firming, dispatchable storage

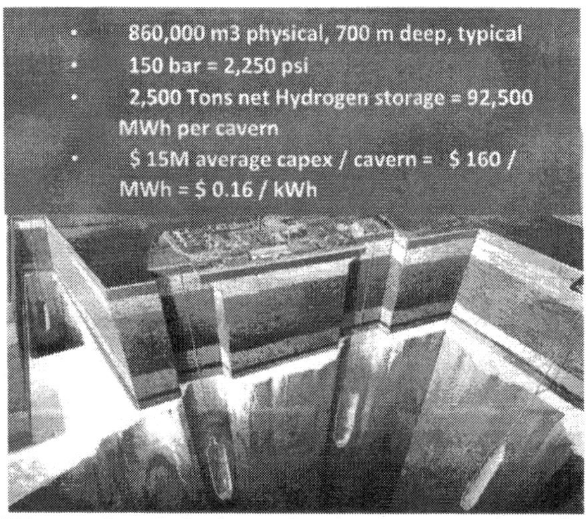

Figure 3. Solution-mined compressed Hydrogen storage caverns in deep domal salt. Multiple caverns manifolded together, share a common surface facility, for very large energy storage.

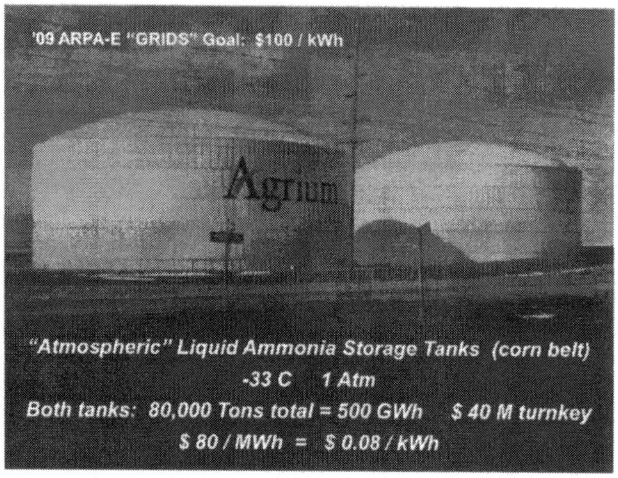

Figure 4. Agrium US, ammonia storage terminal at Early, Iowa, supplying N-fertilizer for corn ag.

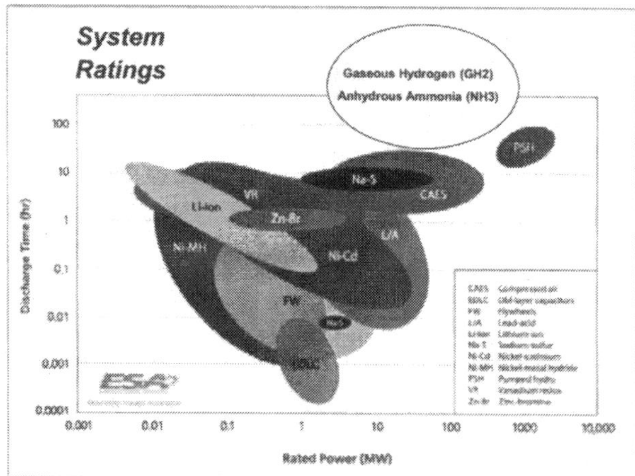

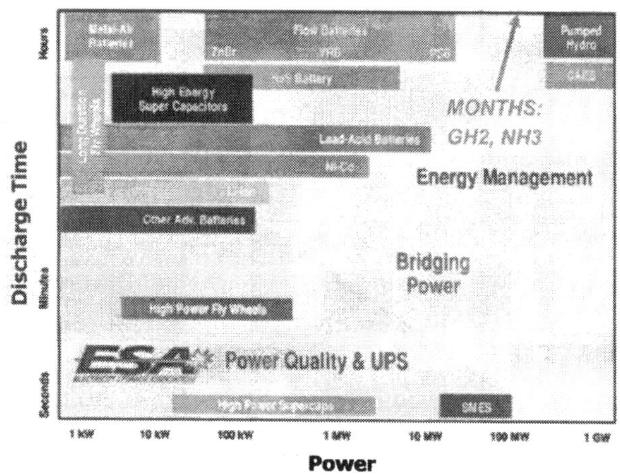

Figure 5. By any measure, the low cost and high capacity of GH2 and NH_3 storage is "off the charts" vis-a-vis the several technologies considered for bulk "electricity" storage as adjuncts to the Grid.

Ni-supported Pd nanoparticles: a new catalyst for low-temperature ammonia cracking

Jaroslaw Polanski[*] Piotr Bartczak[*#], Rafal Sitko[*], Tomasz Siudyga[**], Andrzej Mianowski[**], Jacek Szade[***], Jozef Lelatko[d ****]

[*] Institute of Chemistry, University of Silesia, 40-006 Katowice, Poland
[**] Silesian University of Technology, 44-100 Gliwice, Poland
[***] Silesian Center for Education and Interdisciplinary Research 41-500 Chorzow, Poland
[****] Institute of Materials Science, University of Silesia, 41-500 Chorzow, Poland
[#] NANO-CHEM-TECH, 40-568 Katowice, Poland

ABSTRACT

Ammonia cracking is a method used in the treatment of flue gases from coal or biomass gasification or for hydrogen generation in chemical or related industries. It is an important issue in hydrogen economies. As a carbon-free compound, ammonia provides a potential source of ecological fuel for mobile and stationary power generation, especially for fuel cells, where CO impurities are unacceptable. However, more efficient catalysts are needed for new technologies in this area. Herein, we report a new nanometallic catalyst for ammonia cracking at temperatures well below 500°C. We have discovered that supporting Pd nanoparticles (Pd NPs) directly on Ni grains provides an active and durable catalyst for this reaction. We also report a new, facile and scalable wet chemical method for supporting Pd on Ni; in this method, Pd NPs initially supported on SiO_2 were placed in contact with Ni grains to yield, after a final SiO_2 digestion, a Pd/Ni unalloyed contact.

Keywords: Bimmetalic nanocatalysis, oxidation, Sonogashira coupling, Au nanoparticles, Ni-supported nanoparticles

1 INTRODUCTION

Nanocatalysts are extremely sensitive toward structure differentiation, and their activity and selectivity depend not only on nano-metal and support type but also on size, shape, and composition [1]. Thus, optimization of such materials is an open issue. In fact, more efficient catalysts are still being sought to run the reactions with higher yields and higher selectivity under mild conditions. Other important problems to be addressed include reducing the fraction of noble metals, facilitating the catalyst separation, improving reusability, and reducing contamination of the final products. Probably the widest appliction gained Au nanoparticles (Au NPs) which catalyze a variety of reactions [2-7]. As they tend to agglomerate, they are usually supported on carriers to form more stable catalytic systems. Generally, Au NPs are available on a variety of supports, from carbon-like graphite to inorganic materials. Basically, the first should be wettable by apolar reagents and solvents, while the latter should be wettable by polar ones. Wettability, and consequently catalyst availability for the reactants, is of crucial importance for the reaction progress; e.g., recently oxidation of cyclohexene and D-glucose over nano-Au/SiO_2 in water has been compared. In this reaction, polar polyhydroxyl D-glucose complying with polar SiO_2 support and a polar solvent (H_2O) reacted smoothly at room temperature, whereas nonpolar cyclohexene violating the polarity rule needed the addition of surfactants to react efficiently [8].

2 BIMETALLICS FOR NANOCATLYSIS

Recently, we developed a new method for transferring SiO_2-supported nano-metals to a variety of supports [8-10]. We optimized this procedure using SiO_2 supported Au NPs, e.g., we prepared bimetallic nano-Au or Pd supported on Cu or Ni grains, to test their performance in a variety of reactins, including oxidation or Sonogashira coupling [9-10]. We tested for the first time the efficiency of SiO_2-, Cu-, and Ni-supported Au in deep glycerol oxidation in a diluted and viscous H_2O_2/H_2O liquid phase. Acetic acid (AA), the C2 oxidate, was preferentially formed in such a system. High conversion (100%) and AA yields (90%) were observed for the sol-gel SiO_2-suppported Au in diluted solutions. Although with the increase of glycerol concentration in the viscous liquid phase these values decreased to ca. 40% (conversion) and 20% (AA yield), the addition of acetonitrile improved the AA yield to ca. 40%, while the surfactants were found to be capable of a many-fold enhancement of the catalyst activity at the room temperature highly-viscous liquid phase. High performances were also observed for the bimetallic Au/Cu and Au/Ni catalysts obtained by nano-Au transfer; however, these catalysts were destroyed during the reaction by the Cu or Ni leaching effect.

A copper supported nanopalladium catalyst obtained by an innovative method of nanoparticle transfer from the intermediate carrier SiO_2 to the target Cu carrier was a highly efficient and selective catalyst, giving as much as quantitative conversion in a series of Sonogashira reactions. Comparisons with other Pd systems, especially Pd/SiO_2, indicated substantial advantages for this novel catalytic system. X-Ray photoelectron spectroscopy (XPS) studies of the prepared Pd/Cu catalyst revealed the development of a PdO species in the Pd/Cu system and illustrated for the first time the importance of the resulting Pd/PdO/Cu system for efficient Sonogashira coupling [10].

During this stydy we investigated possible applications of Ni supported Pd NPs.

3 RESULTS AND DISCUSSION

Ammonia cracking is a method used in the treatment of flue gases from coal or biomass gasification or for hydrogen generation in chemical or related industries. It is an important issue in hydrogen economies. As a carbon-free compound, ammonia provides a potential source of ecological fuel for mobile and stationary power generation, especially for fuel cells, where CO impurities are unacceptable. However, more efficient catalysts are needed for new technologies in this area. Herein, we report a new nanometallic catalyst for ammonia cracking at temperatures well below 500°C. We have discovered that supporting Pd nanoparticles (Pd NPs) directly on Ni grains provides an active and durable catalyst for this reaction. We also report a new, facile and scalable wet chemical method for supporting Pd on Ni; in this method, Pd NPs initially supported on SiO_2 were placed in contact with Ni grains to yield, after a final SiO_2 digestion, a Pd/Ni unalloyed contact.

This method provides interesting capabilities for the potential production of industrial catalysts where relatively low-cost Ni, e.g., in the form of wire meshes, provides not only a catalytic moiety but also a flexible catalyst support. This discovery is significant because hydrogen production from ammonia [11] appears to be a difficult problem. Ammonia decomposition is a complex process that proceeds through stepwise dehydrogenation (NH_3 - $H \rightarrow NH_2$ - $H \rightarrow NH$ – $H \rightarrow N$) and yields H and N, which recombine to H_2 and N_2, respectively. Although the binding energy of the nitrogen should be sufficiently strong for dehydrogenation, it should not be too high to block the recombination step [12-15].

Among potential catalysts for this process, Ru has been observed to be the most active; [16] however, *the effects of the Ru particle size and shape can be very difficult to elucidate, and a simultaneous comparison of these factors is almost impossible* [17].

In Figure 1 we presented representative SEM and TEM images of the Pd/Ni catalyst.

Hydrogen productivity of the new catalyst [nanoPd/Ni (1% Pd)] if compared with the literature data showed the evident advantage of the new system.

In Figure 2 we illustrated temperature efficiency of the system.

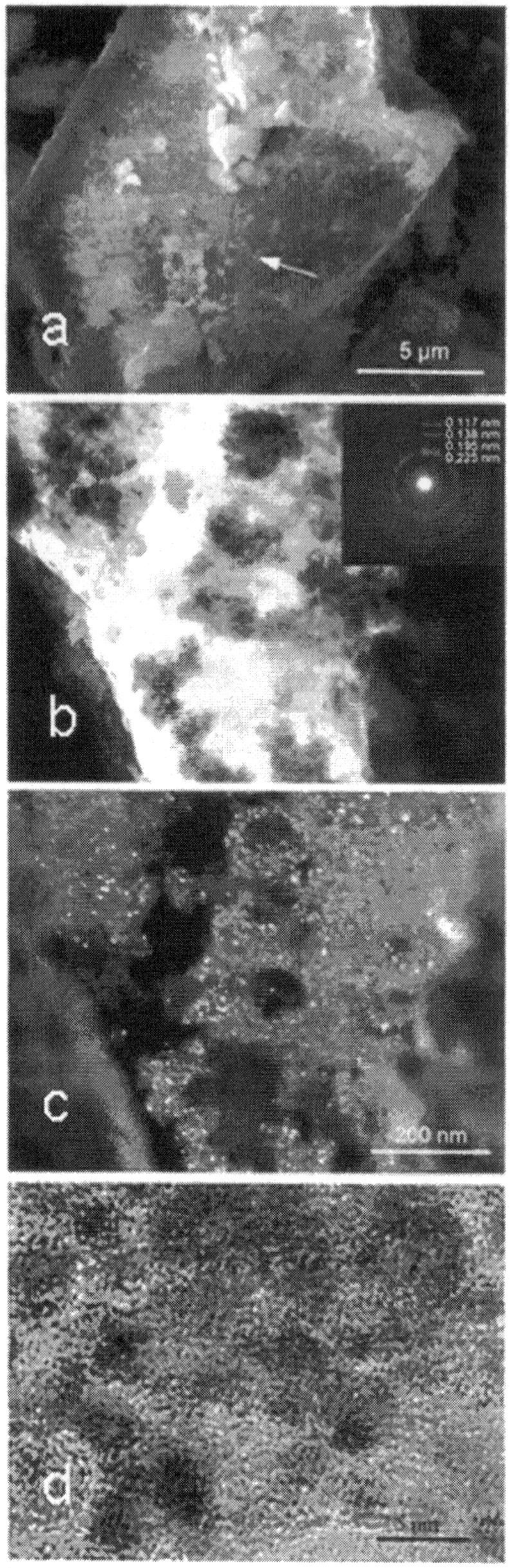

Figure 1: Representative SEM and TEM images of the Pd/Ni catalyst. a - SEM image of the Pd NPs aggregates on the Ni surface; b, c - TEM bright- and dark-field images of the Pd nanocrystalline aggregates on the Ni surface; d - HRTEM image of the Pd NPs with diameters less than 5 nm on the Ni surface.

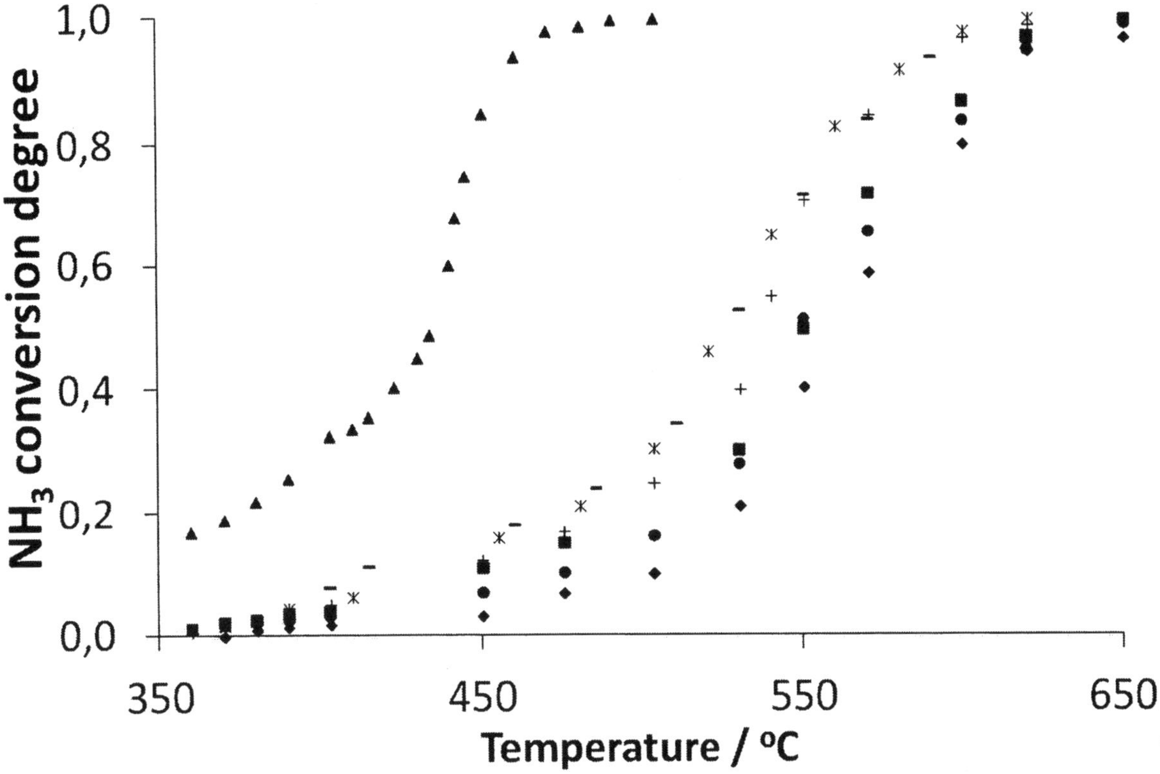

Figure 2: Ammonia conversion on the Pd/Ni catalyst (triangles) compared with that on the analogously processed control Ni carrier (black squares), unprocessed Ni (circles), PdO (crosses), SiO_2 (dashes), Ni/SiO_2 (diamonds) and Pd/SiO_2 (asterisks) at a flow rate of 2 dm^3/h.

4 EXPERIMENTAL

4.1 Preparation of metal NPs on sol-gel silica

The series of metal/silica nanocatalysts were prepared according to the procedure optimized. The carrier is obtained using the Stöber method with tetraethyl orthosilicate (TEOS), which is added to a mixture of ethanol and water with an aqueous ammonia solution.

A mixture was sonicated, then concentrated, dried, and reduced under hydrogen at 500 °C. In a typical procedure, 800 mL of anhydrous ethanol and 135 mL of 25 wt.% solution of ammonia were mixed with 78 mL of deionized water. After 10 min of stirring 60 mL of tetraethyl orthosilicate was added to the reaction mixture which was next stirred for 3 h at room temperature. The colloidal silica suspension was centrifuged, placed in an ultrasound bath and stirred for 90 min. A solution containing, e.g., gold precursor (0.07 g 40% (wt.%) chloroauric acid for 0,1% Au/SiO2) in deionized water (8 mL) was added dropwise into the obtained carrier, i.e., colloidal silica, and mixed in an ultrasound bath for 30 min. Next, it was dried at 60–90 °C for about 12 h in dark, ground and sieved. The reduction was conducted in an oven under hydrogen at 500 °C for 4 h.

4.2 Preparation of bimetallics

Bimetallic catalysts were prepared using a novel facile approach involving the transfer of nanoparticles from the intermediate carrier, i.e., SiO_2, to the target carrier. The general method includes several steps. The target carrier i.e., Cu or Ni, (0.99 g) and Au (Pd) NPs of a low polydispersity deposited on the intermediate carrier i.e., 1,0% Au/SiO_2, (1.00 g) were suspended in deionized water (100 mL) under mechanical stirring and sonication. After 10 minutes of vigorous stirring sodium hydroxide (40 mL 40% w/w) was added to the suspension and stirring was continued for 4 h at 80 °C, whereupon the suspension was allowed to stand for about 18 h until the suspended solids sedimented. The suspension was centrifuged and the supernatant was decanted, the precipitate was washed five times with deionized water and centrifuged again to achieve neutral pH of the supernatant. Precipitate was washed with deionized water, centrifuged and supernatant was removed. The obtained catalyst was dried in an electric dryer to a constant weight at 110 °C.

The resulted preparations of silica and bimetallic catalysts were examined by X-ray photoelectron spectroscopy (XPS) using PHI 5700 spectrometer with monochromatized AlKα a radiation. Alternatively, the samples were suspended in ethanol, sonicated for 15 min and the resulted materials were deposited on carbon adhesive tape for the preparation of the samples for TEM analyses. The transmission electron microscopy (TEM) images of the resultant composites were obtained on a JEOL 2000 FX operating at 200 kV or high resolution (HRTEM) JEM 3010 microscopes (both with EDS systems for chemical composition microanalysis). The scanning electron microscopy (SEM) JSM6480 or a PHILIPS XL 30 was used for morphology investigation of composite powders.

The energy-dispersive X-ray fluorescence (EDXRF) analysis was performed on the Epsilon 3 spectrometer (Panalytical, Almelo, The Netherlands) with a Rh target X-ray tube with 50μm Be window and max. power of 9W. The spectrometer is equipped with thermoelectrically cooled silicon drift detector (SDD) with 8μm Be window and resolution of 135 eV at 5.9 keV. The system is also equipped with five primary filters that can be selected to improve measuring conditions for determined elements. The quantitative analysis of Au or Pd/SiO_2, Pd/Cu or Pd/Ni catalysts was performed using Omnian software based on fundamental parameter method.

ACKNOWLEDGMENTS

The research was co-financed by the National Research and Development Center (NCBiR) under Grant ORGANOMET No: PBS2/A5/40/2014.

REFERENCES

[1] S. Philippe, P. Karine (Eds.), "Nanomaterials in Catalysis Introduction", in: S. Philippe, P. Karine (Eds.), Concepts in Nanocatalysis, Viley-VCH, Weinheim, 1-5, 2013,

[2] Y. Zhang, X. Cui, F. Shi, Y. Deng, Chem. Rev. 112 2467, 2012.

[3] C.D. Pina, E. Falletta, M. Rossi, Chem. Soc. Rev. 41 350, 2012.

[4] M. D. Hughes, Y. J. Xu, P. Jenkins, P. Mcmorn, P. Landon, D. I. Enache, F. Carley, G. A. Attard, G. J. Hutchings, F. King, E. H. Stitt, P. Johnston, K. Griffin, C. J. Kiely, Nature 437, 1132, 2005.

[5] A. S. K. Hashmi, G. J. Hutchings, Angew. Chem. Int. Ed. 45, 7896, 2006.

[6] M. D. Hughes, Y. J. Xu, P. Jenkins, P. Mcmorn, P. Landon, D. I. Enache, F. Carley, G. A. Attard, G. J. Hutchings, F. King, E. H. Stitt, P. Johnston, K. Griffin, C. J. Kiely, Nature 437, 1132, 2005.

[7] A. S. K. Hashmi, G. J. Hutchings, Angew. Chem. Int. Ed. 45, 7896, 2006.

[8] P. Bujak, P. Bartczak, J. Polanski, J. Catal. 295 15–21, 2012.

[9] M. Korzec, P. Bartczak, A. Niemczyk, J. Szade, M. Kapkowski, P. Zenderowska, K. Balin, J. Lelątko, J. Polanski, J. Catal. 313, 1, 2014.

[10] M. Kapkowski, P. Bartczak, M. Korzec, R. Sitko, J. Szade, K. Balin, J. Lelatko, J. Polanski, J. Catal., 319 (2014) 110-118.

[11] F. Schueth, R. Palkovits, R. Schloegl, D.S. Su, Energy & Environmental Science 5, 6278, 2012.

[12] C.J.H. Jacobsen, S. Dahl, B.S. Clausen, S. Bahn, A. Logadottir, J.K. Norskov, J. Am. Chem. Soc. 123 8404, 2001.

[13] A.Boisen, S. Dahl, J.K. Norskov, C.H. Christensen, J. Catal. 230 309, 2005.

[14] D.A. Hansgen, D.G. Vlachos, J.G. Chen, Nature Chemistry 2 484, 2010.

[15] J.C.Ganley, F.S. Thomas, E.G. Seebauer, R.I. Masel, Catalysis Letters 96, 117, 2004.

[16] X. Duan, X. Zhou, D. Chen, Catalysis 25, 118, 2013.

Effects of Size, Composition, and Architecture on the CO Tolerance of Bimetallic Co_xPt_y Nanoparticle Electrocatalysts

C. M. Sims and B. W. Eichhorn

Department of Chemistry and Biochemistry
University of Maryland, College Park
College Park, MD, USA, eichhorn@umd.edu

ABSTRACT

$CoPt_3$, CoPt, and Co_3Pt (Co_xPt_y) alloy and intermetallic nanoparticle (NP) electrocatalysts were synthesized and characterized by a combination of powder X-ray diffraction (XRD), transmission electron microscopy (TEM), and energy dispersive spectroscopy (EDS) studies. Electrochemical carbon monoxide (CO) stripping and rotating disk electrochemical (RDE) experiments showed that the Co_xPt_y alloy NPs have CO-tolerant hydrogen oxidation reaction (HOR) performances similar to monometallic Pt catalysts, but the structural and electronic modifications induced by intermetallic formation were found to be detrimental to their ability to tolerate CO. However, the decreased Pt content makes them more cost-effective as electrocatalysts for proton exchange membrane fuel cells (PEMFCs).

Keywords: bimetallic nanoparticles, electrocatalysis, fuel cells, heterogenous catalysis, intermetallics

1 INTRODUCTION

PEMFCs are highly regarded as a clean alternative energy technology for motor vehicles, power generators, and many other applications. Unfortunately, several factors limit their commercial viability, including the high expense of their catalytic materials, the poisoning of the HOR electrocatalyst at the anode by CO, and the inefficiency of the oxygen reduction reaction (ORR) electrocatalyst at the cathode. To mitigate these issues, researchers have developed new Pt-based bimetallic nanoparticles (NPs) by alloying Pt with cheaper metals such as Fe,[1] or Cu.[2] Use of these bimetallic NPs has led to reduced catalyst costs, improved CO tolerance for the HOR, and increased catalytic efficiencies for the ORR.

Although the Co-Pt bimetallic system has been extensively studied for ORR catalysis, with high catalytic activities predicted by theoretical models and demonstrated in the literature,[3-8] research into the use of Co-Pt NPs for CO-tolerant HOR catalysis is quite limited.[9,10] While these previous studies reported mild improvements in CO tolerance from Co-alloying, the Pt-rich $CoPt_3$ alloy was the focus of research,[9,10] with little discussion of other compositions or architectures.[9] In our previous work, the CO tolerance of the analogous Fe-Pt bimetallic system was found to vary significantly with respect to differing NP architectures, despite the particles' having similar elemental compositions.[11] Studies such as this highlight the importance of compositional *and* structural control of bimetallic NPs, due to their essential roles in determining the ultimate performance of a catalytic system.

Here, we present the synthesis of $CoPt_3$, CoPt, and Co_3Pt (Co_xPt_y) alloy and intermetallic NPs, which were extensively characterized by powder X-ray diffraction (XRD), transmission electron microscopy (TEM), and energy dispersive spectroscopy (EDS) analyses. We conduct a full systematic evaluation of the well-characterized Co_xPt_y NPs for CO-tolerant HOR catalysis to probe the influences of size, composition, and architecture on this energy conversion process.

2 EXPERIMENTAL

2.1 Reagents and Materials

Cobalt acetylacetonate ($Co(acac)_3$, 99.99%), platinum acetylacetonate ($Pt(acac)_2$, 97%), 1-octadecene (90%), sodium triethylborohydride solution ($NaBEt_3H$, 1.0 M in toluene), oleic acid (OA, 90%), and acetone (99.5%) were purchased from Aldrich. Methanol (MeOH, 99.8%) was purchased from VWR. Oleylamine (OAm, >70%) and Nafion® solution (5%) were purchased from Fluka. Isopropanol (iPrOH, 99%) was purchased from Pharmco-AAPER. Hexanes (99.9%) and sulfuric acid (H_2SO_4, 96.4%) were purchased from Fisher. Vulcan XC-72 carbon black powder (CB) was purchased from Cabot Corporation. E-TEK Pt (30% HP Pt on Vulcan XC-72) was purchased from BASF. Ultra-pure water was obtained from deionized water using a Millipore Academic Milli-Q A10 purifier system. All materials were used as received without further purification.

2.2 Synthesis of Co_xPt_y Alloy and Intermetallic Nanoparticles

In a typical synthesis, $Co(acac)_3$ and $Pt(acac)_2$ in the appropriate molar ratios (0.075 mmol total metal, Co:Pt ratios: 3:1, 1:1, 1:3), OA (50 μL), and OAm (50 μL) were mixed in 1-octadecene (5 mL) in a 50 mL round-bottom flask. The resultant mixture was degassed at 80 °C under vacuum, then placed under a flowing N_2 atmosphere and held at 100 °C for 60 min to ensure full dissolution of the precursors. In a separate Schlenk flask, 1-octadecene (15 mL) was added and degassed at 80 °C under vacuum, placed under flowing N_2 and heated to 200 °C. At this temperature, $NaBEt_3H$ solution (3 mL, 1.0 M in toluene) was injected into the Schlenk flask, followed by the immediate injection of the Co_xPt_y precursor solution into the Schlenk flask. The reaction solution instantly turned black with the precursor injection, and the resultant mixture was returned to 200 °C. After being held at 200 °C for 60 min, the heating source was removed and the reaction was allowed to cool down to room temperature. The black colloidal solution was transferred to a 50 mL conical centrifuge tube, where a 1:1 MeOH/acetone mixture (40 mL) was added to the tube, which was sonicated in a sonication bath (FS30H, Fisher Scientific) for 10 min before centrifugation at 6000 rpm for 15 min. The supernatant was discarded and the washing procedure was repeated an additional two times. The black solid was dried overnight, before re-suspension in hexanes (20 mL).

To make carbon-supported Co_xPt_y electrocatalysts with 30 wt % total metal loading, the NP suspensions were mixed with a suitable amount of CB powder and sonicated in a sonication bath (FS30H, Fisher Scientific) for 180 min. The hexanes were evaporated in air overnight and the resultant CB-supported NPs were vacuum-dried. The dried solids were placed in a ceramic boat, which was introduced into a quartz glass tube and heated in a horizontal tube furnace (Thermolyne F21135, Thermo Scientific) under a 5% H_2/95% Ar atmosphere with a flow rate of 90 SCCM for 90 min at 400 °C, 550 °C, or 700 °C to remove the surfactants from the NPs (400 °C, 550 °C, 700 °C) and/or convert the alloy particles into their intermetallic counterparts (550 °C and 700 °C).

2.3 Characterization Methods

Powder X-ray diffraction (XRD) patterns of samples were obtained on a Bruker C2 Discover diffractometer equipped with a VÅNTEC-500 detector using a monochromatic Cu Kα radiation source biased at 40 kV and 40 mA. Transmission electron microscope (TEM) images were obtained on a JEM 2100F Field Emission TEM operating at 200 kV. Energy dispersive spectroscopy (EDS) data were collected on the same TEM operating in the scanning (STEM) mode.

2.4 Electrochemical Analysis

iPrOH (159.2 mL), ultra-pure water (40.0 mL), and Nafion® solution (0.80 mL, 5%) were mixed and stored as a stock solution. The catalyst ink was prepared by mixing the supported NP powder with the above stock solution such that the concentration was 1.0 mg/mL of powder in solution. The resultant mixture was sonicated in a sonication bath (FS30H, Fisher Scientific) for 120 min. The catalyst ink (20 μL) was cast on a glassy carbon (GC) electrode (Pine Instruments, 5.0 mm diameter) using a micropipette and allowed to dry in air overnight while covered. Electrochemical experiments were performed on a potentiostat (Autolab PGSTAT30) with a standard three-electrode electrochemical cell. The rotating GC disk electrode with dried catalyst ink on its surface was used as the working electrode. Pt wire was used as the counter electrode and a saturated calomel electrode (SCE) was used as the reference electrode. All potentials were recorded with respect to the SCE. H_2SO_4 in ultra-pure water (0.5 M) was used as the electrolyte. The catalysts were subjected to 50 potential scan cycles at 50 mV/s between -0.2 V and 0.8 V while Ar-saturated as a conditioning procedure. After catalyst conditioning, fresh electrolyte was used during the collection of experimental data. To perform the CO stripping experiment, the catalyst was saturated with CO by bubbling CO (99.5% pure, Al tank) in the electrolyte for 20 min, followed by Ar-purging for 40 min to remove excess CO. The CO-saturated catalysts were then subjected to a potential scan cycle at 20 mV/s between -0.2 V and 0.8 V. To obtain polarization curves for the electrooxidation of CO-contaminated H_2, the electrolyte was bubbled with a 1000 ppm CO/balance H_2 gas mixture (99% pure, Al tank) for 90 min with the electrode potential held at -0.2 V, followed by a potential scan (1 mV/s) at a rotation rate of 1600 rpm.

3 RESULTS AND DISCUSSION

3.1 Synthesis and Characterization of Co_xPt_y Nanoparticles

Briefly, bimetallic Co_xPt_y ($CoPt_3$, CoPt, and Co_3Pt) NPs were synthesized by using $NaBEt_3H$ to co-reduce appropriate amounts of $Co(acac)_3$ and $Pt(acac)_2$ in 1-octadecene in the presence of an OA/OAm surfactant mixture at 200 °C similarly to previously reported methods.[12,13] The three different as prepared (AP) Co_xPt_y samples ($CoPt_3$, CoPt, and Co_3Pt) were annealed at three different temperatures (400 °C, 550 °C, and 700 °C) to convert the disordered alloys into ordered intermetallic phases. The results of our experiments are in good agreement with previous reports.[12,13] A summary of the experimental data, as determined from XRD, TEM, and EDS analyses of the NPs, is shown in Table 1.

Representative XRD patterns of the Co_xPt_y NPs are shown in Figure 1. Relative to monometallic Pt, the diffraction peaks shift to increasingly higher 2θ angles with increasing Co concentration due to increasing lattice contraction resulting from Co having a smaller lattice constant than Pt. Each of the Co_xPt_y NPs samples also were analyzed with EDS (graphical data not shown for brevity), with the $CoPt_3$, CoPt, and Co_3Pt NPs having average Co:Pt composition ratios of 24:76, 45:55, and 70:30, respectively, in good agreement with the precursor ratios used and the XRD analysis. The composition ratios do not change significantly with respect to annealing temperature.

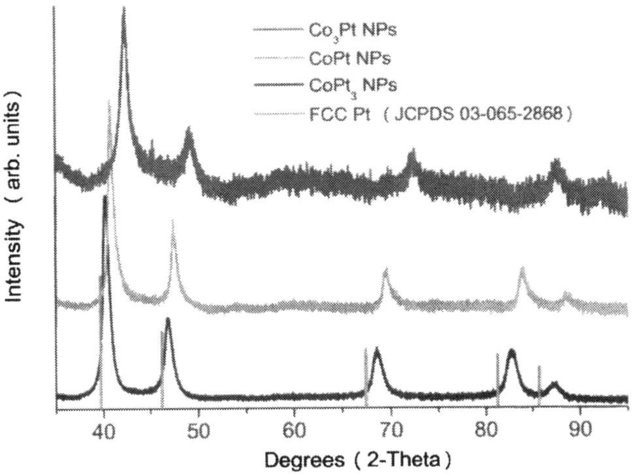

Figure 1. XRD patterns of 550 °C annealed $CoPt_3$ (black), CoPt (magenta), and Co_3Pt (blue) NPs. Red lines indicate the peak positions for FCC phase Pt.

Comparing the different Co_xPt_y NPs shows that the particles generally increase in size with increasing Pt content (Table 1). Each Co_xPt_y synthesis contained the same amount of surfactant with equal moles of total metal, with only the Co:Pt ratios changing between the different Co_xPt_y syntheses. Increasing the Pt content (but maintaining equal moles of metal) within a synthesis results in slightly increased surface area due to the larger size of Pt atoms relative to Co. Since surfactants control NP size with respect to surface area, the increased surface area effectively increases the metal-surfactant ratio, which leads to a slight increase in NP size.

While the Co-Pt phase diagram shows that the $CoPt_3$ and CoPt NPs should exist as intermetallics, the rapid reduction and nucleation of Co and Pt results in the formation of Co_xPt_y alloy particles. Annealing the $CoPt_3$ and CoPt NPs induces their transformation into their thermodynamically stable intermetallic phases, while the Co_3Pt NPs maintain the expected alloy structure. High temperature annealing is necessary to induce the transformation from alloy to intermetallic, but it also results in the growth and agglomeration of the NPs (Table 1). A representative TEM image of the Co_xPt_y NPs is shown in Figure 2.

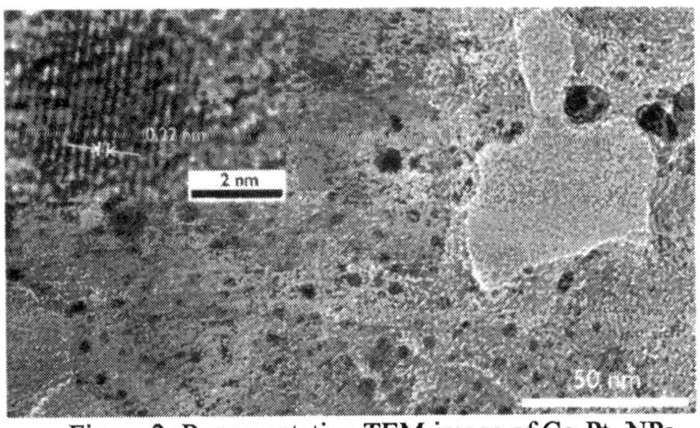

Figure 2: Representative TEM image of Co_xPt_y NPs (550 °C annealed CoPt NPs shown)

3.2 Electrochemical Characterization of Co_xPt_y Alloy and Intermetallic Nanoparticles

To assess the effects of size, composition and architecture on the NPs' catalytic activities, we evaluated the catalytic activities of all nine annealed (400 °C, 550 °C, and 700 °C) Co_xPt_y catalysts for CO oxidation (CO stripping) and CO-tolerant HOR. As a reference, the commercially available E-TEK Pt catalyst was tested under identical conditions. All catalysts were prepared with a 30% total metal loading by weight based on the initial amounts of reagents. Given the similar metal loadings of the Co_xPt_y NP catalysts, we focus our analysis on how the different sizes, compositions and architectures of the NPs impact their resultant catalytic activities. Enhanced catalytic activity is associated with *lower* onset potential (E_{Onset}) values. A summary of the relevant structural and electrochemical data for the different NP catalysts is given in Table 1.

The onset potentials for CO oxidation on the Co_xPt_y catalysts are all in the range of 0.4 - 0.55 V (Table 1), which is similar to the 0.53 V onset potential for the E-TEK Pt reference. Comparing the composition sets ($CoPt_3$ vs. CoPt vs. Co_3Pt), reveals that the onset potential tends to decrease with increasing Co content. Previous studies have demonstrated that the electronic structure modifications induced by alloying Pt catalysts with Co improves their ability to oxidize CO in the absence of H_2.[14,15] The results of these previous CO stripping experiments are reproduced here.

For the electrooxidation of CO/H_2 mixtures (1000 ppm CO), the onset potentials range from 0.27 - 0.4 V, which is lower than the pure CO oxidation onset potentials as expected (Table 1). The $CoPt_3$-400 °C and all three Co_3Pt catalysts have CO/H_2 onset potentials very similar to E-TEK Pt, while the 550 °C and 700 °C annealed CoPt and $CoPt_3$ catalysts have much lower CO tolerances. Unlike other studies where lower CO stripping onset potentials are correlated with lower CO/H_2 onset potentials and increased CO tolerance, there is no clear correlation between the CO oxidation and onset potentials in the Co_xPt_y catalysts.[16]

This lack of correlation suggests the diminished CO tolerance in the intermetallic NPs is most likely due to modified electronic structures that negatively impact their ability to tolerate CO relative to the alloy particles. As previously seen in $FePt_3$ catalysts, this loss of CO tolerance would most likely manifest through reduced H_2 adsorption and activity for HOR.[11] This explanation would also account for the CO tolerance of the CoPt-400 °C catalyst, which is of mixed alloy/intermetallic state and behaves intermediate of the two architectures.

Table 1. Summary of representative experimental data for NPs

Sample	Space Group (Crystal Structure)[a]	Avg. NP Size[b]	Avg. NP Composition[c]	CO Stripping E_{Onset}[d]	CO/H_2 Oxidation E_{Onset}[d]
E-TEK Pt	Fm-3m (FCC)	3.2 ± 0.8 nm	--	0.53	0.29
$CoPt_3$-400 °C	Fm-3m (FCC)	2.8 ± 1.3 nm	Co:Pt 24:76	0.50	0.29
$CoPt_3$-550 °C	Pm-3m (Primitive)	6.0 ± 3.8 nm		0.47	0.38
$CoPt_3$-700 °C	Pm-3m (Primitive)	6.9 ± 4.0 nm		0.54	0.40
CoPt-400 °C	Fm-3m (FCC) P4/mmm (FCT)	2.7 ± 0.7 nm	Co:Pt 45:55	0.48	0.34
CoPt-550 °C	P4/mmm (FCT)	4.7 ± 3.2 nm		0.45	0.37
CoPt-700 °C	P4/mmm (FCT)	6.0 ± 3.9 nm		0.45	0.39
Co_3Pt-400 °C	Fm-3m (FCC)	2.2 ± 0.6 nm	Co:Pt 70:30	0.47	0.29
Co_3Pt-550 °C	Fm-3m (FCC)	3.9 ± 2.5 nm		0.42	0.31
Co_3Pt-700 °C	Fm-3m (FCC)	4.9 ± 3.9 nm		0.41	0.27

[a] Determined from XRD analysis. [b] Calculated by counting 100 small particles and agglomerates.
[c] Determined from EDS analysis. [d] Potential vs. SCE.

4 CONCLUSION

In summary, a series of Co_xPt_y NPs with controlled compositions and architectures were synthesized through a simple co-reduction method. The Co_xPt_y electrocatalysts were tested for their activities towards CO-tolerant HOR to probe the influences of size, composition, and architecture. The Co_xPt_y catalysts were found to have similar or less CO tolerance relative to monometallic Pt, with the structural and electronic modifications induced by the atomic ordering of Co and Pt found to negatively impact their activity for the HOR. Overall, the Co_3Pt NPs have HOR catalytic activities comparable to monometallic Pt, but utilize far less Pt to achieve this result, leading to a significant reduction in the overall cost of the electrocatalyst.

5 ACKNOWLEDGEMENTS

We thank the Air Force Office of Scientific Research for financial support of this work through AFOSR grant FA9550-09-1-0523 and AFOSR grant 4789-UM-AFOSR-0004. C. S. acknowledges the support of the Ann G. Wylie Dissertation Fellowship. We also acknowledge the support of the Maryland NanoCenter and its NispLab. The NispLab is supported in part by the National Science Foundation as a MRSEC Shared Experimental Facility.

REFERENCES

[1] E. Antolini, J.R.C. Salgado, E.R. Gonzalez, J. Power Sources 160 (2006) 957.

[2] M. Oezaslan, P. Strasser, J. Power Sources (2011).

[3] N.M. Marković, T.J. Schmidt, V. Stamenkovic, P.N. Ross, Fuel Cells 1 (2001) 105.

[4] J. Greeley, I.E.L. Stephens, A.S. Bondarenko, T.P. Johansson, H.A. Hansen, T.F. Jaramillo, J. Rossmeisl, I. Chorkendorff, J.K. Nørskov, Nat. Chem. 1 (2009) 552.

[5] V.R. Stamenkovic, B.S. Mun, M. Arenz, K.J.J. Mayrhofer, C.A. Lucas, G. Wang, P.N. Ross, N.M. Markovic, Nat. Mater. 6 (2007) 241.

[6] Q. He, S. Mukerjee, Electrochim. Acta 55 (2010) 1709.

[7] S. Koh, M.F. Toney, P. Strasser, Electrochim. Acta 52 (2007) 2765.

[8] D. Wang, H.L. Xin, R. Hovden, H. Wang, Y. Yu, D.A. Muller, F.J. DiSalvo, H.D. Abruna, Nat. Mater. 12 (2012) 81.

[9] S.K. Cheah, O. Sicardy, M. Marinova, L. Guetaz, O. Lemaire, P. Gélin, A.A. Franco, J. Electrochem. Soc. 158 (2011) B1358.

[10] S.M.M. Ehteshami, Q. Jia, A. Halder, S.H. Chan, S. Mukerjee, Electrochim. Acta 107 (2013) 155.

[11] Z. Liu, G.S. Jackson, B.W. Eichhorn, Energy Environ. Sci. 4 (2011) 1900.

[12] E. Favry, D. Wang, D. Fantauzzi, J. Anton, D.S. Su, T. Jacob, N. Alonso-Vante, Phys. Chem. Chem. Phys. 13 (2011) 9201.

[13] E.V. Shevchenko, D.V. Talapin, H. Schnablegger, A. Kornowski, Ö. Festin, P. Svedlindh, M. Haase, H. Weller, J. Am. Chem. Soc. 125 (2003) 9090.

[14] H. Igarashi, T. Fujino, Y. Zhu, H. Uchida, M. Watanabe, Phys. Chem. Chem. Phys. 3 (2001) 306.

[15] C. Xu, J. Hou, X. Pang, X. Li, M. Zhu, B. Tang, Int. J. Hydrogen Energ. 37 (2012) 10489.

[16] Z. Liu, J.E. Hu, Q. Wang, K. Gaskell, A.I. Frenkel, G.S. Jackson, B. Eichhorn, J. Am. Chem. Soc. 131 (2009) 6924.

Biomimetic Nanomaterials via in situ Synthesis of Atomically-Thin Gold Sheets and Platinum Nanoclusters

M. McTaggart[**], M. Jugroot[*] and C. Malardier-Jugroot[**]

[*]Royal Military College of Canada, Dept. of Aero. and Mech. Eng. Kingston, ON
[**]Royal Military College of Canada, Dept. of. Chem. and Chem. Eng.
Kingston, ON, cecile.malardier-jugroot@rmc.ca

ABSTRACT

This paper presents the synthesis and activity of broadly stable polymer-metal nanoreactors capable of confinement-enhanced catalysis. Poly(styrene-*alt*-maleic acid), SMA, is known to self-assemble in water into highly organized supramolecular nanostructures containing polar and non-polar reaction spaces. We use this polymer architecture to direct the reduction of hydrophobic salt precursors into metal nanoclusters under abient conditions and without extraneous reductant. Small-angle neutron scatering and transmission electron microscopy reveal the gold and platinum nanocatalysts remain embedded within respective hydrophobic and hydrophilic confinement spaces.

Keywords: biomimetics, catalysis, gold nanocrystals, self-assembly, supramolecular materials

1 INTRODUCTION

Many important chemical processes and advanced material classes remain impractical or expensive outside of laboratory scales due to their thermodynamic or kinetic unfavourability [1]. The fixation of atmospheric gases into useable chemicals or the controlled synthesis of functionally complex molecules are two examples of industrially relevant and technologically challenging research areas [1-3]. Yet even the simplest living organisms efficiently promote these processes and indeed make them among the most common chemical reactions on Earth.

A promising approach to understanding Nature's success is to examine the structures responsible for the high reaction efficiency, namely enzymes [4,5]. Enzymes are biological nanoreactors which comprise a macromolecular superstructure and a catalytic active centre. The two elements work in concert: the protein structure reduces the entropic cost of synthetic reactions by increasing the concentration of target substrates within confined reaction sites, and the active center increases reaction kinetics by catalytic control of substrate geometry[5].

Biomimicry can be attempted either through the direct imitation of enzyme structure using chains of amino acids, or by reproducing the physical and chemical properties in other materials. By taking the latter approach, robust nanoreactors could be produced that, unlike natural enzymes, are tolerant to changes in pH, ionic strength, and temperature that may be encountered in industrial environments [5].

The amphiphilic nanostructure selected for development of a confinement system is composed of self-assembling chains of poly(styrene-*alt*-maleic acid), SMA. In neutral water, internal hydrogen bonding along the SMA polymer produces a rigidly linear backbone chain [6]. Self assembly occurs because the enthalpy gain from minimizing the contact area of the styrene groups with water is greater than the entropic cost of association [7].

Small angle neutron scattering (SANS) provides a unique approach for the investigation of the association of polymers because this method allows for characterization in solution. A very high contrast between the deuterated solvent and the hydrogenated polymer allows the size and shape of the nanoarchitechture to be elucidated by fitting the SANS spectrum with characteristic models. A hollow cylinder form factor revealed that SMA self-assembles into nanotubes with 28Å interior and 41Å exterior diameters.

Characterization of the association of high molecular weight SMA shows that the individual chain conformation remains the same as that observed in previous [8] studies of SMA self-assembly but while low molecular weight chains form nanotubes, longer chains instead associate into bi-layer sheets with similar interior dimension [6,7,9]. In both low and high molecular weight cases, the structures' interior spaces have primarily hydrophobic surfaces while the exterior is primarily hydrophilic [10].

This paper presents the synthesis and characterization of noble metal nanocrystalization into active centres within confined reaction sites of a supramolecular, self-assembling polymeric nanostructure. Metal species-specific nanocrystal formation is observed within hydrophobic and hydrophilic confinement areas, making for a highly tunable and versatile reactor system.

2 EXPERIMENTAL

Poly(styrene-*alt*-maleic anhydride), partial methyl-ester (350, 000 mw avg, Sigma-Alrich) was disolved in water to form a 1% wt/wt solution and neutralized to pH 7 with NaOH. Noble metal SMA solutions were produced by adding 0.0358g platinum(II) chloride ($PtCl_2$, 98%, Sigma-Aldrich) or 0.0347g gold(I) chloride (AuCl, 99.9%, Sigma-Aldrich) to 3mL SMA solution and mixing by sonication for 1.5 hours. D_2O and NaOD were used for small angle neutron scattering (SANS) experiments.

Molecular modeling was completed using Gaussian09 software [11]. Metal-polymer interactions were simulated with a three part ONIOM model: the outer monomers were modeled at the semi-empirical PM6 level of theory; the central monomer was modeled using density functional theory (DFT) B3PW91 functional and 6-31G basis set; metal atoms were modeled using DFT B3PW91/ LANL2DZ. This splitting has been shown to give high quality results for simulations of transition metal interactions with oxygen and benzene complexes [12].

SANS experiments were performed on the NG3-30m instrument at the NIST Center for Neutron Research, Gaithersburg, MD, using an incident wavelength of 6Å at sample to detector distances of 1.00, 6.00, and 13.00m. Raw data was reduced at NIST-CNR using the Igor Macros package. Analysis of the reduced data was performed using SasView 3.0.0 (http://www.sasview.org).

Transmission electron microscopy (TEM) experiments were performed at the Canadian Centre for Electron Microscopy, McMaster University, Hamilton, ON, using a high-resolution FEI HR-TEM/STEM Titan 80-300LB microscope operating at 300kV.

3 RESULTS AND DISCUSSION

An overview of the noble metal and SMA nanoreacture system, including its catalytic activity, is detailed in Section 3.1. Section 3.2 contains a thorough characterizationof the Pt-SMA system through molecular modeling, SANS, and TEM. The metal salt reduction method is expanded to gold(I) chloride in Section 3.3.

3.1 Nanoreactor structure and function

The supramolecular SMA polymer provides the structure for the nanoreactors under development. Non-polar molecules, including hydrophobic metal salts, have been found to preferentially accumulate within the SMA template [13]. Pyrrole is a small, heterocyclic, non-polar molecule which is insoluble in water. By dissolving SMA at 1% wt/wt in neutral water, pyrrole was found to dissolve within the hydrophobic domains of SMA nanostructures suspended in aqueous solution. In time, a colour change indicated a reaction had occurred within the solution: pyrrole had polymerized into polypyrrole [14]. Following the polymerization by UV/vis showed the growth of a peak that red-shifted from 310nm to 480nm in proportion to polymer length.

Pyrrole polymerization is not a spontaneous reaction in the bulk, nor in pure water. Density functional theory molecular modeling shows that dimerization requires 93.6 kJ/mol; it isn't until the hexamer is reached that the polymer is favoured (by -191.6kJ/mol) over an equal number of independent monomers [14]. Two experiments were made to distinguish between the effects of physical confinement and the interior chemistry of SMA22. First, the reaction was attempted in 1% solution of poly(isobutylene-*alt*-maleic acid), IMA. Like SMA, this polymer also self-assembles into a bi-layer sheet with a 2nm hydrophobic interior space in neutral water [15]. Pyrrole was found to spontaneously polymerize in this system in the same manner as in SMA solution. Second, a block copolymer polystyrene-*b*-poly(acrylic acid), PS-*b*-PAA, was used to produce 50nm micelles whose internal and external surfaces were chemically equivalent to those of SMA. Pyrrole was found to solubilize into the interior spaces of the micelles, but no polymerization occurred [14].

In the case of Platinum(II) chloride, it was found that the SMA template would also promote the reductant-free synthesis of mono-disperse platinum nanocrystals [13]. The reduction of platinum chloride to platinum metal without a reducing agent had never before been reported. The pyrrole polymerization experiment was repeated using platinum nanoparticle loaded SMA structures. Not only was the reaction rate increased by a factor of three, but the polypyrrole peak under UV/vis spectroscopy became sharper [13]. Since absorbance by polypyrrole is dependent on polymer length, the sharper peak indicates an increase in the frequency of occurrence of a particular chain length.

The confinement effect is thought to be a consequence of surface effects on local water phases. Since the mobility of the SMA structure is much lower than a water molecule, those water molecules that bond to the surface must lose velocity and, subsequently, kinetic energy. Since the water within the hydrophobic space is free to move (in fact much more free to move than bulk water due to the lack of restrictive hydrogen bonds [8]), the velocity of internal water molecules would increase in proportion to the energy lost from the frozen surface waters. The low water density and high velocity produces a supercritical phase of water within the SMA confinement space [8,16].

Pyrrole's hydrophobicity implies that its interaction with water is negligible and so would experience the potential energy field produced by external surface water as a curvature in the space that produces an apparent inward acceleration, i.e. compression, as observed. There is also a growing body of literature describing the formation of very small, monodisperse crystals in supercritical water [17], explaining the synthesis and morphology of the platinum nanocrystals previously observed.

3.2 Platinum in SMA solution

TEM images in the original study of platinum nanocrystal synthesis in SMA showed the crystals embedded within the polymer structure [13]. Investigation of the metal - polymer interactions are therefore important to establish the thermodynamically favoured site for platinum bonding and crystal nucleation. Density functional theory modeling was completed to probe the most stable geometries of ideal systems. A test platinum atom placed equidistant from the styrene (hydrophobic interior) and

Table 1. SANS model fitting parameters.

	Parameter	SMA-Au pH7	SMA-Pt pH7	SMA-Pt pH5
Lamina	SLD_{solid}	1.87e-06	1.87e-06	1.87e-06
	SLD_{solv}	6.38e-06	6.38e-06	6.38e-06
	Spacing s (Å)	8.04	6.66	6.30
	Thickness t (Å)	95.64	14.59	12.61
	Layers n (N)	7.23	1.01	5.38
Ellipsoid	SLD_{solid}	4.66e-06	6.34e-06	6.34e-06
	SLD_{solv}	6.38e-06	1.87e-06	1.87e-06
	Radius a (Å)	2.68	16.76	7.97
	Radius b (Å)	162.04	7.87	14.61
	Radius c (Å)	266.78	13.69	14.74
Quality of fit (χ^2/n)		3.749	18.483	85.763

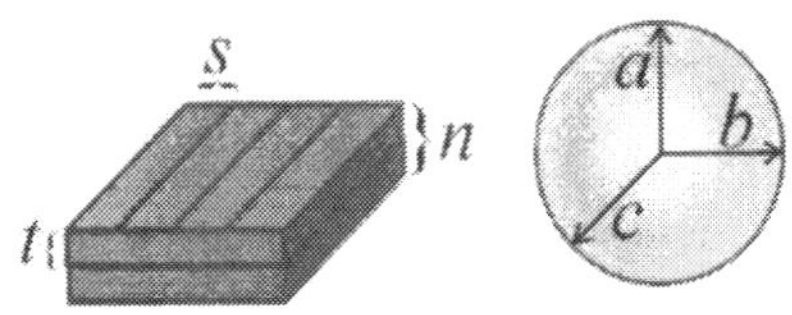

Schematic representations of the form factors used in SANS fitting for lamina and tri-axial ellipsoid.

maleic acid (hydrophilic exterior) moieties optimized to form bonds with two carbons of the styrene ring. The thermodynamic stability granted by this decrease in system energy indicates that the platinum - styrene bond is likely to be the nucleation site for crystal formation. This confirms that the clusters are embedded within the SMA interior hydrophobic spaces and corresponds with the 3nm limit of nanocrystal growth observed in TEM images.

To characterize the Pt-SMA system unsupported in aqueous solution, small angle neuton scattering (SANS) experiments were performed. SANS spectra give the intensity of q which, as a reciprocal of length, depicts the presence of smaller objects in its higher range. Further analysis is completed by fitting a function P(q) to the intensity curve whose parameters precisely relate to the shape and dimensions of suspended materials.

The form factor used in the current study is a combination of a term to model the lamellar nanostructure of the SMA and a second term triaxial ellipsoid model to describe the scattering behaviour of the metal nanocrystals:

$$P(q) = 2\pi(\Delta\rho)^2\Gamma_m \frac{P_{bil}(q)}{q^2} Z_n(q) + \frac{scale}{V_{ell}} \int_0^1 \int_0^1 \phi^2 \left\{q \left[a^2 \cos^2\left(\frac{\pi x}{2}\right) + b \sin^2\left(\frac{\pi x}{2}\right)(1-y^2) + c^2y^2\right]\right\} dxdy$$

where

$$P_{bil}(q) = \left(\frac{sin\left(\frac{qt}{2}\right)}{\frac{qt}{2}}\right)^2$$

$$\phi^2(x) = 9\left(\frac{\sin x - x\cos x}{x^3}\right)^2$$

$Z_n(q)$ accounts for interference from n layers, and a, b, and c describe the radius of the three particle axes.

This model was applied to the Pt-SMA system at pH 5 and pH 7. The parameters used to produce best fit are listed in Table 1. The average diameter of nanocrystals in both samples is 2.5 nm, in good agreement with the previous study [13].

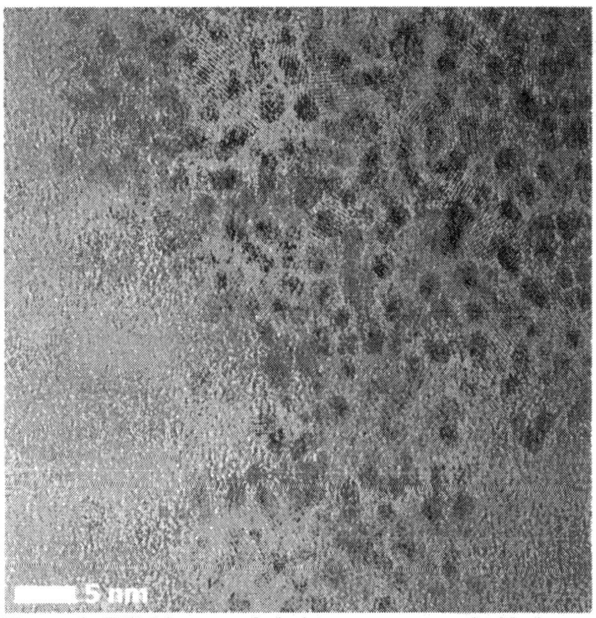

Figure 1: TEM image of platinum nanocrystals (dark circles) and laminar SMA structure (hatched areas) at pH 5. Scale is 5nm.

3.3 Gold in SMA solution

Given the success and reliability of the platinum nanocrystal synthesis in SMA solution, the method was expanded to gold(I) chloride. While gold is generally inert and poor bulk catalyst, nanocrystaline gold demonstrates high catalytic reactivity for oxidation reactions [18]. As with platinum, the investigation into the reaction began with molecular modeling. Unlike platinum, however, DFT simulation showed that the reduced gold atom preferentially interacts with the hydrophilic maleic anhydride group, meaning that crystal formation may occur on the outer surface of the SMA nanostructure.

The same lamina - ellipsoid model described above was applied to SANS data for the Au-SMA system. The radial parameters that describe the metal particles fit where one dimension is atomically thin (2.68 Å) and the others are quite broad (16.2 and 26.6 nm). Furthermore, the solvent adjacent to the gold is not SMA, as in the platinum sample, but D_2O. The parameters used to fit the model to SANS data are listed in Table 1.

TEM confirms the gold crystal morphology described by the SANS data. Images reveal atomically-thin gold nanosheets on the outer surface of the SMA lamina, shown in Figure 2. The regular and consistent 120° corner angles give the nanosheets a distictly hexagonal geometry.

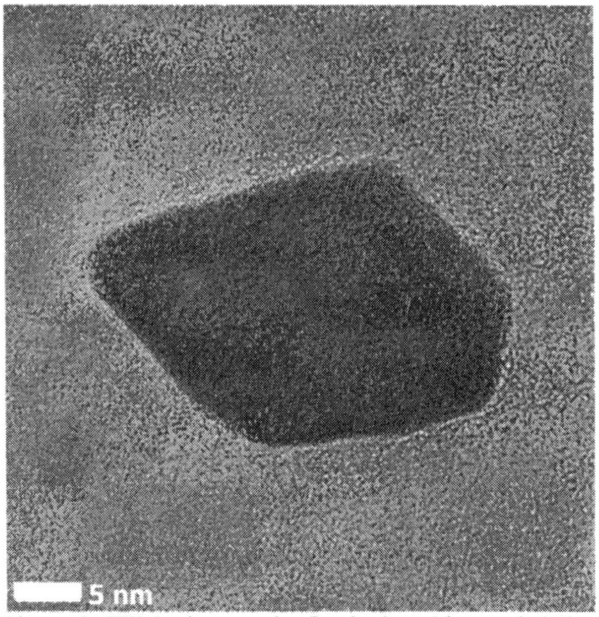

Figure 2: TEM micrograph of a single gold monolayer crystal. Scale is 5 nm.

4 CONCLUSION

Platinum(II) chloride will spontaneously concentrate within the hydrophobic spaces of SMA nanostructures to form monodisperse nanocrystals under 3 nm. The reaction is stable over a broad range of pH and the Pt-SMA nanoreactor it produces is likewise robust. Gold(I) chloride also spontaneously reduces in SMA solution but crystalizes on the surface of the structure rather than within it. These atomically thin hexagonal close-packing nanocrystals grow to hundreds of square nanometers in surface area and may initiate stacking of the Au-SMA system. The structure may then support confinement enhanced catalysis of hydrophilic oxidation reactions. Biomimetic metal-SMA nanoreactor sytems hold great promise as tunable and versatile catalytic agents for systems as diverse as carbon and nitrogen capture, proton exchange fuel cells, or water detoxification.

REFERENCES

[1] D. Gust, T.A. Moore, and A.L. Moore, Acc. Chem. Res., 42, 1890-1898, 2009.

[2] A. Magnuson, M. Anderlund, O. Johansson, P. Lindblad et al. Acc. Chem. Res., 42, 1899-09, 2009.

[3] M. Müllner, T. Lunkenbein, M. Scheider, A.H. Gröschel et al. Macromolecules, 45, 6981-88, 2012.

[4] J. Kang and J. Rebek Jr, Nature, 385, 50-52, 1997.

[5] M. Raynal, P. Ballester, A. Vidal-Ferran, and P.W.N.M. Van Leeuwen, Chem. Soc. Rev., 43, 1734-1787, 2014.

[6] C. Malardier-Jugroot, T.G.M. Van de Ven, T. Cosgrove et al. Langmuir., 21, 10179-10187, 2005.

[7] C. Malardier-Jugroot, T.G.M. Van de Ven, M. Whitehead. J. Phys. Chem. B, 109, 7022-7032, 2005.

[8] M.E. Johnson, C. Malardier-Jugroot, R. Murarka et al. J. Phys Chem. B, 4082-4092, 2008.

[9] G. Garnier, M. Duskova-Smrckova, R. Vyhnalkova et al. Langmuir, 16, 3757-3763, 2000.

[10] M. McTaggart, M. Jugroot, C. Malardier-Jugroot, C. Chem. Phys. Lett., Submitted, 2015.

[11] Gaussian 09, Rev. B.01, M.J. Frisch et al. Gaussian, Inc., Wallingford CT, 2009.

[12] E.S. Kryachko, A.V. Arbuznikov, M.F.A. Hendrickx, J. Phys. Chem. B, 105, 3557-3566, 2001.

[13] M.N. Groves, C. Malardier-Jugroot, and M. Jugroot, Chem. Phys. Lett., 612, 309-312, 2014.

[14] X. Li and C. Malardier-Jugroot, Macromol, 46, 2258-2266, 2013.

[15] A.S.W. Chan, M. Groves, C. Malardier-Jugroot. Mol. Simul., 37, 701-709, 2011.

[16] A.J. Patel. J. Phys. Chem. B., 1632-1637, 2010.

[17] H. Hayashi, Y. Hakuta. Materials (Basel)., 3, 3794–3817, 2010.

[18] T. Fujita, P. Guan, K. McKenna, X. Lang et al. Nat. Mater., 11, 775-780, 2012.

[19] X. Huang, S. Li, Y. Huang, S. Wu, X. Zhou et al. Nat. Comms., 2, 292, 2011.

Colloidal Palladium Nanoparticles versus Commercial Palladium Catalysts for Suzuki Cross Coupling Reactions – The Influence of Surface Functionalization

G. Collins[*,**], C. O'Dwyer[*,***], and J. D. Holmes[*,**]

[*]Chemistry Department and Tyndall National Institute, University College Cork, Cork, Ireland, g.collins@ucc.ie
[**]CRANN@AMBER, Trinity College Dublin, Dublin, Ireland.
[***]Materials and Surface Science Institute, University of Limerick, Limerick, Ireland.

ABSTRAST

The effect of nanoparticle (NP) surface chemistry and functionalization on the catalytic performance in Suzuki coupling is assessed. Capping ligands commonly used in colloidal synthesis such as oleylamine, dodecanethiol, propylene glycol and PVP are compared with commercial Pd on carbon. The catalysts are analyzed pre- and post-reaction with transmission electron microscopy (TEM) and X-ray photoelectron spectroscopy (XPS). Capping ligands are found to have a positive or negative effect on the catalytic performance depending on the chemistry of the functional group. Capping ligands further impact on the reusability of the catalyst. Changes in the NP surface chemistry and morphology during the reaction are key deactivation pathways for catalysts, even a room temperature. In particular PVP, was observed to give excellent stability to colloidal catalysts preventing significant NP aggregation and facilitating excellent reusability.

Keywords: palladium, nanoparticles, capping ligands, surface chemistry, Suzuki

1 INTRODUCTION

The benefits of colloidal nanoparticles (NPs) in catalytic applications include the ability to tune the particle size, shape and surface chemistry. This versatility is not only advantageous for optimising catalyst design but also in understanding structure-property relationships in catalytic systems. The Suzuki-Miyaura reaction is a highly utilized reaction for aryl-aryl bond formation in organic synthesis and represents a key reaction used in the chemicals industry. These reactions are typically associated with homogeneous catalysis but there has been an enormous surge in the use of colloidal NPs for applications in Suzuki coupling reactions.[1-2] Colloidal NP catalysts can boast impressive turn over frequencies (10^5), and recyclability of colloidal catalysts over multiple reaction cycles catalyst have also been demonstrated.[3] Recently, we have shown that the Suzuki reaction to be highly sensitive to the size[4] and shape[5] of the Pd catalyst. Furthermore, we found the mechanism for this activity stems, in part, from the leaching susceptibility of surface atoms.[6] Mechanistic studies identify leaching processes to occur in many NP catalyzed reactions.[7-9] While leaching of soluble Pd species has been shown to catalyze the Suzuki reaction it can also lead to aggregation and deactivation of the catalyst, limiting recyclability and resulting in possible metal contamination of the reaction product. Therefore, the chemical nature of the surface can have a marked impact on the catalytic activity, stability and reusability. In this study we compare the catalytic activity of colloidal NPs capped by common stabilizing ligands typically used in colloidal synthesis with commercial Pd on carbon catalysts.

2 EXPERIMENTAL

Polyvinylpyrrolidone (PVP)[10], dodecanethiol (DDT)[11], oleylamine (OAm)[4] and propylene glycol (PG)[12] stabilized Pd NPs were prepared by methods previously described. Pd NPs were deposited onto activated carbon by stirring solutions overnight. The catalysts were collected by filtration and washed thoroughly with water for PVP, PG NPs and toluene for DDT and OAm catalysts. Commercial Pd/C was purchased from Sigma-Aldrich. The catalytic performance of the colloidal catalysts was compared in the cross coupling of 4- methoxybromobenzene or 4-methoxyiodobenzene (2 mmol) and phenylboronic (2.2 mmol) in $EtOH/H_2O$ (3:1), with 2 equivalents of potassium carbonate as base. Reaction profiles were monitored by gas chromatography. Products were identified against authenticated standards and quantified by calibration to obtain response factors against the known internal standard. X-ray Photoelectron Spectroscopy (XPS) was acquired using a KRATOS AXIS 165 monochromatized X-ray photoelectron spectrometer equipped with an Al Kα (hv = 1486.6 eV) X-ray source. Spectra were collected at a take-off angle of 90° and all spectra were reference to the C 1s peak at 284.6 eV. Transmission electron microscopy (TEM) analysis was performed using a Jeol 2100 electron microscope at an operating voltage of 200 kV.

RESULTS

Figure 1 shows representative TEM images of the prepared NPs before and after deposition onto the catalyst support. Figure 1 (a) shows OAm capped and DDT capped NPS before immobilisation onto activated carbon. Figure 1(c) and (d) show the PVP-catalysts and commercial Pd/C. The mean diameters of the PVP, OAm, PG and DDT NPs were 3.4 nm 3.3 nm 4.2 nm and 3.1 nm, respectively. Commercial Pd/C is typically characterized by a broad range of NP diameters including small (< 5 nm), well dispersed NPs such as those shown in Figure 1 (d) and larger (>100 nm) metal aggregates, as shown in Figure 1 (d) inset.

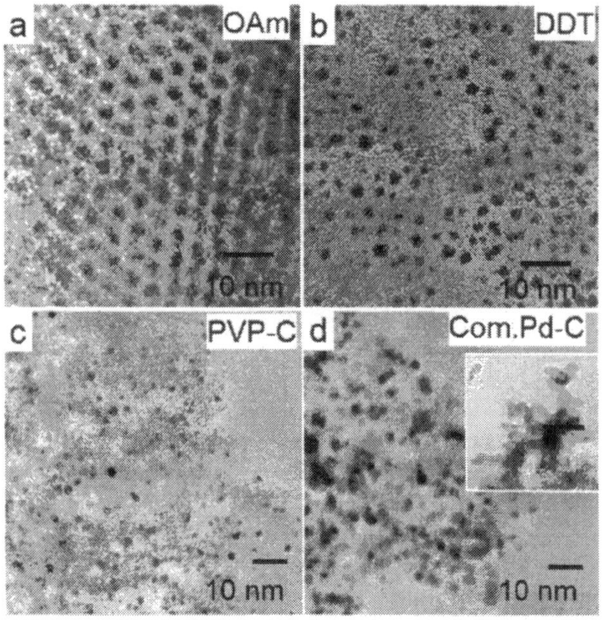

Figure 1. TEM images of unsupported (a) OAm and (b) DDT capped Pd NPs. Carbon supported (c) PVP capped NPs and (d) typical commercial Pd/C.

The catalytic performance of the colloidal catalysts was compared in the cross coupling of 4-methoxybromobenzene (**1**) acid and phenylboronic (**2**) in EtOH/H_2O, as illustrated in the reaction scheme shown in Figure 2 inset. Figure 2 compares the reaction rate profiles for the various colloidal catalysts, using a Pd concentration of 0.5 mol % in the cross coupling for aryl bromides. All the catalysts turn over the reaction to completion at room temperature, yet there are considerable differences in the rates for each catalyst. From Figure 2, it can be seen that the reactivity trend follows PVP > OAm > PG > commercial Pd/C ~ DDT. For PVP and OAm catalysts, the initial reaction is fast as indicated by the steep reaction profile, with no induction period observed at the start of the reaction. After 45 min the reaction catalyzed by PVP-NPs is 71%, while in contrast the DDT NPs displayed a much slower conversion. A 45 min induction period was observed for the DDT catalysts before any product was formed. The large difference in activity between the thiol capped and amine capped NPs is likely due to the strongly binding thiol functional groups at the NP surface, which block access of the reagents. Furthermore, the straight chain alkyl groups exhibit a high packing density compared to the other stabilizing ligands. The PG catalysts had a similar initial reaction rate as PVP and OAm NPs, however the reaction slowed as it progressed. The smaller size of the PG ligands ($C_3H_8O_2$) can less effectively stabilized Pd NPs compared to the long chain OAm and bulky PVP polymer ligands. The catalytic activity of colloidal catalyst were generally superior to commercial carbon. It is worth noting that the reproducibility of commercial Pd on carbon also varied due to variations in NP dispersion.

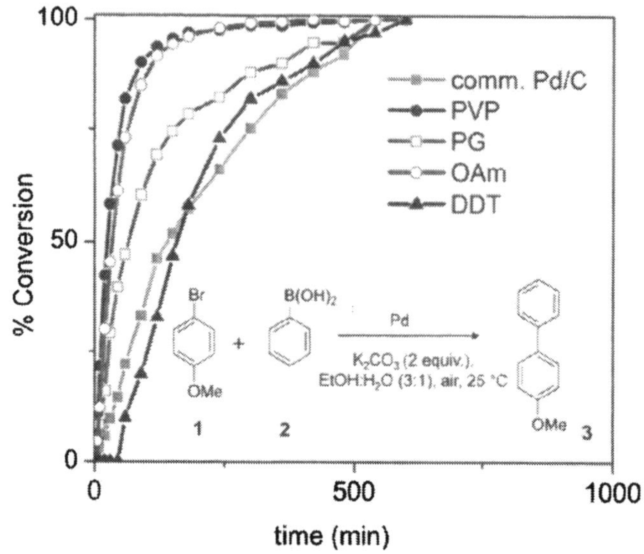

Figure 2. Reaction profile of colloidal and commercial catalysts comparing catalytic performance in Suzuki cross coupling reactions carried out a room temperature.

XPS analysis of the catalysts were carried out to evaluate the surface chemistry of the Pd NPs. Figure 3 displays the Pd 3d core level of PG-capped NP and commercial Pd/C. Both spectra show a doublet at binding energies of 335.5 eV and 341 eV, which can be assigned to Pd $3d_{5/2}$ and Pd $3d_{3/2}$, respectively. Metallic Pd(0) is indicated by the peak at a binding energy of 335.5 eV. *In situ* XPS measurements of Pd determined that peaks shifted to binding energies greater than +1.5 eV surface oxide can be assigned to bulk oxide phases, while smaller shifts are attributed to surface oxide and sub-surface oxide species.[13-14] For the PG catalysts, the NPs are metallic Pd with smaller amounts of surface and bulk oxide at a binding energy of 336.3 eV and 337.1 eV, respectively. For commercial Pd/C, a large intensity peak at a binding energy of 337.1 eV can be assigned to bulk PdO. The presence of higher Pd oxidation states are indicated by the peak at a binding energy of 339 eV, which can be assigned to PdO_2 or $Pd(OH)_4$. From the XPS analysis it can be seen that colloidal NPs are primarily metallic Pd(0) and commercial Pd/C is comprised mainly of PdO, due to surface oxidation. TEM analysis of the catalysis post reaction revealed that the size and morphology of the PG catalysts were well maintained during the reaction, as shown in figure 3 (c). The mean diameter and polydispersity increased slightly to 4.5 nm ± 2.2 nm. In comparison the commercial catalysts underwent significant aggregation with large diameter particles (>10 nm) readily observed by TEM analysis, as shown in Figure 3 (b). It is interesting to note that reduction of commercial Pd/C under a flow of H_2 gas before the reaction lowers the activity of the catalyst. This can be attributed to the reduction of the oxide to metallic Pd. For commercial Pd/C the large portion of surface oxide can be readily dissolved during the reaction. The effectiveness of commercial Pd/C stems from its pseudo-homogeneous nature with an easily leachable source of Pd. The oxide is more easily leached before H_2 reduction compared to leaching of metallic Pd(0).

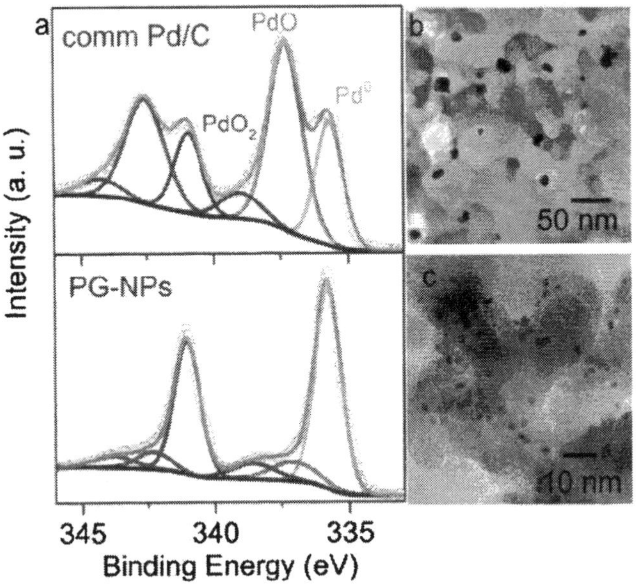

Figure 3. (a) Pd 3d core level XPS spectra of PG and commercial Pd/C. (b) TEM image of commercial Pd/C and (d) PG-catalysts after Suzuki coupling reaction.

To investigate how removing the capping ligand effects catalytic performance, OAm ligands were removed by refluxing in acetic acid.[15] XPS analysis confirmed the removal of the ligands by the absence of the N 1s peak in the survey spectra. Evaluation of catalytic performance in the Suzuki coupling reaction found that ligand removal was beneficial to catalytic activity, with the reaction going to completion faster than the capped catalysts. After 45 min the surface cleaned catalyst displayed a yield of 82% compared to a yield of 71% with the ligands present. This increase in yield can be attributed to the more exposed NP surface available for reaction when the ligands are removed. While a modest increase in catalytic performance was observed after surface cleaning, the impact on recyclability was more pronounced. The OAm capped NPs demonstrated good reusability with 8% drop in yield in the second cycle, under the same reaction conditions. A 100% yield could still be achieved by increasing the reaction time by 60 min. In contrast a significant loss in catalytic performance was observed for the uncapped NPs with a final yield of only 56%, under the same conditions. Increasing the reaction time by a further 60 min resulted in a final yield of 63%. This observation demonstrates the positive effect of capping ligands on stabilizing the Pd during the reaction leading to improved catalytic performance on recycling the catalyst.

XPS analysis was conducted to further investigate the effect of surface chemistry and cleaning in the OAm capped catalysts. Figure 4 compares the Pd 3d XPS core levels for OAm capped NPs before and after acetic acid treatment and show the NPs undergo changes to their surface chemistry as a result of the surface cleaning treatment. The as synthesized Pd NPs are mainly metallic Pd(0) as indicated by the peak at 335.5 eV. A second peak at a binding energy of 336.3 eV is attributed to surface oxide and a smaller shoulder peak at 337.9 eV indicates the presence of bulk PdO. Analysis of the Pd 3d core level

(Figure 4(b)) after surface cleaning show an increase in surface oxidation as indicated by the greater intensity of the Pd(II) peak. The absence of capping ligands and surface oxidation of the NPs leads to greater leaching and loss of Pd during the reaction. While the leached Pd contributes to the catalyzing the reaction, it concurrently leads to catalyst deactivation, therefore significantly lowering the activity of the recycled catalyst.

The reusability of the catalysts was found to follow the trend PVP > OAm > PG > DDT > commercial Pd/C. PVP capped NPs, displayed excellent activity and recyclability with quantitative conversion being maintained over three consecutive cycles, although a longer reaction time was needed to achieve 100% yield. Lower yields on the second cycle were observed for PG and DDT catalysts, of 82% and 75%, respectively. Increasing the reaction temperature to 50 °C increases the activity of the reused and 100% yield can be achieved at evaluated reaction temperatures. To further investigate the excellent recyclability of PVP-capped catalysts changes to the NP surface chemistry and morphology were evaluated by comparing XPS and TEM of the catalyst before and after the reaction. Figure 5(a) shows the Pd 3d core level of carbon supported PVP catalyst before and after reaction. The NPs are primarily metallic with a shoulder peak attributed to some surface oxide that forms after immobilisation onto activated carbon. The surface chemistry of the NPs is also largely preserved during the reaction with no significant changes the oxidization state. A slight increase in the peak intensity ratio of Pd(0):Pd(II) is observed for the used catalyst. This can be attributed to the dissolution of the surface oxide during the reaction.[5] Analysis of the N 1s core level after the reaction displayed in Figure 5(c), show the presence of a peak at a binding energy of 400 eV, which is characteristic of the prrolidone N group.[16] XPS confirms the presence of the PVP capping ligands remaining intact on the surface during the reaction. TEM of the catalyst post reaction shown in Figure 5 (b), reveal that the morphology and size distribution of the NPs is essentially preserved during the reaction.

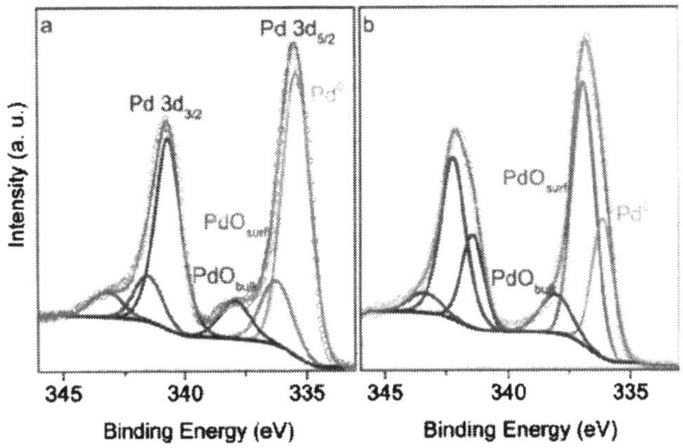

Figure 4. Pd 3d XPS core level of OAm catalysts (a) before and (b) after ligand removing with acetic acid.

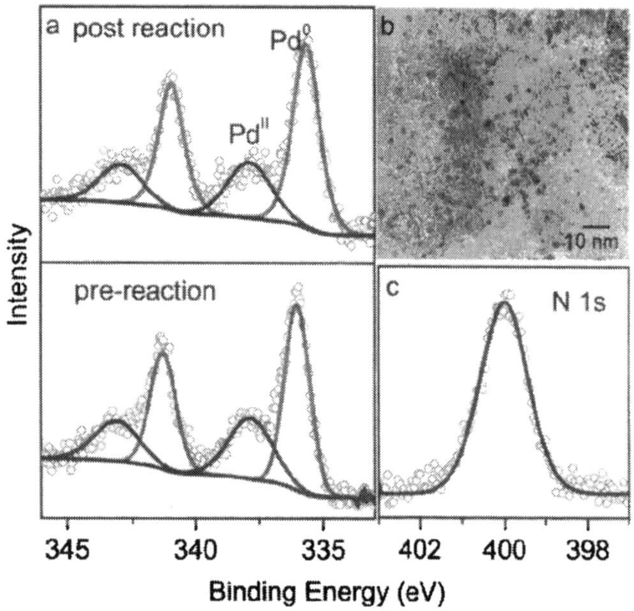

Figure 5. (a) Pd 3d XPS core level of PVP-capped catalysts before and after Suzuki coupling reaction. (b) TEM image of catalyst after the reaction and (d) N1s spectra of catalyst after the reaction.

CONCLUSIONS

The influence of surface functionalization on the activity and recyclability of Pd NPs in Suzuki coupling reaction is evaluated. Capping ligands are beneficial to catalytic activity by stabilizing the NP during the reaction, limiting the degree of metal leaching and aggregation. The presence of capping ligands, in particular PVP, was observed to give excellent stability to colloidal catalysts preventing significant NP aggregation and facilitating excellent reusability. The work highlights the potential of modifying the surface chemistry of colloidal NPs and commercial catalysts to improve the recyclability of NPs during the reaction. Overall, a complex interplay between surface chemistry and stability under the reaction conditions influence catalytic performance and recyclability.

REFERENCES

1. Scheuermann, G. M.; Rumi, L.; Steurer, P.; Bannwarth, W.; Muelhaupt, R., *J. Am. Chem. Soc.* **2009,** *131*, 8262-8270.
2. Balanta, A.; Godard, C.; Claver, C., *Chem. Soc. Rev.* **2011,** *40*, 4973-4985.
3. Santra, S.; Hota, P. K.; Bhattacharyya, R.; Bera, P.; Ghosh, P.; Mandal, S. K., *ACS Cat.* **2013,** *3*, 2776-2789.
4. Collins, G.; Blomker, M.; Osaik, M.; Holmes, J. D.; Bredol, M.; O'Dwyer, C., *Chem. Mater.* **2013,** *25*, 4312-4320.
5. Collins, G.; Schmidt, M.; O'Dwyer, C.; Holmes, J. D.; McGlacken, G. P., *Angew. Chem. Int. Ed.* **2014,** *53*, 4142-4145.
6. Collins, G.; Schmidt, M.; O'Dwyer, C.; McGlacken, G.; Holmes, J. D., *ACS Catalysis* **2014,** *4*, 3105-3111.
7. Pachon, L. D.; Rothenberg, G., *Appl. Organometal. Chem.* **2008,** *22*, 288-299.
8. Niu, Z.; Peng, Q.; Zhuang, Z.; He, W.; Li, Y., *Chem. Eur. J.* **2012,** *18*, 9813-9817.
9. Gaikwad, A. V.; Holuigue, A.; Thathagar, M. B.; ten Elshof, J. E.; Rothenberg, G., *Chem. Eur. J.* **2007,** *13*, 6908-6913.
10. Xian, J.; Hua, Q.; Jiang, Z.; Ma, Y.; Huang, W., *Langmuir* **2012,** *28*, 6736-6741.
11. Cargnello, M.; Wieder, N. L.; Canton, P.; Montini, T.; Giambastiani, G.; Benedetti, A.; Gorte, R. J.; Fornasiero, P., *Chem. Mater.* **2011,** *23*, 3961-3969.
12. Jeon, T.-Y.; Yoo, S. J.; Park, H.-Y.; Kim, S.-K.; Lim, S.; Peck, D.; Jung, D.-H.; Sung, Y.-E., *Langmuir* **2012,** *28*, 3664-3670.
13. Ketteler, G.; Ogletree, D. F.; Bluhm, H.; Liu, H. J.; Hebenstreit, E. L. D.; Salmeron, M., *J. Am. Chem. Soc.* **2005,** *127*, 18269-18273.
14. Lundgren, E.; Kresse, G.; Klein, C.; Borg, M.; Andersen, J. N.; De Santis, M.; Gauthier, Y.; Konvicka, C.; Schmid, M.; Varga, P., *Phys. Rev. Lett.* **2002,** *88*, 246103-246104.
15. Niu, Z.; Li, Y., *Chem. Mater.* **2014,** *26*, 72-83.
16. Collins, G.; Schmidt, M.; McGlacken, G. P.; O'Dwyer, C.; Holmes, J. D., *J. Phys. Chem. C* **2014,** *118*, 6522-6530.

Optimization of Catalyst Nano-structures through Alloying and De-alloying Process---Research on Ag-Co Catalyst for Catalytic Oxygen Reduction in Alkaline Media

Rongzhong Jiang, Deryn Chu, Joshua P. McClure, and Dat T. Tran

Sensors and Electron Devices Directorate, U.S. Army Research Laboratory

2800 Powder Mill Road, Adelphi, MD 20783-1197. Email: Rongzhong.jiang.civ@mail.mil, Phone: +1 301-394-0295, Fax: +1 301-394-0273

ABSTRACT

A highly active and durable Ag-Co catalyst was synthesized by optimizing the catalyst nano-structures for realization of fast electron-transfer and ion-transfer. De-alloying process was used to remove excess of cobalt in Ag-Co sample and form nano-dendrimers. The activity of the Ag-Co materials is dependent on the Co content; and the Co/Ag ratio in 0.10 (a/a) gives the best performance for oxygen reduction reaction (ORR). Further enhancement of activity is achieved by addition of graphene nano-platelets (GNP) as a catalyst support into the Ag-Co material, and followed by a process of heat-treatment. Oxygen reduction at the Ag-Co nano-catalyst belongs to a 4-electron reaction process to produce water in alkaline media. Finally, the catalytic activity was compared with a commercially available Ag nano material (20 – 40 nm). There is a 170 mV decrease in over potential; and 6-times enhancement in current density at 0.75V (versus RHE) and 1600 rpm rotation rate for ORR in oxygen saturated 0.1M KOH solution.

Keywords: Catalyst nanostructure, ORR, Ag-Co, graphene nano-platelets, Heat-treatment.

1 INTRODUCTION

The state of the art catalysts for fuel cell are expensive noble metals, including platinum black and platinum-based metal alloys. For a polymer electrolyte membrane fuel cell (PEMFC), the catalyst covers about 60% of the cost of a fuel cell stack in automobile application [1]. The quest for more efficient fuel cell technologies necessitates the development of low cost, highly active and stable catalyst materials. The fast improvement of catalytic activity in recent years is benefit from the emerging of nanotechnology for material synthesis and processing [2-4], such as carbon nanotube, nanofiber, mesoporous carbon], and graphene. These nano materials can either be used as a catalyst support or directly used as a catalyst. In recent years, a lot of nano-structured noble metals and alloys [5-8], such as metal nanoparticles, nanowires, nanoframes, nanohollows, and nano-coreshells, have been successfully synthesized, and tested for fuel cell catalysts. Optimization of particle size and particle's nano-structure at the atomic level can precisely and effectively tune catalytic properties of materials, and enabling enhancement in both catalytic activity and stability. It is well known that silver is used for catalytic oxygen reduction in alkaline media [9-10]. The cost of silver is around 1% of Pt, and the abundance of it is approximately 1500% of Pt. After alloying with transition metals, such as Co, it gives better catalytic activity [11]. Holewinski et al [12] synthesized an Ag-Co surface alloy nano particles at high temperature (500 °C), having activity around fivefold improvement over pure silver nanoparticles. In this article we report a different approach to synthesize Ag-Co nanoparticles at low temperature. Interestingly, we obtained Ag-Co nano-dendrimers by precisely controlling the size, shape, architecture, and composition. The dendritic nanoparticles showed much higher activity and stability than pure Ag nano particles for catalytic oxygen reduction reaction (ORR) in alkaline media.

2 EXPERIMENTAL

2.1 Synthesis of nanostructured catalysts

Graphene nano-platelets (GNP) with a BET surface area between 600 – 650 m^2/g were purchased from Cheap Tubes Inc. Some commercially available catalysts were purchased from Johnson Matthey, including 40% Pt nano catalyst supported on 60% active carbon (JM's 40% PtC), and Ag-nanopowder (JM's Ag-nano, particle size 20–40 nm). Nano-structured Ag-Co was synthesized by reducing a mixture of silver and cobalt compounds in the presence of GNP. Silver ammonia complex was formed by adding ~ 1g $AgNO_3$ and 5 ml ammonia hydroxide, and moved in a 250 ml flask containing 80 ml water. On the other hand, different quantity of $CoC_2O_4{\cdot}2H_2O$ was suspended in 40 ml water and 5 ml ammonia hydroxide, and moved in the same flask. The reducing reagent was $NaBH_4$ solution by dissolving $NaBH_4$ in 40 ml water. The Ag-Co was synthesized by slowly dropping $NaBH_4$ solution into the flask and fast stirring. The amount of Co in the Ag-Co alloy was adjusted by a de-alloying process in 1M HCl and controlling Co/Ag ratio by a specific time period. The Ag content in the Ag-Co alloy was determined by gravimetrical analysis, in which hydrochloric acid was used to thoroughly remove the cobalt existed in the Ag-Co alloy, and weighting the dry Ag powder. The pure Ag fine power was synthesized with the same method of making Ag-Co but without adding cobalt oxalate in the reaction solution. GNP supported Ag-Co (Ag-Co/GNP) was synthesized with the same method of making Ag-Co and controlling Co/Ag atomic ratio 0.10. A quantitative amount of GNP was added in the reaction solution before dropping $NaBH_4$ solution in the flask. The Ag-Co/GNP contained 30% Ag-Co alloy and 70% GNP. The Ag-Co/GNP was further heat-treated to enhance the catalytic activity at 500 °C in argon atmosphere. We called the heat-treated catalyst as HT-Ag-Co/GNP.

2.2 Instrumental Analysis

Transmission electron microscope (TEM) was obtained with a high-resolution JEOL 2100 FE instrument equipped with an EDAX X-ray detector. Selected area electron diffraction (SAED) patterns and energy-dispersive X-ray spectroscopy (EDS) analyses were also obtained during the measurements. Scanning electron microscope (SEM) image was obtained with FEI QUANTA 200F instrument.

X-Ray diffraction analysis (XRD) results were obtained from a Rigaku Ultima III instrument with Cu Kα radiation ($\lambda = 1.5418$ Å) using a Bragg-Brentano Configuration. The measurements were conducted with scan rate of 1.0 degree (2θ) per minute and each diffraction data point was collected at the interval of 0.02 degree (2θ) for the total spectrum range of 10 to 90 degree (2θ).

2.3 Electrochemical Analysis

A Pine Bipotentiostat RDE4 was used for electrochemically analyzing the synthesized catalysts by depositing the catalyst samples from ink mixtures onto a glassy carbon (GC) disk electrode. The catalyst ink in 1:1 water/1-propanol solution was prepared by ultrasonically treatment with a Branson Sonifier 450 at a duty cycle 40 and out-put control 8 for 10 minutes. The final content of Ag or Ag-Co ink was 4 mg/ml after excluding the mass of supporting GNP. Next, 5 μl of catalyst ink was coated on the GC disk working electrode (0.196 cm^2), and dried at 40°C for ~60 minutes. The final coating on the GC disk electrode contained 95% total catalyst and 5% Nafion with an Ag or Ag-Co loading of 0.1 mg/cm^2. The inks for JM's 40% PtC and JM's Ag-nano were prepared separately with the same procedures. The final coating on the GC electrode was 95% Nafion and 5% catalyst with a Pt loading 0.02 mg/cm^2, and Ag loading 0.10 mg/cm^2. All voltammetrical experiments were performed in argon (Ar.) or O_2 saturated 0.1M potassium hydroxide (KOH) solution at room temperature (20 ± 1°C). A three electrode glass cell was used for electrochemical study containing three compartments separated by a porous ceramic layer. A platinum wire counter electrode and a saturated calomel reference electrode (SCE) were used in the cell. The experimental results are reported based on reversible hydrogen electrode (RHE).

3 RESULTS AND DISCUSSION

3.1 TEM and SEM results

Fig. 1 shows TEM and SEM images of various catalysts obtained by chemical synthesis. The particle size of Ag-Co is significantly reduced in comparison of pure Ag without alloying, although they are synthesized in the same experimental conditions. The particles of graphene supported Ag-Co/GNP are dispersed on the large surface of GNP. After heat-treatment, the particles of HT-Ag-Co/GNP seem embedded in the GNP, and less Ag-Co particles can be seen. Interestingly, the unsupported Ag-Co looks like a dendritic material, where the particles are first joined together to form short wires, and then, the wires are connected to form dendrimers. The SAED image indicates that the dendrimers have good crystal structure, which is judged from the multiple clear rings and the bright dots on the rings in the dark field. The Ag and Co elements are detected by EDS analysis (not shown in the figure). The elemental maps (also not shown in the figure) demonstrate a uniform distribution of Ag and Co. The elemental map of Co has the same shape with that of the Ag, which is an evidence of Ag-Co alloy, instead of an Ag and Co mixture.

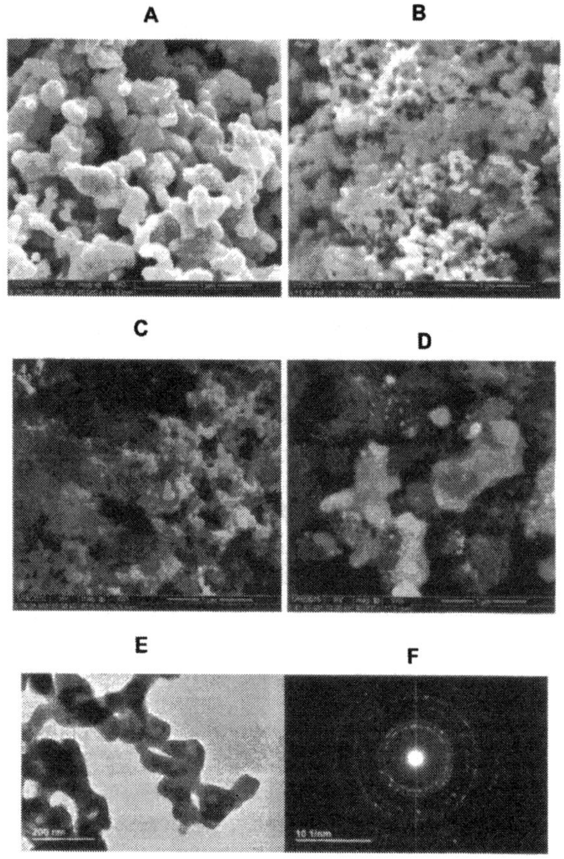

Fig.1 TEM and SEM images of nanostructured catalysts prepared at different conditions. A. SEM of Ag-nano particles by chemical synthesis; (B) SEM of Ag-Co with Co/Ag ratio 0.10 (atomic); (C) GNP (70%) supported Ag-Co (30%, Co/Ag=0.10); D. HT-Ag-Co/GNP (HT at 500 °C); (E) TEM of Ag-Co (Co/Ag = 0.10); and (F) SAED of Ag-Co.

3.2. XRD results

Fig. 2A shows XRD patterns of unsupported Ag-Co alloys with different Co/Ag atomic ratios. Five typical peaks of pure Ag can be seen. When Ag is alloyed with Co, the positions of these peaks have only a slight movement; but all of the peaks become wider. The parameters of the highest peak with Miller index at plane (111) are examined. With increasing the content of Co in the alloy, the half-wave width of the peaks increases, which is a sign of decrease in crystallite size; and the peak position moves to higher degree (2θ). Simultaneously, the perpendicular distance between adjacent lattice planes (d-spacing) becomes decreased, which implies that the Ag lattice structure shrinks after alloying with Co. The shrink in Ag lattice structure may cause a defect of Ag crystal, and change the catalytic property. Fig. 2B shows XRD pattern of Ag-Co supported on GNP. The GNP was introduced in the synthetic process. A small new peak appears at 26.76 degree (2θ) for Ag-Co/GNP, which is a typical peak of graphite structure. For the Ag-Co/GNP, the peak position at plane (111) further increases to higher 2θ value, and lower d-spacing.

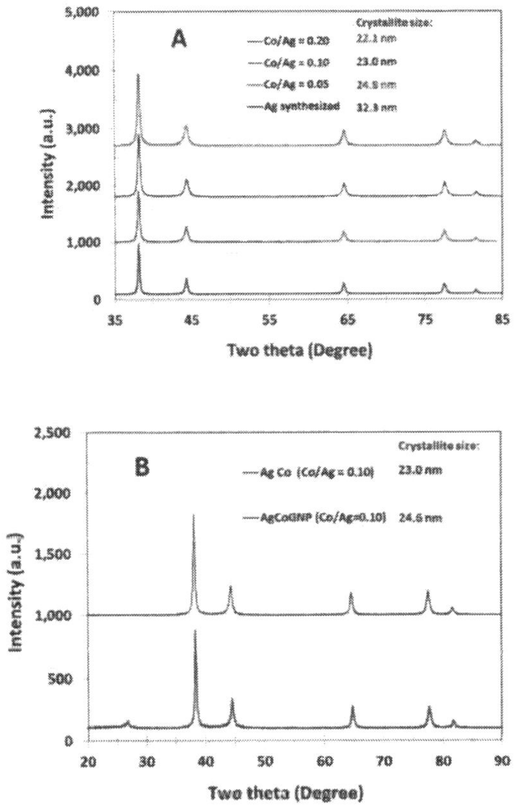

Fig. 2 XRD patterns of various catalysts. (A) XRD of Ag-Co with different Co/Ag ratios; and (B) XRD of Ag-Co with and without GNP support.

3.3 Electrochemical results

The activity of original Ag-Co alloy was not satisfied. Higher activity was achieved by a de-alloying process acidic solution to remove excess of cobalt in the Ag-Co. Fig. 3 shows electrochemical behaviors of Ag-Co coated electrodes and their catalytic activity for oxygen reduction. In argon saturated solution (Fig. 3A), the oxygen was removed from the electrolyte; and the cyclic voltammograms (CVs) show typical electrochemcial surface waves. The original Ag-Co alloy has a small cathodic peak and two small anodic peaks. After de-alloyed, only one anodic peak is seen; and both the cathodic and anodic peaks become about two times larger. About 70–100 mV positive potential shift is observed versus pure Ag nano catalyst (JM's Ag-nano). The catalytic activity was measured with a rotating disk electrode in O_2 saturated electrolyte solution (Fig. 3B). The highest catalytic activity was obtained from the de-alloyed Ag-Co catalyst, and its polarization curve has the most positive half-wave potential (0.74V vs. RHE). Unfortunately, the Co content in the de-alloyed Ag-Co is difficult to control accurately. Therefore, we changed the Co content in the synthetic process by more accurately control the Co/Ag ratio. When the Co/Ag atomic ratio as 0.10, the best activity was obtained, which has the same electrochemical performance as that of the de-alloyed Ag-Co.

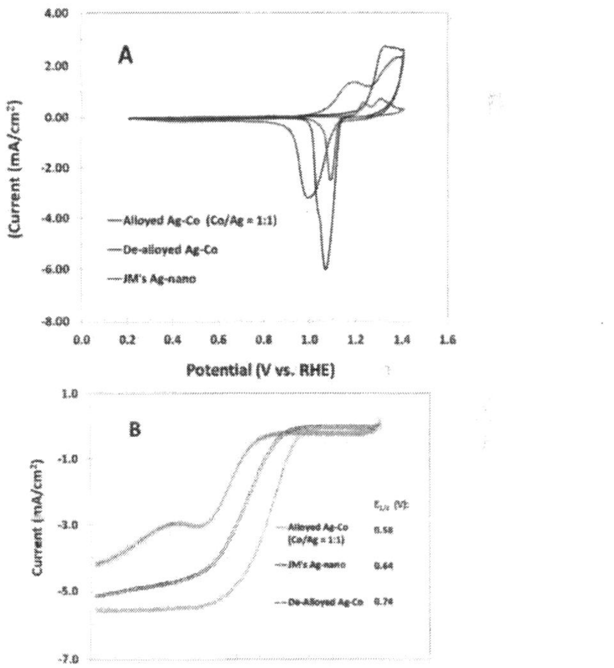

Fig. 3. Electrochemical behaviors of Ag-Co coated electrodes and their catalytic activity for oxygen reduction. 10 mV/s. 1600 rpm. (A) Cyclic voltammograms in argon saturated 0.1M KOH; (B) Polarization curves in O_2 saturated 0.1M KOH;

Further enhancement of activity is achieved by addition of graphene nano-platelets (GNP) as a catalyst support into the Ag-Co material, and followed by a process of heat-treatment. The performance of the best catalyst is evaluated by comparison with commercially available catalysts. Fig.4A shows polarization curves of HT-Ag-Co/GNP and some commercial catalysts coated electrodes for catalytic oxygen reduction. In comparison of JM's Ag-nano, the HT-Ag-Co/GNP catalyst has raised 170 mV in half wave potential. In comparison of JM's 40% PtC, the HT-Ag-Co/GNP catalyst has only 20 mV lower in half wave potential. Fig. 4B shows Tafel plots of the polarization curves. The HT-Ag-Co/GNP catalyst has even lower Tafel slope than that of the JM's PtC. The quantitative comparison of catalytic activity for a series of catalysts is shown in Fig. 4C. There is 170 mV decrease in half wave potential, and 6 times improvement in current density at 0.75V (vs. RHE).

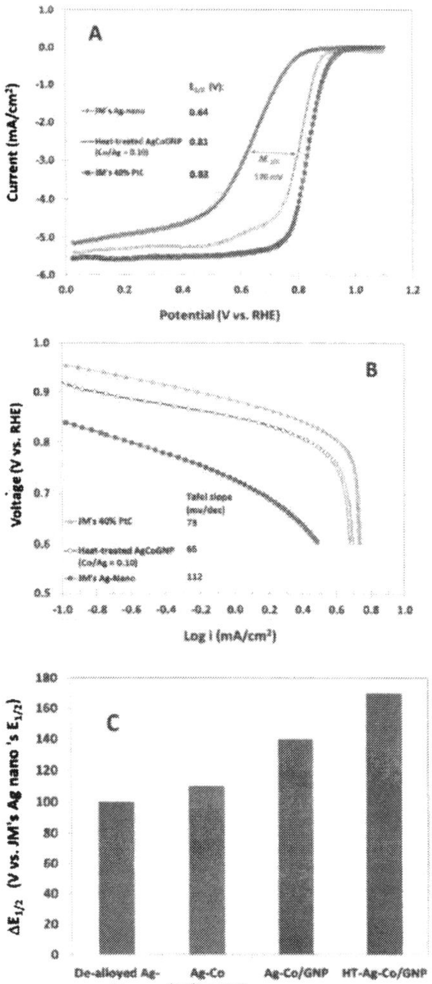

Fig. 4 (A) Polarization curves of HT-Ag-Co/GNP catalyst and commercial catalysts coated electrodes in O_2 saturated 0.1M KOH. 10 mv/s, 1600 rpm. (B) Tafel plots of the data in A. (C) Comparison of catalytic activity versus JM's Ag nano-particles (20-30 μm).

REFERENCES

[1] Marcinkoski, J.; James, B. D.; Kalinoski, J. A.; Podolski, W.; Benjamin, T.; Kopasz, J., "Manufacturing process assumptions used in fuel cell system cost anal-yses", J. Power Sources, 196, 5282–5292, 2011.

[2] Ebbesen, T. W.; Ajayan, P. M., "Large-scale synthesis of carbon nanotubes", Nature, 358, 220-222, 1992.

[3] Che, G. L.; Lakshmi, B. B.; Fisher, E. R.; Martin, C. R., "Carbon nanotube membrane for electrochemical energy storage and production", Nature 8, 393, 346-349, 1998.

[4] Andersen, S. M.; Borghei, M.; Lund, P.; Elina, Y. R.; Pasanen, A. Esko Kauppinen, E.; Ruiz, V.; Kauranen, P.; Skou, E. M., "Durability of carbon nanofiber (CNF) & carbon nanotube (CNT) as catalyst support for Proton Exchange Mem-brane Fuel Cells" Solid State Ionics, 3, 231, 94–101, 2013.

[5] Rong, C.; Jiang, R. Z.; Sarney, W.; Chu, D., "Ultrasound-assisted microemulsion for synthesis of Pt and PtCo nano-particles", Electrochimica Acta, 55, 6872–6878, 2010

[6] Chen, C; Kang, Y. J.; Huo, Z. Y.; Zhu, Z. W.; Wenyu Huang, W. Y.; Xin, H. L.; Snyder, J. D.; Li, D. G.; Herron, J. A.; Mavrikakis, M.; Chi, M. F.; More, K. L.; Li, Y. D.; Markovic, N. M.; Gabor A. Somorjai, G. A.; Yang, P. D.; Stamen-kovic, V. R., "Highly Crystalline Multimetallic Nanoframes with Three-Dimensional Electrocatalytic Surfaces", Science, 343, 1339-1343, 2014.

[7] Chen, H. M.; Liu, R. S.; Lo, M. Y. Chang, S. H.; Tsai, L D.; Peng, Y. M.; Lee, J. F., "Hollow Platinum Spheres with Nano-Channels: Synthesis and Enhanced Catal-ysis for Oxygen Reduction", J. Phys. Chem. C , 112, 7522–7526, 2008.

[8] Tuaev,X.; Rudi, S.; Petkov, V.; Hoell, A.; Strasser, P., "In Situ Study of Atomic Structure Trans-formations of Pt-Ni Nanoparticle Catalysts during Electrochemical Potential Cycling", ACS Nano 2013, 7, 5666–5674, 2013.

[9] Beer, S. Z.; Sandier, Y. L., "Oxygen Reduction at Silver and Silver Based Alloy Electrodes", J. Electrochem. Soc., 112, 1133, 1965.

[10] Yang Y. F.; Zhou, Y. H., "Particle size ef-fects for oxygen reduction on dispersed silver + carbon electrodes in alkaline solution", J. Electroanal. Chem., 397, 271, 1995.

[11] Lima, F. H. B.; de Castro, J. F. R.; Tician-elli, E. A., "Silver-cobalt bimetallic parti-cles for oxygen reduction in alkaline me-dia", J. Power Sources 2006, 161, 806–812.

[12] Holewinski, A.; Idrobo, J. C.; Linic, S., "High-performance Ag–Co alloy catalysts for elec-trochemical oxygen reduction", Nature Chemistry, 6, 828-834, 2014.

Green Fenton-like catalysts for the removal of water pollutants

C. Nadejde[*], M. Neamtu[*], V.-D. Hodoroaba[**], R. J. Schneider[**], A. Paul[**], G. Ababei[***] and U. Panne[**,****]

[*]Interdisciplinary Research Department – Field Science, 'Alexandru Ioan Cuza' University, Lascar Catargi 54, Iasi, Romania, claudia.nadejde@uaic.ro, mariana.neamtu@uaic.ro
[**]BAM Federal Institute for Materials Research and Testing, Unter den Eichen 87, Berlin, Germany
[***]National Institute of Research and Development for Technical Physics, Dimitrie Mangeron Blvd. 47, Iasi, Romania
[****]Department of Chemistry, Humboldt-Universität zu Berlin, Brook-Taylor-Str. 2, Berlin, Germany

ABSTRACT

Two types of surface modified magnetite (Fe_3O_4) nanoparticles, coated with either tannic acid (TA) or dissolved natural organic matter (NOM), were evaluated as magnetic heterogeneous catalysts. Simple and efficient procedures for the synthesis of the magnetic catalysts were employed, their properties being fully assessed by various characterization techniques. The catalytic activity in heterogeneous oxidation of aqueous solutions containing Bisphenol A (BPA) over the catalysts was comparatively studied. The optimum experimental parameters were: 1g/L of catalysts, 10 mM H_2O_2, under UV irradiation. The highest mineralization rates were observed for Fe_3O_4-TA catalyst. More than 80% of BPA were removed after 30 minutes of reaction time under the specified experimental conditions. The results showed that the obtained catalysts are suitable candidates for the removal of pollutants in wastewaters by means of heterogeneous reaction.

Keywords: nanocatalysts, characterization, Fenton oxidation, wastewater, bisphenol A degradation

1 INTRODUCTION

Since nowadays water pollution became an issue of increasing concern, novel greener alternatives for water decontamination are continuously sought in order to protect the environment. The use of magnetic materials as adsorbents or heterogeneous catalysts in water treatment has attracted increasing attention in the last years [1-3].

Magnetite (Fe_3O_4) nanoparticles are good candidates for these procedures due to their benign nature, and also, their properties can be tailored according to their purpose; moreover, the nanostructures' surface can be functionalized with a wide variety of non-toxic materials containing (photo)active groups, that act both as nanoparticle stabilizers and also as catalysts. The magnetic properties allow facile re-collection and possible re-use of such nanomaterials from water. All these advantages significantly contribute in reducing costs and increasing availability for large-scale applications. Thus, for the present study, tannic acid (TA) and dissolved natural organic matter (NOM) were the coatings of choice for stabilizing magnetite nanoparticles (obtained by co-precipitation method), due to their known strong binding to iron ions. Moreover, they are widly-spread, safe and inexpensive natural materials. The further use of these functionalized magnetic iron oxide nanoparticles as Fenton-like catalysts for the removal of persistent contaminants from wastewaters was the aim of the present investigation. Although other studies reported the use of these type of catalysts in the degradation of pigments or adsorption of heavy metals from water [4-6], the knowledge on their performance in the removal of hazardous water pollutants such as endocrine disruptors (i.e. Bisphenol A) is limited. Thus, the properties and catalytic activity of the above mentioned nanomaterials in heterogeneous oxidation of aqueous solutions containing BPA are discussed.

2 EXPERIMENTAL

2.1 Materials

The catalyst synthesis procedures were performed using high purity reageants (Sigma-Aldrich), while the dissolved natural organic matter (NOM), was isolated from surface water (Fuchskuhlelake, Brandenburg, Germany), up-concentrated by reverse osmosis and later freeze-dried. Deionized water (18.2 MΩ/cm), obtained with Barnstead EASYpure™ II Ultrapure Water System was used throughout all experimental protocols.

2.2 Catalyst synthesis

The magnetic nanocatalysts were fabricated by employing a facile cost-effective synthesis procedure (chemical co-precipitation method) as described in detail in [3]. Briefly, the iron oxide (Fe_3O_4) magnetic nanoparticles were obtained by thoroughly mixing together ferric and ferrous salt precursors (2:1 stoichiometric ratio) in the presence of a strong alkaline precipitation agent (1.7 M NaOH); the method was performed at relatively high temperature (80 °C) with continuous stirring, under normal atmospheric conditions. The as-obtained magnetic black precipitate was subsequently purified, then coated with either TA or NOM, taking in consideration a molar ratio of 0.05 between the magnetic core and stabilizing shell. After 2 h of mechanical vigurous stirring at 40 °C, the samples were separated from suspension, repeatedly washed with water and ethanol, then dried in a vacuum oven at ~ 40 °C.

2.3 Catalyst characterization

The nanocatalysts were fully characterized by X-ray diffraction (XRD, Shimadzu LabX XRD-6000 diffractometer), nitrogen adsorption-desorption isotherms (Micromeritics ASAP 2020™ Physisorption system - Norcross, USA), energy dispersive X-ray spectroscopy (EDX, Bruker XFlash® 5030 SDD X-ray spectrometer - Berlin, Germany), high-resolution scanning electron microscopy (SEM, Supra 40 Carl Zeiss - Oberkochen, Germany) and high resolution transmission electron microscopy (HRTEM, Zeiss Libra 200 MC TEM/STEM electron microscope), Fourier transform infrared spectroscopy (FTIR, Jasco 660 Plus spectrometer) and vibrating sample magnetometry (VSM, MicroMag Model 2900/3900 magnetometer).

3 RESULTS AND DISCUSSION

Both samples exhibit good cristallinity as confirmed by XRD analysis. Fig. 1 shows the XRD patterns of TA and respectively NOM coated iron oxide nanoparticles, which were found to be typical for cubic Fe_3O_4 phase with spinel structure, the peak positions being consistent with those from the standard data for magnetite.

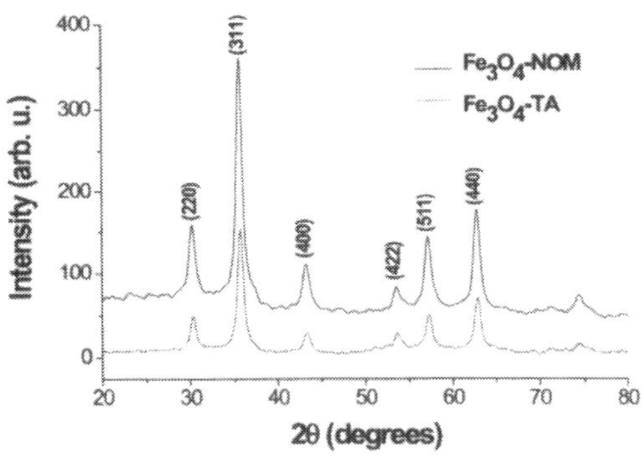

Figure 1: The XRD patterns of Fe_3O_4-TA and Fe_3O_4-NOM

BET measurements revealed similar values of the surface area of both sensitized nanomaterials (Table 1) consistent with those reported in the literature. The plots of pore size distribution (Fig. 2 a,b), determined by the BJH method, show the presence of a dominant peak in the mesoporous range (2–50 nm); the mesopores on spheres can be attributed to the interspaces of constituent particles. The pore diameter and volume for both catalysts were also determined (see Table 1). Overall, the Fe_3O_4-TA catalyst is characterized by a larger surface area with larger pores than those of Fe_3O_4-NOM sample.

Catalyst type	BET surface area (m^2/g)	BJH pore size (nm)	Pore volume (cm^3/g)
Fe_3O_4-TA	52.971	24.373	0.3023
Fe_3O_4-NOM	49.250	20.297	0.2499

Table 1: Surface properties of the studied nanocatalysts.

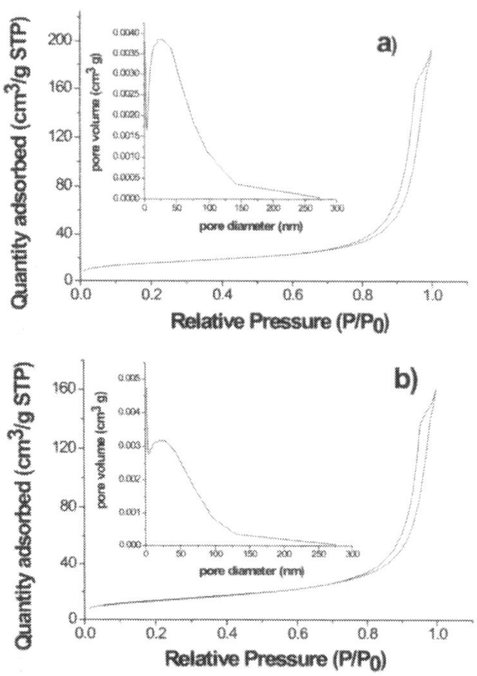

Figure 2: The nitrogen adsorption-desorption isotherm and pore size distribution curve (inset) of Fe_3O_4-TA catalyst (a) and Fe_3O_4-NOM catalyst (b).

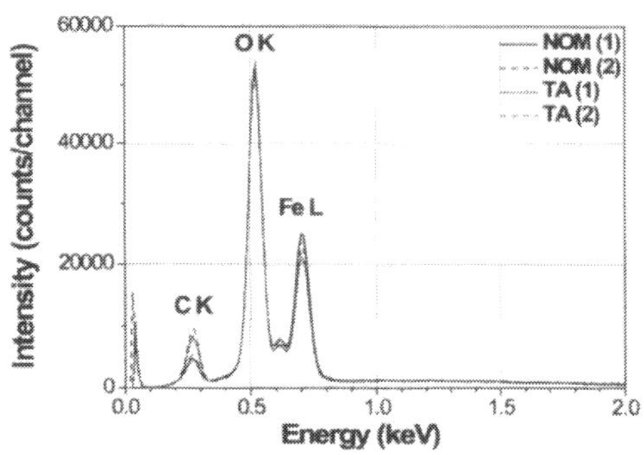

Figure 3: 5 kV EDX spectra measured at two locations on each sample, Fe_3O_4-NOM (black) and Fe_3O_4-TA (red); the EDX spectra were normalized on the O K line intensity.

The EDX investigation at 5 kV provided more surface sensitive analysis. The results at 5 kV were validated by EDX measurements at 20 kV acceleration voltage. As it can be seen from Fig. 3, the Fe_3O_4-TA sample contains significantly more C and Fe (relative to O) than the Fe_3O_4-NOM sample.

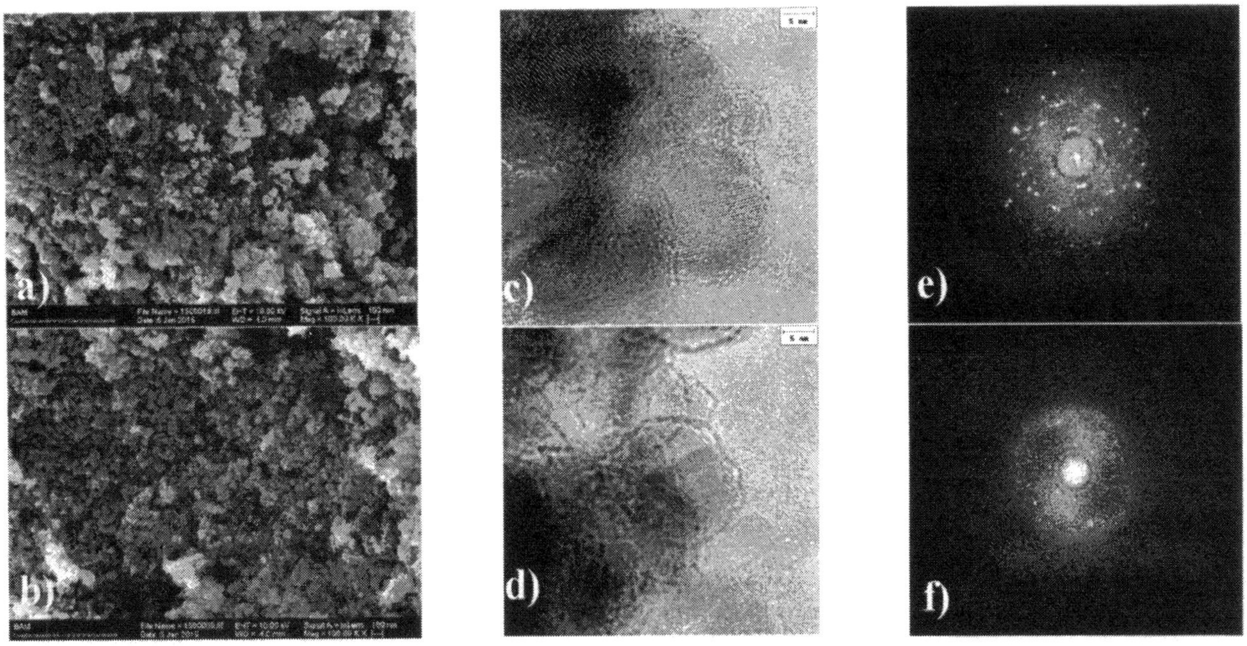

Figure 4: SEM micrographs (left), HRTEM images (middle) together with the corresponding SAED patterns (right) of the Fe_3O_4-TA (up) and Fe_3O_4-NOM (down) catalysts

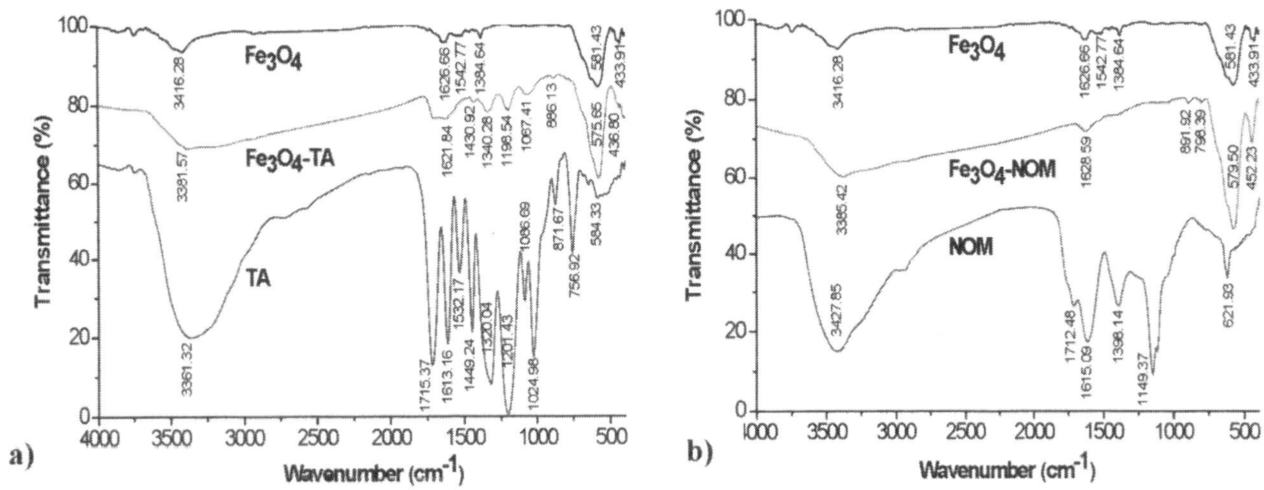

Figure 5: FTIR spectra of bare Fe_3O_4, of Fe_3O_4-TA and TA (a); FTIR spectra of bare Fe_3O_4, of Fe_3O_4-NOM and NOM (b);

The morphology of the samples' surface and the size of the primary particles were examined by high-resolution SEM (Fig. 4 a,b) and TEM (Fig. 4 c,d). The SEM images show the primary nanoparticles of rather spherical-like shape having sizes between 10 and 20 nm; the data is in agreement with the ones obtained by TEM; the high-magnification TEM images evidence the thin surrounding layer (below 1 nm) of the coating material (Fig. 4 c,d). The diffraction rings indicate that the magnetite cores are polycrystalline as revealed by the SAED pattern (Fig. 4 e,f).

FTIR measurements were further carried out in order to assess the binding of the active material onto the iron oxide surface. Fig. 5 a and b show the FTIR spectra of Fe_3O_4-TA, and Fe_3O_4-NOM respectively, compared to those of pure TA and NOM, and also compared to the uncoated magnetite spectral data. The FTIR bands of coated and uncoated Fe_3O_4 at wavenumbers lower than 600 cm^{-1} are ascribed to the Fe–O vibrations of iron oxide. In the Fe_3O_4-TA spectrum (Fig. 5a), the phenolic groups of TA are seen in the region ~1000–1343 cm^{-1}, while the absorption bands between 1400 cm^{-1} and 1650 cm^{-1} are related to aromatic –C=C– bonds [7]. In the case of pure NOM spectrum (Fig. 5b), the peaks at 1616.cm^{-1} and 1712 cm^{-1} correspond to the stretching vibration of C=O bond in carboxylic salt and free carboxylic acid respectively [5]; these are also present in the Fe_3O_4-NOM catalyst spectrum at ~1628 cm^{-1} suggesting the interaction of carboxilate group with the FeO surfaces. The FTIR measurements also reveal that TA formed stronger interactions with Fe_3O_4, compared to NOM.

The above findings can also explain the higher saturation magnetization (Ms = 68.0 emu/g) of Fe_3O_4-NOM than that of Fe_3O_4-TA (Ms = 58.2 emu/g), as depicted in Fig. 6. The obtained results are in agreement with those reported in literature for the magnetization curves of coated Fe_3O_4 recorded at room temperature.

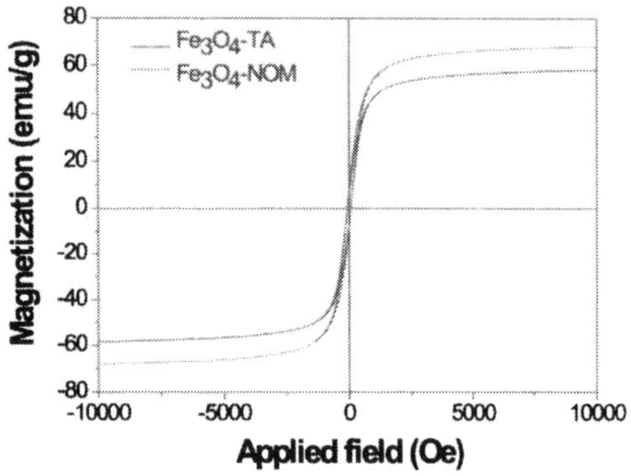

Figure 6: Magnetization curves of the two nanocatalysts

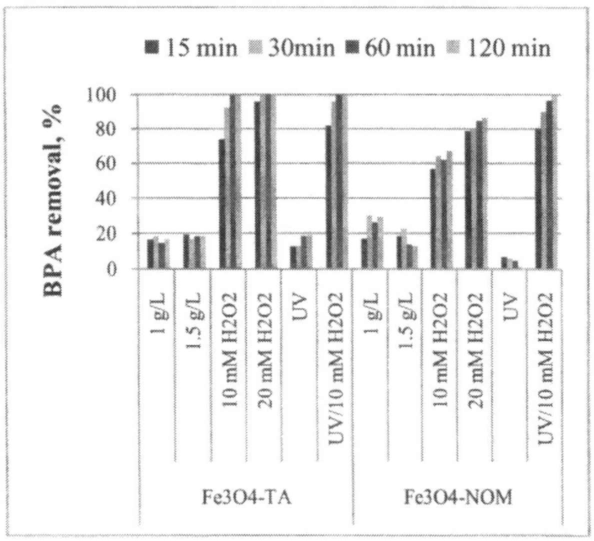

Figure 7: The effect of catalyst loading, H_2O_2 dosage and UV light on BPA removal over the two catalysts. Initial conditions: 0.5 μM of pollutant, 1 g/L catalyst.

Following catalysts' characterization, the catalytic activity in heterogeneous oxidation of aqueous solutions containing a model pollutant (Bisphenol A, BPA) was comparatively studied. BPA, used as a raw material for polycarbonate and epoxy resins, is known to accumulate in nature without decomposition and has been found to be weakly estrogenic [8].

The effect of catalyst loading, H_2O_2 dosage and UV light on the degradation performance of the oxidation process was investigated. Moreover, the stability and reusability of the catalysts were also established (data not shown).

The optimum experimental parameters were found to be 1 g/L of catalyst, 10 mM H_2O_2, under UV irradiation. More than 80% of BPA were removed by each catalyst after 30 minutes of reaction time under the experimental conditions given above (Fig. 7). The higher overall performance of the Fe_3O_4-TA material can be observed, while Fe_3O_4-NOM exhibited a lower reactivity compared to Fe_3O_4-TA.

4 CONCLUSION

The synthesis, characterization and performance of two green Fenton-like catalysts were investigated at the laboratory scale in the present study. Tannic acid and dissolved natural organic matter were the reactive materials of choice grafted on the surface of magnetite cores. The results obtained in catalytic Fenton heterogeneous reactions showed the suitability of both catalysts for the efficient degradation of persistent pollutants in wastewaters.

Acknowledgement The authors acknowledge the financial support of the Romanian Ministry of National Education through the national grant PN-II-ID-PCE-2012-4-0477.

To whom all correspondence should be addressed[1].

REFERENCES

[1] T. Shahwan, S. Abu Sirriah, M. Nairat, E. Boyaci, A.E. Eroglu, T.B. Scott, K.R. Hallam, Chem. Eng. J., 172, 258–266, 2011.

[2] S.-P. Sun, A.T. Lemley, J. Molecular Catal. A: Chem., 349, 71–79, 2011.

[3] C. Nadejde, M. Neamtu, R.J. Schneider, V.-D. Hodoroaba, G. Ababei, U. Panne, Appl. Surf. Sci. 2015, DOI: 10.1016/j.apsusc.2015.01.036.

[4] H.C. Genuino, N. Mazrui, M.S. Seraji, Z. Luo and G.E. Hoag, Ch. 3. Green Synthesis of Iron Nanomaterials for Oxidative Catalysis of Organic Environmental Pollutants, in New and Future Developments in Catalysis, 41–61, 2013.

[5] L. Peng, P. Qin, M. Lei, Q. Zeng, H. Song, J. Yang, J. Shao, B. Liao, J. Gu, J. Haz. Mat., 209–210, 193–198, 2012.

[6] J.-F. Liu, Z.-S.Zhao, G.-B. Jiang, Environ. Sci. Technol., 42, 6949–6954, 2008.

[7] K. Atacan, M. Özacar, Coll. Surf. B: Biointerfaces, 128, 227–236, 2015.

[8] M. Neamţu, D. Grandjean, A. Sienkiewicz, S. Le Faucheur, V. Slaveykova, J. Velez Colmenares, C. Pulgarín, F.L. De Alencastro, Appl. Catal. B: Environ., 158–159, 30–37, 2014.

[1] C. Nadejde and M. Neamtu, 'Alexandru Ioan Cuza' University, 54 Lascar Catargi, Iasi, Romania, Ph: +40(232)201102, Fax: +40(232)201102, claudia.nadejde@uaic.ro, mariana.neamtu@uaic.ro.

Controlled synthesis of metal@fullerene nanoarchitectures for catalysis

F. Leng, M. R. Axet and P. Serp*

* Laboratoire de Chimie de Coordination UPR CNRS 8241
composante ENSIACET, Universite de Toulouse UPS-INP-LCC
4 allee Emile Monso BP 44362, 31030 Toulouse Cedex 4 (France)
faqiang.leng@ensiacet.fr

ABSTRACT

Ru@fullerene C_{60} nanostructures have been synthesized by decomposition of an organometallic precursor ([Ru-(COD)(COT)] (COD= 1, 5 cyclooctadiene, COT= 1, 3, 5-cyclooctatriene)) in the presence of C_{60} using mild conditions. The nanoarchictetures obtained display small Ru nanoparticles (1.5 nm of mean diameter). The shape of the nanostructures is controlled by the use of different solvents. The Ru@C_{60} nanoarchitectures have a good catalytic activity and selectivity in the hydrogennation of nitrobenzene.

Keywords: ruthenium nanoparicles, C_{60} fullerene, nanostructures, catalysis

The association of metallic or semi-metallic nanopartiles with carbon materials provides unique materials that are useful in many applications including catalysis, gas storage, sensors, fuel cells and batteries.[1] The use of nano-structured carbon materials to build such assemblies is particularly attractive since they are unique and ideal templates onto which nanoparticles can be grafted, allowing the construction of designed nanoarchitectures.[2] A plethora of reports therefore exists on the functionalization of nano-structured carbon materials with metal particles. Metallic nanoparticles have been associated to carbon nanotubes,[3] graphene,[4] fullerenes[5] and even nano-diamonds.[6] However, the main limitation of these assemblies is the complete absence of a control of their organization, *i.e.* the metallic nanoparticles are randomly distributed on the surface of the nanocarbons. Consequently, it is extremely difficult to achieve a high metal loading with small metal nanoparticle size, which is detrimental to many applications such as fuel cells, sensors or gas storage. Additionally, the distance between nanoparticles being not controlled, their properties are not optimized. Indeed, it has been shown that the proximity of nanopartic-les may affect their catalytic performances.[7]

Fullerenes are an important member in the carbon material family, which have been extensively investigated up to now. As far as we know, most of metal nano-particles@fullerene C_{60} structures have been synthesized by the following methods: a) deposition of pre-synthesized metal nanopaticles on fullerene C_{60};[8] b) thermal decom-postion of an organometallic complex in the presence of fullerene C_{60};[9] c) co-evaporation of the metal and fullerene C_{60};[10] and d) electrochemically.[11]

Fullerene C_{60} may produce metal polymer fullerenes.[12] The first metal fullerene polymer was described by Nagashima,[13a] which was synthesized from C_{60} and $[Pd_2(dba)_3]$ (equation (1)). Pt fullerene polymer has been also synthesized in the same way.[13b] The polymer presents an amphorous structrue and no metal nanoparticles are observ-ed. Metal nanoparticles can be obtained by thermal decom-position of the polymer at 650 K.[14] Ru_nC_{60} species were prepared by the reaction of $[Ru_3(CO)_{12}]$ and C_{60} in refluxing toluene (equation (2)).[15] Compared to the Pd-, and Pt-C_{60} polymers, the amorphous matrix showed small ruthenium particles (2-5 nm mean diameter) in the Ru fullerene compound.

$$[M_x(dba)_y] + C_{60} \xrightarrow[\text{toluene}]{\text{reflux}} M_nC_{60} \quad (1)$$

$x = 0, 2, y= 2, 3$ $\qquad$ $n<7$

$M = Pd, Pt$

$$[Ru_3(CO)_{12}] + C_{60} \xrightarrow[\text{toluene}]{\text{reflux}} Ru_3C_{60} \quad (2)$$

In this context and highly interested in the development of nanostructured hybrid materials, we associate here fullerene C_{60} with ruthenium. The Ru@C_{60} nanoarchitectures were synthesized by decomposition of [Ru(COD)(COT)] in the presence of C_{60} using an organometallic approach.[16] Scheme 1 depicts the procedure of preparation of Ru@C_{60} nanostructures. [Ru(COD)(COT)] was decomposed by molecular H_2 (3 bar) in the presence of fullerene C_{60}. The solvent used as well as the Ru/C_{60} ratio play an important role in the synthesized nanostructures.

$$[Ru(COD)(COT)] + C_{60} \xrightarrow[\text{r.t.}]{\text{3 bar } H_2\text{, solvent}} Ru@C_{60}$$

Scheme 1. Synthesis of Ru@C_{60} nanoarchitectures.

The effect of different solvents (toluene, chlorobenzene, *o*-dichlorobenzene, dichloromethane, and decalin) was investigated first with a Ru/C_{60} ratio of 2. TEM images (Figure 1) show that several structures can be obtained depending on the solvent used. Chlorinated solvents produced Ru@C_{60} polymeric particles with mean size of: 285 ± 3 nm for chlorobenzene, 200 ± 3 nm for *o*-dichlorobenzene and 39.6 ± 0.7 nm for dichloromethane. On the other hand, decalin and toluene produced amorphous structures. In addition, non coordinating solvents such as decalin and dichloromethane allowed to obtain organometallic polymers decorated with Ru nanoparticles.

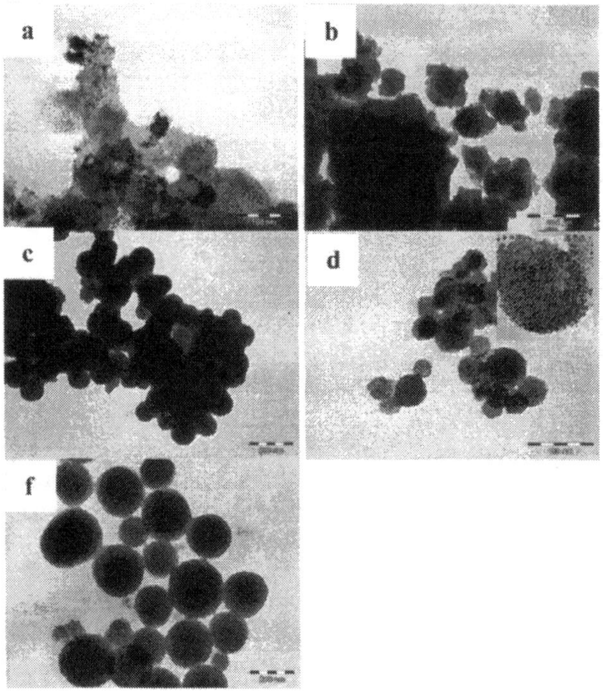

Figure 1. TEM images of Ru/C_{60}=2/1 synthesized in different solvents: a) decalin (scale bar 100 nm), b) toluene (scale bar 500 nm), c) chlorobenzene (scale bar 500 nm), d) dichloromethane (scale bar 100 nm), and (f) *o*-dichorobenzene (scale bar 200 nm).

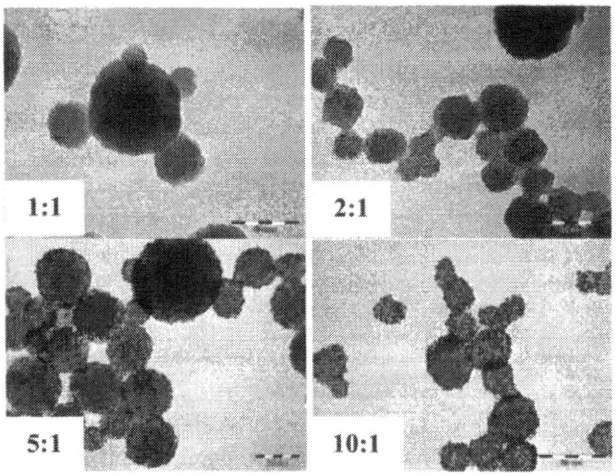

Figure 2. TEM images of nanostructures produced in dichloromethane using different Ru/C_{60} ratio, scale bar are 50 nm.

Using dichloromethane as solvent, which allow to obtain well defined polymeric nanospheres, we examined the effect of the Ru/C_{60} ratio (Figure 2). Ru/C_{60}=1/1 display polymeric nanospheres (37.8 ± 1.0 nm) with no Ru nanoparticles. The increase of Ru content allow to obtain polymeric nanospheres decorated with Ru nanoparticles. The size of the nansopheres being constant and the Ru nanoparicles size showing a narrow distribution (1 nm).

Table 1. Hydrogenation of nitrobenzene using Ru@C_{60}.[a]

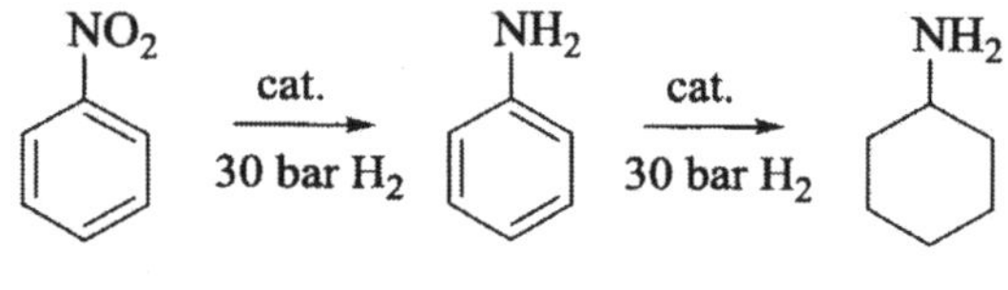

1 2 3

Ru/C_{60}	2 Conversion (selectivity)%[b], Time	3 Conversion (selectivity)%[b], Time
1:1	21(84), 24h	n. d., 48h
2:1	39(6), 24h	Trace, 48h
5:1	98 (96), 24h	Trace, 48h
10:1	91(92), 4h	81 (80), 3h

[a] Reaction conditions: 5.0 mg of catalyst, 4.07mmol of $PhNO_2$, 0.58 mmol of dodecane as internal standard, 30 mL of EtOH, 30bar of H_2 at 80°C. [b] Conversion was determined by GC-MS using dodecane as internal standard. n.d.= not detected.

The Ru@C_{60} nanoarchitectures, were tested as catalyst in the hydrogenation of nitrobenzene (Table 1). Ru/C_{60} polymer show a low conversion to aniline (**2**) after 24 h of reaction, however, increasing the ruthenium content (*i.e.* the nanoparticles on the nanosphere surface) the catalysts were more active reaching total conversion to aniline in 4h with Ru/C_{60}=10/1 nanocatalyst. Furthermore, cyclohexylamine (**3**) was produced in high yields with Ru/C_{60}=10/1 catalyst under mild conditions. Indeed, only a few papers report the direct synthesis of cyclohexylamine from nitrobenzene,[17] always using harsh conditions (170 °C).

In summary, this work presents the synthesis of well defined nanoarchitectures Ru@C_{60} through decomposition of an organometallic complex [Ru(COD)(COT)] in the presence of C_{60} at room temperature. According to the experiment data, the solvent is the key to control the nanostructures, and different Ru@C_{60} composites were obatined controlling the Ru@C_{60} ratio. It is also revealed that the Ru@C_{60} nanoarchitecture exhibited high catalytic activity in the hydrogenation of nitrobenzene. We foresee that our method would open the way to design special morphologies and structures of metal@C_{60} nanostructure, which could find use in various applications.

EXPERIMENT SECTION

General methods

All operations were carried out under argon atmosphere using standard Schlenk techniques or in an MBraun glove-box. Solvents were purified by an MBraun SPS-800 solvent purification system. [Ru(COD)(COT)] was purchased from Nanomeps (Toulouse) and fullerene C_{60} from Aldrich.

All Ru@C_{60} nanostructures were synthesised using the same procedure. We describe here a typical experiment:

Synthesis of the Ru/C_{60} = 2/1. [Ru(COD)(COT)] (20 mg, 0.0635 mmol) was introduced in a Fischer-Porter bottle in the glovebox. A solution of C_{60} (23 mg, 0.0317 mmol) in 50 mL CH_2Cl_2, was then canulatted to the Fischer-Porter bottle. The resulting purple solution was stirred for 30 min at room temperature, after which the bottle was pressurized under 3 bar of H_2. A black precipitate formed quickly indicating ruthenium particles were produced, then the solution was allowed to react for 12 h. After elimination of excess H_2, the resulting solution was reduced to 15 mL, after which addition of 30 mL pentane led to a dark precipitate. The solid was filtered, washed with pentane three times, and dried in vacuum pump.

General method for hydrogenation of nitrobenzene:

Ru/C_{60} catalyst (5 mg), dodecane (100 mg) and nitro-benzene were added to a schlenked tube (100 mL) with 30 mL of degased ethanol in the glovebox. The resulting disp-ersion was ultralsonicated for 15 min and then introduced to the autoclave. After, the mixture was heated to the desired temperature (80°C) and autoclave pressurized to 30 bar of H_2 pressure. Samples of the liquid phase were taken during the reaction and analized by GC-MS.

Characterization

The TEM specimens were prepared by slow evaporation in a glovebox of droplets of diluted solutions of the differe-nt colloids deposited on a high-resolution carbon-supported copper grid. TEM analyses were performed at the *Centre de microcaracterisation Raimond Castaing*, UMS 3623, Toul-ouse by using a JEOL JEM 1011 CX-T electron microscope operating at 100 kV with a point resolution of 4.5 Å or a JEOL JEM 1400 electron microscope operating at 120 kV. The approximation of the particles mean size was made through a manual analysis of enlarged micrographs by mea-suring at least 200 particles on a given grid. The ruthenium content was established by inductively coupled plasma opti-cal emission spectroscopy (ICP-OES) performed in a Ther-mo Scientific ICAP 6300. GC-MS samples were detected using instructment of Perkin Elmer Clarus 500 MS.

ACKNOWLEDGEMENT

This work was supported by the Centre National de la Recherche Scientifique (CNRS), which we gratefully acknowledge. The authors acknowledge financial support from the program of China Scholarships Council(CSC) for F. L. grant.

REFERENCES

[1] G. G. Wildgoose, C. E. Banks, R. G. Compton, Small 2, 182-193, 2006.

[2] R. R. Bacsa, P. Serp, in Carbon Meta-Nanotubes, John Wiley & Sons, Ltd, pp. 163-221, 2011.

[3] V. Georgakilas, D. Gournis, V. Tzitzios, L. Pasquato, D. M. Guldi, M. Prato, J. Mater. Chem. 17, 2679-2694, 2007.

[4] P. T. Yin, T.-H. Kim, J.-W. Choi, K.-B Lee, Phys. Chem. Chem. Phys. 15, 12785-12799, 2013.

[5] I. I. S. Lim, J.; Luo, J. Ouyang, L. Wang, S. Zhou, C.-J. Zhong, Chem. Mater. 17, 6528-6531, 2005.

[6] P. Subramanian, Y. Coffinier, D. Steinmüller-Nethl, J. Foord, R. Boukherroub, S. Szunerits, Electrochim. Acta. 110, 4-8, 2013.

[7] a) M. Nesselberger, M. Roefzaad, R. F. Hamou, P. U. Biedermann, F. F. Schweinberger, S. Kunz, K. Schloegl, G. K. H. Wiberg, S. Ashton, U. Heiz, K. J. J. Mayrhofer, M. Arenz, Nat. Mater. 12, 919-924, 2013; b) J. Speder, L. Altmann, M. Baumer, J. J. K. Kirkensgaard, K. Mortensen, M. Arenz, RSC Adv. 4, 14971-14978, 2014; c) G. Prieto, J. Zečević, H. Friedrich, K. P. de Jong, P. E. de Jongh, Nat. Mater. 12, 34-39, 2013; d) G. Prieto, J. D. Meeldijk, K. P. de Jong, P. E. de Jongh, J. Catal. 303, 31-40, 2013.

[8] a) Z.Q. Tan, A. Masuhara, S. Ohara, H. Kasai, H. Nakanishi, H. Oikawa, J. Nanopar. Res. 15, 2029, 2013; b) Y. P. Ren, P. Paira, T. R. Nayak, W. H. Angc, G. Pastorin, Chem. Commun. 47, 7710-7712, 2011; c) M. Brust, C. J. Kiely, D. Bethell, D. J. Schiffrin, J. Am. Chem. Soc. 120, 12367-12368, 1998.

[9] a)D. Saha, S.G. Deng, Langmiur 27, 6780-6786, 2011.; b)G. Lee, J. H. Shim, H. Kang, K. M. Nam, H. Song, J. T. Park, Chem. Commun. 5036-5038, 2009.

[10] a) S. Sakai, H. Naramoto, V. Lavrentiev, K. Narumi, M. Maekawa, A. Kawasuso, T. Yaita, Y. Baba, Mater. Trans. 46, 765-768, 2005; b) U. Zimmermann, N. Maljnowski, A. Burkhardt, T. P. Martin, Carbon 33, 995-1006, 1995; c) J. G. Hou, X. Li, Y. Q. Li, H. Q. Wang, Adv. Mater. 11, 1124-1126, 1999.

[11] a) A.L. Balch, D. A. Costa and K. Winkler, J. Am. Chem. Soc. 120, 9614-9620, 1998; b) A. Hayashi, A. de Bettencourt-Dias, K. Winkler, A. L. Balch, J. Mater. Chem. 12, 2116, 2002; c) E. Brancewicz, E. G.adzka, A. Basa, K. Winkler, Electrochim. Acta 128 91-10, 2014.

[12] a) F. Giacalone and N. Martín, Chem. Rev. 106, 5136-5190, 2006; b) A. L. Balch, M. M. Olmstead, Chem. Rev. 98, 2123-2165, 1998; c)N. F. Goldshleger, Fullerene Sci. Techn., 9, 225-280, 2001.

[13] a) H. Nagashima, A. Nakaoka, Y. Saito, M. Kato, T. Kawanishi, K. Itoh, J. Chem. Soc., Chem. Commun. 377, 1992; b) H. Nagashima, Y.Kato, H.

Yamaguchi, E. Kimura, T. Kawanishi, M. Kato, Y. Saito, M. Haga, K. Itoh, Chem. Lett. 7, 1207,1994.
[14] A.V. Talyzin , A. Dzwilewski, M. Pudelko, Carbon 45, 2564-2569, 2007.
[15] M. Rasinkangas, T. T. Pakkanen, T. A. Pakkanen, J. Organomet. Chem. 476, 6-8, 1994.
[16] B. Chaudret, C. R. Phys. 6, 117-131, 2005.
[17] a) P. Tomkins, E. Gebauer-Henke, W. Leitner, T. E. Muller, ACS Catal. 5, 203-209, 2015; b) S. G. Oh, V. Mishra, J. K. Cho, B.-J. Kim, H. S. Kim, Y.-W. Suh, H. Lee, H. S. Park, Y. Kim, J. Catal. Commun. 43, 79-83, 2014.

Tailoring Particle Behavior for Phase Transfer Catalysis: Recent Developments by the Center for Interfacial Reaction Engineering

Clint P. Aichele[1], Javen Weston[2], Deepika Venkataramani[3], Nicholas Briggs[4], Steven Crossley[5], Jeff Harwell[6], Jeff White[7], and Daniel Resasco[8]

[1]School of Chemical Engineering, Oklahoma State University, Stillwater, OK, USA, clint.aichele@okstate.edu
[2]School of Chemical, Biological, and Materials Engineering, University of Oklahoma, Norman, OK, USA, javen@ou.edu
[3]School of Chemical Engineering, Oklahoma State University, Stillwater, OK, USA, deepika.venkataramani@okstate.edu
[4]School of Chemical, Biological, and Materials Engineering, University of Oklahoma, Norman, OK, USA, nbriggs@ou.edu
[5]School of Chemical, Biological, and Materials Engineering, University of Oklahoma, Norman, OK, USA, stevencrossley@ou.edu
[6]School of Chemical, Biological, and Materials Engineering, University of Oklahoma, Norman, OK, USA, jharwell@ou.edu
[7]Department of Chemistry, Oklahoma State University, Stillwater, OK, USA, jeff.white@okstate.edu
[8]School of Chemical, Biological, and Materials Engineering, University of Oklahoma, Norman, OK, USA, ouresasco@ou.edu

Center for Interfacial Reaction Engineering (CIRE)

ABSTRACT

This presentation highlights recent advances by the Center for Interfacial Reaction Engineering (CIRE) in designing and tailoring catalytic particles for use in phase-transfer catalysis. CIRE consists of researchers across several disciplines including nanotechnology, thermodynamics, interfacial phenomena, and catalysis from three different universities. The results highlight the importance of particle wetting on dynamic interfacial and overall emulsion behavior and has potential importance for a variety of applications in the biofuels, pharmaceuticals, and energy sectors.

Keywords: emulsions, carbon nanotubes, nanparticles, catalysis, interfacial behavior

1 INTRODUCTION

Phase-transfer catalysis involves carrying out chemical reactions in a biphasic mixture of two immiscible liquids, and it is a potentially useful technique that combines standard heterogeneous catalysis with the equivalent of a solvent extraction separation. Figure 1 illustrates the four central themes of CIRE that address phase-transfer catalysis. The first theme of CIRE focuses on synthesizing and characterizing nanoparticles that optimize the reactivity, emulsion stability, mass transfer, and interfacial area of phase-transfer catalysis processes. Second, CIRE focuses on understanding how catalytic reaction mechanisms and kinetics change when the reactions take place at a liquid-liquid interface. Third, CIRE uses spectroscopy and molecular modeling to understand the reaction pathways, competitive adsorption, and thermodynamics of model reactions on interfacial catalysts. Finally, CIRE combines experimentation, theory, and modeling to understand the kinetics of mass transfer and phase equilibrium in dynamic, phase-transfer environments and how altering particle properties can be used to optimize phase transfer.

Phase-transfer catalysis does more than just combine the reaction and separation steps common in numerous industrial applications. It allows engineers and scientists to tune the reactivity and selectivity of specific reactions by controlling solvent/catalyst interactions and taking advantage of differences in reactant/product solubilities in the two liquid phases. The technique is also extremely useful when the desired reaction product would be unstable under reaction conditions in one solvent, but can quickly partition into the other solvent where it is more stable. Much early work into phase-transfer catalysis has used molecular surfactants to create emulsions, which increase the interfacial area and increase total interphase mass transfer. However, molecular surfactants can be very difficult to separate from the reaction product mixtures, a problem that can be mediated by using interfacially active nanoparticles as both catalyst supports and emulsion stabilizers since the nanoparticles can be more easily recovered via centrifugation or filtration.

This presentation highlights recent advances by the Center for Interfacial Reaction Engineering in designing phase-transfer catalytic systems that show improved selectivity and reactivity when compared to standard heterogeneous catalysis. Our most recent work involves studying photocatalytic partial oxidation of cyclohexane on a titania catalyst and hydrogenation of alkenes using a palladium catalyst supported on multi-walled carbon nanotubes.

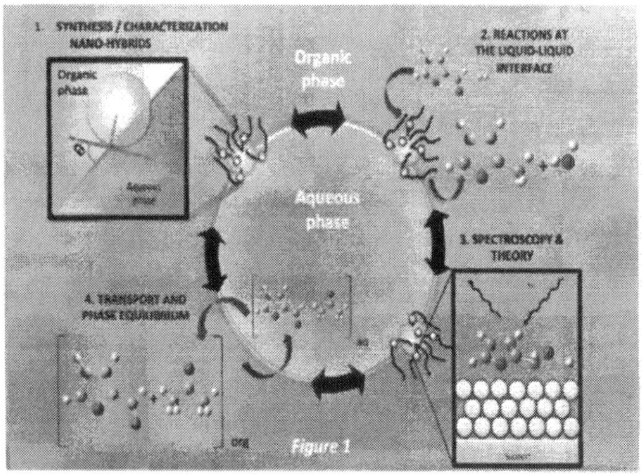

Figure 1. Four central research thrusts of CIRE.

2 EXPERIMENTAL TECHNIQUES

2.1 Nuclear Magnetic Resonance (NMR)

Nuclear Magnetic Resonance (NMR) is a core tool used by CIRE to probe emulsion morphology, speciation, and the impact of water and other species on catalyst performance. CIRE recently employed NMR to provide fundamental insight regarding the impact of water on zeolite catalysis [1]. Diffusion NMR techniques [2] are currently being employed by CIRE to quantify the impact of reactions on emulsion morphology and separation efficiency. NMR has the unique ability to simultaneously quantify droplet size distributions and speciation, thereby providing critical information about the impact of reactions on emulsion morphology. NMR is not dependent on the optical properties of the emulsions, so it is ideally suited for concentrated emulsion characterization where the systems are generally opaque.

2.2 Electron Microscopy

To determine the location of MWNTs at the interface MWNT Pickering emulsion wax droplet stabilized with oxidized MWNTs, pristine MWNTs with alkyl glucoside, or pristine MWNTs with 2-hydroxyethyl cellulose were filtered using a Whatman 1 filter and rinsed with water to remove any unattached particles. After this 3.5 wt% of the wax droplets covered with MWNTs was mixed with LR White resin, a hydrophilic resin. The solution was then hand shook for several minutes to disperse the wax droplets in the resin. After mixing the solution was put in a gel capsule and then placed in an oven at 40°C to allow the LR White Resin to polymerize. The gel capsule was removed and the polymerized LR White was then thin sectioned using an ultramicrotome, making 100 nm thick sections. These sections were then placed on a TEM gird. TEM of the samples was then performed to see the location of the MWNTs at the interface using a JEOL 2000-FX with an acceleration voltage of 200 kV.

2.3 Immersion Microcalorimetry

A Setaram C80 micro-calorimeter with a specially designed pair of powder cells was used to measure the heat of immersion for all of the modified nanoparticles. The heat of immersion is related to the particle contact angle using the theoretical approach described by Yan.29 The method uses the enthalpy of immersion, a result of the change in Gibbs free energy that results from replacing a solid-gas interface with a solid-liquid interface, and relates it to the solid-liquid contact angle using the Young-Laplace equation. The resulting contact angle equation yields:

$$cos\theta = \frac{-KT - h_i - \pi_e}{\gamma_{lv}} \quad (1)$$

Where θ is the solid-liquid contact angle, K is the difference between the temperature dependence of solid-liquid interfacial tension and solid surface tension, T is the absolute temperature, hi is the enthalpy of immersion, γlv is the liquid-vapor surface tension, and πe is the difference between the solid surface tension and the solid-vapor surface tension (γs − γsv), which is assumed to be negligible for systems with reasonably large contact angles. In this case, K is assumed to be 7.0 • 10-5 J/m2•K, corresponding to a low-energy surface and a γsl that is temperature independent, following the assumptions made by Yan. The powder cells consist of a stainless steel cylinder and a pair of brass sealing rings that hold a

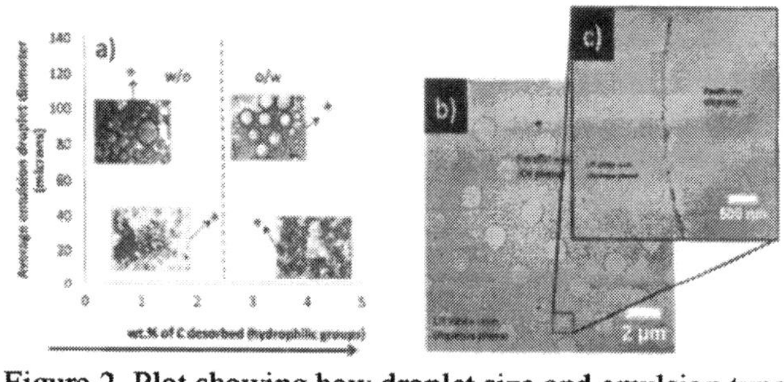

Figure 2. Plot showing how droplet size and emulsion type change with CNT hydrophilicity, and two micrographs that that show nanotubes adsorbed at the interface of an emulsion droplet after thin-sectioning the droplet with an ultramicrotome.

membrane in place and seal each end of the cell; one cell is used as the sample cell, while the other is used as a reference cell. The calorimeter cells are designed by Setaram with a breaking rod that has been modified in-house with a sharp, pointed tip which is used to puncture the powder cell's membrane. Prior to each experiment the modified nanoparticles are dried in a vacuum oven at 1200C for at least 12 hours. Then, a known quantity of sample is placed into each powder cell, and the cell is sealed with a nitrile rubber membrane. Both powder cells are then placed in the larger stainless steel, calorimetry cells (manufactured by Setaram) with approximately 8 mL of the wetting liquid. Both calorimeter cells are then installed into the microcalorimeter and allowed to equilibrate at 40°C. After the heat signal and cell temperatures stabilize, both membranes of the sample cell are punctured with the breaking rod, allowing the liquid to enter the cell and immerse the nanoparticles. The thermal energy released or absorbed by the sample immersion is then measured and recorded by the microcalorimeter. A series of eight blank punctures of sealed, empty powder cells were used to estimate the heat released by only puncturing the nitrile rubber membranes and subtracted from the measured immersion enthalpy as a constant background. The heat of immersion is measured at least four times for each sample and the average was used to calculate the particle contact angle.

2.4 Chemical Reactions

Hydrogenation reactions were carried out using 1-dodecene as a oil soluble reactant and cis-2-butene-1,4-diol as a water soluble reactant. Both of these molecuels have high solubility in either the water or oil. When operating under mass transfer limitations the reactants will have limited mass transfer through the other phase.

Reactions were carried out in oil, water, and emulsions. Reactants 1-dodecene and cis-2-butene-1,4-diol were used when the solvent which the reactant was soluble in was used.

Emulsions were created by first dispersing Pd/MWNTs in either oil or water by horn sonication. Then adding the opposite phase and horn sonicating to form a emulsion. After emuslification the catalyst was reduced for 3 hours under H_2. Then the reactant injected and allowed to react for 1 hour. Reduction and reaction were both carried out at 45°C or 70°C.

Chemical reactions were carried out using a glass reflux system with a three neck round bottom flask. One neck was connected to the reflux condenser, another to a gas line to inject hydrogen, and the last a septum seal for injecting the reactants once reduction of the catalyst was completed.

3 RESULTS

Sample	Heat of Immersion, mJ/m^2	Contact Angle
Bare Ludox	-110.2 ± 6.7	<0
Low Coverage SAPTMS	-135.5 ± 28.8	<0
High Coverage SAPTMS	-170.1 ± 23.9	<0
Low Coverage MPTMS	-60.2 ± 2.6	58
High Coverage MPTMS	-65.8 ± 16.4	53
Low Coverage PhTCS	-65.8 ± 5.1	52
High Coverage PhTCS	-47.5 ± 3.3	69
Low Coverage TMCS	-5.1 ± 7.3	104
High Coverage TMCS	1.9 ± 3.9	109
Low Coverage OTCS	-0.4 ± 7.9	107
High Coverage OTCS	13.0 ± 2.1	120

Table 1. Heats of immersion and calculated water contact angles of surface-modified silica nanoparticles measured using microcalorimetry

Multi-walled carbon nanotubes have been found to adsorb at the interface of emulsion droplets and prevent droplet coalescence, as shown in the TEM micrographs shown in Figure 2. Oxidizing the multi-walled carbon nanotubes and adding surfactants or polymers to the nanotube dispersion can control the number of nanotubes at the interface and the thickness of the interface. Droplet size and emulsion type (oil-in-water vs. water-in-oil) can also be controlled by tuning the hydrophobicity of the carbon nanotube via oxidation or surfactant addition, as illustrated in Figure 2. Furthermore, droplet size and emulsion type can be further tuned by varying the particle ratio when stabilizing emulsions with mixtures carbon nanotubes with different levels of oxidation.

The wettability and contact angle of spherical silica nanoparticles can be controlled by reacting alkyl silanes with hydroxyl groups on the nanoparticle surface. The final contact angle, as measured with a variety of methods, was found to strongly depend on the type of alkyl group grafted onto the nanoparticle, and a less pronounced dependence on the level of surface coverage. Phenyl, trimethyl, and octadecyl chlorosilanes were grafted onto the silica surface and the particle contact angle was evaluated using a heat of immersion method, the results of which are shown in Table 1 [3]. Heat of immersion tests also confirmed the results of Oh [4] in regards to the creation of superhydrophilic surface moieties when mercaptopropyl trimethoxy silane-modified silica is successively exposed to hydrogen peroxide and sodium hydroxide solutions.

Silanization can also be used to control the properties of Pickering emulsions stabilized with the modified silica nanoparticles. Emulsions stabilized by the various hydrophobically modified silica nanoparticles were shown to follow four general trends: 1) low surface coverage nanoparticles stabilized oil-in-water emulsions while high surface coverage nanoparticles stabilized water-in-oil emulsions in almost all cases; 2) emulsions stabilized by high surface coverage nanoparticles have slightly smaller average droplet sizes than emulsions stabilized by low surface coverage nanoparticles in every case but the oil-dispersed trimethyl chlorosilane-modified particles; 3) when the particles are initially dispersed in water, the emulsions stabilized by more hydrophobic nanoparticles had smaller average droplet sizes than emulsions stabilized by more hydrophilic nanoparticles, regardless of surface coverage; and 4) when the particles are initially dispersed in decane, the emulsions stabilized by more hydrophilic nanoparticles had smaller average droplet sizes than emulsions stabilized by more hydrophilic nanoparticles, in most cases.

Pickering emulsions offer a variety of methods to control reaction selectivity. One way is by controlling the location of the catalytically active portion of the particles relative to the oil and water phases. Janus particles made with a hydrophobic and hydrophilic side offer this ability. By depositing palladium particles on one side of the Janus particles the reaction in the aqueous or oil phase of the emulsion can be highly favored [5]. Another way to control selectivity in emulsions is by operating under surface reaction, temperature of 45°C, or hydrogen diffusion limitations, temperature of 70°C. Operating under hydrogen diffusion limitations utilizes the speed of hydrogen to diffuse through the two phases. By switching between these two limiting steps, the rate of reaction normalized to a single aqueous phase system can be cut in half or doubled for the aqueous phase of an emulsion system, as shown in Figure 3.

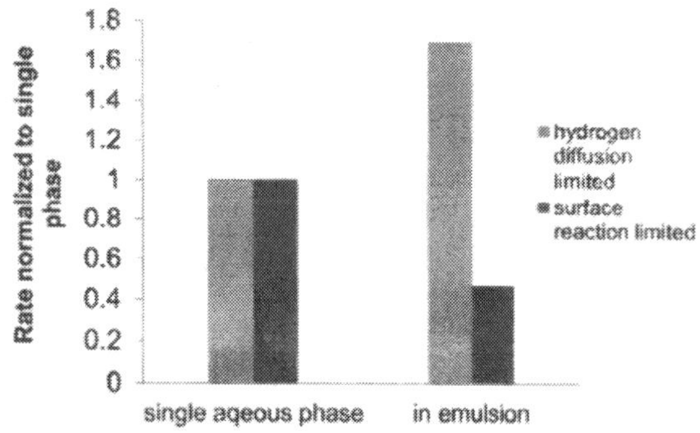

Figure 3. Use of an emulsion to control reaction selectivity when operating under surface reaction limited or hydrogen diffusion limitations. A Pd/MWNT catalyst was used.

4 CONCLUSIONS

The Center for Interfacial Reaction Engineering (CIRE) leverages core experimental capabilities to probe fundamental interfacial properties and phase transfer catalysis. Recent work by the center has shown the importance of emulsion morphology and particle behavior on the overall performance of bi-phasic, catalytic systems. Future work will focus on developing advanced catalysts and catalyst supports to improve the efficiency of the production of desired products while leveraging the advantageous properties of emulsions.

ACKNOWLEDGEMENTS

The authors thank the Department of Energy (DOE/EPSCOR) program for funding this work (Grant DE-SC0004600).

REFERENCES

[1] K. Chen, J. Damron, C. Pearson, D. Resasco, L. Zhang, and J. White, "Zeolite Catalysis: Water can Dramatically Increase or Suppress Alkane C-H Bond Activation," *ACS Catalysis*, 4, 3039-3044, 2014.

[2] C. Aichele, M. Flaum, T. Jiang, G. Hirasaki, and W. Chapman, "Water in Oil Emulsion Droplet Size Characterization using a Pulsed Field Gradient with Diffusion Editing (PFG-DE) NMR Technique," *J. of Coll. And Interf. Sci.*, 315, 607-619, 2007.

[3] N. Yan, Y. Maham, J. H. Masliyah, M. R. Gray, and A. E. Mather, "Measurement of Contact Angles for Fumed Silica Nanospheres Using Enthalpy of Immersion Data," *J. of Coll. And Interf. Sci.*, 228, 1-6, 2000.

[4] Y.-K. Oh, L.-Y. Hong, Y. Asthana, and D.-P. Kim, "Synthesis of Super-hydrophilic Mesoporous Silica via a Sulfonation Route," *J. Ind. Eng. Chem.*, 6, 911-917, 2006.

[5] J. Faria, M. Pilar Ruiz, D. Resasco, "Phase-Selective Catalysis in Emulsions Stabilized by Janus Silica Nanoparticles," *Advanced Synthesis & Catalysis*, 352, 2359-2364, 2010.

Low Temperature Plasma Surface Treatment of Catalytic Material

George H. Miley*, Shailendra N. Srivastava**, Kyujung Kim*, George Chen**

*NPL Associates, Inc.
912 W Armory Ave, Champaign, Illinois 61821
**Illinois Applied Research Institute
2100 S Oak St Suite 206, Champaign, Illinois 61820

ABSTRACT

Enhancing the performance of heterogeneous catalysts through plasma treatment has many applications in industry. One way to improve catalytic activity is through plasma treatment of surface "structure-sensitive" catalysts to create defects which can serve as possible catalytic reaction sites. The purpose of the plasma treatment process is to increase catalytic performance of various catalytic material.

Keywords: plasma, heterogeneous catalysts, plasma treatment / modification / irradiation

1 INTRODUCTION

In this paper we emphasize the application of the plasma treatment method to improve a variety of commercial processes that employ catalysts. However in a closely related paper at this meeting we discuss using the plasma treatment method to improve PEM fuel cell catalysts [1]. Heterogeneous catalysts are in a different phase than the reactants. Homogeneous catalysts are in the same phase as the reactants. Homogeneous catalysts have high activity and selectivity [2] compared to heterogeneous catalyst. As with most real world situations there are tradeoffs; for example heterogeneous catalyst which produce good reaction rates are sometimes composed of noble metals (e.g. platinum, palladium, etc.) which are scarce and expensive [3]. There are a variety ways to try to circumvent these tradeoffs [2] [4], but none have been fully successful to date. The process under study here uses a low temperature plasma to enhance catalytic performance of heterogeneous catalysts.

2 BACKGROUND

Heterogeneous catalyst can be classified by surface "structure-sensitive" and surface "structure-insensitive." These classifications partially explain the catalytic reaction mechanism of the heterogeneous catalyst. The catalytic activity of structure-sensitive catalyst as the name suggest depends on the structure of the surface. That is certain sites on the surface of the catalysts enhance the rate of the chemical reactions [4] [5] [6] [7]. Finding, predicting, and understanding the mechanism by which these certain reaction sites promote chemical reactions is still not well understood [6] [7]. This is because measuring or observing chemical reactions on the catalytic reaction site while it is taking place has never been performed. What is known are some qualitative observations such as, for structure-sensitive catalyst the reaction rate greatly increases with the number of surface defects (e.g. steps, kinks, etc.) [6] [7] [8]. This observation is based on experiments comparing reaction rates of catalyst with single-crystal surfaces and catalyst containing many surface defects [7].

In addition to surface defects other factors that enhance catalytic activity include the surface area, increasing the surface area of the heterogeneous catalyst also increase the catalytic reaction rate [9]. Furthermore catalytic reactions of materials with equal surface areas may also differ depending on their actual form [9]. For example, materials that are normally inert in macroscopic solid form (e.g. gold), demonstrate high catalytic activity in submicron solid form [10].

Low temperature plasma discharges are used in a variety of processes; for example the manufacture of semiconductor products. In the manufacturing process the plasma is used to manipulate the surface of solid materials. The effects of plasmas on surfaces are numerous, plasmas can damage the surfaces of materials through ion bombardment which can eject surface atoms a process called sputtering. Along with surface defects and damage from exposure to plasma, submicron structures have been shown to form on material surfaces under exposure to certain types of plasmas.

The effects caused by plasma-surface interaction (notably surface damage and defects, as well as the formation of submicron structures) are utilized to possibly alter the heterogeneous catalysts surface so the catalytic activity can be enhanced. The plasma-catalysts treatment process has been studied for some time [11].

3 EXPERIMENT

Figure 1 shows a schematic of a DC plasma discharge system with the cathode (exposed to the plasma) made out of a conducting heterogeneous catalysts. Exposing the catalysts to the plasma allows the plasma to modify the catalysts surface.

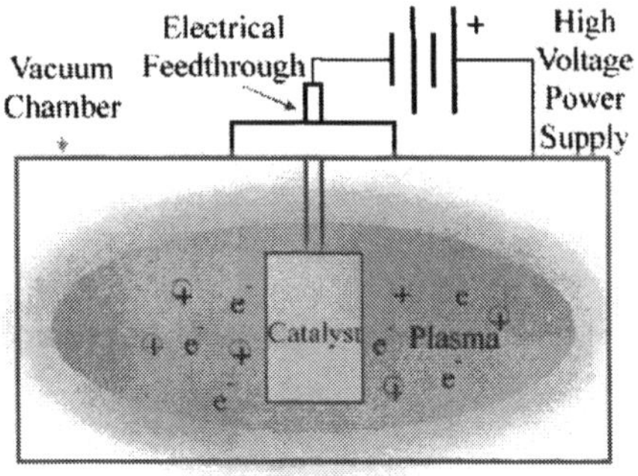

Figure 1: Schematic of DC plasma source modifying the surface of a conducting heterogeneous catalysts.

Figure 1 is a schematic of one possible system and is not the only configuration possible. A DC system is presented here because DC plasma systems are relatively simple compared to other plasma systems. However many heterogeneous catalysts are not electrically conducting. In such a case an AC discharge may be more appropriate to treat the surfaces of such catalysts. AC plasma discharges include DBD, RF, and microwave. RF, microwave are classified as AC, the distinction that separates them are the frequencies. As mentioned before using an AC discharge allows for the treatment of a greater variety of heterogeneous catalysts, however AC systems are more complicated usually requiring impedance matching which increases costs.

Heterogeneous catalytic material is being plasma-treated to modify the surface. The plasma-surface irradiation coarsens the surface of the catalytic material creating very pack submicron structures and defects [12] [13] [14] [15] [16] [17]. These submicron formations and defects on the catalytic material can serve as reactant adsorption and reaction sites [4] [5] [6] [7] [18] [19] [20]. The plasma modification can potentially increase the number reaction sites this can greatly improve the performance of the catalytic material. The plasma-surface treatment is relatively simple and can possibly enhance the catalytic activity of many less expensive heterogeneous catalysts. This then holds promise for lowering costs and/or raising the performance of catalysts for a variety of industrially and commercially useful chemical reactions. This could allow plasma treated catalysts to supplant the more expensive catalysts currently used. Furthermore plasma irradiating some of the more expensive catalysts can also increase their catalytic activity, thereby increasing reaction rates and ultimately improving the economics without changing the catalysts employed in the process. The plasma treatment is a simple method to increase heterogeneous catalyst performance; these catalysts are ubiquitous in many industries.

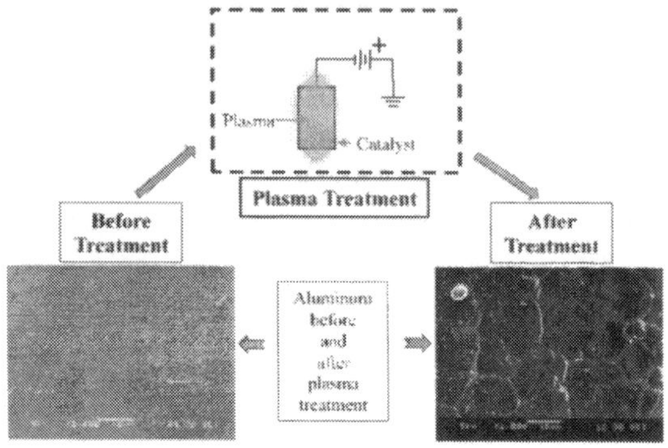

Figure 2: Plasma processing of aluminum foil demonstrating the effects of plasma on material surfaces.

Figure 2 illustrates a DC plasma treatment system currently set up to study and optimize this catalysts treatment process. As shown, initial measurements using SEM photos of surfaces before and after treatment are uses to identify the nature of the surface changes. The case shown is for an aluminum foil. In addition to SEM scans, changes in the catalysts reactivity will be measured.

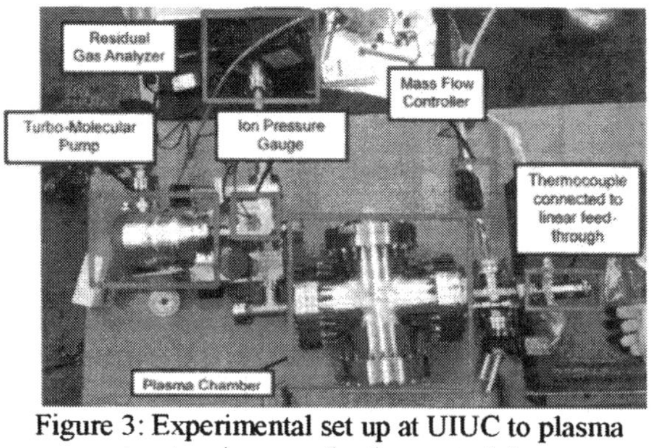

Figure 3: Experimental set up at UIUC to plasma treat various conducting materials.

Figure 3 shows a setup at the University of Illinois at Urbana-Champaign used to treat various conducting materials. The setup is a DC discharge similar to the one shown in figure 1. Various diagnostics are connected to the chamber such as thermocouples and a RGA. The pressure in the system can be very low by using the turbo-molecular pump; this low pressure helps reduce contamination and allows better process control.

4 CONCLUSION

This plasma treatment technique can also be applied to other electrically insulating catalysts through the use of other types of plasma discharges (e.g. AC discharges). The ability to increase the performance of catalysts can also have a profound impact on numerous industries such as energy and pharmaceuticals just to name a few. Catalysts are ubiquitous in chemistry and enhancing the activity of catalyst through simple plasma treatment can ripple through the commercial sector ultimately lowering cost of products and services for the public. This approach can also allow the introduction of previous cost prohibitive technologies (e.g. fuel cell vehicles) to the market. According to FRPT Research the global market value for catalyst will be worth over $20-billion by 2021 [21]. Even small improvements can result in large financial savings.

REFERENCES

[1] G. H. Miley, S. N. Srivastava and G. Chen, "Catalytic Performance Enhancement by Low Temperature Plasma Treatment," in *TechConnect World 2015*, Washington, DC, 2015.

[2] M. P. de Almeida and S. A. C. Carabineiro, "The Best of Two Worlds from the Gold Catalysis Universe Making Homogeneous Heterogeneous," *CHEMCATCHEM*, vol. 4, pp. 18-29, 2012.

[3] M. Armbruster, K. Kovnir, M. Friedrich, D. Teschner, G. Wowsnick, M. Hahne, P. Gille, L. Szentmiklosi, M. Feuerbacher, M. Heggen, F. Girgsdies, D. Rosenthal, R. Schlogl and Y. Grin, "Al13Fe4 as a low-cost alternative for palladium in heterogeneous hydrogenation," *Nature Materials*, vol. 11, pp. 690-693, 2012.

[4] S. He, C. Li, H. Chen, D. Su, B. Zhang, X. Cao, B. Wang, M. Wei, D. G. Evans and X. Duan, "A Surface Defect-Promoted Ni Nanocatalyst with Simultaneously Enhanced Activity and Stability," *Chemistry of Materials*, vol. 25, pp. 1040-1046, 2013.

[5] M. Boudart, "Heterogeneous Catalysis By Metals," *Journal of Molecular Catalysis*, vol. 30, pp. 27-38, 1985.

[6] T. Zambelli, J. Wintterlin, J. Trost and G. Ertl, "Identification of the "Active Sites" of a Surface-Catalyzed Reaction," *Science*, vol. 273, pp. 1688-1690, 1996.

[7] Z.-P. Liu and P. Hu, "General Rules for Predicting Where a Catalytic Reaction Should Occur on Metal Surfaces: A Density Functional Theory Study of C-H and C-O Bond Breaking/Making on Flat, Stepped, And Kinked Metal Surfaces," *Journal of the American Chemical Society*, vol. 125, pp. 1958-1967, 2003.

[8] R. T. Vang, K. Honkala, S. Dahl, E. K. Vestergaard, J. Schnadt, E. Laegsgaard, B. S. Clasusen, J. K. Norskov and F. Besenbacher, "Controlling the catalytic bond-breaking selectivity of Ni surfaces by step blocking," *Nature Materials*, vol. 4, pp. 160-162, 2005.

[9] U. Heiz, F. Vanolli, A. Sanchez and W. D. Schneider, "Size Dependent Molecular Dissociation on Mass-Selected, Supported Metal Clusters," *Journal of the American Chemical Society*, vol. 120, pp. 9668-9671, 1998.

[10] A. Sanchez, S. Abbet, U. Heiz, W. D. Schneider, H. Hakkinen, R. N. Barnett and U. Landman, "When Gold Is Not Noble: Nanoscale Gold Catalysts," *Journal of Physical Chemistry A*, vol. 103, pp. 9573-9578, 1999.

[11] C.-j. Liu, G. P. Vissokov and B. W.-L. Jang, "Catalyst preparation using plasma technologies," *Catalysis Today*, vol. 72, pp. 173-184, 2002.

[12] Y. Zayachuk, A. Manhard, M. H. J. 't Hoen, W. Jacob, P. A. Zeijlmans can Emmichoven and G. van Oost, "Depth profiling of the modification induced by high-flux deuterium plasma in tungsten and tungsten-tantalum alloys," *Nuclear Fusion*, vol. 54, no. 123013, 2014.

[13] H. Y. Xu, G. N. Luo, H. Schut, Y. Yuan, B. Q. Fu, A. Godfrey, W. Liu and G. D. Temmerman, "Enhanced modification of tungsten surface by nanostructure formation during high flux deuterium plasma exposure," *Journal of Nuclear Materials*, vol. 447, pp. 22-27, 2014.

[14] M. Laurent-Brocq, N. Job, D. Eskenazi and J.-J. Pireaux, "Pt/C catalyst for PEM fuel cells: Control of Pt nanoparticles characteristics through a novel plasma deposition method," *Applied Catalysis B: Environmental*, vol. 147, pp. 453-463, 2014.

[15] S. M. M. Coelho, F. D. Auret, P. J. Janse van Rensburg and J. M. Nel, "Unexpected properties of the inductively coupled plasma induced defect in germanium," *Physica B,* vol. 439, pp. 97-100, 2014.

[16] O. V. Ogorodnikova, J. Roth and M. Mayer, "Deuterium retention in tungsten in dependence of the surface conditions," *Journal of Nuclear Materials,* Vols. 313-316, pp. 469-477, 2003.

[17] T. Takimoto, K. Takeshita, S. Nakamura and T. Okumura, "Effects of plasma-induced defects on electrical characteristics of AlGaN/BaN heterostructure before and after low-temperature annealing," *Thin Solid Films,* vol. 557, pp. 212-215, 2014.

[18] R. C. Egeberg, S. Ullmann, I. Alstrup, C. B. Mullins and I. Chorkendorff, "Dissociation of CH4 on Ni(1 1 1) and Ru(0 0 0 1)," *Surface Science,* vol. 497, pp. 183-193, 2002.

[19] H. Uetsuka, K. Watanabe, H. Kimpara and K. Kunimori, "Structure Sensitivity in the Kinetics and the Dynamics of CO Oxidation over Stepped Pd(335) Studied by the Molecular Beam Infrared Chemiluminescence Technique: Determination of Working Sites during the Steady-State Reaction," *Langmuir,* vol. 15, pp. 5795-5799, 1999.

[20] B. Hammer, "Bond Activation at Monatomic Steps: NO Dissociation at Corrugated Ru(0001)," *Physical Review Letters,* vol. 83, no. 18, pp. 3681-3684, 1999.

[21] FRPT Research, "Global market for catalysts to be worth over $20-bn by 2021: Report," FRPT Research, 2015.

Effect of Big-Cube Zr_2Ni on the Absorption and Desorption Kinetics of Nanocrystalline MgH_2 Powders

M. Sherif El-Eskandarany, H. Al-Matrouk, Ahmed Al-Duweesh and Ehab Shaban

Nanotechnology and Advanced Materials Program,
Energy and Building Research Center, Kuwait Institute for Scientific Research
Safat 13109, Kuwait - State of Kuwait, msherif@kisr.edu.kw

ABSTRACT

A new hydrogen storage system of nanocomposite MgH_2/10 wt.% big-cube Zr_2Ni nanocrystalline powders were synthesized. A β-MgH_2 phase coexisted with small mole fractions of γ-MgH_2 nanocrystalline ultrafine powders were obtained after reactive ball milling (RBM) technique of pure Mg powders of metallic Mg powders under a pressure of 50 bar of hydrogen gas. Mechanical alloying technique, using a high-energy ball milling was employed to fabricate big-cube phase of Zr_2Ni nanocrystalline powders starting from tetragonal-Zr_2Ni alloy prepared by arc melting technique. The hydrogenation/dehydrogenation properties of nanocomposite MgH_2/10 big-cube Zr_2Ni powders possessed fast sorption and desorption kinetics at 250 °C. The system absorbed 4.9 wt.% H_2 within 92 s, whereas it released the absorbed hydrogen within 320 s. At 250 °C, the synthesized nanocomposite system in the present study showed extraordinary high absorption/desorption cycles for any MgH_2 system, as far as the authors know, which reached to 2546 complete cycles covered in 1240 h.

Keywords: reactive ball milling, mechanical alloying, metal hydrides, Zr-based metastable alloys, gas sorption/desorption kinetics, cycle-life-time

1 INTRODUCTION

However, the hydrogenation/dehydrogenation processes (absorption/desorption) kinetics of MgH_2 is very slow, Mg metal with its low cost and high capability to store a high hydrogen content (7.60 wt.%) is considered to be the most promising materials for solid-state hydrogen storage in light vehicles [1]. The absorption and desorption kinetics of MgH_2 powders prepared by RBM process [2] can be improved by subjecting the powders to long-term of milling [3]. Further improvement of can be also achieved by mechanical alloying of MgH_2 with 3d transition metals [4] and their oxides [5].

Although there are many reported efforts performed to improve the hydrogenation/dehydrogenation properties of MgH_2 system [6], there is a still space for new attempts that aims to find a suitable catalytic system used to improve the absorption and desorption kinetics of this promising metal hydride system. The present study was tackled in order to investigate the effect of a new nanocatalyst system (big-Zr_2Ni) used for the first time on improving the hydrogenation and dehydrogenation kinetics of MgH_2 system synthesized by RBM technique.

2 EXPERIMENTAL PROCEDURE

Hydrogen gas (99.999 %), elemental Mg powders (~ 80 μm, 99.8 %), bulk Zr (99.7%) and bulk Ni metals were used in the present study. Small amount of the Mg powders (5 g) was balanced inside a helium (He) gas atmosphere (99.99%) - glove box and then were sealed together with twenty five hardened steel balls into a hardened steel vial (220 ml in volume), using a gas-temperature-monitoring system (GST). The ball-to-powder weight ratio was 40:1. The vial was then evacuated to the level of 10^{-3} bar before introducing H_2 gas to fill the vial with a pressure of 50 bar. The milling process was carried out at room temperature using high energy ball mill operated at a rotation speed of 250 rpm. After 25 h of RBM the powders were discharged from the vial inside the glove box and then mixed with 10 wt.% of previously prepared big-cube-Zr_2Ni nanocrystalline powders [7]. The mixed powders were then ball milled for 50 h, using a high energy ball mill. The structure, morphology, elemental analysis, thermal stability, composition, and the hydrogenation/dehydrogenation kinetics of the all samples obtained after different ball milling time were invistigated. X-ray diffraction (XRD) with CuKα radiation, 200 kV-field emission high resolution transmission electron microscopy/scanning transmission electron microscopy (FE-HRTEM/STEM), equipped with energy-dispersive X-ray spectroscopy (EDS), 15 kV-field emission scanning electron microscope (FE-SEM/EDS), differential scanning calorimetery (DSC), and Sievert's approach were employed to characterize and investigate the different properties of the ball milled powders.

3 RESULTS AND DISCUSSIONS

Figure 1 The XRD patterns of the staring reactant materials of Mg powders is shown in Fig. 1(a). At this initial stage of RBM time, the powders were polycrystalline hcp-Mg phase with large grains, suggested by the sharp diffraction Bragg's peaks shown in Fig. 1(a). After 1 h of the reactive ball milling (RBM) time, a small mole fraction of a new phase could be detected, as suggested by the low intensity Bragg-peaks denoted by squared symbols in Fig. 1(b). Comparing the Bragg-peak positions of this new phase with those reported for tetragonal β-MgH_2 confirmed the formation of β-MgH_2 phase after this early stage of RBM. After 3 h of RBM time, the mole fraction of the existed β-MgH_2 phase in the milled powders was increased, suggested by the appearance of pronounced and sharp Bragg peaks corresponding to the β-MgH_2 phase, as presented in Fig. 1(c). Whereas, the Bragg's peaks corresponding to the β-MgH_2 phase become very sharp and more pronounced after 6 h of the RBM time (Fig. 1(d)), the Bragg's lines for non-reacted Mg crystals can hardly be seen, implying a continuous increasing in the mole fractions of the formed hydride phase against the unreacted Mg powders, respectively. After 25 h of the RBM time, nanocrystalline β-MgH_2 coexisting with small mole fraction of γ-MgH_2 phases were obtained, suggested by the obvious broadening in the Bragg's lines for both phases, as shown in Fig. 1(e). Low-intensity peaks (less than 40 cps) of cubic-MgO phase, which has a lattice constant of a_o=0.4216 nm, were detected in sample. This oxide phase appeared as a result of surface oxidation of the fine MgH_2 powders during the sample preparations for XRD.

The image of the high resolution transmission electron microscope (HRTEM) for the powders obtained after 25 h of RBM time is presented in Fig. 2. The formation of ultrafine grains, ranging from 4 to 8 nm, of β-MgH_2 and γ-MgH_2 phases can be obviously seen, as denoted by the Miller indexed zones (Fig. 2).

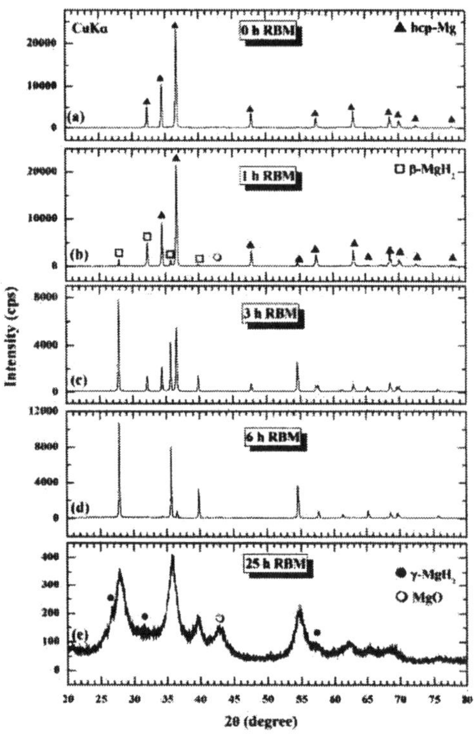

Figure 1: XRD patterns of Mg powders after RBM time for (a) 0 h, (b) 1 h, (c) 3 h, (d) 6 h and (e) 25 h.

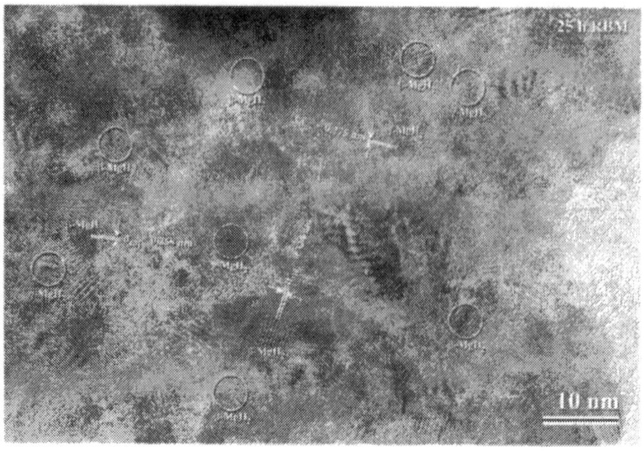

Figure 2: HRTEM image of Mg powders obtained after RBM time for 25 h.

The thermal stability of synthesized MgH_2 powders after different RMB time was investigated by DSC analysis at a constant heating rate of 20 °C/min. The DSC thermograms of the powders obtained after selected RBM times are shown in Fig. 3. The decomposition event for the sample obtained after 6 h of RBM time took place through a single-endothermic peak (Fig. 3(a)), suggesting the existence of a single phase of β-MgH_2. The onset and peak decomposition temperatures were 438 °C and 459 °C, respectively. These temperatures tended to be shifted to the low temperature side after 12.5 h of the RBM time (Fig. 3(b)) and found to be 431 °C (onset temperature) and 451 °C (peak temperature), suggesting a destabilization on the β-MgH_2 phase.

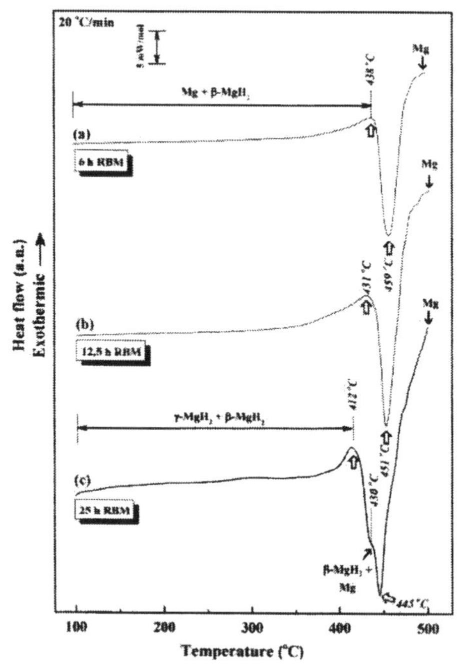

Figure 3: DSC curves of Mg powders obtained after RBM time for (a) 6 h, (b) 12.5 h and (c) 25 h.

The DSC trace for the sample, which was obtained after 25 h of the RBM time revealed a shoulder-like endothermic peak, suggesting two-decomposition stages, as shown in Fig. 3(c). The first decomposition event refers to the decomposition of the γ-MgH_2 into β-MgH_2 and Mg powders, as confirmed by the XRD analysis (not shown here). The first endothermic peak, which characterized by an onset temperature of 412 °C, vanished at an end-set temperature of about 430 °C, where the second decomposition peak (β-MgH_2 $\rightarrow$Mg) starts. The peak temperature of the second decomposition event is recorded and found to be 445 °C, suggesting a continuous destabilization of the formed hydride phase, as shown in Fig. 3(c).

According to the poor hydrogenation/dehydrogenation kinetics of as-synthesized MgH_2 powders and in part to propose a new nanocatalyst system, we prepared big-cube Zr_2Ni nanopowders [7] for using as catalysts to improve the kinetic of absorption and desorption of MgH_2 powders and to maximize its cycle-life time. The as-synthesized MgH_2 powders obtained after 25 h of the RBM were mixed with 10 wt.% of the big-cube and then ball milled under hydrogen gas pressure for 50 h. The XRD patterns of the big-cube phase obtained after 150 h of ball milling is shown in Fig. 4(a). Obviously, very broad Bragg peaks were appeared, suggesting the formation of nanocrystalline phase. The analysis of the diffracted lines presented in Fig. 4(a) suggests that the powders at this stage of ball milling reveal polycrystalline structure corresponding to Ti_2Ni [8] structure ($E9_3$ structure, space group Fd3m).

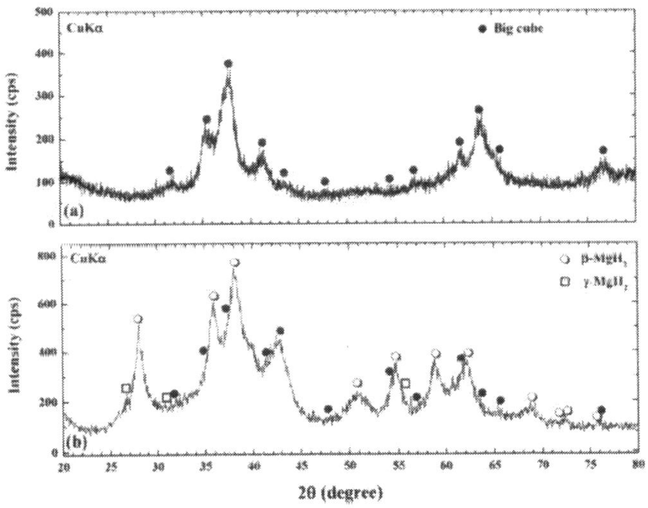

Figure 4: (a) XRD patterns of big-cube Zr_2Ni nanocrystalline powders obtained after 150 h of ball milling time. The XRD pattern of as-prepared MgH_2 powders and then ball milled with the big cube phase for 50 h under a hydrogen gas atmosphere is shown in (b).

The STEM-BFI and of the nanocomposite MgH_2 doped with 10 wt.% Zr_2Ni obtained after 50 h of ball milling is shown in Fig. 5. The light gray matrix shown in the figure is for MgH_2 matrix whereas the spherical dispersoids nanoparticles (~3 – 11 nm in diameter) embedded into the matrix are nano-cells big cube phase.

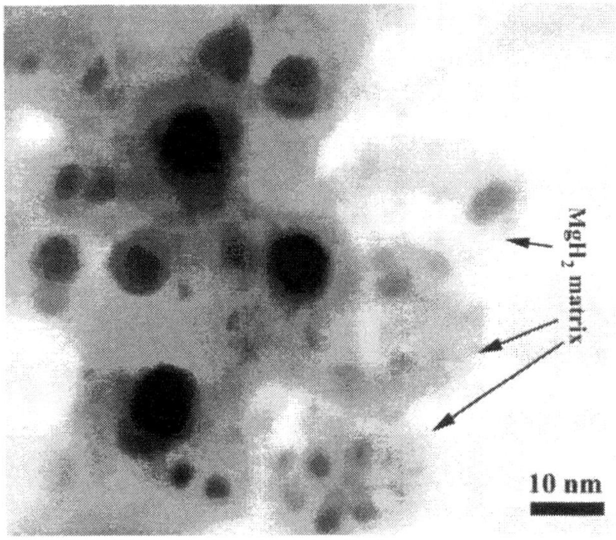

Figure 5: STEM-BFI and of MgH_2/10 Zr_2Ni nanocomposite powders obtained after 50 h of ball milling time.

The DSC analysis was performed in order to investigate the effect of big-cube Zr_2Ni additive on the thermal stability of MgH_2 powders. The DSC trace of the nanocomposite powders obtained after 50 h of the ball milling time is displayed in Fig. 6(a). There are two opposite reactions shown in the micrograph. The first reaction is an endothermic and refers to the decomposition of β-MgH_2 phase into hcp-Mg that taken place at an onset temperature of 288 °C, as shown in Fig. 6(a). Comparing this value with that one for pure MgH_2 powders (431 °C; Fig. 3(c)) one can be noted that adding the big-cube phase leads to decrease the decomposition temperature of β-MgH_2 phase by a value of 143 °C and this indicates a crucial effect of the big-cube phase on destabilizing of the MgH_2 phase. The XRD pattern of the sample heated up to 324 °C (just above the endothermic reaction) reveals Bragg-peaks related to hcp-Mg and big-cube Zr_2Ni phase, without any evidence for the formation of any other phases, as displayed in Fig. 6(b).

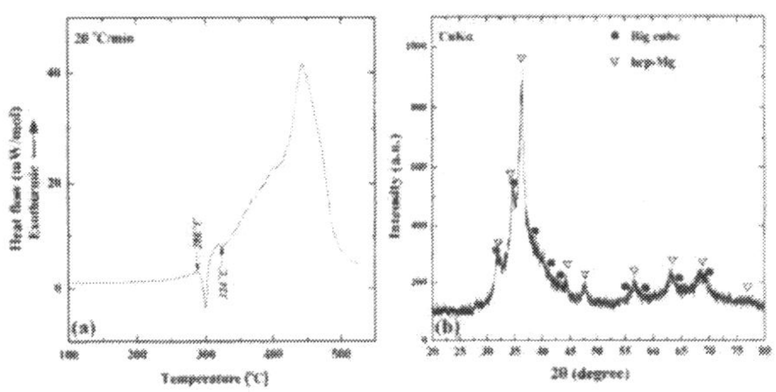

Figure 6: (a) DSC curves of MgH_2 powders obtained after RBM for 25 h and then ball milled with 10 wt.% of big-cube Zr_2Ni for 50 h. The XRD pattern of a sample heated up to 324 °C is shown in (b).

The second reaction displayed in Fig. 5(a), which took place at a rather high temperature of about 440 °C (peak temperature), is an exothermic reaction and refers to a big cube→fcc phase transformation, as was suggested by El-Eskandarany and Inoue [9].

In order to understand the catalytic effect of big-cube phase on improving the hydrogenation/dehydrogenation properties of nanocomposite MgH_2, the absorption (Fig. 7(a)) and desorption (Fig. 7(b)) kinetics achieved at 250 °C were investigated. The sample absorbed about 4.7 wt.% H_2 within 80 s and reached to saturated at about 4.9 wt.% H_2 after 92 s, as displayed in Fig. 7(a). Then, the nanocomposite powders desorbed about 4 wt.% H_2 within 150 s and completely released about 4.9 wt.% H_2 after 320 s (Fig. 7(b)). Comparing these results with those obtained for pure MgH_2 system (see for example [3]) one can conclude that the synthesized nanocomposite MgH_2/10 big-cube Zr_2Ni system enjoys fast absorption and desorption kinetics, suggested by the superior role of big-cube nanocatalysts on improving the hydrogenation/ dehydrogenation behavior of MgH_2 system.

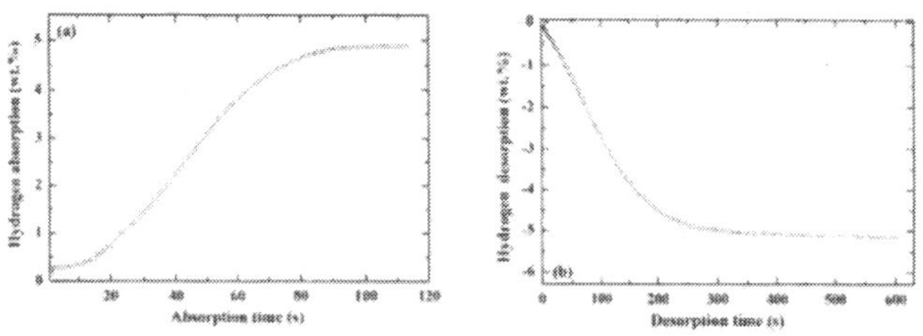

Figure 7: (a) absorption and (b) desorption kinetics developed at 250°C for MgH_2 powders doped with big cube Zr_2Ni and then ball milled for 50 h.

The hydrogen chargeability and cyclic stability of nanocomposite MgH_2/10big-cube Zr_2Ni system, was carried out at 250°C. Figure 8 displays the cyclic-life-time curves of the examined nanocomposite powders prepared in the present study. The powders possessed extraordinary high absorption/desorption cycles (2546 complete cycles) covered within 1240 h, as elucidated in Fig. 8. No remarkable degradation in the hydrogen storage capacity could be observed during the first 250 h where the sample maintained its hydrogen chargeability at a constant value of 5 wt.%. A slight monotonical decreasing in the hydrogen storage capacity (about 0.5 wt.%) can be seen when the sample was subjected to further cycle-life-time (above 250 h). This slight degradation occurred due to the agglomeration of the powder particles and the grain growth happened in the big-cube and Mg grains, as was checked by SEM and TEM analysis (not shown here).

4 CONCLUSIONS

Reactive ball milling technique was employed to fabricate nanocrystalline MgH_2 powders, using high energy

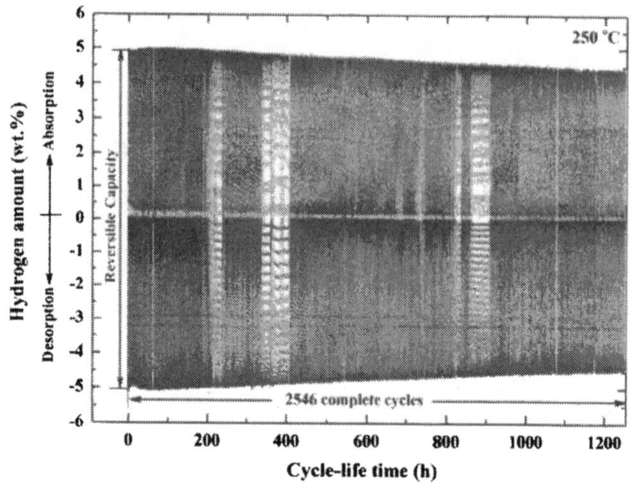

Figure 7: (a) absorption and (b) desorption kinetics measured at 250°C for for MgH_2 powders doped with big cube Zr_2Ni and then ball milled for 50 h.

ball milling technique. The synthesized powders were mechanically ball milled for 50 h together with 10 wt.% of synthesized big-cube Zr_2Ni nanocrystalline powders to form uniform nanocomposite powders. This new nanocomposite binary system showed fast hydrogenation and dehydrogenation kinetics at 250°C. Moreover, the synthesized nanocomposite system possessed extraordinary high absorption/desorption cycles of 2546 cycles with a measured cycle-life-time of 1240.

REFERENCES

[1] M Sherif El-Eskandarany, "Mechanical Alloying for Nanotechnology, Materials Science and Powder Metallurgy," 2nd ed., Elsevier Publishing, Oxford-UK, 2015, in press.

[2] M. Sherif El-Eskandarany, K. Sumiyama, K. Aoki, and K. Suzuki, Mater. Sci. Forum, 88, 801-808, 1992.

[3] M. Sherif El-Eskandarany, Ehab Shaban, and Badryiah Al-Halaili, Int. J. Hydrogen Energy, 39, 12727-12740, 2014.

[4] G. Liang, J. Huot, S. Boily, Neste A. Van and R. Schulz, J. Alloys Compd., 292, 247–452, 1999.

[5] T. Sadhasivama, M. Sterlin Leo Hudson, K. Sunita Pandey, Ashish Bhatnagar, K. Singh Milind, K. Gurunathan, and O.N. Srivastav, J. Hydrogen Energy, 38, 7353-7362, 2013.

[6] J. Huot, D.B. Ravnsbæk, J. Zhang, F. Cuevas, M. Latroche, T.R. Jensen, Prog Mater Sci, 58, 30-75, 2013.

[7] M. Sherif El-Eskandarany, Intermetallics, 61, 2015, in press.

[8] Z. Altounian, E. Batalla, J.O. Strom-Olsen, and W.L. J. Walter, J. Appl. Phys., 61, 149-154, 1987.

[9] M. Sherif El-Eskandarany, A. Inoue, Physical Review B, 75, 224109 – 1 to 224109-9, 2007.

Cellulose Fiber Immobilized Silver Nanoparticles and Their Applications in Catalysis, SERS and Antimicrobial Activity

B. Baruah*

*Kennesaw State University, Department of Chemistry and Biochemistry, Science Laboratory Building, Room SL4015, 105 Marietta Drive NW, Kennesaw, GA 30144, USA, bbaruah@kennesaw.edu

ABSTRACT

This work involves *in-situ* preparation of silver nanoparticles on nonwoven fibers in baby wipes (a mixture of synthetic fiber and cellulose fiber) by impregnating Ag^+ ions. The Ag^+ ions bind to the hydroxyl groups in the cellulose fibers and subsequently reduced to silver nanoparticles (AgNPs) with ascorbic acid. The hydroxyl groups in the cellulose fibers provided the affinity for the immobilization of Ag^+ ions first and then nanoparticles after reduction. The structure and properties of the composite wipes were verified by scanning electron microscopy (SEM), energy dispersive X-ray analysis (EDX), and Fourier transformed infrared spectroscopy (FTIR). We have tested this AgNP@wipe composite for catalytic activity, surface enhanced Raman scattering (SERS) and antimicrobial activity. The composite exhibited a very good catalytic activity for the reduction of 4-nitrophenol to 4-aminophenol in presence of sodium borohydride, significant SERS activity for Raman active analyte molecules.

Keywords: Cellulose fiber, silver nanoparticles, catalysis, SERS, and SEM

1 INTRODUCTION

Silver nanoparticles (AgNPs) uniquely respond to incident light's electromagnetic field due to the presence and collective oscillation of the free conduction band electrons [1]. Such interaction give rise to localized surface plasmon resonance (LSPR) [1]. The LSPR in AgNPs induces light absorption and near-field enhancement of incident electromagnetic fields by several order of magnitude [2]. These properties are primarily dependent on the size and shape of AgNPs [3] and this draws special attention to AgNPs as size and shape dictate applications in catalysis [4], SERS [5], sensors [6] and antimicrobial applications [7]. Silver is highly active metal for SERS applications compared to gold. However, silver gets oxidized under ambient conditions and colloidal stability of AgNPs is not satisfactory, it is not ideal to develop solution dispersed substrates with silver [8]. In order to minimize these drawbacks efforts have been made to fabricate AgNPs in to various composite materials by coating with titania [9] and silica [10]. However, for SERS application sample preparation is still a challenge with these AgNPs containing composite materials. Similarly, AgNPs has been gaining enormous interest in heterogeneous catalysis [11]. However, such catalysts, where solution dispersed AgNPs are used, have been facing some real challenges as there is no way to separate the products and catalyst from the reaction mixture using traditional methods [12]. This reduces the reusability of the catalyst. Among some of the recently adopted strategies is the immobilization of metal nanoparticles on to solid supports, such as polymer [13] or inorganic microspheres [12]. The type of support or substrate and immobilization of AgNPs play critical role in determining the efficiency as heterogeneous catalyst [14] and SERS substrates.

In the current study we have immobilized AgNPs on cellulose based fibers by *in-situ* synthesis. We tested the SERS activity of AgNP@wipe composite utilizing a Raman active food contaminant, melamine and tested for the catalytic activity and recyclability with the reduction reaction of 4-nitrophenol to 4-aminophenol in presence of sodium borohydride. AgNP@wipe composites were characterized by FTIR, SEM, and EDX line analysis and tested for SERS activity with Raman spectroscopy and catalytic activity with UV-visible spectroscopy. It was also found that the fibers dipped in higher concentration of precursor $AgNO_3$ solution would show higher activity in SERS detection of analyte. The AgNP@wipe conjugates show very promising recyclability in the catalytic reaction.

2 EXPERIMENTAL

2.1 Materials

$AgNO_3$ (99%) and sodium borohydride (99%) were purchased from Sigma-Aldrich; ascorbic acid (98%), 4-nitrophenol (98%) and melamine (98.0%) were purchased from Across Organic. All chemicals and solvents were used without further purification. Kroger brand baby wipes are used as a source of nonwoven fiber. The DI water in all experiments was Milli-Q water (18 MΩ cm, Millipore).

2.2 *In-situ* Synthesis of AgNPs on Nonwoven Fibers

Dried baby wipes were used as source of cellulose based nonwoven fiber. As shown in Scheme 1, such wipes were first thoroughly washed with ethanol in order to get rid of hydrophobic chemicals (in baby wipes) and then dried in the oven at 60°C. These were further pre-treated with 10% HCl followed by 10% NaOH and then dried after rinsing with excess water. Such nonwoven fibers contain hydroxyl groups and that allow the Ag^+ ions to bind to the fiber [15]. Sixteen pieces of dry pre-treated wipes (3.5 cm by 3.5 cm) were immersed in 50 mL of $AgNO_3$ solutions with $AgNO_3$ concentration of 10 mM or 25 mM or 50 mM for 12 hours. Silver ions were impregnated into the cellulose fibers through the hydroxyl groups. Then the wipes were rinsed with ethanol to remove un-bound Ag^+ ions. After reduction in aqueous ascorbic acid (40 mM) solution for 15 min and rinsing with DI water, the original colorless wipe turned dark yellow (shown in Scheme 1)

2.3 UV-visible and FTIR Studies of AgNPs

The absorption spectra of 4-nitrophenol reduction were recorded using a Cary 50 UV-visible spectrophotometer equipped with a dip probe. FTIR spectroscopy was performed using Perkin-Elmer FTIR Spectra 100 spectrometer fitted with diamond ATR. The spectra were further processed and plotted with Origin 8.0 software.

2.4 Raman Spectroscopic Measurement

One piece of dried AgNP@wipe composite was immersed in 1.0 mM aqueous solution of melamine for 30 min to form an adsorbed layer on the metal substrate. After rinsing off the un-adsorbed melamine molecules with water SERS spectra was measured. The analyte adsorbed AgNP@wipe composite was stuffed in a glass sample vials for the SERS measurement. Raman spectroscopic measurements are carried out on a DeltaNu Advantage 200A Raman spectrometer equipped with a HeNe laser set at 632.8 nm with an integration time of 5.0 s.

2.5 SEM and EDX analysis

A SEM was used for the analysis of empty wipe and AgNP@wipe composite. The SEM was equipped with an EDX spectrometer. The SEM/EDX system was operated in low-vacuum mode at 20.0 kV, back-scattered detector was used for imaging at a working distance of 10-12 mm. The EDX has software with a database of reference spectra for elemental analysis, compositional nanoanalysis, and mapping. A small piece of AgNP@wipe composite or empty wipe was fixed on a aluminum stage and introduced into the chamber. For both measurements a Hitachi HD 2000 Scanning Transmission Electron Microscope (STEM) was used with Field Emission source of resolution 0.24 nm at 200 kV.

3 RESULTS AND DISCUSSION

In this study we report *in-situ* synthesis of AgNPs on baby wipes. In a typical synthesis pre-treated baby wipes (source of nonwoven fiber) were socked in an aqueous solution of silver nitrate of known concentration. Later wipes were rinsed with ethanol to wash off the unbound Ag^+ ions followed by reduction of the bound Ag^+ ions into AgNPs with ascorbic acids. The color of the wipes turned yellowish-brown form white after formation of the AgNPs. AgNP@wipe composite formation route is shown in Scheme 1.

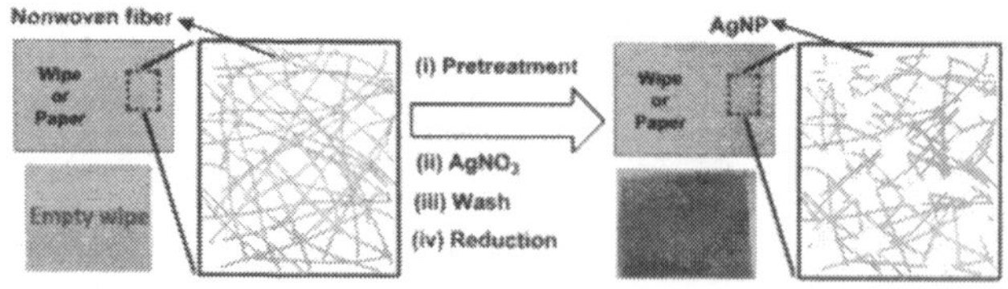

Scheme 1: Schematic representation of *in-situ* synthesis of AgNPs on nonwoven fiber. Digital photograph of empty wipe and AgNP@wipe composites are also shown.

3.1 Characterization of the AgNP@wipe composites by FTIR Spectroscopy

FTIR spectroscopic data suggests the presence of AgNPs bound to the nonwoven fibers. We plot the FTIR spectra of empty wipe, AgNP@wipe-1, AgNP@wipe-2 and AgNP@wipe-3. For the later three precursor $AgNO_3$ concentrations were 10.0 mM, 25.0 mM and 50.0 mM, respectively, during the synthesis. All the spectra were shown in Figure 1. The empty wipe gives signals at 3304, 2917/2848, 1466 and 1012 cm^{-1} that are assigned to O-H stretch, C-H stretch, asymmetric C-H bending and O-H bending. Similar signals in AgNP@wipe-1, AgNP@wipe-2 and AgNP@wipe-3 are assigned at 3320, 2917/2848, 1466 and 1020 cm^{-1}. Clearly the C-H stretch and bend are not affected by AgNPs immobilization on the wipe. However, the signals for both O-H stretch and bend are shifted 16 and 8 cm^{-1}, respectively due to direct interaction of the O-H group with silver metal. The characteristic weak signals at 1315-1375 cm^{-1} were associated with $-CH_2-$ symmetric bonding and wagging vibrations and signals at 1160-1204 cm^{-1} were associated with C-O-C bond stretching.

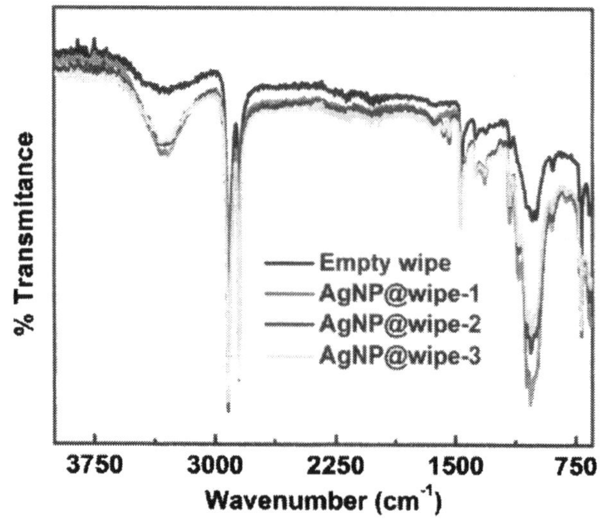

Figure 1: FTIR spectra of empty wipe (black line), AgNP@wipe-1 (red line), AgNP@wipe-2 (blue line) and AgNP@wipe-3 (green line).

3.2 SEM and EDX Analysis of AgNP@wipe Composites

Typical SEM images and digital photographs of wipes fabricated with different amount of silver are presented in Figure 2. It shows that the nonwoven fibers exhibited a porous microfibrous structure with a smooth surface regardless of the $AgNO_3$ concentrations, similar to that of the empty wipe's microfibers. From both digital photographs and SEM images it is clear that the *in-situ* synthesis of AgNPs was successful and AgNPs immobilized onto the surface of fibrous material (Figure 2). The elemental silver signals in the EDX spectra (Figure 3) demonstrated that AgNPs were immobilized onto cellulose microfibers. The presence of oxygen and carbon in the EDX spectra was attributed to cellulose. The absorption peak of 3.0 keV in AgNP@wipe samples clearly indicate that AgNPs were indeed deposited on the wipe.

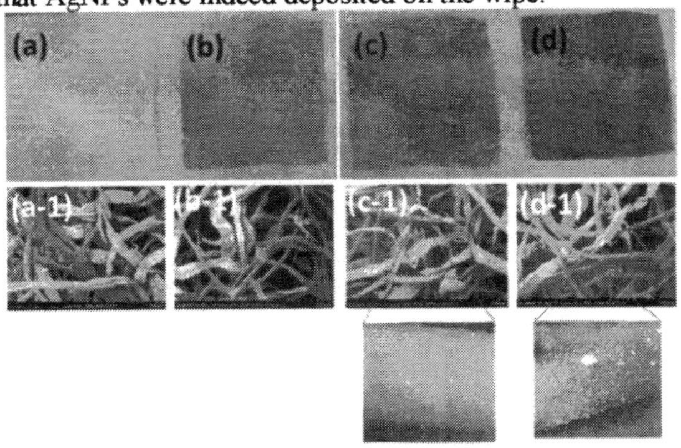

Figure 2: Digital photographs of empty wipe (a), AgNP@wipe-1 (b), AgNP@wipe-2 (c) and AgNP@wipe-3 (d) are shown. Corresponding SEM images of empty wipe (a-1), AgNP@wipe-1 (b-1), AgNP@wipe-2 (c-1) and AgNP@wipe-3 (d-1) are also shown (scale bar 200 µm). Images were taken with a Hitachi HD-2000 STEM at an operating voltage of 15-20 kV.

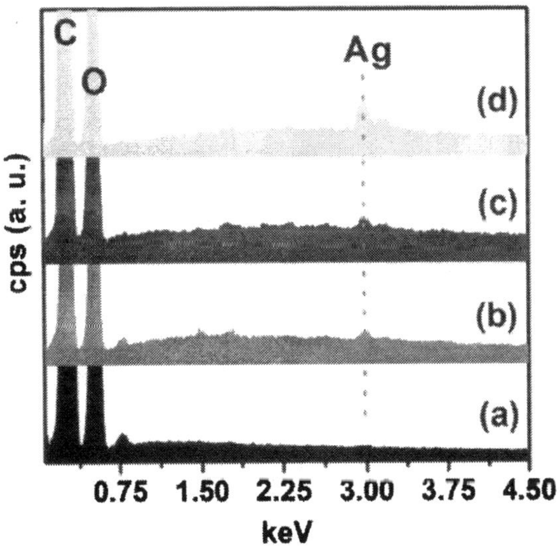

Figure 3: The EDX spectra of empty wipe (a), AgNP@wipe-1 (b), AgNP@wipe-2 (c) and AgNP@wipe-3 (d) are shown in here and confirms the presence of silver in the AgNP@wipe. Analysis was done using a Hitachi HD-2000 STEM at an operating voltage of 20 kV.

3.3 AgNPs@wipe as SERS Substrate

We studied the Raman spectrum of solid melamine and SERS spectra of melamine adsorbed in empty wipe and AgNP@wipe composites from a 1.0 mM stock solution. Figure 4 shows the Raman spectrum of solid melamine and melamine spectra while adsorbed on wipes. Melamine has two characteristic trigonal ring breathing modes at 678 cm^{-1} (>C-NH_2) and 985 cm^{-1} (>C-N=C<). Melamine upon adsorbed on empty wipe do not show any signal in the Raman spectrum. Whereas, when adsorbed on to the wipes containing immobilized AgNPs a significantly enhanced signal appear at 687 cm^{-1}. The signal intensity of the peak at 687 cm^{-1} increases in the following order AgNP@wipe-1 < AgNP@wipe-2 < AgNP@wipe-3. Clearly, wipes with higher AgNPs concentration, AgNP@wipe-3 was the one has highest intensity. It is also noticeable that in the SERS spectra only one of the two signals is showing up which is due to only >C-NH_2 ring breathing. The convincing explanation could be because the melamine molecules most probably bind the surface of the AgNPs present in wipe via the lone pair present in –NH_2 groups. This is also indicative of the fact that the peak position has been shifted by 9 cm^{-1} going from solid melamine to AgNP@wipe composite.

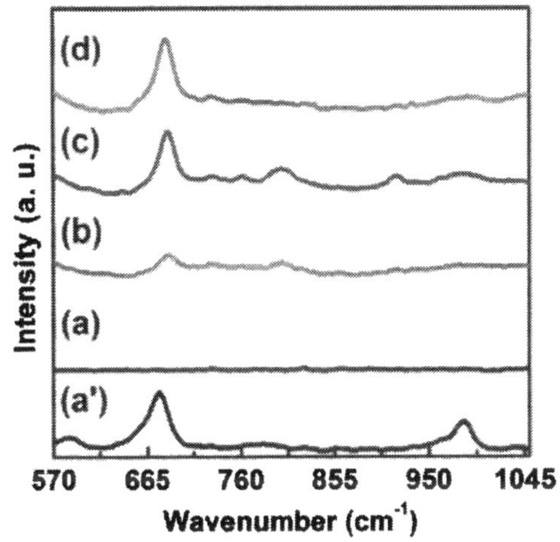

Figure 4: Raman spectra of solid melamine (a') and 1 mM melamine adsorbed in empty wipe (a) and SERS spectra of 1 mM melamine adsorbed in AgNP@wipe-1 (b), 1 mM melamine adsorbed in AgNP@wipe-2 (c) and 1 mM melamine adsorbed in AgNP@wipe-3 (d).

3.4 Catalytic Reduction of 4-Nitrophenol to 4-Aminophenol with AgNP@wipe

In the current study we report the reduction of 4-nitrophenol to 4-aminophenol as a model system [16] in order to evaluate the catalytic activity of AgNP@wipe composites. The reduction of 4-nitrophenol ($E^0_{(4\text{-NP/4-AP})}$ = −0.76 V) by sodium borohydride ($NaBH_4$) ($E^0_{(H_3BO_3/BH_4^-)}$ = −1.33 V) is thermodynamically feasible but kinetically restricted in the absence of a catalyst. Addition of $NaBH_4$ to a 4-nitrophenol solution changes the light yellow color of the solution to intense yellow due to formation of the 4-nitrophenolate ion due to the basic pH [16]. The catalytic reduction was monitored by UV–visible spectroscopy, as depicted in Figure 5A. In the presence of AgNP@wipe-3 and $NaBH_4$ the 4-nitrophenol was reduced, and the intensity of the absorption peak at 400 nm gradually decreased with time and after 7 min it disappeared completely (Figure 5A). In the meantime, a new absorption peak appeared at 298 nm and progressively increased in intensity (Figure 5A). This new peak is attributed to the typical absorption of 4-aminophenol. The UV–visible spectra in Figure 5A showed two isosbestic points at 279 and 314 nm. This result suggests that the catalytic reduction of 4-nitrophenol exclusively yielded 4-aminophenol, without any other side products [17].

In the reduction reaction, $NaBH_4$ concentration is significantly higher than that of 4-nitrophenol. Hence, the pseudo-first-order kinetics was applied in order to determine the rate constant, k. The absorbance of 4-nitrophenol is proportional to its concentration in solution; the absorbance at time t (A_t) and time $t = 0$ (A_0) are equivalent to the concentration at time t (C_t) and time $t = 0$ (C_0). The rate constant (k) was determined from the linear plot of $\ln(A_t/A_0)$ versus reduction time in seconds, and the constant was estimated to be 8.63×10^{-3} s^{-1} (Figure 5B).

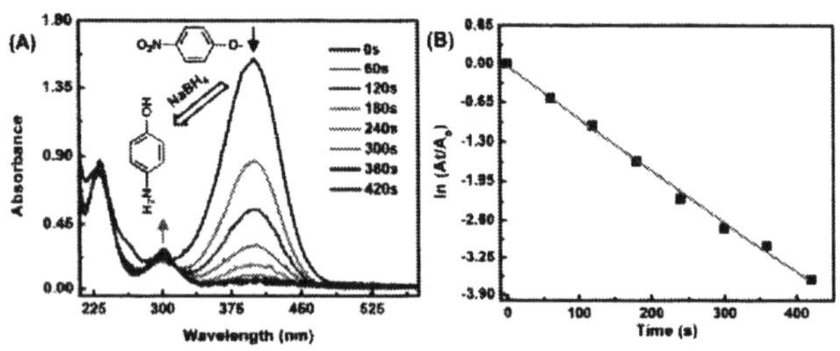

Figure 5: (A) Plot of 0.12 mM 4-nitrophenol with 12 mM $NaBH_4$ in the presence of AgNP@wipe-3 as catalyst and (B) plot of $\ln(A_t/A_0)$ against the reaction time for pseudo-first-order reduction kinetics of 4-nitrophenol in the presence of excess $NaBH_4$ in aqueous solutions.

4 CONCLUSIONS

In summary, *in-situ* synthesis of AgNPs have been reported in presence of nonwoven fibers present in baby wipes. Three different concentrations of silver nitrate ($AgNO_3$) was used as a source of silver ions and that leaded to three different AgNP@wipe composites. Comparison of SEM images of empty wipe and AgNP@wipe composites indicate absence and presence of AgNPs, respectively. EDX line analysis support the presence of Ag in AgNP@wipe composites. This report also shows that AgNP@wipe significantly enhances the Raman signal of melamine when adsorbed on the AgNP@wipe composites compared to the neat melamine. The catalytic activity of AgNP@wipe composites was studied for the reduction reaction of 4-nitrophenol to 4-aminophenol in presence of $NaBH_4$. The pseudo-first order rate constant for AgNP@wipe-3 was 8.63×10^{-3} s^{-1}. Further, antimicrobial studies of AgNP@wipe composites are currently in progress.

5 ACKNOWLEDGEMENTS

This work was supported by Kennesaw State University, Department of Chemistry and Biochemistry, KSU CSM Mentor Protégé fund (BBaruah-01-FY2015-01) awarded to Bharat Baruah. BB would like to thank Dr. Mark Mitchell, Department Chair; Dr. Mark Anderson, Dean of the College of Science and Mathematics for their enormous support. We acknowledge Advanced Materials Research Laboratories (AMRL) at Clemson University for STEM and EDX analysis.

REFERENCES

[1] B.J. Wiley, S.H. Im, Z.-Y. Li, J. McLellan, A. Siekkinen, and Y. Xia, J. Phys. Chem. B 110 (2006) 15666-15675.

[2] F. Jiang, and Y.-L. Hsieh, Biomacromolecules 15 (2014) 3608-3616.

[3] J.E. Millstone, S.J. Hurst, G.S. Métraux, J.I. Cutler, and C.A. Mirkin, Small 5 (2009) 646-664.

[4] Y. Fang, Y. Li, H. Xu, and M. Sun, Langmuir 26 (2010) 7737-7746.

[5] A.M. Michaels, M. Nirmal, and L.E. Brus, J. Am. Chem. Soc. 121 (1999) 9932-9939.

[6] A.D. McFarland, and R.P. Van Duyne, Nano Lett. 3 (2003) 1057-1062.

[7] J.R. Morones, J.L. Elechiguerra, A. Camacho, K. Holt, J.B. Kouri, J.T. Ramirez, and M.J. Yacaman, Nanotechnology 16 (2005) 2346.

[8] Y. Han, R. Lupitskyy, T.-M. Chou, C.M. Stafford, H. Du, and S. Sukhishvili, Anal. Chem. 83 (2011) 5873-5880.

[9] Y.-C. Liu, C.-C. Yu, and T.-C. Hsu, J. Phys. Chem. C 112 (2008) 16022-16027.

[10] M.S. Akin, M. Yilmaz, E. Babur, B. Ozdemir, H. Erdogan, U. Tamer, and G. Demirel, J. Mater. Chem. B 2 (2014) 4894-4900.

[11] B. Naik, S. Hazra, V.S. Prasad, and N.N. Ghosh, Catal. Commun. 12 (2011) 1104-1108.

[12] K. Shin, J.-Y. Choi, C. Park, H. Jang, and K. Kim, Catal. Lett. 133 (2009) 1-7.

[13] K. Kuroda, T. Ishida, and M. Haruta, J. Mol. Catal. A 298 (2009) 7-11.

[14] S. Besson, T. Gacoin, C. Ricolleau, and J.-P. Boilot, Chem. Commun. (2003) 360-361.

[15] J. He, T. Kunitake, and A. Nakao, Chem. Mater. 15 (2003) 4401-4406.

[16] J. Liu, G. Qin, P. Raveendran, and Y. Ikushima, Chem. Eur. J. 12 (2006) 2131-2138.

[17] Q. An, M. Yu, Y. Zhang, W. Ma, J. Guo, and C. Wang, J. Phys. Chem. C 116 (2012) 22432-22440.

Cavitation Disintegration of Microparticles and Nanoparticles in Dense Liquid Dispersions

Richard Dvorsky[*,***,****], L. Svoboda[**], J. Trojková[*], J. Bednář[****], P. Mančík[****] and B. Kostura[**]

VŠB-Technical University of Ostrava, Czech Republic
[*]Department of Physics, [**]Department of Chemistry,
[***]Regional Materials Science and Technology Centre, [****]Nanotechnology centre
richard.dvorsky@vsb.cz

ABSTRACT

With the current rapid development of nanotechnology, top-down preparation of nanoparticles by various disintegration techniques became an important industrial sector. In addition to the presently common high speed bead mills working with particles in liquid dispersions, jet mills are also frequently used. They operate in the dry disintegration mode of extreme aerosol flow with cyclonic separation of the fine fraction. The common characteristic of these both mill types is the submicron size of disintegrated output particles. Very promising alternative to these techniques is the use of the disintegrating effect of cavitation implosion. In this work, we present experimental results and technical experience with a new experimental device for cavitation disintegration of microparticles and nanoparticles in dense liquid dispersions. In first test experiments, the mean size of the primary silicon particles of 140 nm was reduced to 40 nm by the cavitation disintegration lasting 200 minutes.

Keywords: nanoparticles, silicon, disintegration, cavitation

1 INTRODUCTION

With the current rapid development of nanotechnology, preparation of nanoparticles becomes a very important technological factor. Bottom-up processes of nanoparticle production are generally based on chemical methods [1-2]. Their basic problems are the effective management of reaction kinetics in nucleation, growth restrictions [3-4] and possible modifications [5-7]. Another technological approach is top-down production of nanoparticles using various disintegration techniques. In addition to the presently common high speed bead mills which operate in the environment of liquid particle dispersions, jet mills become also frequently used [8]. They operate in the dry disintegration mode of extreme aerosol flow with cyclonic separation of fine fractions. The common characteristic of these both mill types is the submicron size of disintegrated output particles.

Very promising alternative to these techniques is the application of the disintegrating effect of cavitation implosion on the particles dispersed in a liquid. Intense cavitation can be generated by high-energy water jet, like in the disintegrator Water Jet Mill [9]. Our experiments confirmed that it is an effective tool for nanoparticle preparation [10-11]. In this paper we present preliminary experience and experimental results of the cavitation disintegration of micro and nano particles in a dense fluid dispersion under intense ultrasonic field.

2 THEORY OF CAVITATION DISINTEGRATION OF PARTICLES IN DENSE LIQUID DISPERSIONS

The analysis [9] of the mechanism of disintegration based on the implosion of cavitation bubbles proved that the impact pressures can reach tens of GPa, sufficient to disintegrate even very hard materials. For this application, we constructed an ultrasonic disintegrator schematically illustrated in Figure 1.

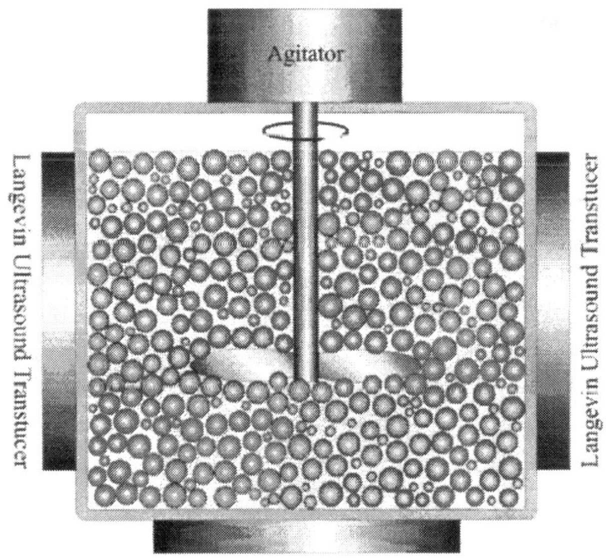

Figure 1: Sketch of the prototype cavitation ultrasonic disintegrator with the stirring of dense particle dispersion (sizes of disintegrated micro particles are significantly magnified to illustrate the required high dispersion density).

Inside the disintegration chamber, the dense liquid dispersion of microparticles is subjected to ultrasonic radiation of Langevin transducers through its metal walls. Simultaneously, the relative positions of microparticles are rapidly changing due to external mixing of dispersion. In these conditions, mutual impacts and subsequent separations of dispersed particles are very frequent. Heterogeneous nucleation of cavitation bubbles on the surface of the particles substantially prevails over homogeneous nucleation. Therefore, cavitation bubbles form mainly between two particles separated by mixing.

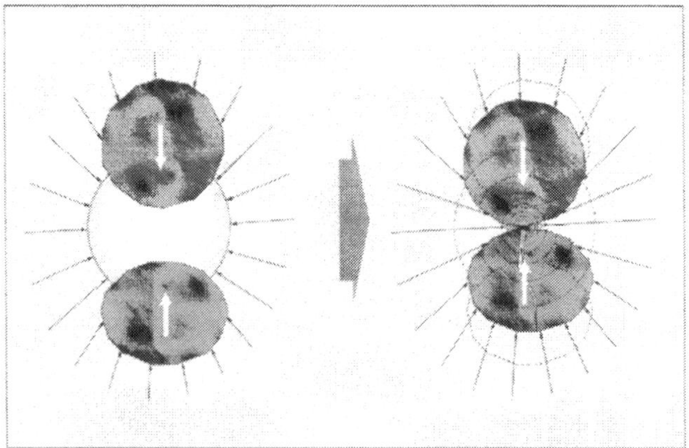

Figure 2: Illustration of the mutual impact disintegration due to the collapse of a cavitation bubble between two particles

This situation is illustrated in Figure 2. On the left, the cavitation bubble reached its maximum size and it started imploding. The two particles form the opposite walls of the collapsing bubble and they are pushed by surrounding liquid into the place of mutual collision, as displayed on the right. Reflecting the analysis of the cavitation disintegration in [9], the following experiments tested the disintegration of semiconductor-grade silicon submicron particles dispersed in degassed water.

3 EXPERIMENTAL VERIFICATION OF ULTRASOUND CAVITATION DISINTEGRATION TECHNOLOGY

The chamber of the ultrasonic cavitation disintegrator was filled with the dense water dispersion (slurry) of semiconductor-grade Si microparticles with the size distribution depicted in Figure 3, with the mean size of 140 nm.

For optimal efficiency of the disintegration it was necessary to provide the volume fraction of dispersed particles in the aqueous medium at least 50%. Such an arrangement in the microscale induces very high frequency of mutual collisions during mixing of the dispersion. Simultaneously, it also ensures plentiful formation of cavitation bubbles between particles after their separation and therefore the high intensity of cavitation disintegration. In the arrangement according to Figure 1, the dense dispersion of the aforementioned primary microparticles was subjected to ultrasonic cavitation disintegration for 200 minutes at the power density of ultrasonic field 50 W/l and the agitator speed of 100 rpm.

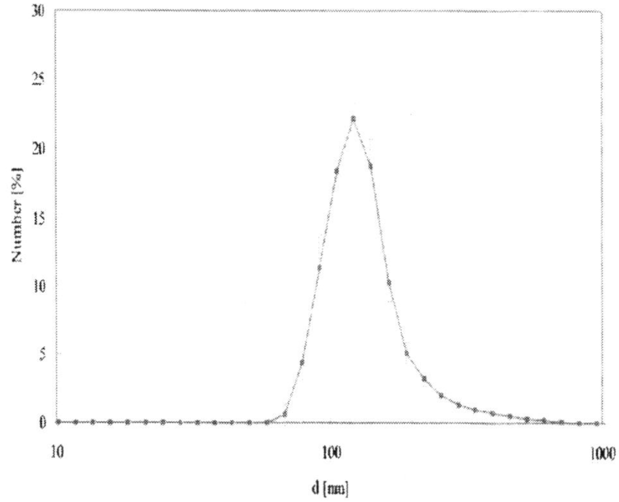

Figure 3: The statistical distribution of diameters *d* of primary silicon microparticles with the mean size of 140 nm, obtained using DLS Malvern Nano ZS considering the correction due to the hydrodynamic coefficient of 0.8. These primary microparticles were produced by milling in a planetary mill and separated by sieve analysis.

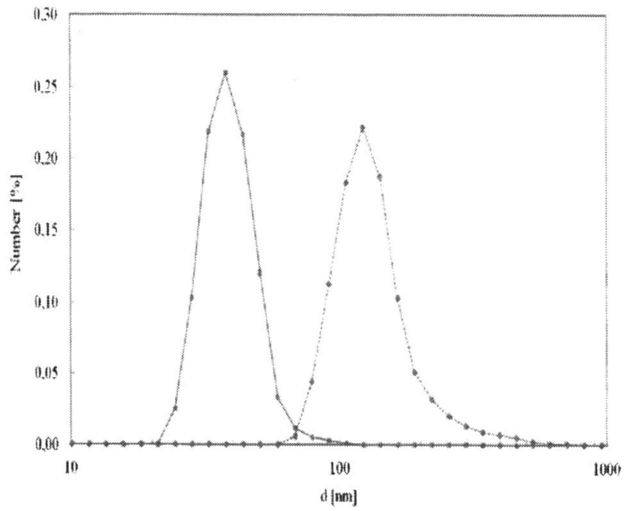

Figure 4: The effect of 200 minutes of ultrasonic cavitation disintegration. The original size distribution of microparticles is depicted in a dashed line, the distribution after the processing in a solid line. The mean diameter is 40 nm. The analysis was obtained using DLS Malvern Nano ZS considering the correction due to the hydrodynamic coefficient of 0.8.

Figure 4 shows the transition in the size distribution of silicon particles due to the disintegration. Primary microparticles (dashed line) transform to resulting nanoparticles (solid line). Following the statistical analysis of [7], the mean size of the particles is reduced to 27%. The resultant nanoparticle liquid dispersion was finally vacuum freeze dried to obtain dry nanopowder for further practical use. This was then analyzed on transmission electron microscope TEM Jeol JEM 1230, 80 kV.

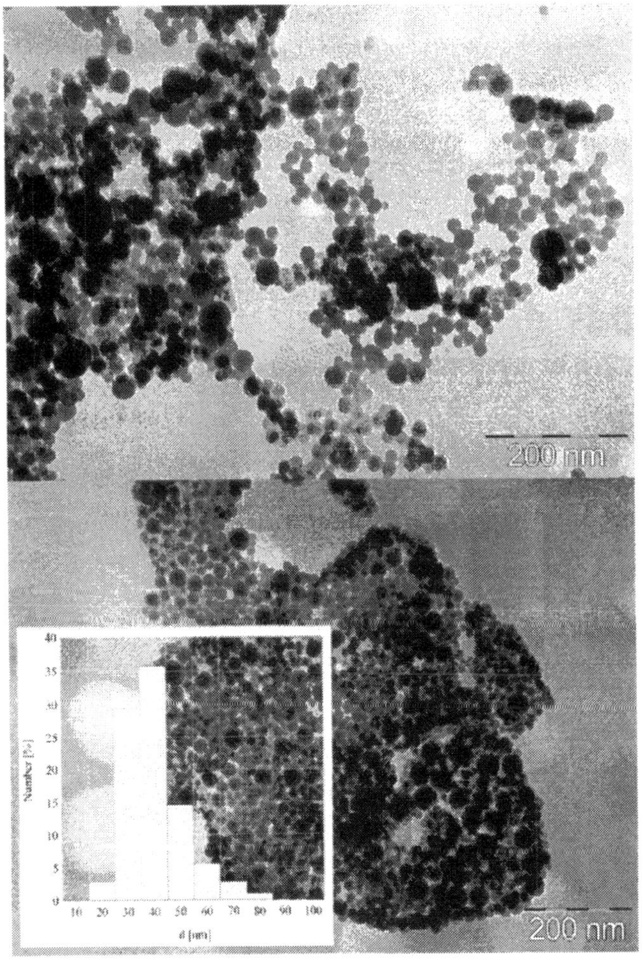

Figure 5: TEM micrograph of a final aggregate of Si nanoparticles after disintegration and vacuum freeze drying (TEM Jeol JEM 1230, 80 kV).

Image analysis of the final silicon nanoparticles gave a statistical ensemble of 100 particles with the size distribution shown in Figure 5. Its mean value of 39 nm is in good agreement with the value of 40 nm obtained by dynamic light scattering method using Malvern Nano ZS.

The disadvantage of our first experimental disintegrator is a low density of acoustic power transferred to the processed dispersion. This naturally reduces the efficiency of the disintegration, increases the time required to achieve the desired grain refinement and restrains disintegration (grinding) limits.

The achieved results prove the described technology of disintegration basically effective, so we can approach the construction of an efficient prototype of the cavitation ultrasonic disintegrator with a density of the transferred acoustic power above 100 W/l.

In the previous works [10-11], the theoretical analysis of maximum attained impact pressures has shown that they are significantly influenced by the external hydraulic pressure of the surrounding liquid. For these reasons, a new prototype of the cavitation ultrasonic disintegrator will operate with higher pressures, up to 15 MPa.

4 CONCLUSIONS AND FUTURE DIRECTIONS

In the first experiments, we tested the response to ultrasonic cavitation disintegration on sparse dispersions, always with a negligible effect. In the light of this outcome, our later results confirm the validity of fundamental theoretical ideas about the mechanism of impact disintegration due to implosion of cavitation bubbles in dense liquid dispersions of microparticles.

With regard to the theoretical calculations of the maximum impact pressures in the previous work [10-11], we estimated the dynamics of a mutual collision of two particles. It follows that the submicron particles collide with the characteristic speed of hundreds of meters per second. Such conditions lead to the destruction of even very hard materials.

Currently, we are finalizing the prototype of a new, more efficient device that will operate with higher hydraulic pressures and higher acoustic power densities. The improved parameters of the new apparatus should increase the intensity of heterogeneous nucleation of cavitation bubbles on solid particles and also the impact velocities in collapse.d

ACKNOWLEDGMENTS

This work was supported by the project Pre-seed VŠB-TUO II - Materials (CZ.1.05/3.1.00/14.0320) and the project "Regional Materials Science and Technology Centre - Feasibility Program" (LO1203).

REFERENCES

[1] Kickelbick G, Liz-Marzán L.M, in: Nalwa H.S. (Eds.) (2009) Encyclopedia of Nanoscience and Nanotechnology 2, American Scientific Publishers, 199–204.

[2] C. Br´echignac, P. Houdy, M. Lahmani: Nanomaterials and Nanochemistry, ISBN 978-3-540-72992-1, Springer Berlin Heidelberg, New York 2007

[3] R. Dvorsky, P. Praus, J. Trojkova: Model of Synthesis of ZnS Nanoparticles Stabilised by CTAB, Chalcogenide Letters 10/2013; 10(10), p.385–391

[4] **Praus P, Dvorsky R, Kovář P, Trojková J** (2012) Stabilization of ZnS nanoparticles in micellar dispersion of cetyltrimethylammonium bromide, Acta Chimica Slovenica 59: 784–788.

[5] Lai S, Suk J.S, Pace A, Wang Y, Yang M, Mert O, Chen J, Kim J, Hanes J (2011) Drug carrier nanoparticles that penetrate human chronic rhinosinusitis mucus, Biomaterials 32: 6285-6290.

[6] P. Praus, M. Turicová, M. Karlíková, L. Kvítek, R.Dvorsky: Nanocomposite of montmorillonite and silver nanoparticles: Characterization and application in catalytic reduction of 4-nitrophenol, Materials Chemistry and Physics, 01/**2013**, 140, p.493-498.

[7] **Dvorsky R, Trojkova J, Praus P, Luňáček J** (2013) Analysis of mechanisms of composite particles modification in liquid dispersions, Materials Research Bulletin 48: 2286-2288

[8] K. Nogi,M., Naito,T., Yokoyama: Nanoparticle Technology Handbook, 1st Edition, ISBN 978-0-56336-1, Elsevier 2012

[9] Dvorský. R., Lunacek, J., Slíva, A., Piksová, K.: Preparation of Silicon Nanoparticles by Means of Disintegration in a Cavitation Water Jet, 2nd International Conference NANOCON 2010, Olomouc 2010 (12-14 October 2010) Conference Proceedings Book (ISBN 978-80-87294-19-2), p.144-150.

[10] **Dvorský, R., Luňáček, J., Slíva, A.: Dynamics Analysis** of Microparticles Cavitation Disintegration During Nanopowder Preparation in New Water Jet Mill (WJM), Advanced Powder Technology, Volume 22, Issue 5, September 2011, Pages 639-643

[11] R. Dvorsky, J. Trojková: Handbook of Mechanical Nanostructuring, Chapter 23 - Cavitation disintegration of powder microparticles, Wiley-VCH (2015)

Comparative Study of Carbonaceous Materials as Catalyst Supports for the Enantioselective Hydrogenation of (E)-α-Phenylcinnamic Acid: Influence of the Support Acidity

Tran Si Bui Trung[*], Yeonwoo Kim[*], Sungho Kang[*], Hangil Lee[**], Sehun Kim[*]

[*]*Molecular-Level Interface Research Center, Department of Chemistry, KAIST, 305-701, Republic of Korea*

[**]*Department of Chemistry, Sookmyung Women's University, Seoul 140-742, Republic of Korea*

ABSTRACT: A systematic comparative study of carbonaceous materials as catalyst supports for the asymmetric heterogeneous hydrogenation of (E)-α-phenylcinnamic acid (PCA) was performed to investigate the influence of the supports on the reaction enantioselectivity. A series of Pd-based catalysts supported on activated carbon, graphene oxide (GO), or carbon nanotubes was prepared using deposition methods, and their physicochemical properties were characterized. The results demonstrated that the textural structure of the supports had a trivial effect on the enantiomeric excess (*ee*). The TGA, FT-IR, and TPD–NH_3 data revealed an abundance of acidic sites present on the Pd/GO catalyst, unlike the other catalysts, and these sites reduced the enantioselective hydrogenation of PCA. The effect of the support acidity on the enantioselectivity was ascribed to the preferential adsorption mode of cinchonidine (CD) on the GO surface *via* electrostatic interactions between the negatively charged oxygen functional groups present on the GO surface and the positively protonated amine groups present in the CD molecule.

1. INTRODUCTION

Heterogeneous catalyst systems have several advantages over homogeneous catalysts in that they are more easily recyclable and offer a higher yield. Heterogeneous systems could potentially provide green alternatives to stoichiometric processes and reduce the environmental impact of manufacturing. In a heterogeneous reaction, the catalyst support is critical to determine the catalytic performance [1–3]. Therefore, a thoroughly understanding the influence of the catalyst support in each application is essential for developing state-of-the-art catalysts with improved catalytic performances.

Activated carbon (AC) is one of the most extensively utilized supports for metal-based catalysts used in low-temperature reducing reactions, such as liquid phase heterogeneous hydrogenation reactions [4]. The recently discovered carbonaceous materials, including carbon nanotubes (CNTs) and graphene oxide (GO), show promise in a wide range of applications seeking to capitalize on their unique and fascinated properties [5,6]. CNTs and GO are broadly employed as heterogeneous catalyst supports because they possess a large surface area, provide good mechanical strength, and stable under harsh conditions [7,8]. Although AC, GO, and CNTs comprise layers of carbon atoms arranged in six-membered rings, their structures are distinct from one another. AC has an amorphous structure that consists of aromatic sheets and strips with slits of a variety of molecular dimensions that form porosity [9], whereas CNTs are composed of curved sp^2-hybridized carbon atoms arranged in an axial alignment of concentric cylindrical planes [10]. On the other hand, GO is a flat monolayer of carbon atoms tightly packed into a two-dimensional honeycomb lattice on which a variety of reactive oxygen functional groups are present [11]. The structural diversity of these carbonaceous supports results in different metallic active site structures and, consequently, induces distinctive catalytic behaviors.

The asymmetric heterogeneous hydrogenation of prochiral α,β-unsaturated carboxylic acids over cinchonidine (CD)–modified Pd-based catalysts has attracted enormous attention in recent decades due to the growing demand for enantiomerically pure compounds for use in various applications [12–14]. Among the most widely studied reactions is the enantioselective hydrogenation of (E)-α-phenylcinnamic acid (PCA), which has been intensively examined by Nitta and others [15–18]. Many factors that affect the enantioselectivity of the reaction, including the solvents [19], additives [20], or catalyst preparation conditions [21] have been investigated with the aim of enhancing the enantiomeric excess (*ee*) values. Activated carbon is a conventional catalyst support used in this type of reaction [22–25], and little effort has been applied toward the use of other carbonaceous materials, such as GO [26] or CNTs [27]. Moreover, no systematic studies have yet examined the effects of these carbonaceous supports on the enantioselectivity of the hydrogenation reaction of prochiral α,β-unsaturated carboxylic acids. Therefore, a comparison of these carbonaceous materials as catalyst supports for the hydrogenation of PCA could reveal the influence of the supports and clarify the mechanism underlying this reaction. These insights could assist the design of novel catalysts with enhanced enantioselectivity.

This work seeks fundamental insights into the role of the catalyst support in the enantioselectivity of the asymmetric hydrogenation of PCA by systematically comparing the properties of carbonaceous material supports for Pd-based catalysts under similar catalyst preparation and hydrogenation reaction conditions. Pd-based catalysts supported on AC, GO, or CNTs were prepared using deposition methods and were characterized by a variety of approaches. The oxygen functional groups present on the supports were found to play a key role in determining the enantiodifferentiation properties.

2. EXPERIMENTAL SECTIONS

2.1. Pd catalyst preparation

Graphene oxide was synthesized using the modified Hummers method [28]. Supported palladium catalysts were prepared by the deposition–precipitation method using a $PdCl_2$ precursor. The $PdCl_2$ solution was prepared by ultrasonicating a mixture containing 25 mg $PdCl_2$ and 2 mL HCl (37%) in 10 mL H_2O for 20 minutes. In a typical preparation, 300 mg of the support (AC, GO, CNTs) were dissolved in 100 mL H_2O with ultrasonic treatment for 1 hour. The pH of the mixture was then adjusted to 10 by the addition of a 1 M Na_2CO_3 solution, followed by 30 minutes stirring. The $PdCl_2$ solution was then added to the suspension with stirring, and the pH of the mixture was adjusted to 10 again. After the pH adjustment, the mixture was stirred for 5 hours and then was heated at 80°C for 2 hours. The heat treatment magnified the acidity of the solution and promoted the hydrolysis process to form $Pd(OH)_2$, which anchored to the support surface [29,30]. The suspension was cooled to room temperature, and $NaBH_4$ dissolved in 20 mL H_2O was slowly dropped into it. The reaction was continued for 3 hours with stirring, and the black precipitate was filtered, washed 5 times with water (50 ml each time), dried overnight under ambient atmosphere in an oven (110°C), and stored for use.

2.2. Hydrogenation reaction

The solvent used in the hydrogenation reactions was wet 1,4-dioxane (containing 2.5% double-distilled water). Each reaction was repeated at least 3 times to obtain the average *ee* value. The hydrogenation reactions were carried out using procedures conducted with or without the pretreatment [31]. The pretreatment procedure involved heating 15 mg of the catalyst with 2 mL solvent in a 25 mL glass bottle with magnetic stirring (1000 rpm) at 80°C for 30 minutes under a H_2 atmosphere supplied by a balloon. After the pretreatment process, a solution of 0.01 mmol CD in 1 mL solvent was added at 23°C. The mixture was stirred for 30 minutes, and then a solution of 0.25 mmol PCA in 2 mL solvent was added, followed by the injection of 0.03 mL BA. The slurry was stirred with magnetic agitation at 1000 rpm at 23°C under the H_2 atmosphere. After the hydrogenation reaction had finished (8–12 hours). The product was esterified to produce the methyl ester, and the *ee* values were determined by HPLC using a chiral Chiralcel OD-H (Daicel) column.

3. RESULTS AND DISCUSSION

3.1. Characterization of the synthesized Pd-based carbonaceous supports

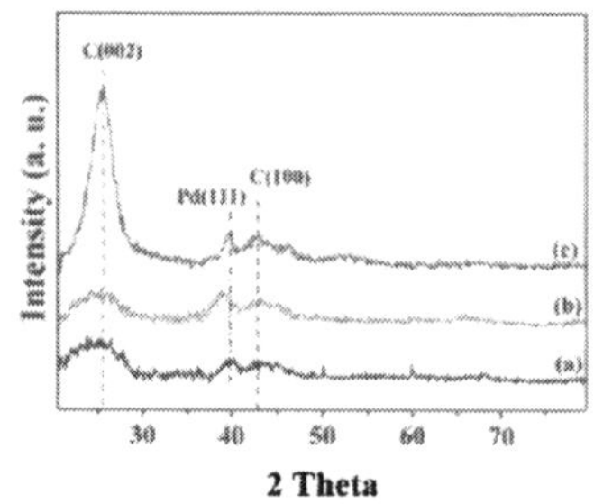

Figure 1. XRD patterns of the prepared catalysts (a) Pd/AC, (b) Pd/GO, and (c) Pd/CNTs.

The X-ray diffraction patterns of the Pd-based catalysts prepared with different carbonaceous supports were collected over the range 20°–80° to study the crystalline structures. As shown in Figure 1, diffraction peaks at 26° and 43° were observed in all catalysts and were assigned to the typical (002) and (100) planes of the hexagonal graphite structure, respectively. In the Pd/GO catalyst, these peaks indicated that GO was reduced partially to graphene sheets during the catalyst preparation steps, and the graphene sheets restacked to form an ordered crystalline structure [32]. The three synthesized catalysts exhibited characteristic diffraction peaks at $2\theta=40°$, which corresponded to the (111) crystalline plane of the *fcc* Pd nanoparticles (JCPDS #46-1043). It should be noted that this peak was broader in the Pd/AC and Pd/GO catalysts and sharper in the Pd/CNT catalyst, suggesting that the Pd nanoparticles present in the Pd/AC and Pd/GO catalysts were smaller than the Pd nanoparticles present in the Pd/CNT catalyst. In addition, this peak in the Pd/GO catalyst was shifted slightly toward lower diffraction angles due to the effects of the lattice strain, which increased the lattice parameter to a small degree [33].

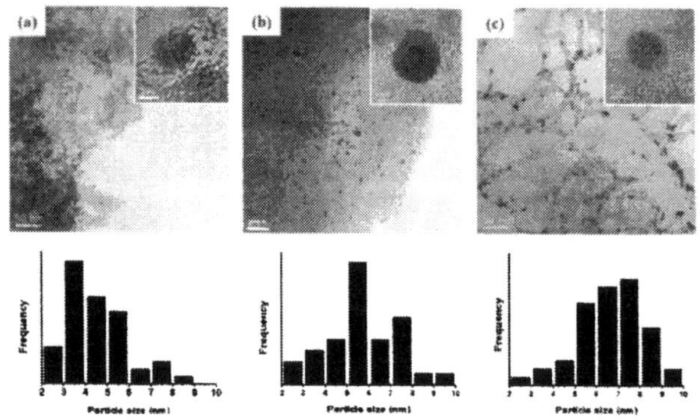

Figure 2. TEM images and (inset) HR-TEM images of the prepared Pd-based catalysts: (a) Pd/AC, (b) Pd/GO, or (c) Pd/CNTs, and their corresponding particle size distributions.

The morphologies and dispersions of the Pd nanoparticles in the prepared Pd-based catalysts were evaluated using bright-field TEM techniques. Representative

TEM images and the corresponding particle size distributions are shown in Figure 2. The insets show HR-TEM images of a randomly selected dark spot on the prepared Pd/AC, Pd/GO, or Pd/CNT catalysts, confirming that the particles were enclosed by facets with interplanar distances of 0.225nm, 0.228 nm, and 0.225 nm, respectively, based on the adjacent lattice fringes. These lattice parameters corresponded to the (111) plane of the *fcc* Pd crystal [34], in good agreement with the observations collected from the XRD data. The inset HR-TEM images also revealed that the Pd nanoparticles assumed sphere-like shapes in all of the catalysts. The Pd nanoparticles in the prepared catalysts were homogeneously decorated across the carbonaceous materials, and the particle size typically did not exceed 10 nm. The particle size distributions revealed that the Pd/AC catalysts included more numerous small (2–6 nm) Pd particles, whereas the Pd particles present on the Pd/CNTs catalysts were predominantly between 5 and 9 nm, in good agreement with the XRD data. The results obtained from the XRD and TEM analyses indicated that the Pd nanoparticles were successfully deposited onto the carbonaceous supports after the catalyst preparation process.

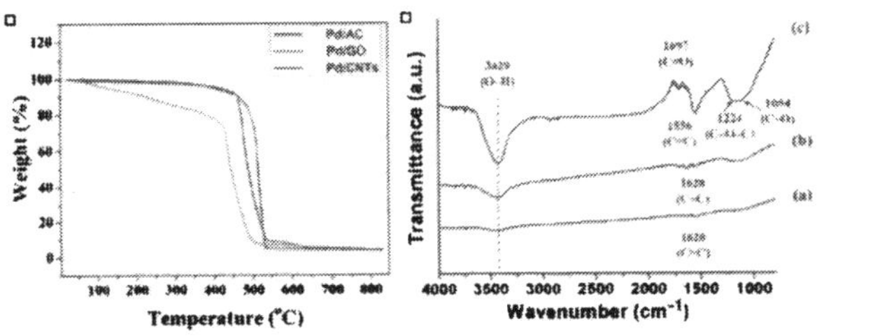

Figure 3. Left panel: TGA study of the prepared Pd-based catalysts: (black) Pd/AC, (red) Pd/GO, and (blue) Pd/CNTs. Right panel: FT-IR data of the prepared Pd-based catalysts: (a) Pd/CNTs, (b) Pd/AC, (c) and Pd/GO.

TGA studies were conducted over the temperature range between room temperature and 850°C to examine the thermal stability of the prepared Pd-based catalysts and to determine the practical Pd wt% loading. As shown in Figure 3 (left panel), the Pd/AC and Pd/CNT samples exhibited similar thermal behaviors. Initially, a small mass loss was observed as the temperature increased to 450°C; however, the TGA curves rapidly dropped to 5%wt within the temperature range 450–600°C, which was attributed to the weight loss due to the removal of the carbon component. By contrast, the Pd/GO catalysts displayed over 15% weight loss at 300°C as a result of the removal of oxygen-functional groups, such as C–O, C=O, and O–H, which remained on GO after catalyst preparation [35]. As the temperature increased to 600°C, more than 70% of the mass was lost due to the decomposition of the remaining stable oxygen functionalities and the carbon component [36]. These results indicated the presence of oxygen functional groups on the Pd/GO after the catalyst preparation. Beyond 650°C, all catalysts remained at a ~5% weight, which was assigned to the wt% of Pd loading on the carbonaceous supports. The FT-IR spectra were also collected to identify the major oxygen functional groups present on the prepared Pd-based catalysts. As shown in Figure 3 (Right panel), the peak assigned to the O–H bond stretching at 3429 cm^{-1} was present in all samples. Peaks corresponding to the C=C bond stretching of the graphite structure at 1620 cm^{-1} were detected in the Pd/AC and Pd/CNT catalysts, whereas a peak at 1556 cm^{-1} in the Pd/GO catalyst spectrum was attributed to the skeletal vibrations of the graphene sheets [37]. Furthermore, the FT-IR spectrum of the Pd/GO catalyst showed additional bands at 1054 cm^{-1}, 1224 cm^{-1}, and 1697 cm^{-1}, which were attributed to carboxylate C–O, epoxide C–O–C, and carboxyl C=O stretching vibrations, respectively. It is worth noting that the absorption bands of the Pd/GO catalyst were more intense than the bands collected from the other samples due to the large number of oxygen functional groups present on the GO after the catalyst preparation steps. The FT-IR spectra indicated that the Pd/GO catalyst contained a larger variety of oxygen functional groups compared to the other samples, consistent with the TGA data.

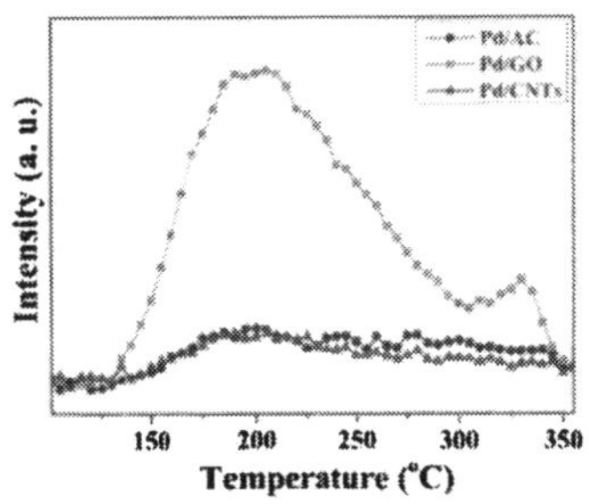

Figure 4. NH_3-TPD analysis of the prepared Pd-based catalysts: (black) Pd/AC, (red) Pd/GO, and (blue) Pd/CNTs.

The oxygen functional groups observed in the TGA and FT-IR analyses were further characterized using the temperature-programmed desorption of ammonia (TPD–NH_3) technique to measure the quantity as well as strength of the acidic sites present on the synthesized Pd-based catalysts. The kinetic diameter of an NH_3 molecule is sufficiently small (0.26 nm) to diffuse through narrow pores, adsorb onto the surfaces of the catalysts, and react with Brønsted and Lewis acid sites. Although the data obtained from TPD–NH_3 analysis could not distinguish between the Brønsted and Lewis acid groups [38], it provided comparative information about the surface acidity of each catalyst. Conventionally, acidic sites are classified based on the desorption temperature range, where weak sites desorb at 120°C–250°C, medium sites desorb at 250°C–400°C, and strong sites at >400°C [39].

In the present work, the TPD–NH_3 studies were conducted only up to 350°C due to the instability of the carbonaceous materials at higher temperatures, as indicated by the TGA data. Figure 4 shows that the Pd/GO catalyst exhibited an intensely broad desorption signal ranging from 120°C to 300°C and a narrower but tailing signal in the range 300°C–350°C. These signals were attributed to desorption from weak and medium acidic sites, respectively. Moreover, the TPD–NH_3 profiles revealed that both the weak and medium acidic sites on the Pd/GO catalyst were present in significant excess of those present on the Pd/AC and Pd/CNTs catalysts. The total concentration of acidic sites present on the Pd/GO catalyst was calculated to be 134.76 μmolg^{-1}, 6.5 times the concentration of acidic sites present on the Pd/AC catalyst and 7.3 times that present on the Pd/CNTs catalyst. The acidic sites on the Pd/GO catalyst originated from the oxygen functional groups that remained after the catalyst preparation steps, further confirm the previous data.

3.2. Asymmetric hydrogenation reaction of PCA

Metal surfaces in asymmetric heterogeneous hydrogenation systems act as the active sites for reactions, whereas chiral modifiers generate chiral sites *via* adsorption onto the metal surface [40]. After protonation, the modifier CD molecule adsorbs onto the Pd surface in a suitable mode to provide a chiral environment and subsequently interacts with the substrate *via* hydrogen bonds to form intermolecular complexes that finally produces products with enantiodifferentiation [41,42]. The pretreatment procedure is performed to remove oxygen and impurities from the Pd surface, which enhances enantioenrichment by increasing the number of adsorbed CD molecules and also favoring the adsorption of CD on the Pd surface in the suitable mode [31].

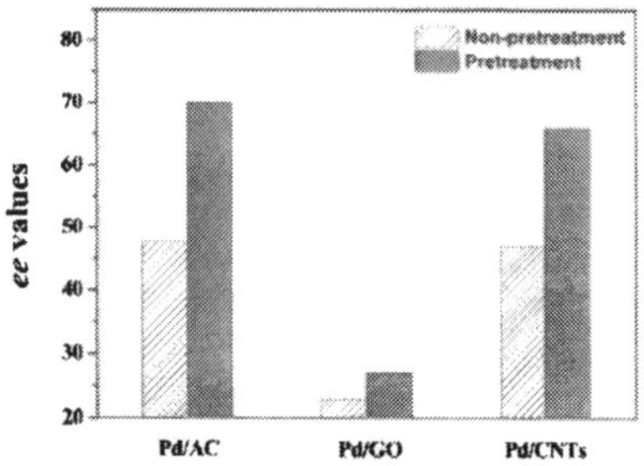

Figure 5. *ee* values obtained from the prepared Pd-based catalysts: Pd/AC, Pd/GO and Pd/CNTs.

As shown in Figure 5, the *ee* values obtained from the Pd/AC and Pd/CNTs catalysts were modest, 48%, in the non-pretreatment case. The pretreatment procedure significantly enhanced the enantioselectivity to values of 70% and 66%, respectively. The *ee* values attained from the Pd/AC and Pd/CNTs catalysts suggested that the structural morphologies of the supports only trivially influenced the enantioselective performance. On the other hand, the prepared Pd/GO catalyst displayed an *ee* of only 23% in the non-pretreatment case, which was very poor compared with the values obtained from the other catalysts. Moreover, the pretreatment procedure only enhanced the *ee* up to 27%. The similar drop of *ee* values obtained from the Pd/GO catalyst submitted to both procedures suggested that the purified Pd nanoparticle surface after the pretreatment procedure was essentially impervious to improving the *ee*. Instead, the quantity of CD molecules adsorbed onto the Pd surface and their adsorption mode were decisive in determining the *ee*. In other words, CD molecules did not favorably adsorb onto the Pd surface, regardless of the removal of contaminants on the surface. Under conditions that did favor adsorption onto the Pd surface, the CD molecules did not adsorb in the suitable mode to induce enantioselectivity.

The support polarity has been shown to affect the catalytic performance in a variety of reactions [43–45]. These effects have been ascribed to variations in the adsorbed reactant quantities due to the presence of oxygen functional groups on the support surface, which promote or reduce the desired products. Protonated amine groups adsorbed onto the GO support appear to incorporate themselves between the negatively charged carboxylate functional groups remaining on the reduced graphene surface after chemical reduction and the positively protonated amine groups *via* attractive electrostatic interactions. This binding mode has been supported by both theoretical calculations [46] and experimental data [47].

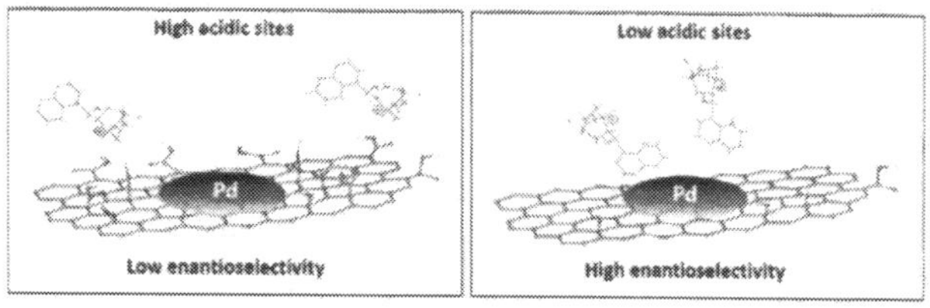

Scheme 1. Schematic diagram showing the influence of oxygen functional groups on the enantioselective hydrogenation of PCA.

The effects of the support polarity and the abundance of acidic sites present on the Pd/GO catalyst suggest that the low *ee* values measured from the Pd/GO catalyst were associated with oxygen-containing groups present on the catalyst surface. In other words, the negatively charged oxygen functional groups present on the GO surface of the prepared Pd/GO catalyst might attract or even interact with the positively protonated quinuclidine nitrogen moiety in the CD molecule *via* electrostatic forces (Scheme 1). These effects could alter the adsorption mode of CD on the Pd surface in the former case, or favor the preferential adsorption of CD on the GO surface in the latter case. This, in turn, decreased the quantity of modifier–substrate intermediate complexes that could produce enantioselectivity, leading to an inevitable drop in the *ee* values.

4. CONCLUSION

We investigated the influence of the catalyst support on the enantioselectivity of the asymmetric heterogeneous hydrogenation of PCA in a comparative study of Pd-based catalysts supported on carbonaceous materials (AC, GO, or CNTs). The Pd-based catalysts were prepared by the deposition method and were characterized using a variety of methods. The results suggested that the morphologies of the catalyst supports did not significantly affect the *ee* value. The Pd/GO catalyst, which contained an abundance of acidic sites relative to the other catalysts due to the presence of oxygen functional groups after the catalyst preparation, induced very low *ee* values in the hydrogenation of PCA. The results suggest that the support acidity reduced the *ee* value by favoring the preferential adsorption mode of CD molecules on the GO surface *via* electrostatic interactions.

REFERENCES

[1] A. Villa, M. Schiavoni, L. Prati, *Catal. Sci. Tech.* **2012**, *2*, 673.

[2] G. Agostini, C. Lamberti, R. Pellegrini, G. Leofanti, F. Giannici, A. Longo, E. Groppo, *ACS Catal.* **2014**, *4*, 187.

[3] F. Cárdenas-Lizana, Y. Hao, M. Crespo-Quesada, I. Yuranov, X. Wang, M. A. Keane, L. Kiwi-Minsker, *ACS Catal.* **2013**, *3*, 1386.

[4] F. Rodríguez-reinoso, *Carbon* **1998**, *36*, 159.

[5] X. Peng, J. Chen, J. A. Misewich, S. S. Wong, *Chem. Soc. Rev.* **2009**, *38*, 1076.

[6] C. N. R. Rao, A. K. Sood, K. S. Subrahmanyam, A. Govindaraj, *Angew. Chem. Int. Ed.* **2009**, *48*, 7752.

[7] B. Wu, Y. Kuang, X. Zhang, J. Chen, *Nano Today* **2011**, *6*, 75.

[8] B. Luo, S. Liu, L. Zhi, *Small* **2012**, *8*, 630.

[9] H. F. Stoeckli, *Carbon* **1990**, *28*, 1.

[10] I. A. A. C. Esteves, F. J. A. L. Cruz, E. A. Müller, S. Agnihotri, J. P. B. Mota, *Carbon* **2009**, *47*, 948.

[11] D. R. Dreyer, S. Park, C. W. Bielawski, R. S. Ruoff, *Chem. Soc. Rev.* **2010**, *39*, 228.

[12] K. Borszeky, T. Mallat, A. Baiker, *Tetrahedron: Asymmetry* **1997**, *8*, 3745.

[13] G. Szöllősi, K. Balázsik, M. Bartók, *Appl. Catal. A* **2007**, *319*, 193.

[14] G. Szöllősi, E. Szabó, M. Bartók, *Adv. Synth. Catal.* **2007**, *349*, 405.

[15] Y. Nitta, Y. Ueda, T. Imanaka, *Chem. Lett.* **1994**, *23*, 1095.

[16] Y. Nitta, K. Kobiro, *Chem. Lett.* **1995**, *24*, 165.

[17] G. Szőllősi, B. Hermán, K. Felföldi, F. Fülöp, M. Bartók, *J. Mol. Catal. A: Chem.* **2008**, *290*, 54.

[18] T. Sugimura, T. Uchida, J. Watanabe, T. Kubota, Y. Okamoto, T. Misaki, T. Okuyama, *J. Catal.* **2009**, *262*, 57.

[19] Y. Nitta, T. Kubota, Y. Okamoto, *J. Mol. Catal. A: Chem.* **2004**, *212*, 155.

[20] T. Y. Kim, T. Sugimura, *J. Mol. Catal. A: Chem.***2010**, *327*, 58. [21] T. Kubota, H. Ogawa, Y. Okamoto, T. Misaki, T. Sugimura, *Appl. Catal. A* **2012**, *437 438*, 18.

[22] T. Sugimura, J. Watanabe, T. Okuyama, Y. Nitta, *Tetrahedron: Asymmetry* **2005**, *16*, 1573.

[23] T. Sugimura, J. Watanabe, T. Uchida, Y. Nitta, T. Okuyama, *Catal. Lett.* **2006**, *112*, 27.

[24] T. Y. Kim, M. Yokota, T.; Uchida, T. Sugimura, *Catal. Lett.* **2009**, *131*, 279.

[25] T. Sugimura, H. Ogawa, *Chem. Lett.* **2010**, *39*, 232.

[26] K. Szőri, R. Puskás, G. Szőllősi, I. Bertóti, J. Szépvölgyi, M. Bartók, M. *Catal. Lett.* **2013**, *143*, 539.

[27] Z. Guan, S. Lu, C. Li, *J. Catal.* **2014**, *311*, 1.

[28] V. C. Tung, M. J. Allen, Y. Yang, R. B. Kaner, *Nat. Nanotechnol.* **2009**, *4*, 25.

[29] M. L. Toebes, J. A. van Dillen, K. P. de Jong, *J. Mol. Catal. A: Chem.* **2001**, *173*, 75.

[30] S. Louis, H. J. Merritt (1958). *"Method of preparing hydrogenation catalysts"* U.S. Patent 2,857,337 A, Oct 21, 1958.

[31] Y. Nitta, J. Watanabe, T. Okuyama, T. Sugimura, *J. Catal.* **2005**, *236*, 164.

[32] S. Yang, C. Shen, Y. Tian, X. Zhang, H. J. Gao, *Nanoscale* **2014**, *6*, 13154.

[33] B. Fultz, J. Howe (2013) *Transmission Electron Microscopy and Diffractometry of Materials*; Springer Berlin Heidelberg, 2013; p 1.

[34] G. J. Thomas, R. W. Siegel, J. A. Eastman, *Mat. Res. Soc. Symp. Proc.* **1989**, *153*, 13.

[35] S. Stankovich, D. A. Dikin, R. D. Piner, K. A. Kohlhaas, A. Kleinhammes, Y. Jia, Y. Wu, S. T. Nguyen, R. S. Ruoff, *Carbon* **2007**, *45*, 1558.

[36] T. Zhou, F. Chen, K. Liu, H. Deng, Q. Zhang, J. Feng, Q. Fu, *Nanotechnol.* **2011**, *22*, 045704.

[37] S. Park, J. An, I. Jung, R. D. Piner, S. J. An, X. Li, A. Velamakanni, R. S. Ruoff, *Nano Lett.* **2009**, *9*, 1593.

[38] M. Niwa, N. Katada, *Chem. Record* **2013**, *13*, 432.

[39] A. Boréave, A. Auroux, C. Guimon, *Microporous Mater.* **1997**, *11*, 275.

[40] T. Mallat, E. Orglmeister, A. Baiker, *Chem. Rev.* **2007**, *107*, 4863.

[41] Y. Nitta, A. Shibata, *Chem. Lett.* **1998**, *27*, 161.

[42] G. Kyriakou, G.; S. K. Beaumont, R. M. Lambert, *Langmuir* **2011**, *27*, 9687.

[43] X. Wang, N. Li, J. A. Webb, L. D. Pfefferle, G. L. Haller, *Appl. Catal. B* **2010**, *101*, 21.

[44] M. L. Toebes, Y. Zhang, J. Hájek, T. Alexander Nijhuis, J. H. Bitter, A. Jos van Dillen, D. Y. Murzin, D. C. Koningsberger, K. P. de Jong, *J. Catal.* **2004**, *226*, 215.

[45] R. W. Gosselink, W. Xia, M. Muhler, K. P. de Jong, J. H. Bitter, *ACS Catal.* **2013**, *3*, 2397.

[46] D. Stauffer, N. Dragneva, W. B. Floriano, R. C. Mawhinney, G. Fanchini, S. French, O. Rubel, *J. Chem. Phys.* **2014**, *141*, 044705.

[47] E. Y. Choi, T. H. Han, J. H. Hong, J. E. Kim, S. H. Lee, H. W. Kim, S. O. Kim, *J. Mater. Chem.* **2010**, *20*, 1907.

Scalable Production of Stabilized Catalysts via Atomic Layer Deposition

A. Goulas*, A. Dokania**, F. Grillo**, J.R. van Ommen** and B. van Limpt*

*Delft IMP, Julianalaan 136, 2628 BL, Delft The Netherlands
**Delft University of Technology, ChemE, Product and Process Engineering, Julianalaan 136, 2628 BL, Delft, The Netherlands, j.r.vanommen@tudelft.nl

ABSTRACT

We report the fabrication of ultrathin aluminum-based overcoats on supported platinum catalysts using atomic layer deposition and hydrid atomic/molecular layer deposition. Tuning the process parameters and applying heat treatment for porosity induction can result to controllably accesible model catalysts. We examined the effect of the number of deposition cycles (1-25), co-reactant choice (water, ethylene glycol, 1,5-pentanediol) and exposure time on the characteristics of the alumina and alucone stabilizing films. The use of fluidized bed reactors for atomic layer deposition applied for active phase deposition and overcoat stabilization paves the way for large-scale gas-phase production of tailored nanostructured catalysts.

Keywords: Atomic Layer Deposition (ALD), heterogeneous catalysis, catalyst stability, platinum (Pt) nanoparticles, alumina (Al_2O_3) overcoat

1 INTRODUCTION

Optimizing catalyst performance at the atomic level can have a tremendous impact on the intensification of chemical processes. Tuning composition and structure parameters of the active species (size/dispersion, shape, position) during synthesis is one of the main objectives of catalyst nanostructuring [1]. As a result, more active, selective and stable catalysts with limited deactivation problems at demanding conditions (temperature, pressure, pH) can be obtained [2]. Additional advantages are minimizing the use of scarce and expensive materials (e.g. noble metals) and reducing the amount of waste related with the catalyst production process. In the past years, atomic layer deposition (ALD) has emerged as a nanostructuring toolbox for the modification of surface properties with atomic precision [3-5]. Already, support modification [6-8], nanoparticle decoration [5-13] and ultrathin-film overcoats *via* ALD have been used to effectively synthesize and stabilize noble [5-9, 13-17] and base metal [18-23] supported catalysts. Molecular layer deposition (MLD), the organic counterpart of ALD, has also been used in hybrid schemes for creating porous catalyst overcoats.[10-12] With these schemes, several enhanced catalysts with metal oxide and hybrid organic/inorganic overcoats have already been reported as shown in table 1.

Feng *et al.* [6] reported the application of 1 or 2 ALD cycles of TMA/H_2O to improve the behavior of Pd-ZnO/Al_2O_3 catalysts in methanol decomposition. The authors initially claimed that TMA would not react with the Pd clusters, thus resulting in selective coating of the ZnO support. The influence of the number of ALD cycles on the activity of the catalyst and the mobility of the metal clusters was further studied [7-8] and the low GPC observed (0.8 Å/cycle) was correlated to non-uniformity of the surface coverage. Liang *et al.* [11] applied alucone MLD overcoats to stabilize Pt clusters up to 800 °C. The overcoats were rendered porous after calcination in air. Moreover, evidence for preferential coverage of the most energetic metal sites (edges, corners) during overcoat growth was indicated. Lu *et al.* [15] demonstrated the applicability of ALD overcoats on conventional catalysts. An accesible film of Al_2O_3 greatly improved the performance of Pd/Al_2O_3 catalysts in ethane oxidative dehydrogenation (ODH) by minimizing coking and sintering of the active phase.

The importance of overcoats for stabilization of catalysts used in liquid-phase reactions was recently discussed [20] further highlighting the importance of the calcination treatment for porosity induction *via* crystallization of the ALD overcoat. Additionally, the porous overoat films enhanced the performance of the catalyst to size-selective reactions due to the ability to tune the pore size [12]. The possiblity of adding acidity modifier such as MgO_x was also demonstrated. More bifunctional catalysts based on the growth of binary oxides have also been reported [18]. Finally, the use of blocking agents for selective growth of protective nanocages and selective nanobowls was introduced [5, 9]. A comprehensive review article [24] on the possibilies of ALD for catalyst design was recently communicated by O'Neill *et al.*

overcoat	reaction	reference
AlO_x	CH_3OH decomposition	[6-8]
alucone	CO oxidation	[11]
AlO_x	C_2H_6 ODH	[14-15]
AlO_x, MgO_x, NbO_x, TiO_x, alucone	furfural, C_4H_6, C_6H_{12} and $(CH_2)_8$ hydrogenation	[12-13, 17-20, 22-23]
alucone	CH_4 dry reforming	[10]
ZrO_x	O_2 reduction	[9]

Table 1: Reported ALD catalyst overcoats for enhancement of catalytic chemical reactions.

TechConnect Briefs 2015, TechConnect.org, ISBN 978-1-4987-4728-8

2 EXPERIMENTAL SECTION

The model catalysts were synthesized on titania nanoparticles (Aeroxide TiO_2 P 25, Evonik Industries, ~10 g) which are non-porous (specific surface area of 52 m^2/g) and have an average primary particle size of 21 nm. Prior using them in the ALD reactor for the Pt deposition, the nanoparticles were dried at 150 °C for 4 hours in a muffle furnace and sieved (45 µm). For the growth of the active metal phase (Pt) and the protective Al-based overcoats a custom-built fluidized bed reactor operating in atmospheric pressure described elsewhere [25] was used. ALD of Pt was performed using alternative exposures to (trimethyl) methylcyclopentadienyl-platinum(IV) ($MeCpPtMe_3$, Strem Chemicals) and pre-heated dry synthetic air (Linde Gas Benelux) at a deposition temperature of 250 °C. The Pt precursor was contained in a stainless steel bubbler and vaporization was carried at 70 °C.For the overcoating experiments, ~1 g of particles were *in-situ* dried in the ALD reactor at 120 °C in order to remove possibly physisorbed water. ALD of Al_2O_3 was performed using alternative exposures to trimethyl aluminium (TMA semiconductor grade, AkzoNobel HPMO) and demineralized water at a deposition temperature of 120 °C. For the hybrid ALD/MLD growth of the alucone layers, ethylene glycol (EG) or 1,5 pentanediol (PD) (both from Sigma-Aldrich) were implemented as co-reactants along with TMA. The TMA and water were contained in stainless steel bubblers and vaporization was carried at 30 °C. The diols were vaporized at 80 °C (EG) and 120 °C (PD) respectively. All the precursor transport lines were heat-traced at a temperature above the vaporization temperature in order to prevent precursor condensation. Dry nitrogen (HiQ 5.0, Linde Gas Benelux) was used for fluidization, purging of the reactor and transport lines and as a precursor carrier gas. The prepared catalysts were calcined in a furnace using a heating rate of 10 K/min and a final temperature of 700 °C or 800 °C with a hold time of 2 hours. Subsequently the catalysts were cooled down to room temperature.

The Pt and Al content of the catalysts was measured by inductively coupled plasma atomic emission spectroscopy (ICP-AES) in a PerkinElmer Optima 4300 spectrometer. Bright-field transmission electron microscopy (TEM) in a JEOL JEM 1400 microscope was used for probing the Pt cluster size and the thicknes of the Al_2O_3 and alucone overcoats. The obtained images were processed with ImageJ in order to derive particle size distributions for the metal clusters. Nitrogen adsorption and BET surface area analysis was performed in a Micromeritics TriStar II 3020. CO adsorption measurements were carried at the reduced samples (H_2, 350 °C) in a Micromeritics ASAP 2020.

3 RESULTS AND DISCUSSION

The model supported Pt catalysts were synthesized on TiO_2 nanoparticles using a reported method [26]. The deposited Pt clusters were 1.5-2.0 nm in diameter for both loadings of 0.45 and 0.74 wt.% Pt obtained by varying the total Pt precursor exposure. The specific surface area of the support nanoparticles remained effectively unchanged (54 m^2/g) after the deposition of the active phase.

For conventional ALD overcoating schemes, variation of the number of applied cycles will result in the growth of ultrathin protective films of different thickness. The growth-per-cycle (GPC) is influenced by certain support- and process-related parameters (e.g. availability of OH active sites, reaction temperature) and in general it is accepted that a full Al_2O_3 monolayer (3.8 Å) can only be achieved after 3-4 cycles [27]. Sequential exposure of TMA and H_2O for 2 cycles is expected to result in the formation of discontinuous Al_2O_3 patches instead of a film [16]. Based on the Al content of the catalysts a GPC of 0.8 Å was estimated after 15 cycles. This is slightly lower than our previous reported value (1.4Å) for growth at ambient conditions [25]. The linear increase of the amount of material deposited, as measured by ICP, versus the number of ALD cycles (5-25) performed is shown in figure 1.

The thickness of the grown protective films was verified by TEM measurements. A thickness of ~1.9 nm was estimated (TEM) for the sample prepared after 15 ALD cycles. The discrepancy between the value estimated from ICP and TEM can be tentatively attributed to assumptions during the conversion of the Al content to film thickness. Deviations from the density value of 2.8 g/cm^3 [28] that was selected based on the growth temperature of 120 °C and the assumption of spherical shape of the support particles can be expected. The overcoats lead to an increase of the primary particle size, as indicated by the measured value of the BET specific surface area of 41 m^2/g for the 15-cycle sample.

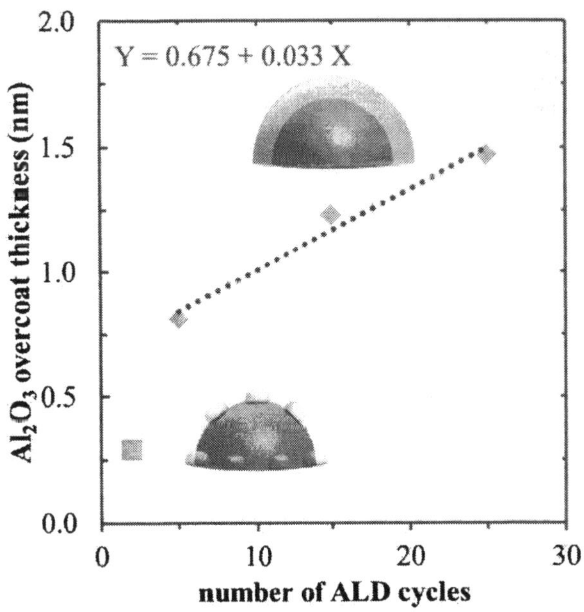

Figure 1: Thickness of Al_2O_3 overcoats versus ALD cycles (square, ■ corresponds to patches deposition; diamonds, ♦ correspond to film deposition).

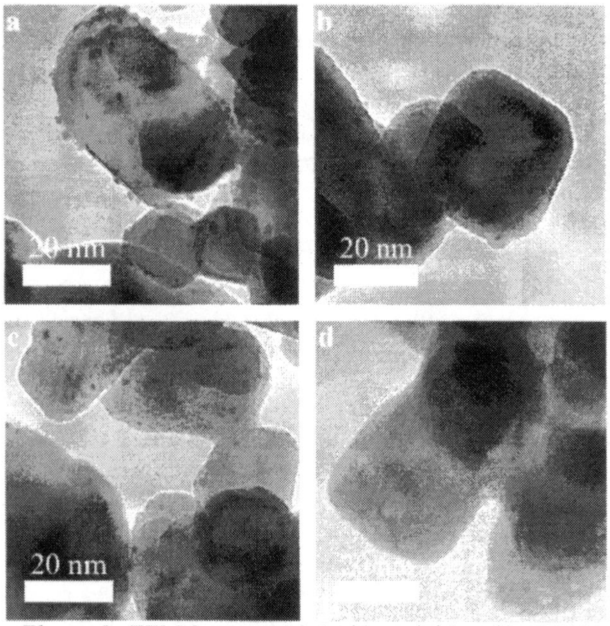

Figure 2: TEM images: (a) $2AlO_x$-Pt/TiO_2; (b) $15AlO_x$-Pt/TiO_2; (c) $15AlO_x$-Pt/TiO_2-800°C; (d) sub-saturated (0.75min H_2O pulse) 0.75sub_$15AlO_x$-Pt/TiO_2.

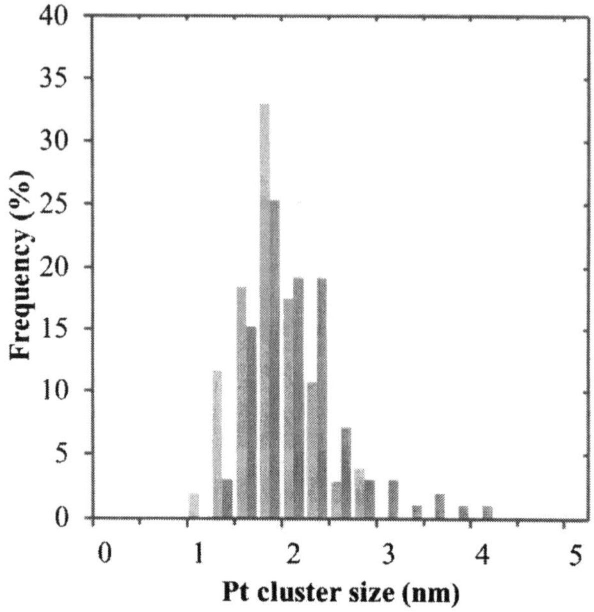

Figure 3: Size distribution histogram of Pt clusters for a sample of 2 ALD overcoat cycles (blue) and a calcined sample of 15 ALD overcoat cycles (red).

Additionally, samples with sub-saturative co-reactant (H_2O) pulses were prepared in order to study the effect of incomplete ligand removal to the subsequent porosity induction during calcination. Two overcoat experiments were performed with exposure time of ½ and ⅓ of the base case (1.5 min). In that way, films of 1.4-2.2 nm were obtained. These samples indicated some non-conformal growth features as shown in figure 2 (d). By utilizing a hybrid ALD/MLD scheme, composite films were also grown following the same pulsing sequence with the base case but for 8 cycles. The use of diols (EG and PD) as co-reactants instead of H_2O lead generally to lower growth.

In order to induce porosity to the coatings, selected samples were exposed to air in high-temperature. This is expected to crack the coatings open due to the removal of trapped H_2O and remaining organic ligands. No significant changes in the BET surface area of the samples (40 m^2/g) were observed after the calcination. The particle size distribution of the Pt species was derived from TEM images [Figure 2 (a) and (c)] before ($2AlO_x$-Pt/TiO_2) and after ($15AlO_x$-Pt/TiO_2-800°C) the calcination process. As shown in figure 3, the average Pt cluster size slightly grew from 1.7 nm to 1.9 nm. The overcoat can thus protect to a certain extent the underlying Pt clusters. Optimizing the overcoat thickness will enhance further the thermal stability of the active metal phase.

The total adsorption of CO indicated that the 15-cycle overcoat [shown in figure 2 (b)] blocked some of the Pt clusters. Calcination of the samples induced accesibility to the clusters (metal surface area increased from 0.32 to 0.83 m^2/g catalyst). However, the results for CO chemisorption (table 2), indicated that for the overcoated sample ~25% of the Pt clusters were still accesible. That amount was decreased to 0.03 m^2/g in the calcined catalyst. A possible explanation is that calcination at the crystallization temperature (800 °C) resulted in densification of the AlO_x phase [29]. Repetition of the measurements under different overcoat thickness values is expected to elucidate the effect of the overcoat in the stability of the metal clusters.

catalyst	Total adsorption [m^2/g catalyst]	Chemisorption [m^2/g catalyst]
Pt/TiO_2	2.47	0.51
$15AlO_x$-Pt/TiO_2	0.32	0.12
$15AlO_x$-Pt/TiO_2-800^{oC}	0.83	0.03

Table 2: Total CO adsorption and chemisorption amounts for i) the as-deposited, ii) overcoated (15 cycles) and iii) calcined (800°C) overcoated (15 cycles) catalysts.

4 CONCLUSIONS

We have demonstrated several nanostructuring concepts (*viz.* alteration of overcoat thickness for 2-25 cycles of conventional ALD, incomplete films by sub-saturation of the co-reactant (H_2O) pulse, hybrid MLD growth of alucone layers) coupled with porosity induction *via* heat treatment (calcination at 700 °C and 800 °C) can result to stabilization of the metal active phase against thermal sintering. Tuning the film characteristics can further improve the stability of the catalysts. ALD/MLD in fluidized bed or spatial ALD reactors is a promising fabrication route for preparing precisely tailored heterogeneous catalysts.

ACKNOWLEDGEMENTS

The research leading to these results has received funding from the European Research Council under the European Union's Seventh Framework Programme (FP/2007-2013)/ERC Grant, Agreement No. 279632.

REFERENCES

[1] F. Zaera, Catalysis Letters, 142, 501, 2012.
[2] M. D. Argyle, C. H. Bartholomew, Catalysts, 5, 145, 2015.
[3] J. Lu, J. W. Elam, P. C. Stair, Accounts of Chemical Research, 2013.
[4] J. Lu, K.-B. Low, Y. Lei, J. A. Libera, A. Nicholls, P. C. Stair, J. W. Elam, Nature Communications, 5, 2014.
[5] N. A. Ray, R. P. Van Duyne, P. C. Stair, The Journal of Physical Chemistry C, 116, 7748, 2012.
[6] H. Feng, J. W. Elam, J. A. Libera, W. Setthapun, P. C. Stair, Chemistry of Materials, 22, 3133, 2010.
[7] H. Feng, J. A. Libera, P. C. Stair, J. T. Miller, J. W. Elam, ACS Catalysis, 1, 665, 2011.
[8] H. Feng, J. Lu, P. C. Stair, J. W. Elam, Catalysis Letters, 141, 512, 2011.
[9] N. Cheng, M. N. Banis, J. Liu, A. Riese, X. Li, R. Li, S. Ye, S. Knights, X. Sun, Advanced Materials, 27, 277, 2015.
[10] T. D. Gould, A. Izar, A. W. Weimer, J. L. Falconer, J. W. Medlin, ACS Catalysis, 4, 2714, 2014.
[11] X. Liang, J. Li, M. Yu, C. N. McMurray, J. L. Falconer, A. W. Weimer, ACS Catalysis, 1, 1162, 2011.
[12] Z. Shang, R. L. Patel, B. W. Evanko, X. Liang, Chemical Communications, 49, 10067, 2013.
[13] H. Yi, H. Du, Y. Hu, H. Yan, H.-L. Jiang, J. Lu, ACS Catalysis, 2735, 2015.
[14] B. Fu, J. Lu, P. C. Stair, G. Xiao, M. C. Kung, H. H. Kung, Journal of Catalysis, 297, 289, 2013.
[15] J. Lu, B. Fu, M. C. Kung, G. Xiao, J. W. Elam, H. H. Kung, P. C. Stair, Science, 335, 1205, 2012.
[16] J. Lu, B. Liu, J. P. Greeley, Z. Feng, J. A. Libera, Y. Lei, M. J. Bedzyk, P. C. Stair, J. W. Elam, Chemistry of Materials, 24, 2047, 2012.
[17] H. Zhang, X.-K. Gu, C. Canlas, A. J. Kropf, P. Aich, J. P. Greeley, J. W. Elam, R. J. Meyers, J. A. Dumesic, P. C. Stair, C. L. Marshall, Angewandte Chemie International Edition, 53, 12132, 2014.
[18] A. C. Alba-Rubio, B. J. O'Neill, F. Shi, C. Akatay, C. Canlas, T. Li, R. Winans, J. W. Elam, E. A. Stach, P. M. Voyles, J. A. Dumesic, ACS Catalysis, 4, 1554, 2014.
[19] J. Lee, D. Jackson, T. Li, R. E. Winans, J. Dumesic, T. Kuech, G. Huber, Energy & Environmental Science, 2014.
[20] B. J. O'Neill, D. H. K. Jackson, A. J. Crisci, C. A. Farberow, F. Shi, A. C. Alba-Rubio, J. Lu, P. J. Dietrich, X. Gu, C. L. Marshall, P. C. Stair, J. W. Elam, J. T. Miller, F. H. Ribeiro, P. M. Voyles, J. Greeley, M. Mavrikakis, S. L. Scott, T. F. Kuech, J. A. Dumesic, Angewandte Chemie International Edition, 52, 13808, 2013.
[21] B. J. O'Neill, J. T. Miller, P. J. Dietrich, F. G. Sollberger, F. H. Ribeiro, J. A. Dumesic, ChemCatChem, 6, 2493, 2014.
[22] B. J. O'Neill, C. Sener, D. H. K. Jackson, T. F. Kuech, J. A. Dumesic, ChemSusChem, 7, 3247, 2014.
[23] H. Zhang, Y. Lei, A. J. Kropf, G. Zhang, J. W. Elam, J. T. Miller, F. Sollberger, F. Ribeiro, M. C. Akatay, E. A. Stach, J. A. Dumesic, C. L. Marshall, Journal of Catalysis, 317, 284, 2014.
[24] B. J. O'Neill, D. H. K. Jackson, J. Lee, C. Canlas, P. C. Stair, C. L. Marshall, J. W. Elam, T. F. Kuech, J. A. Dumesic, G. W. Huber, ACS Catalysis, 5, 1804, 2015.
[25] D. Valdesueiro, G. Meesters, M. Kreutzer, J. van Ommen, Materials, 8, 1249, 2015.
[26] A. Goulas, R. J. van Ommen, Journal of Materials Chemistry A, 1, 4647, 2013.
[27] S. M. George, Chemical Reviews, 110, 111, 2010.
[28] M. D. Groner, F. H. Fabreguette, J. W. Elam, S. M. George, Chemistry of Materials, 16, 639, 2004.
[29] V. Cimalla, M. Baeumler, L. Kirste, M. Prescher, B. Christian, F. Passow, F. Benkhelifa, F. Bernhardt, G. Eichapfel, M. Himmerlich, S. Krischok, J. Pezoldt, Materials Sciences and Applications, 5, 628, 2014.

Nano-oxyhydride catalysts for hydrogen production from bioethanol at room temperature

W. Fang*, C. Pirez*, S. Paul**, H. Jobic***, F. Dumeignil**** and L. Jalowiecki-Duhamel*

*Université Lille Nord de France, 59000 Lille, France
CNRS UMR8181, Unité de Catalyse et Chimie du Solide, UCCS, 59655 Villeneuve d'Ascq, France, louise.duhamel@univ-lille1.fr
**Ecole Centrale de Lille, 59655 Villeneuve d'Ascq, France, Sebastien.Paul@ec-lille.fr
***IRCELyon Institut de Recherches sur la Catalyse et l'Environnement de Lyon, 69626 Villeurbanne Cedex, France, herve.jobic@ircelyon.univ-lyon1.fr
****Institut Universitaire de France, 75005 Paris, France, Franck.Dumeignil@univ-lille1.fr

ABSTRACT

Well tuning the preparation, formulation and conditions applied, different series of nickel based nano-oxyhydride (Ni-Ce-O and Ni-Mg-Al-O based) catalysts were developed for the highly efficient and sustainable H_2 formation from ethanol in the presence of oxygen and water at room temperature. Once *in situ* treated in H_2 at adequate temperature, the nickel based mixed oxides become oxyhydrides, with the presence of hydrogen species of hydride nature in the anionic vacancies of the mixed oxides. The insertion into the solid of hydride species is well shown by Inelastic Neutron Scattering (INS). By taking advantage of the chemical energy released from the reaction between nano-oxyhydride catalysts and O_2, continuous complete conversion of ethanol specifically at 50°C (oven temperature), with simultaneously production of H_2, can be obtained. The sustainable process can be explained by a mechanism involving a heterolytic abstraction of hydride species from ethanol regenerating the oxyhydride catalyst.

Keywords: Hydrogen, oxyhydride, catalyst, ethanol, nickel.

1 INTRODUCTION

The potential benefits of a hydrogen economy coming from renewable energy sources are creating a large consensus, it is important to find a way for obtaining "green" energy. Globally, the steam reforming of natural gas, especially methane, is still the most commonly used and the most economically competitive method for H_2 production. However, H_2 economy could be clean and sustainable only if it is produced from renewable sources and bioethanol obtained through biomass is a promising renewable source of hydrogen. In theory, hydrogen production from biomass or biomass derived liquids can be a carbon-emission free process and because of its low toxicity and ease of deliverability, ethanol lends itself very well to a distributed-production strategy. Besides, the wide-spread application of fuel cells become closer to reality, so increased attention is focused on hydrogen production technologies.

So far, hydrogen has been produced from ethanol by steam reforming (SRE, Equation (1)), partial oxidation (POE, Equation (2)), oxidative steam reforming (OSRE), and autothermal reforming reactions [1-4]. The endothermic SRE reaction extracts more hydrogen atoms from ethanol and water, however, additional energy supply is needed. One alternative way of supplying heat is to add oxygen or air to the feedstock (OSRE) and simultaneously to burn a portion of ethanol.

$$C_2H_5OH + 3\ H_2O \rightarrow 2\ CO_2 + 6\ H_2 \qquad (1)$$
$$\Delta H^0_{298K} = +\ 174\ KJ\ mol^{-1}$$

$$C_2H_5OH + 1.5\ O_2 \rightarrow 2\ CO_2 + 3\ H_2 \qquad (2)$$
$$\Delta H^0_{298K} = -\ 545\ KJ\ mol^{-1}$$

The OSRE reaction is generally studied in the catalytic fixed-bed reactors [1-5] or the micro-channel membrane reactors [6]. The crucial factor is the development of highly efficient, stable and low-cost catalysts, that is, the design of the appropriate active sites which fulfil the above features of the OSRE reaction, especially the design of the non-noble metal catalysts. An ideal OSRE process desires to develop a catalyst working at low temperatures in order to build long life and safe reformers which can be employed for the portable applications.

Recently, we reported a novel route for the room temperature H_2 production over $CeNiH_ZO_Y$ nano-oxyhydride catalyst [5]. This promising technology can dramatically save energy by means of a smart combination of the exothermic reaction between the hydride species in the catalyst and O_2 (chemical energy) and the exothermic reaction between ethanol and O_2 (POE reaction). The $CeNiH_ZO_Y$ nano-oxyhydride catalyst was able to simultaneously activate ethanol with an oven fixed at 60°C, produce hydrogen, and provide hydride species to sustain the chemical reaction with O_2. It has to be precised that the reaction must be started at about 200°C. Moreover, it was shown that the effect of Ni content brings also interesting catalytic behaviors for the OSRE reaction. As a matter of fact, we reported the $CeNi_XH_ZO_Y$ oxyhydrides as efficient and stable catalysts for H_2 production from ethanol with an oven fixed at 60°C [7].

It is today a major challenge to provide low cost catalysts able to efficiently break the C-C bond from bio-ethanol. Materials, such as catalysts or electrocatalysts that are able to activate fuels like H_2, and alcohols are absolutely required for the development of new technologies [8, 9].

In the present study, we report H_2 formation on different nickel based oxyhydride catalysts from ethanol in the presence of oxygen and water. Correlations between catalytic activity and physico-chemical characterizations allow participating to the open debate on the active site and mechanism.

2 EXPERIMENTAL

The catalyst precursors were prepared by the co-precipitation method. The nano-oxides were obtained by calcination of the precursors at 500°C in the presence of air flow for 4 h [5, 7, 10, 11]. The metal loadings were analyzed by ICP-MS (CNRS–Service Central d'Analyses) (Table 1).

Inelastic neutron scattering (INS) experiments were carried out on the IN1 Lagrange spectrometer at the Institut Laue Langevin, Grenoble. 36 g of the solid was placed in a stainless steel container and *in-situ* treated in high purity H_2 (10 h).

The catalytic reaction was performed under atmospheric pressure in a fixed-bed up-flow quartz reactor (inner diameter 4 mm) fitted in a programmable oven [5, 7]. The catalyst (0.03 g) was previously *in situ* treated in pure H_2 at 250°C for Ni-Ce-O catalyst and at 450°C for Ni-Mg-Al-O catalyst for 10 h. The reactant mixture (molar ratio: $EtOH/H_2O = 1/3$) was injected into a heating chamber (ca. 140°C) by HPLC pump to the flow of O_2-N_2 mixture (60 mL min^{-1}). The gas stream of $EtOH/H_2O/O_2/N_2$=1/3/1.6/1.3 was then fed to the reactor. The outlet gases were analyzed online by GC (TRACE GC ULTRA) equipped with a TCD and a FID detector. To be noticed, the reactants were introduced in a specific order, and O_2 must be added in the last to avoid the immediate re-oxidation of the oxyhydride. After all the reactants were introduced, the reaction was activated at a certain temperature, depending on the ability of the oxyhydride catalyst to react with O_2. The heat of the oven was reduced as soon as the measured reaction temperature started to increase drastically, and the variation in the reaction temperature was recorded.

3 RESULTS AND DISCUSSION

Different nickel based nano-oxyhydride (Ni-Ce-O and Ni-Mg-Al-O based) catalysts are developped. The selected pre-treatment temperature in H_2 is important for generation of an oxyhydride. The influence of the pretreatment temperature (T_T) on ethanol conversion in SRE conditions has been studied previously. On cerium nickel based compounds an optimum T_T has been reported at 250°C [12, 13], while on Ni-Mg-Al-O compound it has been shown that when increasing T_T an increase of the conversion can be obtained however, carbon formation increases also, therefore a good compromise was obtained for a T_T at 450°C [10, 11].

Catalyst	Ni [wt%]	S_{BET} [m^2 g^{-1}]
Ni-Ce-O	21	110
Ni-Mg-Al-O	62	196

Table 1. Ni content and specific area of the catalysts

Once *in situ* treated in H_2 at adequate temperature, the nickel based mixed oxides (Table 1) become oxyhydrides, with the presence of hydrogen species of hydride nature located in the anionic vacancies formed in the mixed oxides. The insertion into the solid of hydride species is well shown here by Inelastic Neutron Scattering (INS) [14, 15]. The H_2 treatment generates an increase of the INS spectrum level as well as the emergence of new peaks associated to hydrogen vibration bands. Importantly, the spectrum level is proportional to the concentration of hydrogen. The hydrogen species with the hydride nature appear clearly on Ni-Ce-O compound after a treatment in H_2 at 250 °C (Figure 1) and on Ni-Mg-Al-O compound after a treatment in H_2 at 450°C (Figure 2). It has to be remarked that after treatment in vaccum at 200°C (to eliminate physisorbed water) some hydrogen species are always present in the compounds, corresponding to hydroxyl groups. The vibration bands obtained at about 30, 50 and 77 meV were assigned to hydroxyl groups on the $CeNi_XO_Y$ nano-compounds [5]. After treatment in H_2, main new peaks are observed at around 60 meV and 110 meV for Ni-Ce-O compound and at around 76, 98 and 208 meV for Ni-Mg-Al-O, even if the spectra are complex contribution from the ternary Mg–Al–Ni system. It has been reported that magnesium hydride presents two peaks in the range between 60-80 meV [16], while aluminum hydride leads to a large spectrum with several peaks between 76 and 117 meV [17], and nickel hydride shows broad bands at around 100 and 200 meV [18].

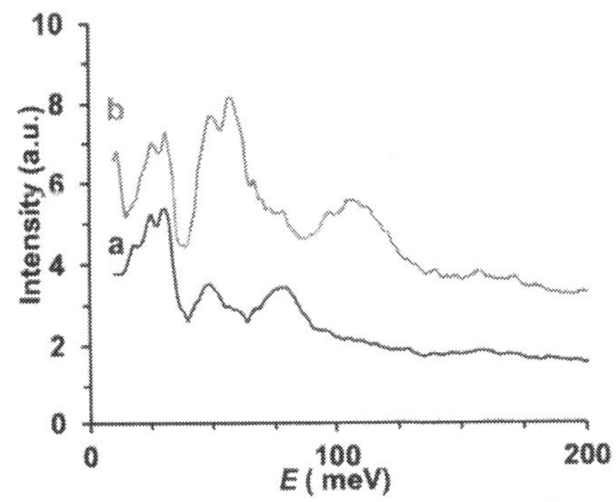

Figure 1: INS spectra of Ni-Ce-O a) treated in vacuum at 200°C for 2 h and b) in H_2 at 250°C for 10 h.

Hydrogen can be mainly heterolytically dissociated on an anionic vacancy and an O^{2-} species of the catalyst [Equation (3)].

$$O^{2-}\,M^{n+}\,\square + H_2 \rightarrow OH^-\,M^{n+}\,H^- \qquad (3)$$

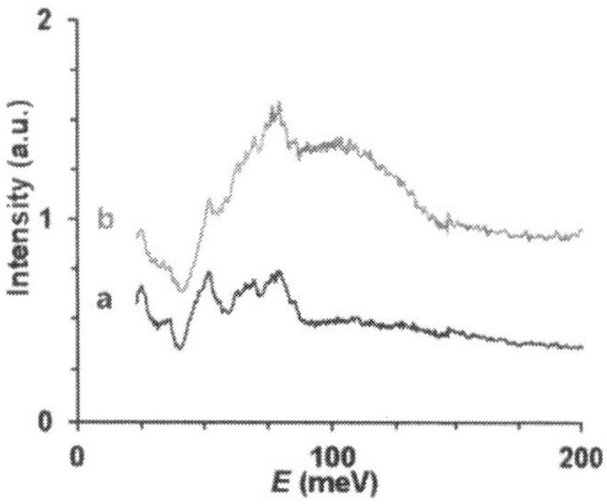

Figure 2: INS spectra of Ni-Mg-Al-O a) treated in vacuum at 200°C for 2 h and b) in H_2 at 450°C for 10 h.

The catalyst is first *in situ* pre-treated in H_2 in order to form the nano-oxyhydride, then it is studied for H_2 production in the presence of ethanol, water and O_2 (molar ratio: $EtOH/H_2O/O_2/N_2$ = 1/3/1.6/1.3). The oven temperature is fixed at only 50°C whereas the reaction temperature is measured and reported. The reaction must be initiated at about 200°C, then the temperature quickly increases during an induction period of a few minutes (around 3 min). The heating can be then lowered and only a small extra energy is needed to realize the reaction. The measured reaction temperature is reported at the stable state.

Ethanol and oxygen are completely converted on the catalysts (Figures 3 and 4). Only small quantities of CH_4 and CH_3CHO are detected in outlet gases (< 1%). In complement to the exothermic partial oxidation reaction of ethanol, by taking advantage of the chemical energy released from the reaction between hydride species stored in the nano-oxyhydride catalysts and O_2, continuous complete conversion of ethanol specifically at 50°C (oven temperature), with simultaneously production of H_2, can be obtained. Figure 3 shows the results obtained on the Ni-Ce-O based nano-oxyhydride, with a H_2 formation of about 46% (mol%) relative to all the gas phase products (dry basis). The other products analyzed are mainly CO_2 (41%) and CO (12%). The oven temperature is fixed at 50°C and the measured reaction temperature is much higher at 330°C. Figure 4 shows that relatively similar results can be obtained on a Ni-Mg-Al-O based nano-oxyhydride catalyst. A H_2 formation of about 45% is obtained with mainly CO_2 (45%) and CO (10%). The oven temperature is fixed at 50°C while the measured reaction temperature is at 260°C.

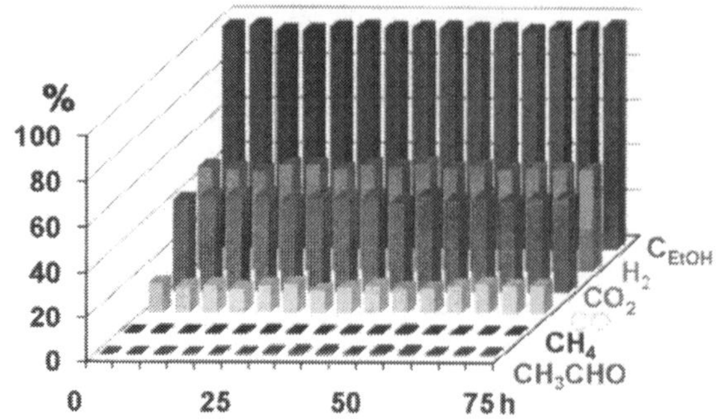

Figure 3: Ethanol conversion and products distribution on Ni-Ce-O based nano-oxyhydride. The reaction mixture is $EtOH/H_2O/O_2/N_2$= 1/3/1.6/1.3. The oven temperature is fixed at 50°C and the measured reaction temperature is at 330°C.

The catalysts exhibit remarkable stability after at least 75 h of reaction, even if filamentous carbon deposition is found after the reaction. The conversion is increased significantly by the addition of O_2 into the reactant mixture. For comparison, we have reported a high and stable activity (5 h) at 250°C in the SRE conditions (14 mol.% EtOH) with ethanol conversion at 50% and H_2 formation at 55% over high Ni content Ni-Mg-Al-O catalyst (0.05 g).

H_2 and CO_2 are the ideal products from SRE and POE processes [Equations (1) and (2)], and CO is the only byproduct. The sum of acetaldehyde and methane is lower than 1%. However, solid carbon is also formed.

Dehydrogenation of ethanol produces acetaldehyde and H_2 [Equation (4)], whereas decomposition of ethanol (and/or acetaldehyde) produces CO, methane and H_2 [Equation (5)] and total decomposition leads to CO, H_2 and carbon [Equation (6)]. Water-gas shift reaction favored at low temperature allows the transformation of CO [Equation (7)]. In the presence of O_2, partial oxidation reactions take place. POE may be considered as a combination of partial oxidation to syngas coupled with CO oxidation [2].

$$C_2H_5OH \rightarrow C_2H_4O + H_2 \qquad (4)$$
$$C_2H_5OH \rightarrow CO + CH_4 + H_2 \qquad (5)$$
$$C_2H_5OH \rightarrow CO + 3\,H_2 + C \qquad (6)$$
$$CO + H_2O \rightarrow CO_2 + H_2 \qquad (7)$$

Acetaldehyde has been proposed as an important reaction intermediate in SRE and OSRE, and the abstraction and dissociation of hydrogen species from ethanol (dehydrogenation) through C–H rupture is usually remarked as the rate-determining step. As we have proposed for the heterolytic dissociation of H_2 [Equation (3)], an active site involving an anionic vacancy and an O^{2-} species of the solid can be envisaged for the heterolytic dissociation of ethanol [Equation (8)].

$$O^{2-}M^{n+}\square + C_2H_5OH \rightarrow OH^-M^{n+}H^- + C_2H_4O \quad (8)$$

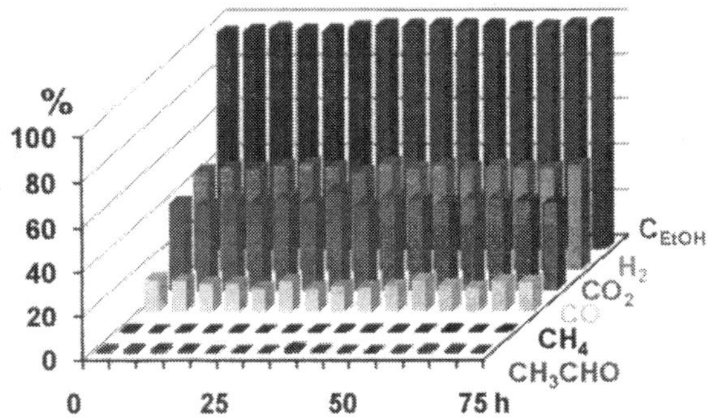

Figure 4: Ethanol conversion and products distribution on Ni-Mg-Al-O based nano-oxyhydride. The reaction mixture is $EtOH/H_2O/O_2/N_2$= 1/3/1.6/1.3. The oven temperature is fixed at 50°C and the measured reaction temperature is at 260°C.

The present result is a major advance in the field, because conversion of concentrated ethanol into H_2 production with low energy consumption is of value in the practical use.

Different physicochemical techniques were used to characterize the catalysts and carbon formed. The active nickel species are found in strong interaction with other cations. Finally, to explain the sustainable process, an active site involving an anionic vacancy and an O^{2-} species of the solid can be envisaged, as proposed in Scheme 1, with abstraction of hydride species from ethanol and regeneration of the oxyhydride catalyst.

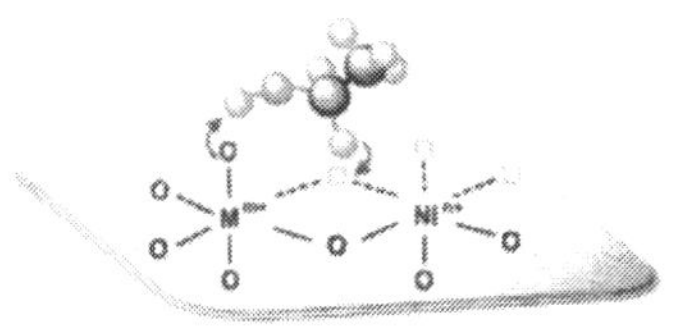

Scheme 1: Activation of ethanol

4 CONCLUSION

Hydrogen has been produced from ethanol in the presence of oxygen and water. Well tuning the catalyst preparation, formulation and Ni content, as well as the activation procedure and the reaction conditions allows obtaining a high and stable activity at room temperature (oven) even if some solid carbon is formed. It appears that the active nickel species are in strong interaction with other cations. An active site involving cations and anionic vacancies is proposed leading to a mechanism involving heterolytic abstraction of hydride species from ethanol.

REFERENCES

[1] G. A. Deluga, J. R. Salge, L. D. Schmidt, X. E. Verykios, Science 30, 5660, 993, 2004.
[2] V. Subramani and C. Song, in Catalysis: Volume 20, The Royal Society of Chemistry, pp. 65, 2007.
[3] K. Sato, K. Kawano, A. Ito, Y. Takita and K. Nagaoka, ChemSusChem, 3, 1364, 2010.
[4] L. V. Mattos, G. Jacobs, B. H. Davis, F. B. Noronha, Chem. Rev. 112, 4094, 2012.
[5] C. Pirez, M. Capron, H. Jobic, F. Dumeignil and L. Jalowiecki-Duhamel, Angew. Chem. Int. Ed 50, 10193, 2011.
[6] N. Zhu, X. Dong, Z. Liu, G. Zhang, W. Jin and N. Xu, Chem. Commun. 48, 7137, 2012.
[7] W. Fang, C. Pirez, S. Paul, M. Capron, H. Jobic, F. Dumeignil and L. Jalowiecki-Duhamel, ChemCatChem 5, 2207, 2013.
[8] H. Qin, Z. Zhu, Q. Liu, Y. Jing, R. Raza, S. Imran, M. Singh, G. Abbas and B. Zhu, Energy Environ. Sci. 4, 1273, 2011.
[9] A. Kowal, M. Li, M. Shao, K. Sasaki, M. B. Vukmirovic, J. Zhang, N. S.Marinkovic, P. Liu, A. I. Frenkel and R.R. Adzic, Nat. Mater. 8, 325, 2009.
[10] W. Fang, S. Paul, M. Capron, F. Dumeignil and L. Jalowiecki-Duhamel, Appl. Catal. B 152–153, 370, 2014.
[11] W. Fang, S. Paul, M. Capron, A. V. Biradar, S. B. Umbarkar, M. K. Dongare, F. Dumeignil and L. Jalowiecki-Duhamel, Appl. Catal. B 166–167, 485, 2015.
[12] L. Jalowiecki-Duhamel, C. Pirez, M. Capron, F. Dumeignil and E. Payen, Int. J. Hydrogen Energy 35, 12741, 2010.
[13] L. Jalowiecki-Duhamel, C. Pirez, M. Capron, F. Dumeignil and E. Payen, Catal. Today 157, 456, 2010.
[14] L. Jalowiecki-Duhamel, F. Dumeignil, W. Fang, M. Jiménez-Ruiz, H Jobic, Institut Laue-Langevin (ILL), 2012, doi:10.5291/ILL-DATA.7-05-397.
[15] H. Jobic, Handbook of Heterogeneous Catalysis, 2nd Ed., Eds. G. Ertl, H. Knözinger, F. Schüth, J. Weitkamp, Wiley-VCH, 971, 2008.
[16] A. I. Kolesnikov, V. E. Antonov, V. S. Efimchenko, G. Granroth, S. N. Klyamkin, A. V. Levchenko, M. K. Sakharov and Y. Ren, J. Alloys Compd. 509, 2, S599, 2011.
[17] A. I. Kolesnikov, V. E. Antonov, Y. E. Markushkin, I. Natkaniec and M. K. Sakharov, Phys. Rev. B 76, 064302, 2007.
[18] V. E. Antonov, V. K. Fedotov, B. A. Gnesin, G. Grosse, A. S. Ivanov, A. I. Kolesnikov and F. E Wagner, Europhys. Lett 51, 140, 2000.

On the behavior of modified TiO_2 nanotubes for a photoanode-driven photoelectrochemical reduction of CO_2

S. Palmas*, M. Mascia*, A. Vacca*, P. Ampudia*, L. Mais*, F. Ferrara** and A. Pettinau**

*Dipartimento di Ingegneria Meccanica Chimica e dei Materiali, Università degli Studi di Cagliari. Via Marengo 2, 09123 Cagliari, Italy, simonetta.palmas@dimcm.unica.it

**Sotacarbo S.p.A., c/o Grande Miniera di Serbariu, 09013 Carbonia, Italy, f.ferrara@sotacarbo.it

ABSTRACT

Chemical reduction of CO_2 is thermodynamically feasible only in the presence of strong reducing agents. Photoelectrochemical (PEC) reduction represents one of the possible alternatives: this technique may be considered "green" when solar energy and water are used to drive the process, like in natural photosynthesis. Different system configurations and different experimental conditions are needed, depending on the use of photosensitive material as cathode, anode or both in the PEC cell. In the present paper we investigated the possibility to use a photoanode-driven PEC cell, in which a metal was used as cathode, while the anode material was a nanotubular structure of TiO_2. The photocatalytic performance of the anode was improved by deposition of tungsten oxide. Electrodeposition was performed both by potential step (PST) and pulse potential technique (PPT). Different morphology of the deposit was obtained and, in turn, different performance was measured in terms of photocurrent at the related samples.

Keywords: titania nanotubes, tungsten oxide electrodeposition, photoelectrocatalysis, pulse and step potential techniques.

1 INTRODUCTION

Carbon dioxide is one of the main greenhouse gases responsible for global warming. Due to the high costs of its disposal, many efforts have recently been addressed to its reuse: converting CO_2 to valuable compounds such as CO, methane and methanol, allows considering it as a feedstock, rather than a waste product [1, 2].

Actually, chemical reduction of carbon dioxide is thermodynamically feasible but only in the presence of strong reducing agents [2]. In alternative, electroreduction of CO_2 has been widely studied. The literature reports excellent overview of possible products obtained by electrochemical CO_2 reduction, at different electrode materials and experimental conditions [1-7]. However, research on this topic is still opened, being high energy efficiency achievable through a combination of high faradaic efficiency for the desired product, and low cell potential [4].

In this contest using a PEC cell, in which at least one electrode is a photosensitive material, could represent a valid solution to reduce the cell voltage.

In the present paper a photoanode-driven PEC cell is considered in which the lowering of the cell voltage is obtained from the photopotential generated at the nanotubular structure of TiO_2 n-type semiconductor, irradiated by a suitable light wavelenght. This anode could be coupled with metals or metal composite oxides, which are indicated in the literature as effective cathodic materials to this aim [3]. Titania nanotubular anodes were already tested in our laboratory for the photoelectrosplitting of water or oxidation of organics [8-11]. The basic properties of the original structure, such as regular architecture, straight path for photogenerated carriers, and high specific surface, confirmed the effectiveness of the samples in the UV range of light. In those cases a nominal 1 cm^2 of active surface was used.

In the present work a scale up of the system has been made to obtain electrodes of 15 cm^2. Moreover, to improve the final photocatalytic performance in the near visible range, surface modification by tungsten oxide was attempted. Electro-deposition was performed both by step potential (SPT) and pulse potential (PPT) techniques. Depending on the experimental technique, different morphologies of the deposit were obtained and in turn, different performances were measured in terms of photocurrent at the related samples.

2 EXPERIMENTAL

2.1 Preparation of nanotubular structures

Ti foils (0.25 mm thickness, 99.7 % metal basis, Aldrich) of 5 cm diameter were used as raw material to obtain the nanotubular structures by electrochemical oxidation (final samples are indicated as NT5 in the rest of the text). Figure 1 shows a schematic view of the electrochemical cell used to obtain NT5 samples. The teflon cell is composed by one cylindrical chamber (d=4.4 x h=5 cm) containing the electrolyte solution. The electrical contact with the Ti foil was realized by means of an aluminium disc (d=5 x h=0.5 cm) located at the bottom side of the cell. The electrode/electrolyte contact area was 15 cm^2. A platinum grid cathode was set at 1 cm distance, in front of the anode. Prior to the electrochemical oxidation,

Ti foils were degreased by sonication in acetone, then in isopropanol and finally in methanol; then, they were rinsed with deionized water and dried in a Nitrogen stream.

The oxidation process was performed in (10%) deionized water / (90%) glycerol organic solution with 0.14 M of NH_4F at room temperature. A potential ramp was imposed from open-circuit-voltage (OCV) at a fixed potential of 20 V with a scan rate of 100 mVs^{-1}. The applied potential was maintained at this fixed value for 4 h.

A final annealing treatment was required in order to transform the amorphous structure into crystalline one. Thermal treatment was performed in air atmosphere at 400°C for 1 h.

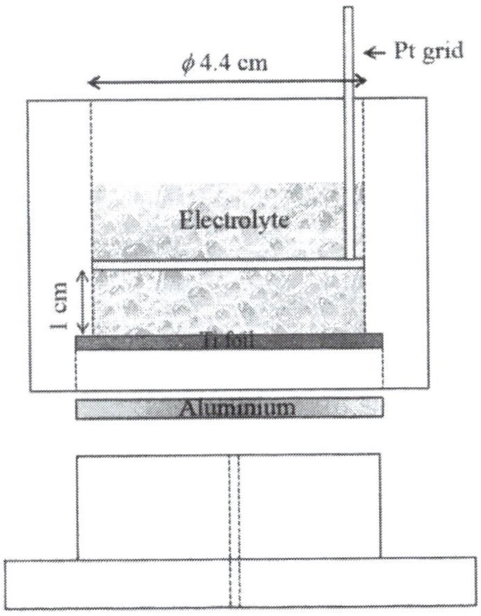

Figure 1: Electrochemical cell for NT5 preparation.

2.2 Electrodeposition of tungsten oxide

The electrodeposition of tungsten oxide was performed at room temperature in a classical three-electrode cell in which a saturated calomel electrode (SCE) was the reference, a platinium grid was the counter electrode, while titania was used as working electrodes. To this aim samples of 3.75 cm^2 nominal area, were derived by cutting an NT5 sample, previously obtained (final samples at which deposit of tungsten oxide was obtained are indicated as NTWx in the rest of the text, x indicating the depisition time). The electrodes were connected to a potentiostat-galvanostat (Amel 7050), controlled by Junior Assist software. A receipt suggested by the literature was adopted to deposit tungsten oxide [12]. Tipically, 25 mL of electrolytic solution, composed by 25 mM Na_2WO_4, 30 mM of concentrated H_2O_2, was used. pH of the solution was adjusted to 1.4 by adding nitric acid.

Preliminary voltammetric study was performed (results not showed) which allowed to individuate the main reductive wave at -0.3 V vs. SCE. This potential value was adopted in the deposition process of tungsten oxide by potentiostatic runs. To this aim both step (SPT) and pulse (PPT) techniques were used in order to obtain the oxide deposits over the nanotubular structures. During SPT, the electrodeposition of tungsten oxide was performed at different times, fixing the potential value at -0.3 V for 8 and 12 min, respectively. During PPT the potential was alternately swiftly between opened circuit potential (OCV = 0.4 V) and -0.3 V. A serie of pulses of equal amplitude but different width was performed. Each pulse consisted of an ON-time of 10 s during which -0.3 V was applied, and an OFF-time of 1 s during which the electrode was maintained at OCV. The number of pulses required to reach the same deposition times adopted for the deposition step technique, was calculated according to the ON-time duration. The current density was recorded during all the deposition experiments. The obtained NTWx samples were annealed in air atmosphere at 400°C for 1h.

2.3 Characterization of the samples

The uniformity of the NT5 structure and its photoactivity was tested in the three-electrode cell: cyclic voltammetries and photo-currents were measured at 9 points of the NT5 sample (see figure 2). Specific runs were performed in which a teflon holed mask allowed one section a time to be exposed to the solution and to the irradiation of a Xenon lamp (nominal power: 300W). The photocurrents were recorded at different wave lenghts; also measurements were done in which the lamp was used without filter, in order to simulate the solar radiation. SEM analysis was performed to investigate on the morphology of the structures. Elemental maps were also collected for NTWx samples to assess the presence of W.

3 RESULTS AND DISCUSSION

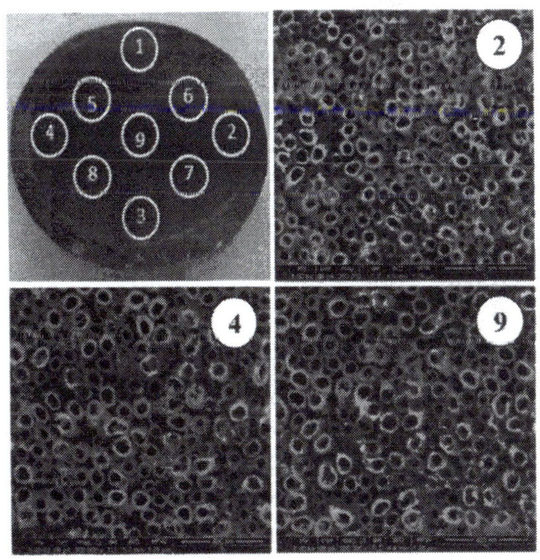

Figure 2: SEM micrographs of NT5 sample at 3 investigated points.

The adopted procedure, described in the experimental section, allowed us to obtain NT5 with an active surface of 15 cm^2.

Figure 2 shows some examples of the results of the SEM analyses performed at different section points in order to verify the uniformity of the NTs.

If we exclude the border line effects (about 5mm corresponding to the thickness of the wall of the cilindrical cell), a almost regular structure of the NTs was observed with a mean tube diameter between 40 and 50 nm measured in different points of the sample.

Also border effects are visible from the results obtained from photocurrent measurements. Figure 3 resumes the values of photocurrent measured at the different points of the active surface. Also in this case we may observe a quite regular trend, most in the central part of the surface, while peripherical points 1,2,3,4 exhibit lower values. However, in this case, the effect could result not only from border effects, but also from a higher concentration of the light irradiation in the central part of the spot emmitted by the Xenon lamp.

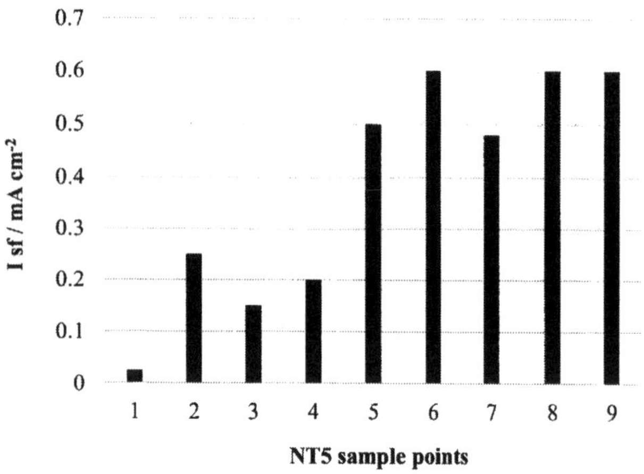

Figure 3: Photocurrents density distribution at NT5 sample obtained by irradiating with the Xenon lamp without filter.

Slight morphological differences observed in different points of the sample also resulted in a slight shift of the peak current densities in the CV, as shown in Figure 4, while no variation of the peak potential position was observed.

Similar results were obtained at the different samples prepared under the same conditions, confirming a good uniformity and, overall, a good reproducibility of the synthesis.

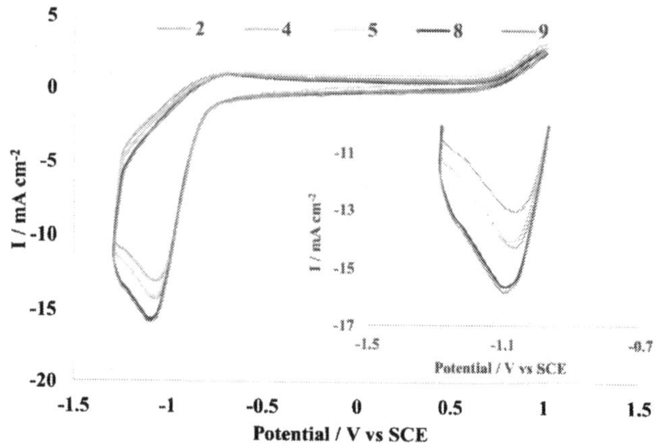

Figure 4: Cyclic voltammetries of NT5 sample recorded in different points of the surface.

Figure 5 shows the results of photoactivity of the NTWx samples, in terms of photocurrents obtained by irradiating the samples with light of 380 and 400 nm wavelenght.

In particular, data obtained with PPT and SPT are compared with those obtained at original NT (no tungsten deposit).

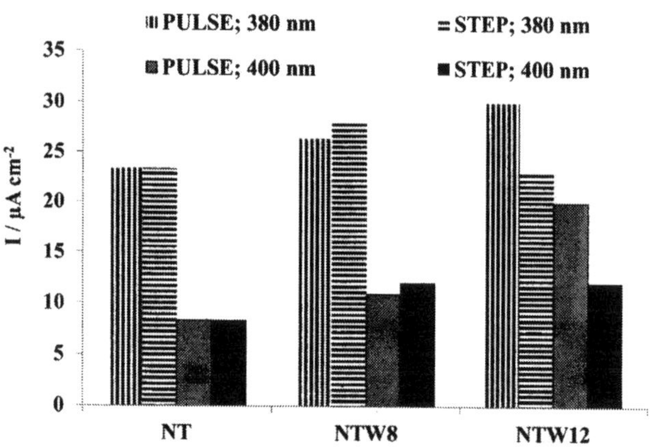

Figure 5: Effect of electrodeposition method on the photocurrents density response at different wave lenghts.

A slight increase of the performance is generally observed at samples where tungsten oxide was deposited. Moreover, as we expected, at the highest wave length lower values of photocurrent are obtained (full bars). If the data related to samples obtained by PPT are considered, at both 380 nm (vertical lines bars) and 400 nm (gray bars), an increasing trend is observed while passing from NT to NTW8 and NTW12. The trend is different when we consider the results related to samples obtained by SPT: actually, a sort of saturation point may be observed at 400 nm (black bars) at which the performance at NTW12 is almost the same as at NTW8, and especially at 380 nm (horizontal lines bars) where the performance of NTW12 is even lowered with respect to that measuered at NTW8.

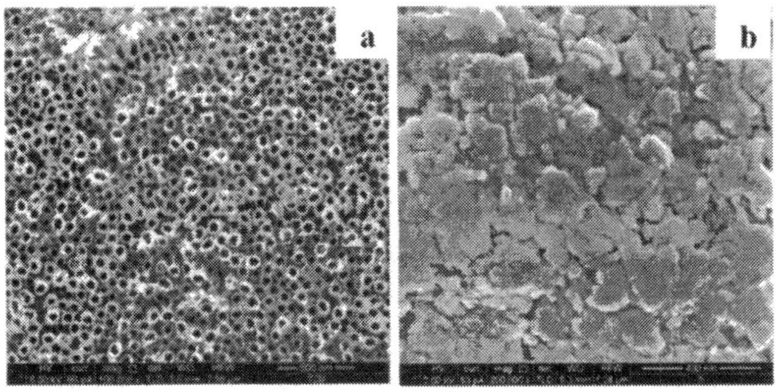

Figure 6: SEM micrographs of electrodeposited tungsten oxide obtained by SPT at a) 8 min; and b) 15 min.

In order to justify these results also the morphology of the tungsten oxide deposit has been considered. Elemental map analysis of the surface (data not shown) indicated a regular distribution of W all over the surface: according to the pH (1.4) and V (-0.3 V), at which the deposit is obtained, WO_3 could be the termodinamically stable form of W [13]. Figure 6 shows as an example, the sample surface after a step potential of 8 minute (left) and 15 minutes (right) was applied. As we can observe, the high deposition time provoks an eccess of deposit which plugs all the NTs so that the performance is strongly affected.

4 CONCLUSIONS

This paper reported the results of an experimental work in which an attempt is made to scale up the process of oxidation of Ti foils, obtaining titania NTs of 15 cm^2 nominal surface to be used in a photoanode-driven PEC cell. The results confirm the quite regular structure of the samples, in terms of morphology and photoactivity.

Samples were also used to derive the substrates to test the tungsten oxide deposition, performed with SPT and PPT. Even if the exact composition of the oxide is not identified, the results obtained up to now indicate the presence of W on the nanotubular structure of the electrodes as well as its positive effect on the photoactivity. Deposition with the PPT seems to give better results. In particular the performance obtained at NTW12 at 400 nn is very similar to that measured at the sample obtained by SPT at 380 nm.

However, it must be observed that the data reported in this paper are not optimised: in particular, an optimum in the times selected for the deposits has to be identified. Further research will be addressed to study this effect both for PPT (in order to evaluate the effect of different times of the ON-OFF states), and especially for samples obtained by SPT, because high deposition times may originate deposits which may shield the nanotubular structure, which in turn loss the good properties, so decreasing the performances of the whole electrode. Moreover, a comparison based on the electrical charge involved in the two techiques will be performed.

REFERENCES

[1] Y.-P. Peng, Y.-T. Yeh, S. I. Shah and C. P. Huang, Appl. Catal., B, 123, 414, 2012.

[2] G. Centi, E. A. Quadrelli and S. Perathoner, Energy Environ. Sci., 6, 1711, 2013.

[3] M. Jitaru, D. A. Lowy, M. Toma, B. C. Toma and L. Oniciu, J. Appl. Electrochem., 27, 875, 1997.

[4] H-R. Jhong, S. Ma, P. JA Kenis, Current Opinion in Chemical Engineering, 2, 191, 2013.

[5] G. Magesh, E.S. Kim, H. J. Kang, M. Banu, J. Y. Kim, J. H. Kimb and J. S. Lee, J. Mater. Chem. A, 2, 2044, 2014.

[6] KP Kuhl, ER Cave, DN Abram, TF Jaramillo, Energy Environ Sci, 5, 7050, 2012

[7] M. Gattrell, N. Gupta, A. C. J. Electroanal. Chem., 594, 1, 2006.

[8] S. Palmas, M. Mascia, A.Vacca, and I. Tredici, Int. J. Photoen., article ID 173760, ISSN: 1110-662X, 2013.

[9] L. Özcan, S. Yurdakal, V. Augugliaro, V. Loddo, S. Palmas, G. Palmisano, L. Palmisano, Appl. Catal. B, Environm., 132-133, 535, 2013.

[10] S. Palmas, A.Da Pozzo, M. Mascia, A. Vacca, P.C. Ricci, R. Matarrese, J. Solid State Electrochem., 16, 2493, 2012.

[11] S. Palmas, A.M. Polcaro, J. Rodriguez Ruiz, A. Da Pozzo, M. Mascia, A. Vacca, Int. J. Hydrogen Energy, 35, 6561, 2010.

[12] K.R. Reyes, D.B. Robinson, ACS Appl. Mater. Interfaces, 5, 12400, 2013.

[13] J. B. Christian, Chem. Eng. Progr., 99, 32, 2003.

Evolutions of self-assembled Bi_2O_3/LDHs nanostructures to mixtures of mixed oxides and their photocatalytic effiecency under UV and solar light

K. Katsumata*, M. Puscasu**, E. Seftel***, E. F. Grosu**, K. Ikeda*, P. Cool*** and G. Carja**

* Materials and Structures Laboratory, Tokyo Institute of Technology, 4259 Nagatsuta, Midori, Yokohama, 226-8503, Tokyo, Japan

** Department of Chemical Engineering, Faculty of Chemical Engineering and Environemntal Protection, Tecnhical University "Gh. Asachi" of Iasi, Bd. Mangeron no.71, Iasi 700554, Romania, carja@uaic.ro

***Laboratory of Adsorption and Catalysis, Department of Chemistry, University of Antwerpen (CDE), Universiteitsplein 1, 2610 Wilrijk, Antwerpen, Belgium

ABSTRACT

In this work we report self-assemblies of nanoparticles of Bi_2O_3 and layered double hydroxides (LDHs) obtained by exploiting the manifestation of the "structural memory effect" of ZnAlLDH and ZnCrLDH anionic clay matrices in $Bi(NO_3)_3$ aqueous solutions. Controlled thermal treatment of Bi_2O_3/ZnAlLDH and Bi_2O_3/ZnCrLDH self-assemblies gives rise to photoresponsive mixtures of mixed oxides, types Bi_2O_3/ZnO-$ZnAl_2O_4$ and Bi_2O_3/ZnO-$ZnCr_2O_4$, respectively. Powder X-ray diffraction (PXRD), X-ray photoelectron spectroscopy (XPS), high resolution transmission electron microscopy (TEM) and UV-Visible spectroscopy (UV-Vis) showed that the LDHs reconstruction in $Bi(NO_3)_3$ solutions gives rise to specific nanoarchitectonics characteristics in which small NPs of Bi_2O_3 (~5 nm) are highly dispersed on the larger NPs (~150 nm) of ZnAlLDH and ZnCrLDH. UV-Vis analysis indicates the cooperative absorbance capabilities of the novel catalytic systems that were evaluated in the degradation process of phenol from aqueous solutions under solar and UV light.

Keywords: Bi_2O_3, LDHs self-assembled nanocomposites, solar irradiation, photocatalyst

1 INTRODUCTION

In recent years increasing attention is given on designing complex nanocomposites systems engineered to act as efficient photocatalysts; the unique nano size-dependent transport properties related to photons are able to establish specific photoresponsive functions that cannot be achieved by their bulk counterparts [1]. In this regard, manipulation of the photocatalytic properties by selective control of the catalyst nanoarchitectonics and by utilizing the collective properties of nanoassembly systems is a continuous challenge [2]. Design of photocatalysts nanoarchitectonics refers not only to their simple construction as nanomaterials but also to ensemble different nanoscaled structural building units in designed and defined ways for creating complex formulations owning collective and synergic properties [3]. Layered double hydroxides (LDHs) are a class of anionic clays. An important property of LDHs is their structural "memory effect"; it refers to the structural reconstruction of the layered clay matrix when the metal mixed oxides, evolved after the calcination around 550°C, are exposed to the aqueous solutions containing anions [4]. Using the manifestation of the LDHs structural "memory effect" in the aqueous solutions of metal salts, we have recently reported a simple fabrication procedure of nanoparticles of metal oxides-LDHs (MexOy/LDHs) as self assemblies [5]. According to this procedure LDH matrix is able to rebuild its structure by using the anions of the solution as interlayer anions and also to adsorb the metal cations of the aqueous solution and organizing them as nanoparticles (NPs) of metal oxides on its surface [6, 7]. MexOy/LDHs self-ensembled composites with a large compositional diversity can be designed by modifying either the nature of the metal cations of the LDHs layers and/or the nature of the NPs of MexOy. NPs of Bi_2O_3 has attracted a great deal of attention in the last years [8]. Hence the joined functionalities of NPs of Bi_2O_3 and LDHs might give rise to unique photoresponsive characteristics.

Herein, we report for the first time the self - assemblies of NPs of Bi_2O_3 with the specific compositions of ZnAlLDH and ZnCrLDH and their evolution, after the thermal treatment at 800°C to nanstructured mixtures of mixed oxides, types Bi_2O_3/ZnO-$ZnAl_2O_4$ and Bi_2O_3/ZnO-$ZnCr_2O_4$. The fabrication procedure of Bi_2O_3/LDHs is simple and cost-effective. It is based on the reconstruction of the LDH clays, due to the "structural memory" in the aqueous solution of $Bi(NO_3)_3$, at room temperature [6]. Results show that nanoarchitectonics of Bi_2O_3/LDHs and the derived mixed oxides systems are active photocatalysts in the process of phenol photodegradation from aqueous solutions.

2 RESULTS AND DISCUSSIONS

2.1. XRD, TEM and XPS analyses

The crystalline structure of Bi_2O_3/ZnAlLDH and Bi_2O_3/ZnCrLDH - as reconstructed in Bi_2O_3 solution and after calcination at 800°C - was first characterized by XRD. Fig. 1A illustrates the characteristics of the obtained powder X-ray diffraction patterns of Bi_2O_3/ZnAlLDH and Bi_2O_3/ZnCrLDH. It reveals that Bi_2O_3/ZnAlLDHs consists of crystalline phases with reflections clearly assigned to the regular layered structure of hydrotalcite-like clay (ICDD file No. 22-700), in which a series of sharp and symmetric basal reflections of the (00ℓ, ℓ = 3, 6, 9) planes and broad, less intense, reflections for the nonbasal (01ℓ, ℓ = 2, 5, 8) planes are easily recognized [5].

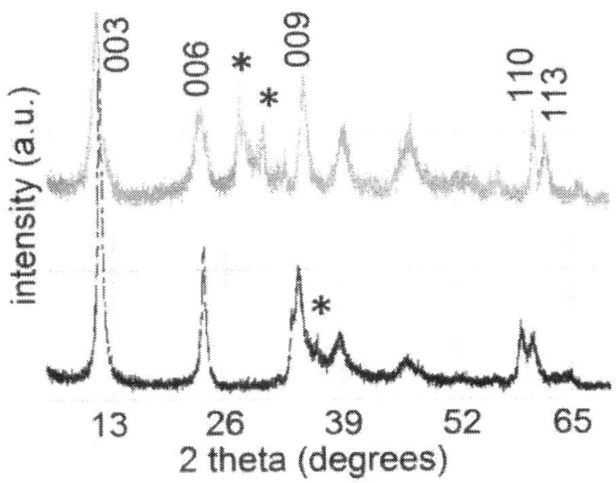

Fig. 1A. The XRD patterns of (-) Bi_2O_3/ZnAlLDH and (-) Bi_2O_3/ZnCrLDH.

On the other hand, besides the reflections assigned to the regular layered structure of LDH some weaker reflections, (marked by (*) in Fig. 1A), with positions which roughly coincide with those of mixed oxides that define the structure of the calcined LDHs can be also identified indicating that the structural reconstruction of the LDHs was not complete. Further, the structural characteristics of the mixtures of mixed oxides obtained after the calcination at 800°C of the as-synthesized and the reconstructed LDHs are shown in Fig. 1B for ZnAlLDH and Fig. 1C for ZnCrLDH, respectively. In Fig 1B some sets of diffraction peaks that can be indexed to hexagonal wurtzite ZnO (JCPDS file no 36-1451), $ZnAl_2O_4$ spinel (JCPDS file No. 5-0669) can be clearly identified while the new peaks observed at 2θ values 27.9, 33.7, 46.6, and 47.1° might correspond to (121), (122), (041) and (104) diffraction planes of α-Bi_2O_3 (JCPDS no. 71−2274). Hence, in this case, a mixture of α-Bi_2O_3/ZnO-$ZnAl_2O_4$ was obtained. In comparison the XRD pattern of Bi_2O_3/ZnCrLDH calcined at 800°C indicates the formation of a mixture of mixed oxides, type α-Bi_2O_3/$ZnCr_2O_4$-ZnO.

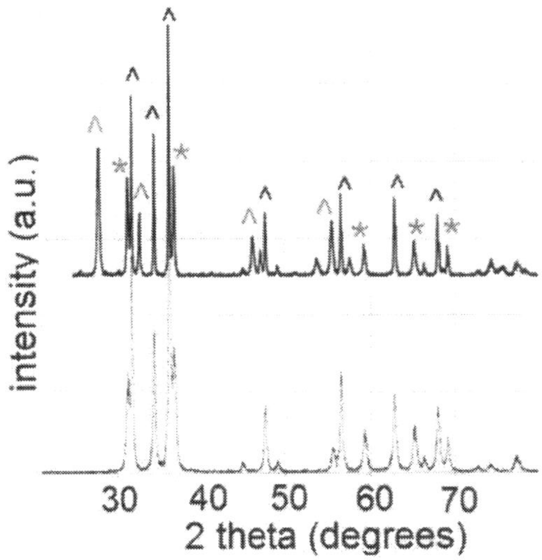

Fig. 1B. The XRD patterns of the mixed oxides derived from - Bi_2O_3/ZnAlLDH and - ZnAlLDH after calcination at 800°C; (^) Bi_2O_3, (^) ZnO, (*) $ZnAl_2O_4$.

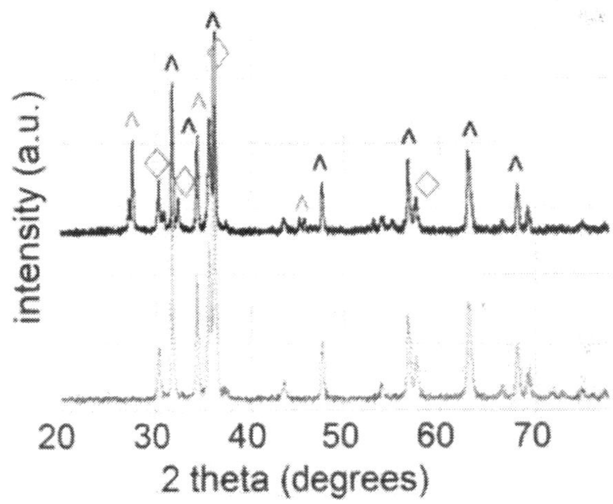

Fig. 1C. The XRD patterns of the mixed oxides derived from (-) Bi_2O_3/ZnCrLDH and (-) ZnCrLDH after calcination at 800°C; (^) Bi_2O_3, (^) ZnO, (◊) $ZnCr_2O_4$.

The detailed crystal structure and morphology was observed by TEM. Fig. 2A-D shows TEM as well as HRTEM images of Bi_2O_3/ZnAlLDH and Bi_2O_3/ZnCrLDH. Fig. 2A illustrates that the reconstructed LDHs shows the conventional LDHs morphology consisting of aggregates of platelet-like particles with average sizes of ~150 nm. In the typical transmission electron microscopy (TEM) image (see Fig. 2B and 2D) very small NPs can be clearly observed as dark spots highly dispersed on the larger particle of the clay; the average size of the loaded NPs is 4.7 nm for

Bi_2O_3/ZnAlLDH and 5.4 nm for Bi_2O_3/ZnCrLDH. The HRTEM image, (see Fig. 2C) indicates that the small NPs are highly crystalline with a well defined spacing of ca. 0.31 nm between adjacent lattice fringes, close to the d-spacing value of the (111) plane of α-Bi_2O_3 [9].

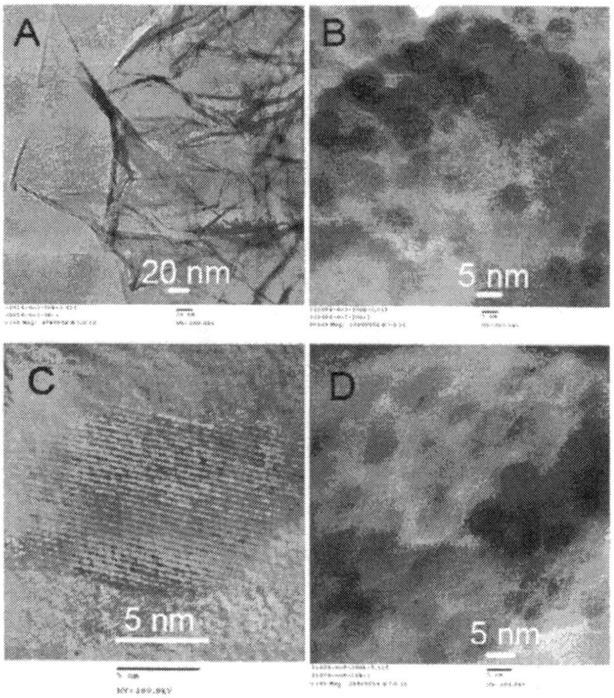

Fig. 2. The TEM and HRTEM images of Bi_2O_3/LDHs.

After calcination at 800°C the evolved mixed oxides presents different nano-sized features. Fig. 3 shows the TEM images of the mixed oxides Bi_2O_3/ZnO-$ZnAl_2O_4$ and α-Bi_2O_3/$ZnCr_2O_4$-ZnO. It reveals that the nanoarchitectonic of the mixed oxides systems might be described by two types of nanoparticles with diameters equal of 9 and 12 nm, respectively. It is interesting to note that in the mixed oxides nanoprticles are interconnected each others revealing a heterojuction between the mixed oxides.

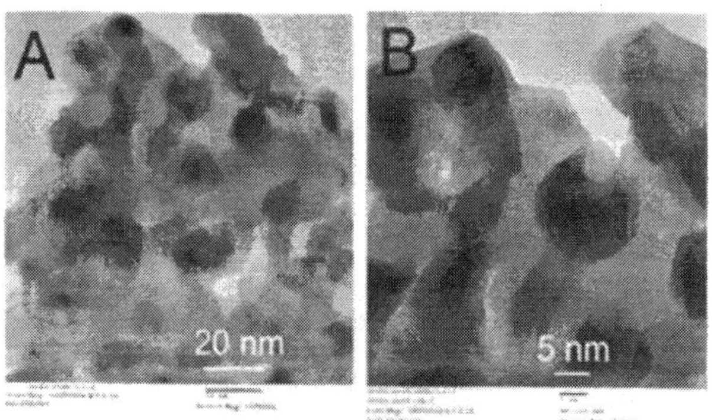

Fig. 3. The TEM images of (A) Bi_2O_3/ZnO-$ZnAl_2O_4$ and (B) Bi_2O_3/$ZnCr_2O_4$-ZnO.

XPS results further confirm the formation of Bi_2O_3/LDHs self-assemblies and indicate that their similar Bi content (equal to ~3%,). For Bi_2O_3/ZnAlLDH the survey EDX-XPS spectra is presented in Fig. 4 while the XPS spectra of Bi_2O_3/ZnCrLDH is given in Fig. 5. The analysis detected peaks from Bi, Zn, Al and O for Bi_2O_3/ZnAlLDH while Bi, Zn, Cr, and O were identified in the surface elemental composition of Bi_2O_3/ZnCrLDH. These results indicate the joined surface composition of AuNPs and specific LDH clay matrix.

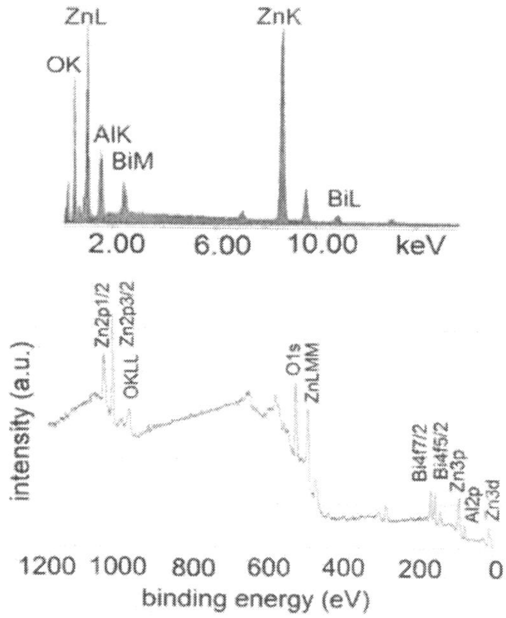

Fig. 4. High resolution XPS and EDX spectra of Bi_2O_3/ZnAlLDH.

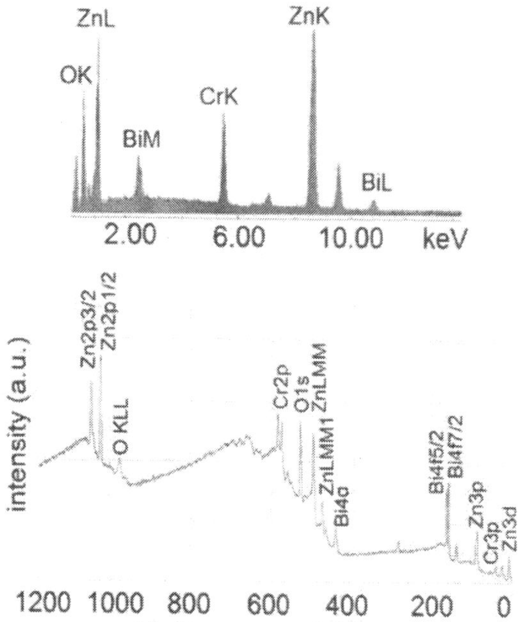

Fig. 5. High resolution XPS and EDX spectra of Bi_2O_3/ZnCrLDH.

2.2. Optical characteristics and photocatalytic properties

Fig. 6 displays the comparison of the room temperature UV-Vis absorption spectra of the studied catalysts. For all the samples the absorption in the UV region is defined by high intensities peaks that can be assigned to the optical responses of the LDHs and their derived mixed oxides.

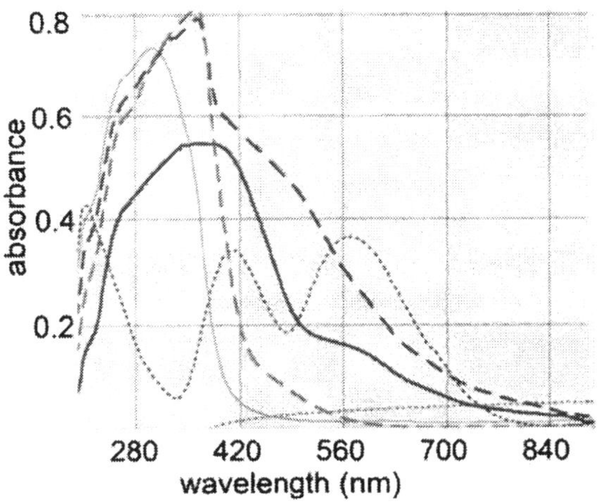

Fig. 6. The UV-Vis absorption spectra of: (-) Bi_2O_3/ZnAlLDH, (-) Bi_2O_3/ZnCrLDH, (---) Bi_2O_3/ZnO-$ZnAl_2O_4$ and (---) Bi_2O_3/$ZnCr_2O_4$-ZnO.

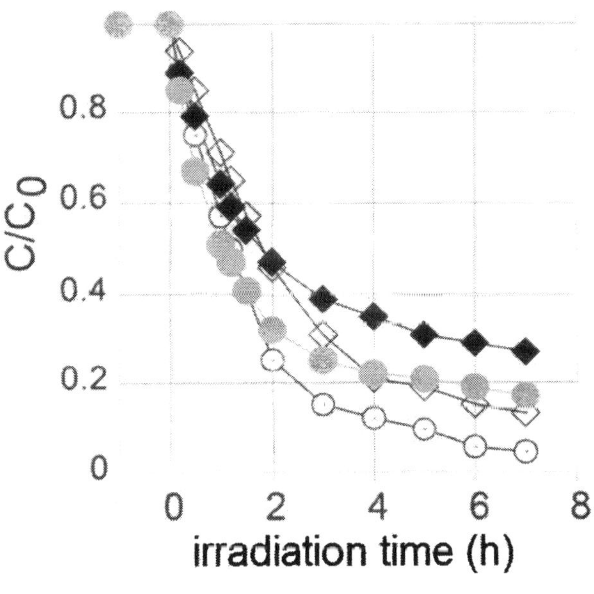

Fig. 7. Comparison of the photocatalytic efficiency of the catalysts during the photodegradation process of phenol over the entire range of wavelength: (•) Bi_2O_3/ZnCrLDH, (♦) Bi_2O_3/ZnAlLDH, (Θ) Bi_2O_3/ZnO-$ZnAl_2O_4$ (◊) Bi_2O_3/$ZnCr_2O_4$-ZnO.

One essential property relevant to the photocatalytic activity of a semiconductor is its energy band configuration. Tuning the energy band gap is a fundamental aspect of the design and fabrication of photocatalysts. The energy band gap configuration will define the absorption of incident photons, the carriers and the redox capabilities of excited-state electrons and holes [11]. Thus, a classical Tauc approach is used to estimate the optical direct band gap (Eg) values of the samples. The obtained Eg values are 3.11 for Bi_2O_3/ZnO-$ZnAl_2O_4$ and 2.27 for Bi_2O_3/$ZnCr_2O_4$-ZnO. For the entire range of wavelength the photocatalytic efficiency of Bi_2O_3/LDHs and the derived mixed oxides are compared in Figure 8. Controlled experiments indicate that that phenol pollutant does not degrade in the absence of photocatalyst. Bi_2O_3/ZnO-$ZnAl_2O_4$ shows the highest catalytic activity with almost complete degradation of the phenol after 6 h under irradiation while in the same conditions Bi_2O_3/ZnAlLDH degrades only 82% of phenol.

REFERENCES

[1] H. Tong, S. Ouyang, Y. Bi, N. Umezawa, M. Oshikiri and J. Ye, *Adv. Mater.*, 2012, 24, 229.

[2] M. Grzelczak, J. Vermant, E. M. Furst and L. M. Liz-Marzaan, *ACS Nano*, 2010, **4**, 3591.

[3] K. Ariga , Q. Ji, M. J. McShane, Y. M. Lvov, A. Vinu and J. P. Hill, *Chem. Mater.*, 2012, 24, 728.

[4] S. Mann, *Nature Mater.*, 2009, 8, 78.

[5] S. Kawamura, M.C Puscasu,, Y. Yoshida, Y. Izumi, G. Carja, Applied Catalysis A: General, 2014, in press.

[6] G. Carja, M. Birsanu, K. Okada, H. Garcia, Journal of Materials Chemistry, 2013, 69, 6875-6883.

[7] E.M. Seftel, M.C. Puscasu, M. Mertens, P. Cool, G. Carja, Applied Catalysis B: Environmental, 2015, 164, 251-260.

[8] J. Prince, F. Tzompantzi, G. Mendoza-Damián, F. Hernández-Beltránc, J.S. Valente, Applied Catalysis B: Environmental, 2015, 163, 352–360.

[9] G. Carja, L. Dartu, K. Okad, E. Fortunato, Chemical Engineering Journal, 2013, 222, 60-66.

[10] S. Kannan, V. Rives and H. Knözinger, J. Sol. St. Chem., 2004, 177, 319.

[11] J. S. Valente, F. Tzompantzi, J. Prince, J. G.H. Cortez, R. Gomez, Appl. Catal. B-Environ., 2009, 90, 330.

Acknowledgment: This work was supported in part by the Kazuchika Okura Memorial Foundation, "Planting Seeds for Research" program from Tokyo Institute of Technology, "Grant-in-Aid for Young Scientists (A; 25708037)" from Japan Society for the Promotion of Science (JSPS), and by the project "Development Base of Advanced Materials Development and Integration of Novel Structured Metallic and Inorganic" of the Ministry of Education, Culture, Sports, Science, and Technology (MEXT), Japan. This work was also supported by the grant of the Romanian National Authority for Scientific Research, CNCS-UEFISCDI, Project number PN-II-ID-PCE-75/2013.

Catalytic Decomposition of Methanol Effected by (Cu-ZnO)@C Nanoreactors

C.-I. Hong*, H.-Y. Kang*, C.-Y. Peng**, J. Wang***, and H. Paul Wang*

*National Cheng Kung University, Tainan 70101, Taiwan, wanghp@mail.ncku.edu.tw
**Tamkang University, New Taipei City 25137, Taiwan
***University of California, San Diego, CA 92037, USA

ABSTRACT

Methanol can be decomposed to yield H_2 that may be the alternative fuels for automobiles and fuel cells. The feasibility study for catalytic decomposition of methanol (DM) to CO and H_2 with (Cu-ZnO)@C core- and yolk-shell nanoreactors was, therefore, carried out. The relatively low activation energy (E_a), low reaction temperature, and high yields of H_2 with the core-shell or yolk-shell nanoreactors were found, which may be associated with the fact of high collision frequency between Cu-ZnO active species and methanol within the confined nanoreactors. The data obtained from the component fitted X-ray absorption near edge structure spectra suggest that metallic copper (Cu) in the nanoreactors can promote the formation of H_2 from the catalytic DM reaction. In addition, the reactivity of DM catalyzed by the nanoreactors increases with an increase of the ZnO fractions in the Cu-ZnO encapsulated in carbon-shell.

Keywords: methanol decomposition, Cu-ZnO@C, XANES

1. INTRODUCTION

Methanol which can be produced commercially from syngas with the $Cu/ZnO/Al_2O_3$ catalysts was used as the feedstock in fuel cells for electrical power generation. The zero-emission vehicles can also be achieved with the feed of the H_2 fuel directly. Although steam reforming of methanol provides more H_2 than DM, DM has the advantages of lower reaction temperatures and simpler processes.

Decomposition of methanol is an endothermic reaction. Methanol-fueled automobiles have more efficient and cleaner than undecomposed methanol and gasoline [1, 2]. The catalysts must be active even at low exhaust-heat temperature and endure rough operating conditions [2].

Decomposition of methanol to methoxy (CH_2OH) on the clean Cu(110) surface takes place at 270 K. Note that Cu(111) and Cu(100) are inactive to DM. It is suggestive that $CH_3OH \rightarrow CH_2OH \rightarrow CH_2O \rightarrow CHO \rightarrow CO$ is the favorable DM reaction pathway on Cu(110). After the initial O-H bond scission of methanol, three successive C-H breaking may occur. The rate-limiting step during DM is the dehydrogenation of CH_2OH to formaldehyde (CH_2O) which is found on the Cu(110) surface. The barrier for the C-H bond breaking of CH_2OH to CH_2O is >0.5 eV that is greater than other steps.

Cu- and ZnO-based catalysts can catalyze the DM reaction at relatively low temperatures. Compared with Ni, Pd and Pt catalysts, Cu is less aggressive in hydrogenation and dehydrogenation reactions [3]. The addition of ZnO can enhance the stability and dispersion of Cu. Recently, the DM reaction with Pt has attracted a great interest taking place on the anodes of the direct methanol fuel cells (DMFCs) which offer direct conversion of methanol to electricity with a greater efficiency than the conventional combustion-based power generation system. However, the performance of DMFCs is limited by the CO poisoning on the Pt-anodes [4].

Nano size-controllable Cu@C core-shell nanoparticles can be prepared in a very simple process [5]. A controlled etching of Cu from the Cu@C core-shell nanoparticles can create desirable space and form the so-called Cu@C yolk-shell nanoparticles [6]. In the present work, zinc was filled into the Cu@C yolk-shell nanoparticles to form the (Cu-ZnO)@C core- and yolk-shell nanoreactors wherein more H_2 can be yielded from the catalytic DM reaction.

2. EXPERIMENTAL

The DM reactions catalyzed by (Cu-ZnO)@C core- and yolk-shell nanoreactors were carried out in a micro-reactor at 473-573 K. Methanol was injected steadily using a syringe pump. The flow rate of the feed containing 21% of methanol and 5% of O_2 in N_2 regulated at 20 mL/min by a mass flow controller. About 0.1 g of the nanoreactor catalyst was used in the catalytic DM experiments.

The concentration of H_2 in the product gases was analyzed by chromatography (Bruker GC-430) equipped with a TCD detector. Molecular sieve 5A column (353 K) was used for separation of H_2 from other gases. Nitrogen at a flow rate of 30 mL/min was used as the carrier gas. CO and CO_2 were analyzed by on-line FT-IR spectroscopy (Varian, FTS-40). A 32-scan data accumulation was conducted at a resolution of 4 cm^{-1}.

The X-ray absorption spectra of copper in the nanoreactors were collected on the Wiggler BL17C beam line at the Taiwan National Synchrotron Radiation Research Center (NSRRC). The energy storage ring was operated under the energy of 1.5 GeV at top-up ring current model (300 mA).

3. RESULTS AND DISCUSSION

H_2 was yielded from the catalytic DM with the (Cu-ZnO)@C core- and yolk-shell nanoreactors at 473-573 K. The Arrhenius parameters such as activation energy (E_a) and pre-exponential factor (A) for the DM reaction catalyzed by the nanoreactors were derived and summarized in Table 1. The E_a for the catalytic DM are in the range of 11-14 kcal/mol. Methanol decomposition may occur in the confined spaces between the core Cu-ZnO and carbon-shell in the nanoreactors which may be associated with the relatively low E_a, high collision frequency and high yields of H_2 in the core- and yolk-shell nanoreactors.

At 573 K, yields of H_2 from the DM reaction catalyzed by (Cu-ZnO)@C core-shell nanoparticles have formation rate of 0.098 mol/gCu-hr. Promotion of H_2 yields in the presence of ZnO in the yolk-shell nanoreactors is also observed.

For more detailed characterization of active copper and zinc species in the nanoreactors, their X-ray absorption spectra were measured. The XANES spectra of copper in the nanoreactors are shown in Fig. 1. The absorption edges of copper and zinc species are observed at 8979 and 9659 eV, respectively. ZnO is the main zinc species in the DM reactions. To determine the copper oxidation states, XANES spectra of copper in the nanoreactors are analyzed by component fitting. After the catalytic DM reactions at 473 K, the main copper species in the nanoreactors are low-oxidation state copper.

Cu/ZnO molar ratios	A (10^5 s^{-1})	E_a (kcal/mol)
Core-shell nanoreactors		
1.5/1.5	7.15	13.3
2.0/1.0	2.30	12.4
2.7/0.3	1.01	11.7
3.0/0.0	0.67	12.3
Yolk-shell nanoreactors		
1.5/1.5	2.10	12.4
3.0/0.0	1.78	12.2

$$\text{Rate} = \gamma_{H_2} = k[CH_3OH],\ k = A\exp(-E_a/RT)$$

Table 1: Arrhenius parameters for catalytic DM reactions with the (Cu-ZnO)@C core- and yolk-shell nanoreactors at 473-573 K.

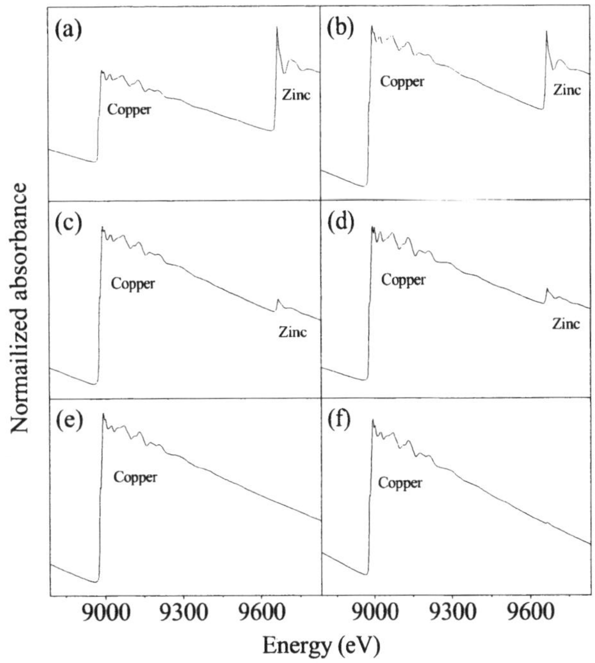

Fig. 1: XANES spectra of (Cu-ZnO)@C core-shell nanoreactors having the Cu/ZnO molar ratios of (a) 1.5/1.5, (b) 2.0/1.0, (c) 2.7/0.3, and (e) 3.0/0 and yolk-shell nanoreactors having the Cu/ZnO molar ratios of (d) 1.5/1.5 and (f) 3.0/0 (having a Cu diameter of 7 nm) after the catalytic DM at 473 K.

4. CONCLUSION

Methanol can be catalytically decomposed to form H_2 effected by the (Cu-ZnO)@C core- and yolk-shell nanoreactors at 473-573 K. By XANES, the main copper and zinc active species in the nanoreactors are Cu and ZnO, respectively. The relatively high reactivity and high yields of H_2 for the catalytic DM reaction may be associated with the fact of high collision frequency between active species (i.e., Cu and ZnO) and methanol in the confined nanoreactors.

ACKNOWLEDGEMENTS

The financial supports of Taiwan National Science Council, Bureau of Energy, and National Synchrotron Radiation Research Center are gratefully acknowledged.

REFERENCES

[1] Y. Usami, K. Kagawa, M. Kawazoe, M. Yasuyuki, H. Sakurai and M. Haruta, Catalytic methanol decomposition at low temperatures over palladium supported on metal oxides. Applied Catalysis A: General, 171, 123-130, 1998.

[2] W.H. Cheng, Deactivation and regeneration of Cu/Cr based methanol decomposition catalysts. Applied Catalysis B: Environmental, 7, 127-136, 1995.
[3] M.V. Twigg and M.S. Spencer, Deactivation of copper metal catalysts for methanol decomposition, methanol steam reforming and methanol synthesis. Topics in Catalysis, 22, 191-203, 2003.
[4] J. Greeley and M. Mavrikakis, Competitive paths for methanol decomposition on Pt(111). Journal of the American Chemical Society, 126, 3910-3919, 2004.
[5] C.H. Huang, H.P. Wang, J.E. Chang and E.M. Eyring, Synthesis of nanosize-controllable copper and its alloys in carbon shells. Chemical Communications, 31, 4663-4665, 2009.
[6] C.I. Hong, H.Y. Kang, H.P. Wang, W.K. Lin, U.S. Jeng and C.H. Su, Cu-ZnO@C nanoreactors studied by in-situ synchrotron SAX spectroscopy. Journal of Electron Spectroscopy and Related Phenomena, 184, 301-303, 2011.

Hydrogen Generation by Ammonia Decomposition Using Co Supported on Mg/La Mixed Oxide: Effect of Calcination Atmosphere of Support

S.Podila[a] , H. Driss[a,b], Y. Alhamed[a,b], A. Al-Zahrani[a,b], L. Petrov[a]
[a]SABIC Chair of Catalysis, King Abdulaziz University, P.O. Box 80204, Jeddah 21589, Saudi Arabia, seetharampodila@gmail.com,
[b]Chemical and Materials Engineering Department, Faculty of Engineering, King Abdulaziz University , P.O. Box 80204, Jeddah 21589, Saudi Arabia, yhamed@kau.edu.sa

ABSTRACT

In this investigation the effect of calcination atmosphere of the support for Co catalyst and its activity in ammonia decomposition (AD) is explored. Mg/La mixed oxide supports (Mg/La = 3) were prepared by co-precipitation from nitrates solutions. The precipitates (support precursor) were either calcined in air, nitrogen or vacuum at 450°C followed by impregnation with 5 wt. % Cobalt. The calcined supports and the supported cobalt catalysts were characterized using nitrogen adsorption, X-ray powder diffraction (XRD), temperature-programmed reduction (H2-TPR) scanning electron microscopy (SEM) and X-ray photoelectron spectroscopy (XPS). Catalyst activity tests were performed in a fixed bed quartz reactor. It was found that the catalyst activity in AD is highest for support calcined in nitrogen followed by that calcined in vacuum while lower activity was exhibited for support calcined in air. This activity pattern is positively correlated with high surface concentration of Co and high Co support interaction as found from XPS and TPR measurements respectively.

Keywords: ammonia decomposition, mixed oxides, hydrogen generation, cobalt catalysts.

1 INTRODUCTION

Generation of CO free hydrogen by ammonia decomposition (AD) is more attractive from practical and economical point of view compared to other process. Much research has been devoted to the investigation of Group VIII metals (Ru, Ni, Ir, Fe, Co and Rh) or metal carbide/nitrides (MoNx, VCx, MoCx, VNx, etc.) as catalytic materials for AD [1-3]. Previous studies showed that Ru supported on multi wall carbon nanotubes (MWCNTs) is the most active catalyst for AD [3]. Unfortunately, MWCNTs as a support suffered from many drawbacks such as high cost, methanation of carbon at high temperatures and low metal support interaction. In addition the high price and limited availability of Ru makes it uneconomical for large scale applications. Therefore, developing a cheap and affordable active catalyst for AD is highly desirable. Many studies reported that support basicity plays a crucial rule in AD thus the tailoring of the basic strength of mixed oxides supports with proper treatment could be used to design better catalysts [4-6] for AD.

In this investigation, the effect of calcination atmosphere of the support on AD for Co supported on Mg/La mixed oxide is explored. Further insight on the effect of such treatment on the morphology and catalyst support interaction of this catalyst is also discussed.

2 EXPERIMENTAL

Mg/La mixed oxide support (Mg/La = 3) were prepared by co-precipitation from Mg and La nitrates solutions under high super saturation conditions using a mixture of KOH and K_2CO_3. The obtained precipitate is then filtered and divided into three equal portions. All portions were calcined at 450 °C but in different atmospheres. The samples were designated as ML-air, ML-N_2 and ML-vac, when calcined in air, N_2 or under vacuum, respectively. For comparison Mg/La mixed oxide support with the same Mg-La ratio was prepared using only KOH for precipitation followed by calcination in air at 450 °C. This support was designated as ML-OH. The prepared supports were impregnated with Co from aqueous cobalt nitrate solution using a rotary evaporator under reduced pressure to obtain 5 wt.% loading of Co. Impregnated precursors of the catalysts were dried in an oven overnight at 100°C followed by calcination at 500°C for 5 hours under nitrogen flow to obtain the final catalyst used in AD. The supported cobalt catalysts were designated as 5CML-air, 5CML-vac, 5CML-N_2 and 5CML-OH.

2.1 Catalyst Characterization

The BET surface areas of the calcined supports and of the supported cobalt catalysts were determined by N_2 adsorption at -196 °C using a Quantachrome Nova200e apparatus. Prior to the experiments, the samples were outgassed at 400°C for 4 h. X-ray powder diffraction (XRD) patterns of the samples were recorded on EQUINOX 1000, Inel (France) using Co Kα radiation (λ=1.789 Å) between 10 -115° at 40kV and 30 mA. Temperature-programmed reduction (H_2-TPR) experiments were performed using the quantachrome pulsar automated chemisorptions analyzer equipped with a thermal conductivity detector (TCD). Prior to measurement, 100 mg of the sample to be tested was loaded into U-tube sample holder and pretreated in a flow of helium at 120°C for 60 minutes to remove the adsorbed impurities and moisture. After cooling down to 50°C, the flow of He was switched to a flow of 5%H_2 in N_2 (15 cm^3 min^{-1}). The temperature of sample was then ramped at a rate of 5°C min^{-1}. The morphologies of all samples were investigated using field emission scanning electron microscopy (FE-SEM, QUANTA FEG450, FEI) using ETD Everhart Thornley Detector (HV mode) and solid-state back scattering electron detector (VCD).

2.2 Activity Studies

Activity tests were performed in a fixed bed quartz reactor under atmospheric pressure (PID system, Eng &Tech, Spain). Prior to the reaction, a 100 mg of catalyst was reduced in a mixture of nitrogen and hydrogen flow at 500°C for 5 hours. After the catalyst activation, pure NH_3 gas was introduced into the reactor at a gs hourly space velocity (GHSV) of 6000 h^{-1}. Experiments were carried out at temperatures ranging from 300 - 550°C with successive rise of 50°C. Products analysis was performed by using an online gas chromatograph (GC-450 Varian, USA) equipped with a thermal conductivity detector and a Poropak Q column.

3. RESULTS AND DISCUSSION

3.1 Catalyst Characterization

The surface area results of supports and corresponding cobalt catalyst are displayed in Table 1. The ML-air, ML-N_2 and ML-vac, supports did not show much difference in their BET surface area while ML-OH support showed the higher surface area. It may be due to the lack of formation mixed oxide where the Mg/La mixed oxide generally have low surface area compared to individual oxides [7]. Impregnation of the support with cobalt resulted decrease in surface area for all supports.

Supports/Catalyst	BET Surface area (m^2/g)	surface cobalt concentration (atom %)
ML-air/5CML-air	59.7/37.2	2.64
ML-N2/5CML -N_2	41.5/32.7	3.32
ML –Vac/5CML -Vac	64.2/34.9	2.96
ML-OH/5CMLa-OH	76.0/45.7	2.56

Table 1: Surface area and XPS results of catalysts

The XRD patterns of freshly calcined supports in different atmospheres are portrayed in Figure1A where it is observed the presence of two types of lanthanum carbonate phases besides MgO phase. Support calcined under vacuum shows one type of lanthanum carbonate with major diffraction lines at 26.66°,34.46°,36.31° and 52.29° corresponding to the La_2CO_5 phase [PDF: 00-023-0320, tetragonal, a= 4.063Å, c = 13.5 Å]. On the other hand, the support calcined under N_2 atmosphere shows diffraction lines at 25.88°, 29.32°, 30.03°, 32.17°, 35.39°, 39.30°, 39.55°, 40.80°, 52.05°, 55.7°, 59.07°, 64.54° and 67.5° corresponding to $La_2O_2CO_3$ phase [PDF: 00-037-0804, hexagonal, a=4.075 Å and c=15.95 Å].

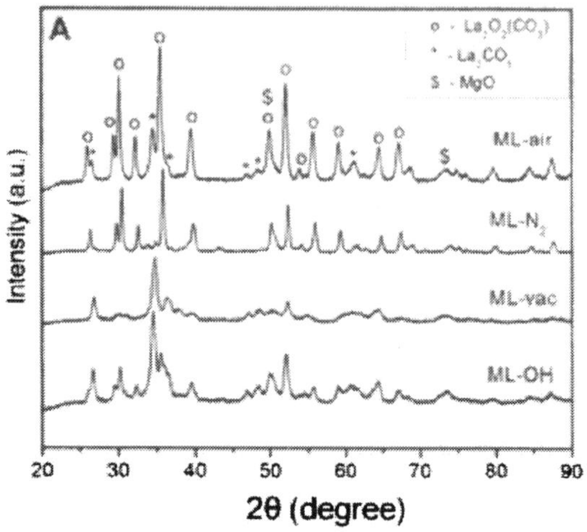

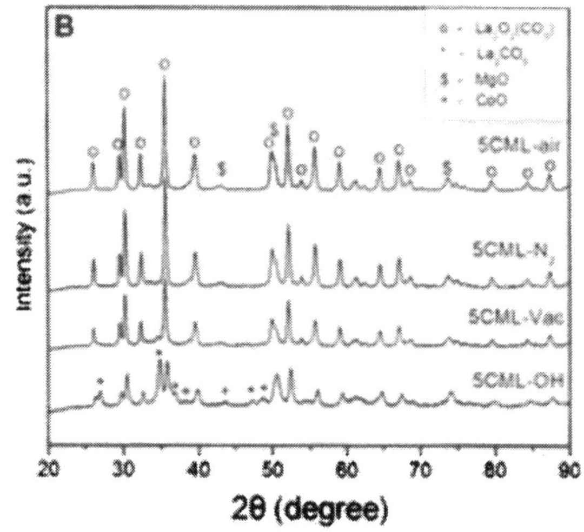

Figure1. XRD patterns of (A) calcined supports, (B) calcined cobalt catalysts

When the support is calcined in air we can clearly see both phases but the hexagonal phase is more dominant compared to tetragonal phase. Conversely the support ML-OH is shows mainly a tetragonal phase as major component. From these results it could be concluded that there are two type lanthanum species, one is showing tetragonal La_2CO_5 phase having cell parameters a=4.06Å c= 3.5Å and the other one is showing hexagonal $La_2O_2(CO_3)$ phase having cell parameters a=4.075Å and c=15.95Å. This might explain the presence of bulk form and mixed oxide form of lanthanum which is forming carbonates by exposure to air.

After impregnation with cobalt all calcined catalysts (in nitrogen) show only hexagonal $La_2O_2CO_3$ phase except

5CML-OH catalyst which is still showing two phases of lanthanum carbonate. No diffraction lines were observed for cobalt except for 5CML-OH. Low loading of and small crystal size of Co species may evade detection cobalt species by XRD for these catalysts.

XPS of Co 2p spectrum of cobalt catalysts are shown in Figure 2. It was found that CoO is the major surface species in calcined cobalt catalysts. The Co 2p3/2 binding energy of CoO in 5CML-N2 and 5CML-Vac catalysts is slightly higher than other catalysts which may indicate a stronger interaction between Co and the support for this catalyst. As seen in Table 1 the surface cobalt concentration (atom %) for the four catalysts follows the order, 5CML-N_2 > 5CML-Vac > 5CML-air > 5CML-OH which is in accordance with the trend observed for catalytic activity where higher surface concentration of Co results in higher catalytic activity.

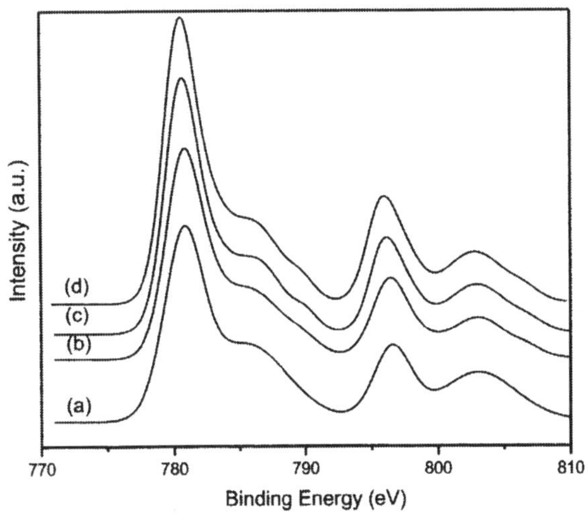

Figure 2: Co 2p XPS spectra of cobalt catalysts with support calcined in different atmosphere (a) 5CML-OH, (b) 5CML-air, (c) 5CML-Vac, (d) 5CML-N_2

TPR patterns are presented in Figure 3. Three major reduction peaks are obtained, the first peak is a relatively narrow and strong and extending from 250-380°C, the second peak is broad and extending from 380 to 650°C and a the third peak starts above 650°C. The first two peaks could be assigned to the reduction of Co_3O_4 to CoO and CoO to metallic cobalt respectively [8]. The very broad tailing nature for the second peak could suggest the reduction of highly heterogeneous CoO species having strong interaction with the support. The third high temperature peak above 700°C is due to reduction of lanthanum carbonate [9]. Analysis of the first peak in figure 3 showed a very clear shift in the maximum of this peak depending on the support calcination conditions. The temperature at which the maxima of this peak is located can be ranked in the following order; 5CML-N_2 (349°C)> 5CML-Vac (334°C) > 5CML-air (322°C). This trend clearly shows that calcination of Mg/La support in N_2 facilitate the process to obtain a better catalyst with enhanced metal support interaction.

Representative SEM images are shown in Figure 4. It is clear from these images that all supports are containing Mg/La mixed oxide which is in grain shape morphology as confirmed by EDAX. The amount of Mg/La mixed oxide phase is more in supports calcined under N_2 and vacuum atmospheres compared to air calcined supports. The SEM and EDAX results showed that air calcined supports are giving big bulk form of MgO and La_2O_3. The Mg/La mixed oxide grains are dispersed on these oxides. Sponge like morphology of MgO is observed in ML-Vac support. Hence the support calcined under different atmospheres not only affects the interaction among the catalyst components, but also affect the morphology of the support.

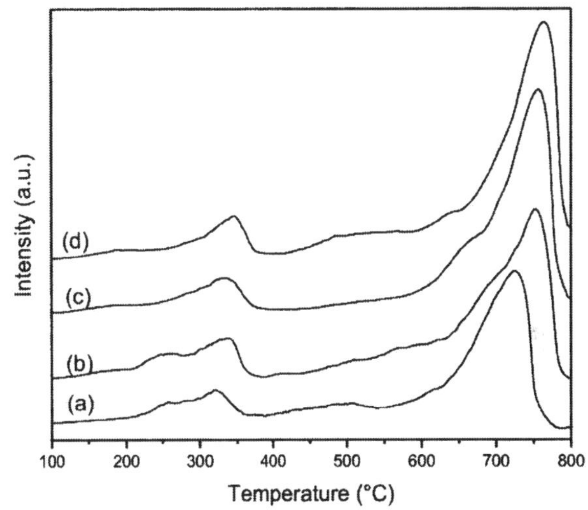

Figure 3: H2 –TPR patterns of cobalt catalysts (a) 5CML-air, (b) 5CML-OH, (c) 5CML-Vac, (d) 5CML-N2

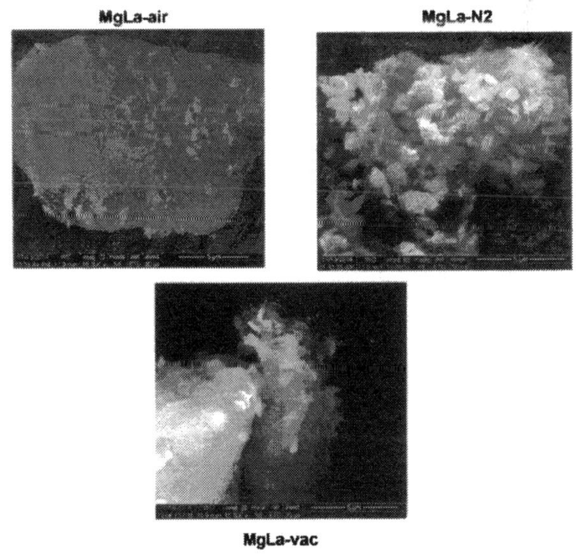

Figure 4: SEM images of Mg/La supports calcined at 450°C under different atmospheres

3.2 Activity Testing

The activity measurements for AD presented in Figure 5.A, clearly shows that 5CML-N2 catalyst is the most active among all catalysts. On the other hand 5CML-OH (prepared by different precipitation procedure) catalyst shows the lowest activity at same reaction conditions. It is clear that the calcination atmosphere of the support greatly influences the final catalytic activity. Studied catalysts are could divided into two groups according to their activity. The first group which has high activity in AD was prepared using supports calcined in non-oxidizing atmosphere (e.g. N_2 or vacuum). The other group of catalysts was prepared by air calcined supports and they exhibited a much lower activity. Thus it could be concluded that support calcination in an oxidizing atmosphere is not beneficial considering the catalyst activity.

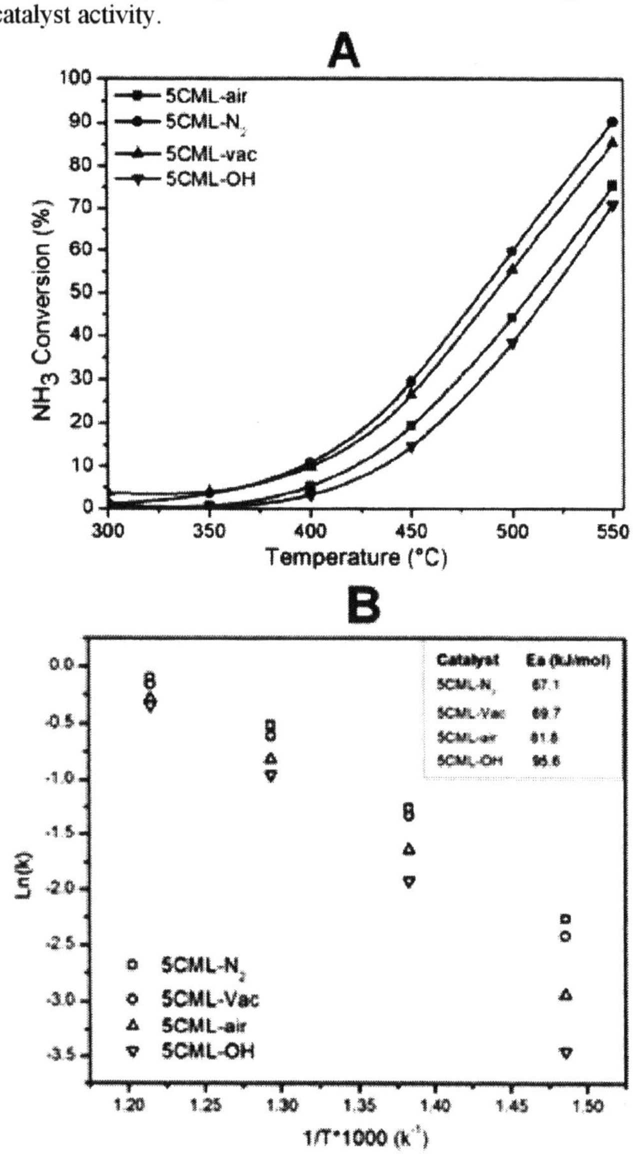

Figure 5: (A) Activity results (B) Arrhenius plots for NH3 decomposition over cobalt catalysts

Figure 5.B illustrates the Arrhenius plots and the estimated apparent activation energies of all catalysts. It is clearly seen that the 5CML-OH catalyst had the highest activation energy (95.6kJ/mol). By calcining the support in N2 atmosphere activation energy greatly decreased to 67.1 kJ/mol and correspondingly NH_3 conversion sharply increased up to 90% at 550 °C. Additional investigation is required to clarify the role of calcinations atmosphere on the activity of these catalysts. Nonetheless, it is very clear that higher activity is positively correlated with high surface concentration of Co and high Co support interaction obtained as a result of calcination of the support in non-oxidizing atmosphere.

3. CONCLUSIONS

Performing calcinations of the support in non-oxidizing atmospheres is a vital precondition for preparation of active cobalt supported catalyst for ammonia decomposition. From the XPS results, it was shown that cobalt surface metal concentration was highest for the most active catalysts for which the support was calcined in nitrogen (5CML-N2 catalysts). The TPR measurements also showed strong metal support interaction for this catalyst. From the SEM results the formation of Mg/La mixed oxide increased in 5CML-N2 catalyst. Further work however is required to better understand the effect support calcinations in different atmosphere on the activity of cobalt catalysts. Nonetheless, it is very clear that higher activity is positively correlated with high surface concentration of Co and high Co support interaction brought about by calcination of the support in non-oxidizing atmosphere.

REFERENCES

[1] J. Zhang, H.Y. Xu, X.L. Jin, Q.J. Ge, W.Z. Li, Appl. Catal., A: General 290, 87-96, 2005.

[2] J. Zhang, H.Y. Xu, W.Z. Li, Appl. Catal., A: General 296, 257-267, 2005.

[3] S.F. Yin, B.Q. Xu, W.X. Zhu, C.F. Ng, X.P. Zhou, C.T. Au, Catal. Today, 27-38, 93, 2004.

[4] H. Zhang, Y. A. Alhamed, A. Al-Zahrani, M. Daous, H. Inokawa, Y. Kojima, L. A. Petrov, Int. J. Hydrogen Energ. 39, 17573-17582, 2014.

[5] H. Zhang, Y. A. Alhamed, W. Chu, Z. Ye, A. AlZahrani, L. Petrov, Appl.Catal. A: General, 464–465, 156-164, 2013.

[6] Zhang J, Xu HY, Jin XL, Ge QJ, Li WZ, Appl. Catal., A: General 290, 87–96, 2005.

[7] V. Bhaskar, J. M. Clacens, F. Figueras, Adv. Synth. Catal. 347, 767-771, 2005.

[8] A. Feller, M. Claeys, and E. van Steen, J. Catal. 185, 120–130, 1999.

[9] G. Jacobs, T. K. Das, Y. Zhang, J. Li, G. Racoillet, B. H. Davis, Appl. Catal A 233, 263–281, 2002.

Computational study of structural dependence of catalytic properties of Pt-Ru nanoalloy catalysts

B. Prasai* and V. Petkov*

* Department of Physics, Central Michigan University, Mt. Pleasant, MI 48858

ABSTRACT

3D structure models of Pt-Ru alloy nanoparticles (NPs) are built by molecular dynamics simulations. Models are refined against the experimental PDF data by reverse Monte Carlo simulations and analysed in terms of prime structural characteristics such as metal-to-metal bond lengths and first coordination numbers for Pt and Ru atoms. Structural characteristics of activated NPs and data for their catalytic activity are compared side by side and strong evidence is found that electronic effects, indicated by significant changes in Pt-Pt and Ru-Ru metal bond lengths at NP surface, and practically unrecognized so far atomic ensemble effects, indicated by the prevalence of particular configurations of Pt and Ru atoms at NP surface and distinctive packing of atomic layers near NP surface, contribute to the observed enhancement of the catalytic activity of composition about 50:50. Implications of so-established relationships between the atomic structure and catalytic activity of Pt-Ru alloy NPs on efforts aimed at improving further the latter by tuning-up the former are discussed and the usefulness of detailed NP structure studies to advancing science and technology of metallic NPs explored for catalytic applications is exemplified.

Keywords: nanoparticle structure modeling, molecular dynamics, fuel cell, Calatysis, sutton-chen, reverse Monte Carlo.

1 INTRODUCTION

In the recent years, metallic nanoparticles (NPs) are being extensively explored for their practical use in various applications such as catalysis, photonics, magnetic storage media, and drug delivery[1, 2, 3, 4]. The exploration of the metallic NPs as catalytic agents has boomed because of their use in fuel cell technologies based on catalysis. Metal nanoalloys, a few nanometer in size, based on Noble metals have shown a great promise as catalyst for several technologically important reactions. He we explored Pt-Ru nanoalloys which is one of the many important nanoalloy catalysts. Pt-Ru was chosen because of its importance for the development of highly pursued current devices for clean energy conversion such as fuel cells, particularly for its remarkable ability to speed up gas-phase CO oxidation and ethanol electro-oxidation reactions. Since most of the studies on Pt-Ru nanoalloys are focused on their catalytic properties with a limited knowledge of the atomic structure, here we pay particular attention on determining the atomic structure of these nanoalloys in fine detail and hence establishing the correlations between atomic structure and catalytic properties, with a hope that, once known, these correlations can provide a feedback loop for streamlining synthesis and optimization of Pt-Ru nanoalloys for fuel cells related applications.

For the purpose we built 3D structure models for catalytically active Pt_xRu_{100-x}; (x=31, 49 and 75) alloy NPs by Molecular Dynamics (MD) simulations and further refined by reverse Monte Carlo (RMC) simulations guided by the experimental atomic pair distribution functions (PDFs). RMC refined models are then analyzed in terms of structural characteristics important for catalysis such as NP phase state, distribution of first atomic neighbor distances, i.e. metal-to-metal bond lengths and first atomic coordination numbers paying particular attention to Pt and Ru atoms at NP surface since it is the NP surface where catalytic reactions take place, and correlated data for catalytic activity with structural characteristics of respective NPs and showed the usefulness of so-established relationships between the atomic structure and catalytic activity of Pt-Ru alloy NPs to efforts aimed at improving further the latter by tuning up the former.

2 COMPUTATIONAL METHODS

MD simulations were carried out with the help of computer program DL-POLY [5] under canonical NVT ensemble in the absence of periodic boundary conditions. Velocity Verlet algorithm with a time step of 2 fs was used. Initial model atomic configurations were equilibrated for 200 ps at 700K, which is just about the temperature at which Pt-Ru NPs were post-synthesis treated as to be activated for catalytic applications. The models were then cooled down to room temperature in steps of 50K and again equilibrated for 100 ps. Sutton-Chen potential parameters used in this study are taken from Ref. [6]. Reverse Monte Carlo (RMC) simulations, details of which are given in Ref.[6], were used to further refine MD generated structure models for Pt-Ru NPs.

3 RESULTS AND DISCUSSION

Accounting for the findings of simplistic empirical modeling, 3D models of post-synthesis treated (catalytically active) Pt-Ru alloy NPs were built by MD simulations based on the archetypal for current metal and alloy atomic structure theory Sutton-Chen (SC) method. MD models featured NPs with the chemical composition, chemical pattern, size (~ 4.5 nm) and shape (spherical) of actual Pt-Ru alloy NPs studied here as previously explained in Ref.[6]. Resulted 3D models are shown in Figure 1a and the atomic PDFs derived from the models are compared with the respective experimental PDFs in Figure 2a. As can be seen in Figure 2a, MD built structure models produce atomic PDFs that capture the essence of experimental ones but fail in reproducing them in fine detail. In particular, MD models fail in reproducing the position, broadening and intensity of several peaks in the experimental PDFs, i.e. the radii, degree of atomic positional disorder and atomic population of several coordination spheres in Pt-Ru alloy NPs. The inability of MD models to reproduce accurately the radii of atomic coordination spheres in Pt_xRu_{100-x} alloy NPs may be expected since SC method employed by MD is, like most theoretical methods currently in use, validated against experimental data for the lattice parameters of respective solids. Results of empirical modeling carried out here, however, showed clearly that NP "lattice parameteres", i.e. the characteristic interatomic distances in Pt-Ru alloy NPs, are different from these of the respective solids. Indeed this is a rather common observation since interatomic distances in metallic NPs can get compressed or stretched out considerably and non uniformly as a result of the inherently increased surface tension of NPs[7] and/or NP surface environment effects such as, for example, exposure to reactive (e.g. H_2) gases [8]. Such an effect is atypical for the respective solids. Besides, charge transfer and optimization of atomic-level stresses in metallic alloy NPs can result in a considerable change in the radii/size of metallic species involved which, again, may not be the case with the respective solids[9, 10]. The inability of MD models to reproduce accurately the degree of local structural disorder in Pt-Ru alloy NPs, in particular its inherent enhancement in coordination spheres involving atoms close to/at the extended NP surface, may also be expected. It is due to the fact that MD treats all atoms in structure models alike whereas atoms inside and at the surface of real-world metallic NPs are not necessarily alike in respect to their immediate atomic neighborhood, and others. The inability of MD models to reproduce accurately the distribution of Pt and Ru atoms across NPs is not a big surprise either. Nevertheless, as discussed in Ref.[9], MD simulations are useful in providing plausible structure models for metallic NPs that can be refined further, when necessary.

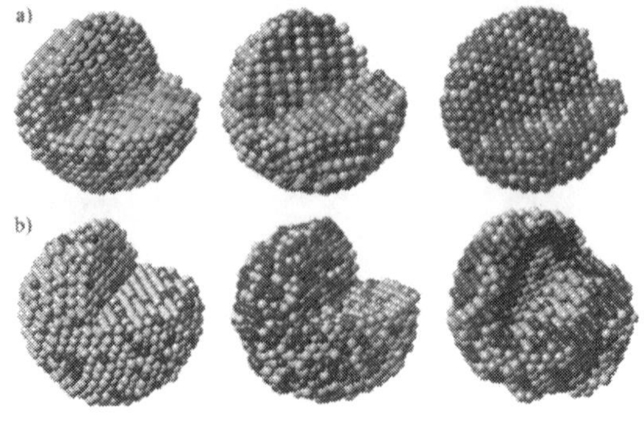

Figure 1: a) MD generated 3D structure models of Pt_xRu_{100-x} alloy NPs (x=75, left); (x=49, middle); (x=31, right), b) RMC refined 3D structure models with the corresponding compositions. The non-uniform distribution of Pt in $Pt_{31}Ru_{69}$ indicates a clear segregation of Pt species at the center and surface of NPs leading to an onion-like chemical pattern. Ru atoms are in green and Pt in gray.

MD models were refined further by RMC simulations. In the simulations positions of atoms in MD models of Figure 1a were adjusted, including switching the positions of nearby Pt and Ru atoms, as that the difference between RMC model derived and respective experimental PDF data is reduced as much as possible. At the same time model's energy was minimized, i.e. model's stability maximized, to the fullest possible extent using reliable pair-wise (Lennard-Jones type) potentials. RMC refined models for Pt-Ru nanoalloys are shown in Figure 1b and the atomic PDFs derived from the corresponding models are compared with the respective experimental PDFs in Figure 2b. As can be seen in Figure 2b, atomic PDFs derived from RMC refined 3D structure models for Pt-Ru alloy NPs reproduce respective experimental atomic PDFs in very good detail (see the reported R_w factors), including the region of higher r-values which is sensitive to structural features of NP surface. Precise accounting for the latter is important for catalysis since it is the NP surface where catalytic reactions take place.

With 3D models of realistic size, shape, chemical composition, chemical species ordering pattern, type of atomic ordering and atomic coordination spheres, including coordination radii and numbers, tuned up against relevant experimental PDF data at hand all important structural characteristics of actual metallic NPs studied can be derived. Here we concentrate on structural characteristics relevant to catalytic properties of metallic NPs such as metal-to-metal bond lengths and coordination numbers (CNs) paying particular attention to metallic species at NP surface. First neighbor Pt-Pt and

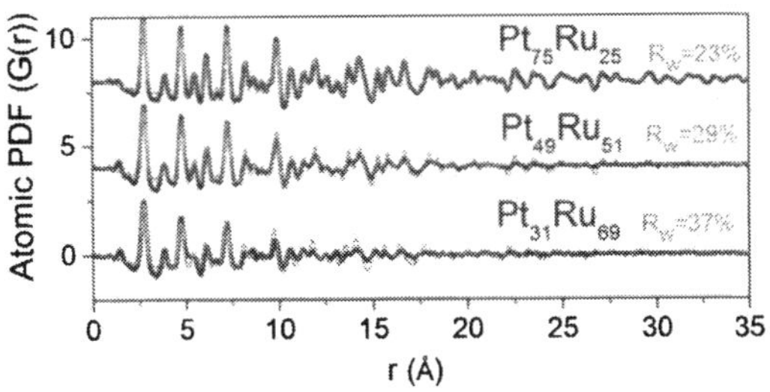

(a)

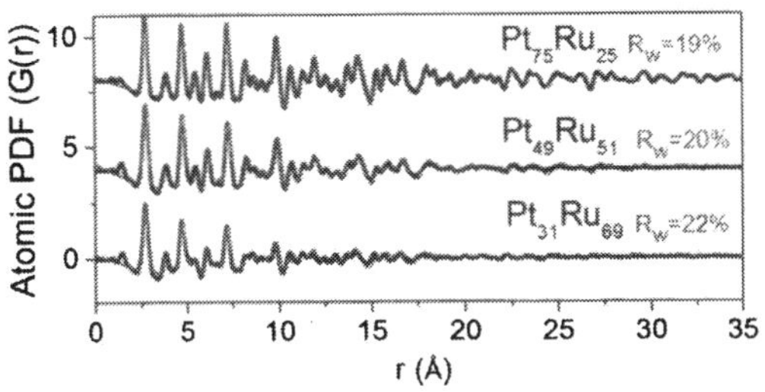

(b)

Figure 2: a) Experimental (symbols) and model (lines in red) PDFs for MD generated Pt_xRu_{100-x} alloy NP (x=31, 49 and 75) atomic configurations, shown in Figure 1a. b) Experimental (symbols) and model (lines in blue) PDFs for RMC refined respective Pt-Ru alloy NP atomic configurations, shown in Figure 1b. Model's quality indicators, R_w are shown for each data set.

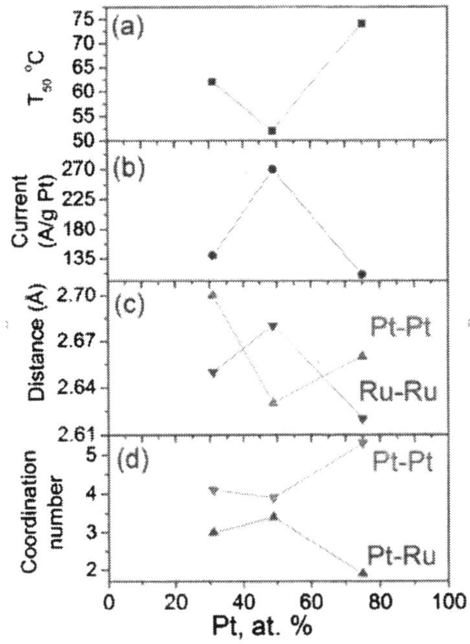

Figure 3: a) Catalytic activity of post-synthesis treated Pt_xRu_{100-x} alloy NPs (x=31, 49 and 75) for gas-phase oxidation of CO given in terms of the temperature T_{50} at which 50 % conversion of CO is achieved, b) Catalytic activity of respective Pt-Ru alloy NPs for electro-oxidation of ethanol given in terms of peak current generated as a result of the reaction, c) Pt-Pt and Ru-Ru bond lengths at the surface of RMC refined 3D atomic structures of Pt_xRu_{100-x} alloy NPs (x=31, 49 and 75) and d) first neighbor Pt-Pt and Pt-Ru CNs at the surface of same NPs.

Ru-Ru distances, i.e. metal-to-metal bond lengths, and first Pt-Pt and Pt-Ru CNs for Pt and Ru atoms at the surface of the Pt-Ru alloy NPs, as derived from their 3D structure models of Figure 1b, are shown in Figure 3. Experimental data for the catalytic activity of same NPs for gas-phase oxidation of CO and electro-oxidation of ethanol, obtained as described in Ref.[6], are also shown in the Figure. Inspection of so-summarized NP atomic structure vs NP property data indicates that the catalytic activity of Pt-Ru alloy NPs studied here peaks when i) Pt-Pt and Ru-Ru bond lengths at NP surface are, respectively, shortened and elongated most and ii) surface Pt-Pt and Pt-Ru first CNs are nearly identical. To comprehend the observed NP atomic structure-catalytic activity relationships we reasoned them within the framework of three most frequently evoked explanations for the higher catalytic activity of metallic alloy NPs as compared to that of respective monometallic NPs. Those explanations argue that i) different metallic species at the surface of alloy NPs do not interact but act cooperatively as to promote particular steps of chemical reactions, ii) different metallic species at the surface of alloy NPs interact by exchanging charge and/or adjusting size both affecting their electronic structure in a way favorable for chemical reactions and iii) different metallic species at NP surface assemble in particular configurations providing active catalytic sites[11, 12].

In the case of Pt-Ru alloy NPs explanation (i) has been put forward in terms of the so-called bi-functional mechanism[13] envisioning formation of surface Ru-hydrate species easing CO oxidation on nearby surface Pt atoms. The mechanism, however, has been found incapable of explaining the improved catalytic activity of Pt-Ru NPs wherein Pt atoms completely cover NP surface thus screening all Ru atoms from chemical species adsorbed at that surface. On the other hand, explanation (ii) has found strong support in numerous experimental and theoretical studies on Pt-Ru alloy NPs indicating the presence of charge transfer between Pt and Ru species[16, 17] accompanied by a considerable shortening of Pt-Pt and elongation of Ru-Ru metal bond lengths[11, 12, 18]. According to Pauling's theory of chemical bonds [19, 20] some charge transfer between Pt and Ru species may be expected due to their somewhat different electronegativity that is 2.3 and 2.2, respectively. Change in Pt-Pt and Ru-Ru bond lengths may also be expected since in order for the atomic-level stresses in Pt-Ru alloy NPs to be optimized the larger in size Pt species (2.77) should shrink and the smaller in size Ru species (2.67) expand as that the ratio of their size/radii becomes as close to one as possible [20, 21]. Changes in metal-to-metal bond length, strength and the electronic structure of respective metal species are known to be largely interrelated[20, 21] and so difficult to take apart. Nevertheless, it has been shown that metal-to-metal bond length changes of the type shown in Figure 3c would induce a strong downshift of the d electron states of surface Pt atoms in Pt_xRu_{100-x} alloy NPs at x$\sim$50. This would modify the energies of bonding of chemical species to surface Pt atoms and so the activation barriers of critical steps of chemical reactions, such as gas-phase oxidation of CO and electro-oxidation of ethanol, in a way enhancing the catalytic activity of Pt-Ru alloy NPs for x$\sim$50[11, 12, 18]. Indeed this is exactly what data in Figure 3(a and b) show. Explanation (iii) has been contemplated in several studies on Pt-Ru alloy NPs [11, 22, 23] but not recognized explicitly so far. Data in Figure 3 provides strong evidence that not only electronic (ii) but also atomic ensemble (iii) effects, signified by the prevalence of atomic configurations where both Pt and Ru metal species have pretty much the same number of unlike first neighbors (Pt-Pt and Pt-Ru first CNs of 3.8 and 3.5, and Ru-Pt and Ru-Ru CNs of 3.5 and 3.8, respectively[6]), contribute to the enhanced catalytic activity for x $\sim$ 50.

4 CONCLUSION

Catalytic activity of metallic NPs is governed by a series of destruction and creation of bonds between NP surface atoms and chemical species adsorbed at NP surface. Complete or partial alloying of metal species at the nanoscale can modify the electronic structure, hence, size (or vice versa) and arrangement of individual metallic species at NP surface thus influencing the catalytic activity of metallic NPs significantly. The arrangement, size, and electronic structure of atoms at NP surface, however, strongly depend on the way atoms in the layer(s) beneath it are positioned. Therefore, detailed knowledge of the atomic-scale structure across metallic alloy NPs is a prerequisite to achieving a better understanding of their catalytic properties and, hence, enabling a rational design approach to producing and optimizing metallic alloy NPs for catalytic applications.

5 ACKNOWLEDGMENTS

Work on this paper was supported by DOE-BES Grant DE-SC0006877.

REFERENCES

[1] J. K. Norskov and C. H. Christensen, Science, 312, 1322, 2006.

[2] S. Link, Z. L. Wang and M. A. El-Sayed, J. Phys. Chem., 103, 3529, 1999.

[3] D. Alloyeau, C. Ricolleau, C. Mottet, T. Oitkawa, C. Langlois, L. Y. Bouar, N. Braidy and A. Loiseau, Nature Mater., 8, 940, 2009.

[4] 9. P. Tartaj, M. d. P. Morales, S. Veintemillas-Verdaguer, T. Gonzalez-Carreno and C. J. Serna, J. Phys. D: Appl. Phys., 36, R182, 2003.

[5] W. Smith, W. C. Yong and P. M. Rodger, Molecular Simulation, 28, 86, 2002.

[6] B. Prasai, Y. Ren, S. Shan, Y. Zhao, H. Cronk, J. Luo, C.-J. Zhong and V. Petkov, Nanoscale [in Press].

[7] W. J. Huang, R. Sun, J. Tao, L. D. Menard, R. G. Nuzzo and J. M. Zuo, Nature Mater., 7, 308, 2008.

[8] V. Petkov, S. Shan, P. Chupas, J. Yin, L. Yang, J. Luo and C. J. Zhong Nanoscale, 5, 7379, 2013.

[9] V. Petkov, B. Prasai, Y. Ren, S. Shan, J. Luo, P. Joseph and Ch. J. Zhong Nanoscale, 6, 10048, 2014.

[10] V. Petkov, S. Shastri, S. Shan, P. Joseph, J. Luo, C. J. Zhong, T. Nakamura, Y. Herbani and S. Sato, J. Phys. Chem. C, 117, 22131, 2013.

[11] B. Hammer and J. K. Norskov, Adv. In Catalysis, 45, 71, 2007.

[12] J. K. Norskov, T. Bigaard, J. Rossmeisl and C. H. Christiansen, Nat. Chemistry, 1, 3746, 2009.

[13] H. A. Gaisteiger, N. Markovic, P. N. Ross, E. J. Cairns, J. Phys. Chem., 98, 617, 1994.

[14] L. Zhang, J. Kim, H. M. Cheng, F. Nah, K. Dude3ck, Ru-Shi liu, G. A. Botton and J. Zhang, J. Power. Sources, 196 9117, 2011.

[15] S. Alayoglu, A. U. Nilekar, M. Mavrikakis, B. Eichhorn. Nature Materials, 7 (4) 333, 2008.

[16] J. McBreen and S. Mukerjee, J. Electrochem. Soc., 142, 3399, 1995.

[17] P. Hu, D. King, M.H. Lee and M.C. Payne, Chem. Phys. Lett., 246, 73, 1995.

[18] A. Schlapka, M. Liska, A. Gros, U. Kasberger and P. Jacob, Phys. Rev. Lett., 91, 016101, 2003.

[19] L. Pauling, Ithaca, NY: Cornell University Press., 1975.

[20] L. Pauling, Proc. Natl. Acad. Sci. U. S. A., 84, 4754, 1987.

[21] Rajasekharam, T.; Seshubai, V., Acta Crystallogr. Sect. A, 68, 156, 2012.

[22] L. Giorgi, A. Pozio, C. Bracchhini, R. Giori and S. Turtu, J. Appl. Electrochem., 31, 325, 2001.

[23] S. Stoupin, E-H. Chung, S. Chattopadhyay, C. U. Segre and E. S. Smotkin, J. Phys. Chem. B., 110, 9932, 2006.

[24] C. Bock, B. MacDougall and Y. LePag, J. Electrochem. Soc., 151, A1269, 2004.

Multifunctional Nanocomposites for Oil and Gas

Andrew Sherman*, Nick Farkas** and Dr. B Werry*

*Powdermet Inc
24112 rockwell dr, Euclid, OH 44117
ajsherman@Powdermetinc.com, bwerry@powdermetinc.com
**Terves Inc, Euclid, OH, 44117 nfarkas@tervesinc.com

ABSTRACT

Currently, wireline, mechnical, and hydraulic mechanisms are used to operate completion tools, and "dumb" materials are used in completion and stimulation operations. "smart" materials, or materials that automatically sense and respond to their environment, are enabled using nanotechnology and hierarchical design principles. These engineered "smart" materials offer potential for new tools and activation mechanisms by responding to changes in temperature, Ph, or fluid environment to provide an engineered response. Response can include shape change, dissolution/reaction, chemical release, and change in state (eg, fluid to solid). Limited applications have been developed in temperature-activated dissolvables, leachable tracer systems, and swellable packers. These, as well as emerging, more advanced materials and applications such as switchable sand control screens and targeted deliver of energy and force will be discussed in this presentation.

Keywords: nanocomposites, multifictional materials, completion tools, materials for oil and gas, magnesium

1 INTRODUCTION

Currently, less than 35% of all known and proven oil and gas reserves are economically recoverable. In fact, the largest source of additional oil and gas ovet the next decade will not be new fields, but enhanced extraction from known oil and gas deposits. Each 1% inrease in oil recovery amounts to over 25 billion barrels of oil. New technologies are needed to enhance completions, stimulation, and oil recovery by controlling crack network formation and pressure distributions during reovery operations. Development of smart, "switchable" materials-materials that have a response triggered by a change in local environment, can be enabling in managing formation and fracture conductivity in different reservoir lithologies.

Controlling structure and chemistry at the nanoscale is an enabling technology for high performance multifictional, or "smart" materials. Nanomaterials have high surface area, so that interface or surface-area driven processes such as catalysis, reaction, or dissolution occur quickly, and can be controlled by controlling the surface or interfacial area by both particle size, and amount. By controlling multiple parameters, multifunctionality (eg strength and dissolution rate) can be controlled with some independence to produce designer materials with a specific set of properties.

Powdermet and its subsidiaries, licensees, and spin-outs such as Terves Inc, has been working to exploit these capabilities by designing multifunctional materials over the last several decades. This has led to a portfolio of currently commercially available and emerging materials being used in aerospace and defense, and more recently oil and gas completion and stimulation applications.

Specific multifunctional materials that are currently in early commerial, or are emerging include; Tervalloy™ dissolvable high strength metals, LocProp™ triggerable sand control screens, ExoProp™, high strength expandable materials, AcoustiFlow™, acoustic enhancement agents.

2 LOCPROP™ PUMPABLE SAND SCREENS

Water injection wells used for secondary and tertiary oil recovery have a high failure rate, requiring frequent intervention or stimulation. The primary failure is a loss of injectability- leading to unacceptable increases in injection pressure and low flow rates. This is particularly cumbersome and costly in offshore wells, where the cost of intervention and workover to restimulate or clean up a well is excessive. A major cause is transfer of fines and sand between zones, and during completions. The problem is exasperated when thermal EOR techniques, such as Huff-n-puff or cyclic steam injection are used. Technology advancement is required to improve performance of single point injection wells, water injection wells, and thermal enhancement wells for offshore application.

Many formations being developed consist of unconsolidated or loosely consolidated sandstone. In these formations, high closure forces an use of high pressure gradients are not achievable. Current techniques for stabilizing these formations, including gravel packing with large proppant, injection of polymer adhesives have proven to be of limited success, or cause secondary issues that create additional problems. Using engineered multifunctional materials, Powdermet has developed a "pumpable screen" consisting of environmentally responsive coatings on the surface of pumpable powders (sand, ceramic beads) that can be easily pumped, and which then crosslinks and forms a rigid screen under the conditions encountered in unconsolidated water injection wells. Previous materials have been primarily developed for higher temperature, or higher closure force applications (eg, procured epoxy's) to prevent flowback of proppants used in stimulation treatments, or are active materials that can set-up and plug the pipeline or during transport (eg, two part epoxy's).

As an alternative, Powdermet has developed a low temperature low pressure crosslinking coating that, when applied to particles, can create a free-flowing system that sts up and cures in mildly acidic conditions at temperatures around 60C. Specifically, LocProp is a low-temp, controlled curable thermoplastic-coated proppant that sets to form an in-situ sand control filter matrix. Designed to prevent proppant flow through in water injection wells, and flowback in production wells, this proppant extends fracture conductivity lifetime with reduction in fines generation and self-healing capability.

This activator free immobile proppant is frac-fluid and breaker compatible, and optimally performs in reservoir temperatures from 100-260F. This patent-pending product is an activated thermo-plastic coating that creeps to form good contact under mild pressure, and acts as a "living polymer", having an environmentally controlled tackiness and crosslinking ability. Typical product features include:

Mesh size: 16/20 and 20/40
Sphericity: 0.9
Roundness: 0.9
Absolute density: 2.7 g/cc
Bulk density: 1.57 g/cc

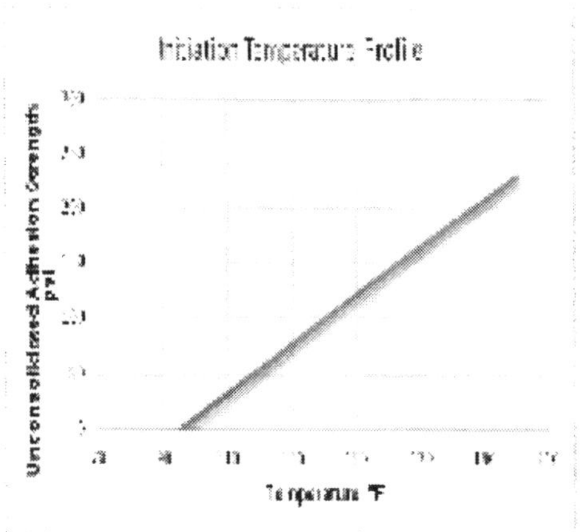

Figure 1: unconstrained cohesive strength as a function of thermal exposue for LocProp pumpable sand screens

3 TERVALLOY™ DISSOLVABLE METALS

In carbonate and shale reservoir lithography, the formations are fractured, or "stimulated" to improve flow. These tight formations have only become economic with the development of horizontal drilling and hydraulic facturing technologies. However, even with stimulation, the formations must be constantly drilled and stimulated, as the wells typicaly loose 80% of their production in the first two years. This pheomena, along with the reduced price of oil, is placing increasing demands on the industry to reduce costs and improve efficiencies in drilling, completion, and stimulation operations.

The development of mproved completion technologies, including sliding sleeve, open hole packers, and swellable packers has driven completion times and costs down dramatically over the last 10 years. This has resulted in a dramatic increase in stages fractued, from only a few in the early 2000's to more than 20 on average, and as many as 150 stages/well currently. One of the innovations driving down completion costs is the development of dissolvable materials that eliminate the need to recover or drill-out flow isolation stages.

Tervalloy is a high strength, nano-engineered magnesium composite that is engineered to degrade in a highly controlled and repeatable manner in completion fluids (brines, slickwater, completion fluid). Use o these dissolvable materials can recuce completion costs be several hundred thousand dolaars or more for each well, by eliminating the need for coiled tubing workover rigs, and reducing completion time and fluid use by 30-50% or more.

Terves Inc licensed Powdermet technology to create a patent pending reactive magnesium metal matrix composite (MMC) that exhibits controlled dissolution rates. These cast and extruded magnesium composites have shown tensile yeild strengths as high as 48ksi with ductilities over 5%. Previous high strength magnesium alloys are based on rare earth additions and rapidly solidified alloys, however these formulations are extremely costly, and do not exhibit the desired controllable dissolution rates. Terves has demonstrated new alloys that have the desired strength and dissolution properties using the addition of fillers through ultrasonic casting and semi-solid mixing. 15 initial active metal formulations were extruded at an 8:1 extrusion ratio at different temperatures and soak times.

Properties of the engineered degradable alloys ranged from 20ksi to 45.6ksi in yield with ultimate tensile strengths from 36ksi to 52.5ksi. Elongation of samples ranged from 5-21% however typical elongation was 5-12%. Dissolution were controlled over a wide range to meet targeted dissolution for common completion brines between 30 and 80 mg/cm2/hr.

Commercial alloys now offered include TA-50x and TA-100x. TA100x has been tested using ASTM E8-

13a for tensile testing and resulted in the following mechanical properties (average results recorded):

Ultimate Tensile Strength: 46.5 ksi
Yield Strength (0.2% offset): 32.5 ksi
Elongation to failure: 12%

This material was also tested for dissolution rate in a 3% KCl solution at 90°C to show that the rate at which extruded product dissolves is consistent with client specifications. These tests showed that the dissolution rate was constant at 50-60 mg/cm^2*hr or about 0.3 mm/hr.

4 TERVALLOY ALLOYS FOR OFFSHORE

As an ilkustration of the tailorability of nanocomposite materials, Terves was requested to develop and alloy having controlled dissolution in calcium containing brines. Calcium in the brinde was found to inhibit the controlled corrosion of active magnesium composites, leading to passivation o standard materials. In response, the galvanic activity was studied extensiveley, and new alloys were designed to couteract the passivation effect of calcium (CaCl2, CaBr2) to enable use in offshore applciations.

Four variations of active corrosion catalysts were varied in a magnesium matrix for a total of nine developmental composites. Each of the nine composites were tested in 3% KCl and 8.9 lb/gal $CaCl_2$ at 90°C for 6 hours. The dissolution results are shown below in Figure 1, demonstrating three potential alloys chemistries that dissolve in both KCl and $CaCl_2$ brines, TA-6, TA-7 and TA-9.

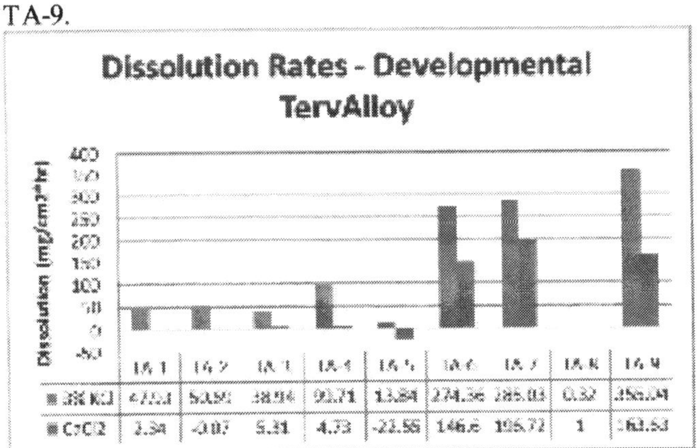

Figure 1. Dissolution Rates of Developmental Alloys

Each of the nine formualtionss were also sent for third party tensile testing. These results are listed in Table 1 below.

Table 1. Tensile Results of Extruded Alloys

ID	UTS (ksi)	YS (ksi)	E %
TA-1	47.1	41.1	6.5
TA-2	45	34.4	2
TA-3	44.6	33.6	12
TA-4	44.5	34.5	8.5
TA-5	50.5	41.7	8.5
TA-6	21.9	21.5	0.5
TA-7	38.4	35.5	3.5
TA-8	40.4	25.9	14
TA-9	35.5	27.6	4.5

5 EMERGING DEVELOPMENTS

Powdermet is currently developing active pumpable materials that can respond to the reservoir environment to generate heat, hydrogen, sound, and force. These "controlled deposition of energy" materials are engineered nanocomposites (how fast, and with what response) that can include control coatings (when activiated). One patent pending material is designed to act as a "mini-jack", applying a large force to extend fracture networks. When added into an engineering polymer resin, this system can be used to provide formation-activiated tools capable of applying tens of thousands of pounds of force to control the development of racture networks- both initiation and extension of fractures.

Additionally, patent-pending zero-valent metal nanocomposites have been engineered and demostrated to provide free hyydrogen, as well as thermal energy *in-situ* in relatively low cost systems. Terves has developed a series of RESPONSE™ coatings that act as thermal or chemical "switches", controlling timing of the activiation of these, as well as dissolution actions of responsive materials.

6 SUMMARY

Nanotechnoloogy and advanced materials processing has been used to develop materials that provide very controled, engineered response to downhole enviroments. Commercially or field-trial level available materials provide controlled agglomeration or degredation to control flow with automated response. Emerging materials provide targeted delivery of energy nand chemicals, including force, thermal energy, and chemical action (reducing, oxidizing, etc.)

A Field Application of Nanoparticle-Based Invert Emulsion Drilling Fluids

Alexey S. Borisov[1], Maen Husein[1], Geir Hareland[2]
[1]Department of Chemical & Petroleum Engineering, University of Calgary, Calgary, AB, Canada.
asboriso@ucalgary.ca; mhusein@ucalgary.ca
[2]Department of Chemical Engineering, Oklahoma State University, OK, USA 74078. geir.hareland@okstate.edu

ABSTRACT

Application of nanotechnology in drilling fluids for the oil and gas industry has been a focus of several recent studies [1-2]. Zakaria et al. [2] developed a process for the in situ synthesis of nanoparticles (NPs) and reported that calcium-based and iron-based nanoparticles (CNP and INP, respectively) with concentrations of 0.5-2.0 wt% can dramatically improve filtration properties of commercial drilling fluids in a lab environment. A modified process for the emulsion-based synthesis of NPs on a large scale has been developed and a NP carrier fluid was employed in full scale field tests. Results showed that total mud losses while drilling with CNP-based invert emulsion were on average 27% lower than in the case of conventional fluids, which falls within the range seen in the laboratory.

Keywords: nanoparticles, drilling fluid, invert emulsion, field application, oil and gas

1 INTRODUCTION

Invasion of drilling fluid filtrate and suspended solids into a near-wellbore region during drilling operations leads to formation damage and can cause wellbore instability, severe permeability impairment and a substantial decline in production [3]. Recently, several studies have shown that drilling muds that combine conventional LCM with particles in the nanometer domain can dramatically reduce fluid loss and form a thinner filter cake, compared to drilling fluids containing LCM alone [1-2]. In this work, six full-scale field tests were conducted in order to provide reliable observations on the performance of NPs under real-life conditions. Field testing employed a 'carrier' approach, whereby a custom invert emulsion of 20 m^3 volume containing 5 wt% in situ CNPs was first prepared in a specialized mixing facility, then delivered to a rig site where it was diluted to the target concentration.

2 MATERIALS AND METHODS

Without delving into proprietary details, industry-scale synthesis of NP-based carrier emulsion followed a bench-scale process developed in our laboratory [4]. Calcium-based NPs (CNP) were produced via a (w/o) microemulsion approach, which served as a mechanism to achieve size control and particle stabilization during chemical co-precipitation of aqueous precursors [5].

In order to determine the performance of the carrier emulsion approach for NP preparation and addition relative to our previous techniques [4], six different virgin and recycled commercial invert emulsion drilling fluids were tested in the laboratory. Drilling fluid testing was carried out using industry-standard equipment and practices [6].
Six full-scale field tests were conducted in horizontal wells in Alberta, Canada in 2014. Upon arrival to the drilling site, 10% vol/vol of carrier were mixed with the circulating drilling fluid to achieve the target concentration of 0.5 wt% CNP, which was maintained thereafter. Mud losses while drilling were estimated via volumetric balance on the total drilling fluid on-site at different depths. The difference between calculated control volumes at a given depth and the initial depth corresponds to cumulative mud losses while drilling. In order to account for differences in measured depth and to standardize the data, cumulative mud losses were converted to losses per 100 m drilled.

3 RESULTS AND DISCUSSION

3.1 Laboratory testing of the carrier emulsion approach

In order to provide a feedback on the effectiveness of the industry-scale mixing and preparation of the carrier fluid, samples from the 20 m^3 carrier fluid were compared with a carrier fluid prepared in the laboratory under controlled environment, e.g. mixing, temperature, etc. The good agreement between the results suggests that the carrier approach was robust even when implemented on a large scale. A laboratory-based carrier emulsion was tested with six commercial invert emulsion drilling fluids to determine its impact on mud weight, electrical stability, plastic viscosity, yield point, gel strength, and volume composition. The results suggested that addition of carrier emulsion did not affect the basic properties of a host fluid significantly.

Once it was established that the carrier approach was compatible with commercial drilling fluids, high pressure high temperature (HPHT) filtration tests at 80 °C and 500 psi were employed to determine laboratory fluid loss in the presence of 0.5 wt% CNP. The results of HPHT experiments are provided in Figure 1 and suggest that CNP reduced the average HPHT fluid loss in virgin Cutter-D and Diesel OBM samples by 52% and 26%, respectively. Filtrate volumes in all three recycled drilling fluids were also reduced in the presence of 0.5 wt% CNP. The respective reduction in HPHT values for the recycled Cutter-D, Distillate 822, and Megadrill samples was 38%, 3%, and 44%, which reproduced previous laboratory results obtained with in situ calcium-based NPs [4].

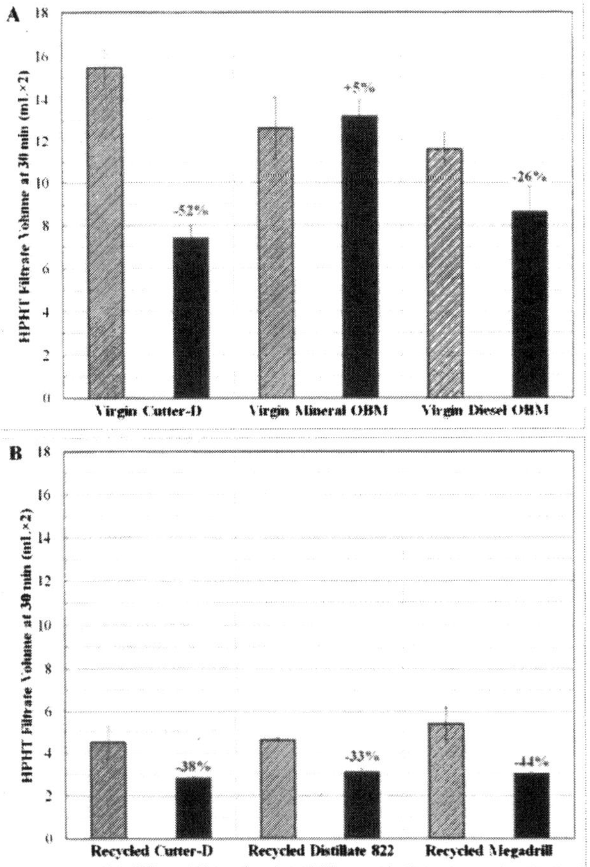

Figure 1: Impact of 10 vol% addition of a carrier emulsion on HPHT fluids loss at 80 °C and 500 psi of commercial virgin (**A**) and recycled (**B**) invert emulsion drilling fluids. Diagonal pattern corresponds to the control samples, while solid black denotes samples with 0.5 wt% CNP.

3.2 Field testing of the carrier emulsion approach

In order to compare the final mud losses in the control and test wells, the average losses per 100 m drilled are plotted in Figure 2.

The average control final losses in groups A, B, and C were in close proximity at 79±3, 77±3, and 82±7 m^3, respectively. Subsequently, same groups showed a 22–34% reduction of mud losses in the presence of 10 vol% carrier emulsion. Group D showed losses of 82 and 72 m^3, which was comparable to the control average. Figure 2-B shows the average mud losses standardized per 100 m drilled. Control wells exhibited losses in the range of 2.5–3 m^3/100 m, which is typical based on mud operator's previous experience in the area. Conversely, losses in the test wells were between 1.9–2.5 m^3/100 m, which again demonstrates reduction in the presence of CNP.

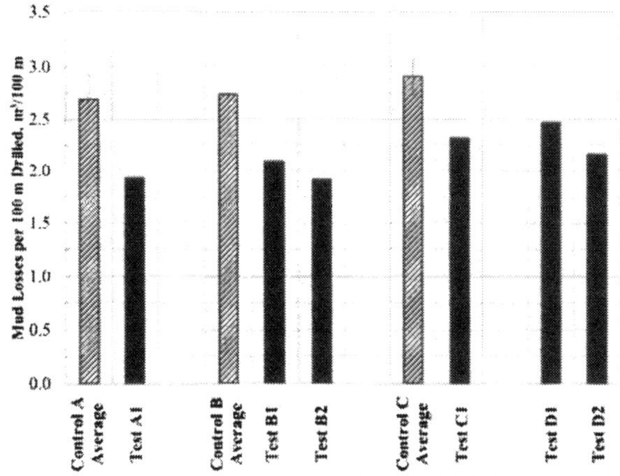

Figure 2: Average losses per 100 m drilled in the control and test wells.

CONCLUSIONS

In conclusion, a novel approach for introducing NPs into drilling fluids using a carrier method was developed and successfully scaled up to an industrial application. Six field tests were conducted using the carrier approach, which was compatible with the existing operations and did not cause any disruptions or drilling problems. The impact of 0.5 wt% CNP on the basic mud properties was within the normal range, while final mud losses in all test wells were 20–30% lower than the control average.

REFERENCES

[1] Contreras, O., Hareland, G., Husein, M., Nygaard, R., and Alsaba, M., "Application of In-House Prepared Nanoparticles as Filtration Control Additive to Reduce Formation Damage", SPE 168116-MS, SPE International Symposium and Exhibition on Formation Damage Control, Lafayette, Louisiana, USA, 26–28 February (2014a)

[2] Zakaria,M., Husein,M., and Hareland,G., "Novel Nanoparticle-Based Drilling Fluid with Improved Characteristics",SPE156992 International Oilfield Nanotechnology Conference and Exhibition, Noordwijk, The Netherlands, June 12–14 (2012)

[3] Jiao, D., and Sharma, M., "Mud Induced Formation Damage in Fractured Reservoirs", SPE 30107, European Formation Damage Control Conference, Hague, The Netherlands, May 15–16 (1996)

[4] Husein, M., Zakaria, M., and Hareland, G., "Novel nanoparticle-containing drilling fluids to mitigate fluid loss", PCT/CA2012/050075 (2012)

[5] Husein, M., and Nassar, N.N., "Nanoparticles preparation using the single microemulsion scheme", Current Nanoscience, 4, pp 370-380 (2008)

[6] API Recommended Practice 13B-2, "Recommended Practice for Field Testing Oil-Based Drilling Fluids", 5th edition (2014)

Development of New Surfactants for CO_2 Foam Enhanced Oil Recovery

Guanwei* Ren and Quoc P. Nguyen**

*Total E&P R&T USA, guangwei.ren@total.com
**Department of Petroleum and Geosystems Engineering, The University of Texas at Austin, Austin, TX, USA,
quoc_p_nguyen@mail.utexas.edu

ABSTRACT

Sweep efficiency in CO_2 enhanced oil recovery processes are major problem owing to the unfavorable mobility of CO_2. One method of controlling CO_2 mobility is through the use of foam. CO_2 foam in the past has predominately utilized commercial surfactant formulations, but with mixed results. This work pertains to the development of a new class of surfactants that can be injected with CO_2 rather than in water to improve foam performance, surfactant efficiency, and reduce water use. The paper describes the methodology used for designing the surfactants and screening the foaming properties. Surfactant solubility and partitioning between supercritical CO_2 (scCO2) and an aqueous phase (brine) were measured as a function of electrolyte concentration and pressure. Foam core flood was conducted to determine foam and surfactant transport. In general, the surfactant solubility in $scCO_2$ increased with pressure. The partitioning of surfactants into $scCO_2$ from brine was almost proportional to pressure, and decreased as temperature increased, where the latter held more sensitivity. Surfactant partition strongly influenced CO_2 displacement of brine in sanstone cores. In-situ foam generation during the injection of CO_2 with dissolved surfactant was remarkable.

Keywords: Supercritical CO2 foam, CO2 soluble surfactants, oil recovery, mobility control.

1 INTRODUCTION

CO_2 flooding is one of the most important methods for enhancing oil recovery (EOR). Miscible CO_2 EOR could recover as high as 20% original oil in place (OOIP) depending on the geological specifics [1.]. Within the United States alone, this EOR method produces 260,000 barrels of oil a day (of 280,000 bbl/d worldwide), representing 5.6 percent of U.S. production. The number of CO_2 injection projects, particularly for carbonate reservoirs, has continuously increased with the importance of CO_2 capture and storage. However, an unfavorable mobility ratio, reservoir heterogeneity and gravity segregation can reduce the macroscopic sweep efficiency. In-situ foaming of CO_2 is the method that has the most potential for improving sweep efficiency based on controlling CO_2 mobility. A dispersion of CO_2 into a surfactant solution has an apparent viscosity several orders of magnitude greater than the viscosity of either phase in the dispersion. In addition, substantial gas trapping in foam can further reduce gas relative permeability as a unique rheological characteristic of dispersed flow in porous media [2,3,4].

Several potential injection strategies for robust foam process have been studied [5,6,7]. Water-alternating-CO_2 injection has been most commonly used in field-scale foam process application [8,9], while simultaneous injection of these two phases has been the main injection scheme for lab-scale foam study. This is because the implementation of the latter strategy has encountered some difficulties such as operational constraints and severe reduction of well injectivity. While conventional foaming surfactants only traverse the reservoir in the aqueous phase, we have conducted an extensive experimental and theoretical study of a novel foam process in which the surfactant can partition between the CO_2 and the aqueous phases [10]. This novel foaming concept exhibits several advantages over conventional foam, of which lower surfactant adsorption, more in-depth robust foam, and higher injectivity are the most important. In addition, it enables the possibility of continuous injection of CO_2 with dissolved surfactant to generate foam in-situ when the injected CO_2 mixes with formation brine.

To increase the solubility of surfactant in scCO2, it was necessary to include fluorinated or silicone based hydrophobes to lower the intermolecular interactions of the surfactants. Although these early CO_2 soluble surfactants were used successfully in supercritical fluid research, some inevitable drawbacks have impeded their use in commercial applications. Expensive fluorous and silicone-based surfactants have environmental and biological persistence issues [11]. Therefore, efforts have been made to obtain low toxicity and less expensive CO_2-soluble hydrocarbon based surfactants, including hydrocarbon polymers [12,13], oxygenated hydrocarbon ionic surfactants [14], and nonionic surfactants [15-18]. Non-fluorinated surfactants is of the most interest.

Previous research has relied almost exclusively on phase behavior, utilizing a stirred high-pressure variable-volume stainless steel view cell [14-17, 19-21], as the standard technique for measuring surfactant solubility. Pressure and temperature induced changes in the turbidity of the solution are noted by visual inspection at each temperature; the phase transition or cloud point pressure is determined by reducing the pressure from a single-phase region until the sample becomes cloudy, i.e., phase separation occurs. This method also does not give any details on the concentration of the solute, thus the conclusions that can be drawn are limited in scope. This article describes an alternative novel method in which the soluble material was directly sampled and characterized to obatin a much more informative data set. This is an important consideration given that many commercial surfactants contain solvents and impurities. To evaluate the effect of surfactant partitioning between CO_2 and brine on the performance of foam process CO_2 foam core flood experiments were conducted on Berea sandstone with the conventional anionic foaming surfactant (CD 1045) and new non-ionic CO_2 soluble surfactant.

2 MATERIALS AND METHODS

The nonionic surfactant (2EH-PO5-EO9) was gift from The Dow Chemical Company and was used as received. The surfactant contained the same base alcohol 2-ethyl-1-hexanol with a linker of 5 PO and 9 EO units, respectively. Sodium chloride (certified ACS, Fisher) was used as received. Brine was made from deionized (DI) water with 3wt% or desired concentration NaCl. Surfactant was used at 0.2 wt%.. Liquid CO_2 was 99.5% purity, industrial grade.

Solubility test. All 2EH-PO5-EO9 solubilities in CO_2 were measured from 1500 psi to 3500 psi and temperature at 35 °C. No water was added. For surfactant concentration analysis, surfactant disolved in each CO_2 sample was trasnfered into a fixed volume of DI water. The surfactant concentration in the DI water was measured directly by HPLC (high-performance liquid chromatography) with an evaporative light scattering detector. The HPLC allows for more accurate determination of the concentration of surfactant and can also give useful information about the distribution of molecules and the partitioning between the phases.

Partitioning test. The method for determining 2EH-PO5-EO9 partition coefficients is quite similar to that used for the surfactant solubility test above. The sampling system and procedure, however, were modified to accommodate the presence of the aqueous phase. In a typical experiment, a high pressure cylinder was charged with CO_2. Surfactant solution was then injected into the cylinder. The system was well equilibrated for 24 hours before recording the equilibrium fluid pressure. Surfactant samples were taken from the water phase for concentration analysis. The brine salinity was fixed at 3 wt% in the studies of pressure effect. The surfactant partition coefficient is calculated using Eq. (1)

$$k = \frac{m_{sCO2} / (m_{sCO2} + m_{fCO2})}{m_{sw} / (m_{sw} + m_{w})} \qquad (1)$$

where m_{sCO2} is the mass of surfactant in the free CO_2, which is the difference between the initial mass of surfactant injected in brine into the cell and the remaining mass of surfactant in the brine at equilibrium (m_{sw}); m_{fCO2} is the mass of free CO_2 determined based on the initial mass of CO_2 charged into the cell and the solubility of CO_2 in the brine at equilibrium pressure and temperature; m_w is the initial mass of water injected into the cell. Note that the solubility of water in CO_2 was neglected.

Core flood test. A Phoenix Hassler-type core holder was mounted in the vertical direction and fluids were injected from the top to the bottom. Two backpressure regulators were used in a series to maintain a constant backpressure of 1500 psig during coreflooding. The core was cleaned and dried in a convection oven at 110 °C for 48 hours. It was then wrapped in three layers of aluminum foil and a thin Teflon heat shrink tube. The wrapped core was placed in the core holder and vacuumed for 10 hours before saturating the core with brine (3 wt% NaCl) in order to measure its porosity. The permeability of the brine saturated core was determined based on Darcy's law.

CO_2 and the surfactant solution were alternatingly injected into the core at 0.1 cc/min (1.4 ft/d) and 0.3 cc/min (4.2 ft/d), respectively. Surfactant solution and gas slug size were 0.1 pore volume (PV) and 0.2 PV respectively. Respective pressure drops over the three sections of the core were recorded. All core floods were conducted at 35 °C and 1500 psi backpressure.

3 RESULTS AND DISCUSSION

The solubility of 2EH-PO5-EO9 increases with increasing pressure (Figure 1) as has been seen with previous cloud point studies. The result suggests that the loss in solvation (CO_2-surfactant interactions) outweighs the benefit gained in decreased surfactant-surfactant interactions, which agrees with previous studies that have shown that the solute-solute interactions are the dominant factor in solubility in CO_2 [22]. This effect becomes more pronounced at higher temperatures and pressures, as the system reaches the limit of compressibility. The EO adds additional polar ether linkages that can increase the solvation by CO_2, however, this is offset by increased interactions between the surfactant molecules. The surfactant-surfactant interactions are not greatly affected by pressure, but are stronger at lower temperatures, whereas the CO_2-surfactant interactions increase with pressure. As a result, the solubility may not significantly decrease at higher pressures as more EO group is added since the CO_2 solvation can compete.

Figure 2 shows the partition coefficient of 2EH-PO5-EO9 as a function of pressure. The result shows that the surfactant was actively transported to the CO_2 phase. Although an increase in CO_2 density modestly raises the solubility of CO_2 in water, this effect has not been found to influence the solvation of the surfactant in the water phase [23].

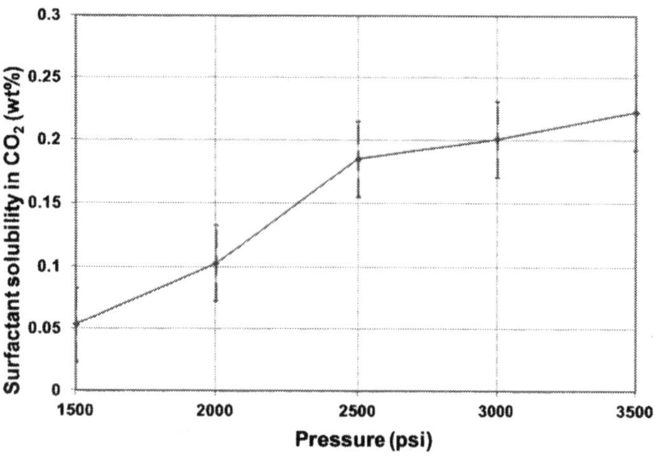

Figure 1: Solubility of 2EH-PO5-EO9 in CO_2 as a function of pressure at 35°C.

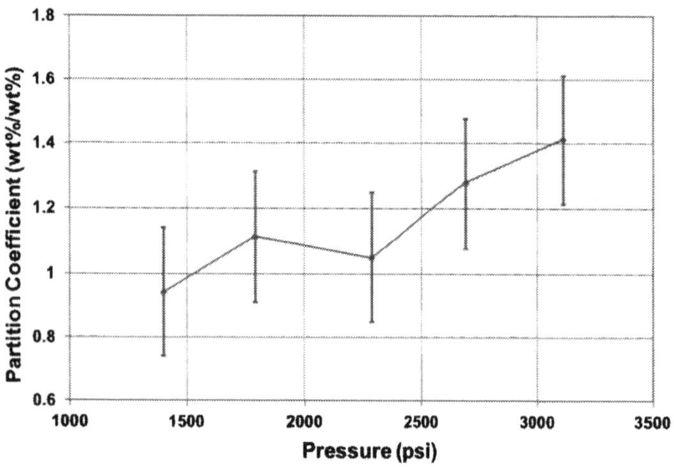

Figure 2: Partition coefficient of 2EH-PO5-EO9 in a CO_2-bine system with 3wt% NaCl at 35°C.

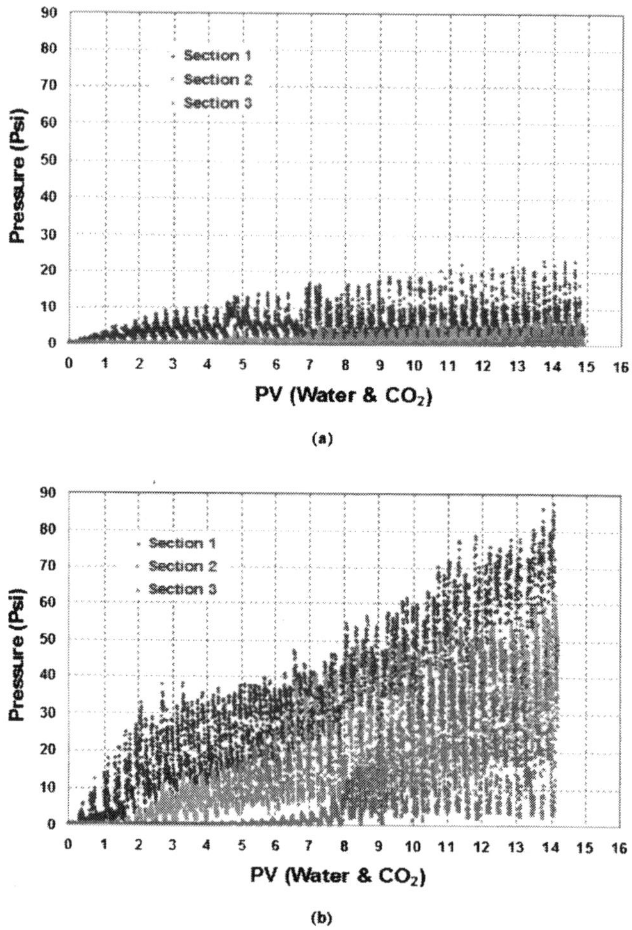

Figure 3: Pressure drop during surfactant alternating gas injection for (a) anionic surfactant CD 1045, and (b) nonioninc surfactant 2EH-PO5-EO9.

Figure 3 shows the sectional pressure drops for 2EH-PO5-EO9 and the anioninc surfactant CD 1045. For both cases, the pressure drops build up during liquid injection and decrease in the chasing gas injection. They tend to fluctuate from cycle to cycle. The fluctuation in pressure drop magnitude becomes more obvious as foam propagates. For 2EH-PO5-EO9, it only took one injected pore volume (IPV) for foam to propagate into Section 2 and then 5 IPV to reach Section 3. On the contrary, a double amount of fluids are required for CD 1045 stabilized foam to yield a significant pressure drop response. It is important to note that CD 1045 and 2EH-PO5-EO9 exhibit almost the same level of adsorption in sandstone. Therefore, the higher foam robustness with 2EH-PO5-EO9 is attributed to the better surfactant transport during gas injection cycle owing to the partitioning of surfactant between CO_2 and water. The magnitude of pressure drop (or CO_2 mobility reduction) for 2EH-PO5-EO9 stabilized foam is 5 to 10 times higher than that for CD 1045.

4 CONCLUSIONS

The solubility of 2EH-PO5-EO9 in CO_2 at 35°C increases up to 0.2 wt% as pressure increases above 2000 psi. The partitioning of the surfactants between CO_2 and water phases increased almost linearly with pressure. The partition coefficient was very sensitive to surfactant impurity (i.e. distribution of EO and PO groups).

Several measurements of surfactant partition coefficient were conducted at 35 °C and 3500 psi with salinity changes up to 9wt% NaCl. The results show little or no effect of electrolyte concentration on the partition coefficient.

Surfactant partitioning strongly affects foam performance in sandstone cores. CO_2 soluble surfactant does improve foam propagation and injectivity. Alternating injection also strenthens the advanatges of CO_2 soluble surfactant stabilized foam with respect to quicker and stronger foam generation. This is attributed to the improved surfactant transport during alternating injection owing to surfactant partitioning between CO_2 phase and aqueous phase.

REFERENCES

[1] Geiger P.2006. CO2 EOR and Sequestration in Wyoming. Proc. 4th Annual EOR Carbon Management Workshop, Houston, TX, U.S.A. Dec 4-5, 2006.

[2] Tang, G. and Kovscek, A.R. 2004. Measurement and Theory of Gas Trapping in Porous Media during Steady-State Foam Flow. Paper SPE 90142 presented at SPE ATCE, Houston, Texas, 26–29 September. doi:10.2118/90142-MS.

[3] Nguyen, Q.P., Rossen, W.R., Zitha, P.L.J., and Currie, P.K. 2009. Determination of Gas Trapping With Foam Using X-Ray CT and Effluent Analysis. SPE J. 14(2): 222-236.

[4] Nguyen, Q.P.; Currie, P.K. Currie; Zitha, P.L.J.: "Determination of Foam Induced Fluid Partitioning in Porous Media using X-ray Computed Tomography," paper SPE 80245 presented at the SPE International Symposium on Oilfield Chemistry, Houston, Texas, U.S.A., 5–8 February 2003.

[5] Renkema, W.J. and Rossen, W.R. 2007. Success of SAG Foam Processes in Heterogeneous Reservoirs. Paper SPE 110408 presented at the 2007 SPE ATCE, Anaheim, CA, 11–14 Nov. doi:10.2118/110408-MS.

[6] Rossen,W.R, Kibodeaux, K.R.,Shi, J.X., and Lim, M.T. 1995. Injectivity and Gravity Override in Surfactant-Alternating-Gas Foam Processes. Paper SPE 30753 presented at the 1995 SPE ATCE, Dallas, TX, 22–25 October. doi:10.2118/30753-MS.

[7] Rossen, W.R., van Duijn, C.J., Nguyen, Q.P. and Chen, C. 2010. Injection Strategies to Overcome Gravity Segregation in Simultaneous Gas and Water Injection into Homogeneous Reservoirs. SPE J. 15(1): 76-90.

[8] Xu, Q.and Rossen, W.R. 2004. Experimental Study of Gas Injection in a Surfactant-Alternating-Gas Foam Process. SPE Res Eval & Eng 7(6): 438-448.

[9] Lawson, J.B. and Reisberg, J. 1980. Alternate Slugs of Gas and Dilute Surfactant for Mobility Control During Chemical Flooding. Paper SPE 8839 presented at the SPE/DOE Enhanced Oil Recovery Symposium, Tulsa, 20–23 April. doi: 10.2118/8839-MS.

[10] Le, V.Q., Nguyen, Q.P., and Sanders, A.W. 2008. A Novel Foam Concept with CO2 Dissolved Surfactants. Paper SPE 113370 presented at 2008 SPE Improved Oil Recovery Symposium, Tulsa, OK, 19–23April. doi:10.2118/113370-MS.

[11] K.P. Johnston, T.W. Randolph, F.V. Bright, S.M. Howdle, Toxicology of a PFPE Surfactant, Science, 272 (1996) 1726.

[12] R. Fink, D. Hancu, R. Valentine, E.J. Beckman, Toward the Development of "CO2-philic" Hydrocarbons. 1. Use of Side-Chain Functionalization to Lower the Miscibility Pressure of Polydimethylsiloxanes in CO2, J. Physical Chemistry, B., 103 (1999) 6441-6444.

[13] T. Sarbu, T. Styranec, E.J. Beckman, Non-flourous polymers with very high solubility in supercritical CO2 down to low pressures, Nature, 405 (2000) 165-168.

[14] X. Fan, Vijay K. Potluri, M.Chandler McLeod, Y.Wang, Oxygenated Hydrocarbon Ionic Surfactants Exhibit CO2 Solubility, J. American Chemical Society, 127 (2005) 11754-11762.

[15] J. Liu, B. Han, G. Li, X. Zhang, Investigation of Nonionic Surfactant Dynol-604 based Reverse Microemulsions Formed in Supercritical Carbon Dioxide, Langmuir, 17 (2001) 8040-8043.

[16] J. Liu, B. Han, Z. Wang, J. Zhang, G, Li and G. Yang, Solubility of Ls-36 and Ls-45 Surfactant in Supercritical CO2 and Loading Water in the Co2/Water/Surfactant System, Langmuir 18 (2002) 3086-3089.

[17] J. Liu, B. Han, Z. Wang, J. Zhang, G. Li and X. Zhang, Chem. Eur. J. 2002, 8, No.6.

[18] G.J. McFann, K.P. Johnston, S.M. Howdle, Solubilization in Nonionic Reverse Micelles in Carbon Dioxide, AIChE J., 40 (1994) 543-555.

[19] A. Dupont, J. Eastoe, M. Murray, L. Martin, F. Guittard, E. Taffin de Givenchy, Hybrid fluorocarbon-hydrocarbon CO2-philic surfactants. 2. formation and properties of water-in-CO2 microemulsions, Langmuir, 20 (2004) 9953-9959.

[20] X. Chen, S. Adkins, A. Sanders, Q. Nguyen, K. Johnston, Interfacial Tension and the Behavior of Microemulsions and Macroemulsions of water and Carbon Dioxide with a Branched Hydrocarbon Nonionic Surfactant, J. Supercritical Fluids, 55 (2010) 712-723.

[21] D. Xing, B. Wei, K. Trickett, A. Mohamed, J. Eastoe, Y. Soong, R. Enick, CO2-soluble Surfactants for Improved Mobility Control. Paper SPE 129907 presented at SPE Improved Oil Recovery Symposium, Tulsa, Oklahoma, USA, 2010, 24 28 April. doi: 10.2118/129907-MS.

[22] S.S. Adkins, X. Chen, I. Chan, E. Torino, Q.P. Nguyen, A.W. Sanders, Keith P. Johnston, Morphology and Stability of CO−2-in-Water foams with Nonionic Hydrocarbon Surfactants, Langmuir, 26 (2010) 5335–5348.

[23] K.R. Wormuth, S. Zushma, Phase behavior of branched surfactants in oil and water, Langmuir, 7 (1991) 2048-2053.

Use of Inert, Corrosion Resistant CVD Coating for Oil and Gas Service

L. Patterson, J.B. Mattzela, D. Smith, M. Yuan, G. Barone and M. Higgins*

*SilcoTek® Corporation, Bellefonte, Pennsylvania, USA, SilcoD@SilcoTek.com

ABSTRACT

As growth in demand for petroleum products continues to drive an increasingly diverse supply, challenges involved with the exploration, extraction, transportation and testing of oil and gas have pushed companies to manage costs, primarily via corrosion control. However, with additional requirements including enhanced wear resistance and chemical sampling inertness, finding a suitable yet economical solution is especially daunting.

The use of advanced, high performance protective coatings is gradually becoming an industry-wide solution for meeting both performance and financial expectations. This paper will introduce a silicon-oxygen-carbon coating (Dursan®) as one such advanced treatment. Following an introduction to proper coating selection criteria, the acidic corrosion resistance, wear resistance, and chemical inertness of this chemical vapor-deposited (CVD) coating will be compared to that of other common surface treatments used in oil and gas sampling service.

Keywords: coating, corrosion control, H_2S, inertness, sampling

1 INTRODUCTION

Sulfur compounds, particularly hydrogen sulfide (H_2S), naturally occur at trace levels in most oil and natural gas reservoirs worldwide. Quantifying the H_2S content i.e. "sweetness" of potential yields is critical to understanding product purity; high H_2S content (generally greater than 0.5%) indicates a "sour" yield that requires additional purification. However, the steels, stainless steels, and exotic alloys which are commonly used in the flow paths of well sampling equipment have proven to adsorb certain sulfur compounds, making accurate, low-level quantification of these molecules impossible [1]. This, coupled with corrosive extraction environments and increasingly strict government regulations, has driven an overwhelming need for the development and use of surface modifications on sample storage and analysis equipment.

Over the years, many solutions have been developed to enhance sampling system compatibility with reactive compounds like H_2S and mercury (Hg), such as polytetrafluoroethylene (PTFE), siloxane, and chemical vapor-deposited silicon-based coatings. The complex sampling requirements discussed above prove that no single solution exists, but pairing a high performance coating with proper base material selection will ensure a reliable quantification of oil and gas yield composition i.e. quality.

2 DISCUSSION

Prior to the full scale development of an oil or natural gas field, samples must be collected and accurately analyzed to ensure the supply is suitable. The substrate material used for most sampling flow path equipment (sample cylinders, valves, filters, etc.) is traditionally stainless steel or other metal alloys as dictated by physical and design parameters like temperature, chemical exposure, and pressure ratings. A variety of these materials meet the physical requirements of a sampling system but fail to provide the chemical inertness necessary for trace level detection of critical compounds such as H_2S.

2.1 Sulfurs (H_2S) and Steels/Exotic Materials

In addition to surface area, temperature and pressure exposure, the chemical composition of a metal substrate will affect the type of reactions that occur with analytes in the sample that move through the flow path. 316L stainless steel is a commonly specified material for sampling componentry due to its ability to form a stable and protective oxide layer. However, highly-reactive free iron (Fe) on the surface poses analytical problems; an ASTM A380 "Ferroxyl Test for Free Iron" demonstrates that the surface's iron and iron oxides are readily attacked, providing avenues for both corrosion and increased surface activity under the harsh conditions typically found in oil and gas sampling service [1]. Over time, due to steady adsorption of sulfurs like H_2S, the surface will reach a near-stable equilibration point and make trace-level quantification impossible.

This phenomenon is even more apparent when copper-containing superalloys like Monel® are used in place of stainless steel to combat harsh corrosives. H_2S and other sulfur compounds are completely scrubbed from sample streams constructed from these materials; Hashem, et al. have proven that samples at 50 parts-per-million (ppm) are scavenged in Monel® tubing lines and no aforementioned equilibration effects occur [2]. It is clear, then, that surface modifications i.e. advanced coatings must be used in conjunction with suitable base materials to ensure the analytical results from oil and gas field qualification tests are exhaustive and error-free.

2.2 Selection of Coatings

The variety of materials available to solve sampling system compatibility problems are listed in Table 1. Some are specific to pH range tolerances while others provide higher thermal stability or specially-tailored surface chemistry specific to certain critical analytes.

Coating	Composition
PTFE or PFA	100% polytetrafluoroethylene or perfluoroalkoxyacetyl
Siloxane	polydimethylsiloxane
a-Si	amorphous silicon
a-Si-R; R=(C-H)	hydrocarbon-functionalized amorphous silicon
a-SiOC (Dursan)	carboxysilicon

Table 1: Common commercial coatings and compositions

Selection of required material properties is often derived from the most significant failure mode in the application. Abrasion and wide pH exposure in oil and gas sampling applications limit the potential coating solutions that can be used. For example, significant wear or vibrations result in poor performance for PTFE coatings, however PTFE is one of only two options discussed here that is suited for pH 0-14. Table 2 summarizes some of the most important physical characteristics for coatings in oil and gas service.

Property	Dursan	a-Si-R	a-Si	Siloxane	PTFE or PFA
Maximum temp.	450° C	450° C	1400 ° C	250° C	260° C
Minimum temp.	-210° C	-210° C	- 210° C	-40° C	-240° C
Low pH limit	0	0	0	2	0
High pH limit	14	7	7	7	14
Thickness	2μm	0.5μm	0.5 μm	0.5μm	25μm
Adhesion	Very good	Very good	Very Good	Good	Poor

Table 2: Physical properties of coatings

The final and most critical criterion to the selection of an appropriate coating for sample stream components is the chemical compatibility of the coating with critical compounds and corrosive media involved in the sample extraction process. Table 3 briefly summarizes the coatings' compatibility in common sampling applications (note: <ppm levels of H_2S and Hg). Inertness in these applications is critical for accurate qualification of oil and gas fields and, additionally, mitigating potential corrosion. These factors have a signficant impact on both the quality of the final product and the project's bottom line.

Coating	H_2S Sampling	Hg Sampling	High pressure sampling	Chloride Containing
PTFE or PFA	X	X		X
Siloxane				
a-Si			X	X
a-Si-R; R=(C-H)	X	X	X	
a-SiOC (Dursan)	X		X	X

Table 3: Compatibility of coatings

It is clear that, specifically for sampling potential oil and natural gas yields, only PTFE/PFA and the amorphous silicon-based coatings are suitable for maintaining an inert barrier between trace levels of H_2S and the metal substrate. Further, only the Dursan coating exhibits the necessary combination of robustness and chemical inertness for repeated, reliable sampling of H_2S in the abrasive conditions present in oil and gas fields. The subsequent sections of this paper will demonstrate Dursan's chemical inertness, corrosion and wear resistance properties.

3 EXPERIMENTAL - DURSAN® COATING

SilcoTek® Corporation's Dursan is a chemical vapor-deposited (CVD) coating comprised of amorphous silicon, oxygen and carbon. Unlike traditional dip or spray coating techniques, CVD is a non-line-of-sight vacuum process which ensures all blind holes and complex geometries are coated. Additionally, unlike PTFE, the coating is physically deposited into the lattice of the base metal, leading to full adhesion and the ability to bend coated parts without risk of flaking. The composition of Dursan and its diffusion into a stainless steel substrate are illustrated in Figure 1 via Auger Electron Spectroscopy (AES).

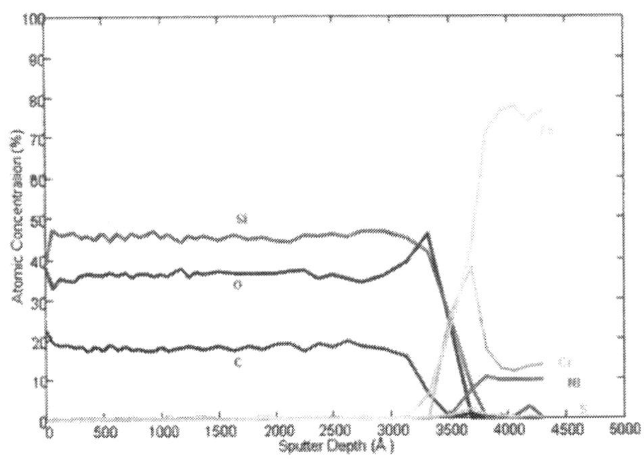

Figure 1: AES profile of Dursan coating

3.1 Inertness to Hydrogen Sulfide (H_2S)

Hydrogen sulfide (H_2S) is the most critical active compound requiring quantification in crude oil and natural gas extraction. Yields containing more than a few ppm of H_2S are considered "sour" and must be purified. Figures 2 and 3 demonstrate the need for an inert coating during the sampling, storage and analysis of this highly active analyte. Non-treated sampling components begin to adsorb the H_2S compounds immediately as described previously in section 2.1. Dursan-coated containers provide sample stability and a realistic time frame for transporting the sample to the laboratory without degradation or loss.

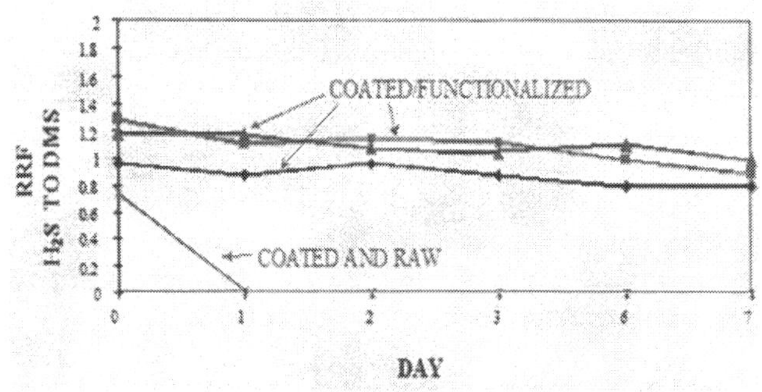

Figure 2: Sulfur compounds at 17 ppbv in Dursan-treated stainless steel containers

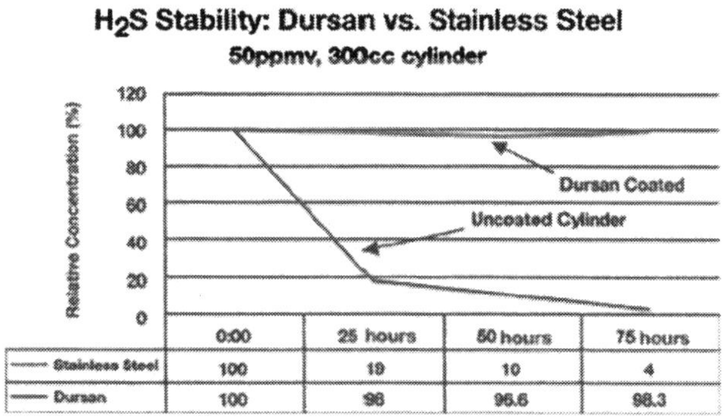

Figure 3: H_2S compounds at 50 ppm in Dursan-coated stainless steel containers versus non-treated containers

As demonstrated, the degradation of H_2S on bare stainless steel is rapid and irreversible. Both at 50 ppm at 17 ppbv levels, H_2S is lost within 24 hours, making an effective analysis practically impossible due to transportation time between field and laboratory. It is worth noting that, in work to test the stability of sulfur compounds during such static sampling, the use of gases like silane (SiH_4) along with multiple day exposure to 5000 ppm H_2S was required to create a passive cylinder for storage [3]. Much of the data in these studies were gathered to demonstrate stability for the use of creating low-level H_2S standards in fuel. Commercially available Dursan-coated components have eliminated this need for passivation and are now recognized as "out of the box" solutions for sulfur sampling and transport. This eliminates the need for working with dangerous materials such as high concentrations of H_2S or pyrophoric gases such as silane.

3.2 Corrosion Resistance

Chloride environments and chloride-containing oil and gas supplies are common in exploration workflows. The aforementioned surface treatments in addition to exotic materials all have been used to increase the lifetime of components in these corrosive applications. Table 4 provides the results obtained from ASTM G31 testing, comparing Dursan to an a-Si coating and bare 304 stainless steel. This method is an immersion test for 24 hours in a 6M HCl (18%) solution at room temperature and pressure. After immersion, differential weighing allows determination of the amount of material loss. The sample size for each configuration was 3 samples. The a-Si coated stainless steel shows greater than 20 times the resistance of non-treated 304 stainless steel in this environment while the Dursan treatment creates greater than a 200 time improvement. Any loss whatsoever in the coated samples occurred as a result of pitting corrosion. The pitting is an indication that there are still pin-holes present in the surface which allowed corrosive attack to initate.

24hr; 6M HCl; 22° C	304 SS	a-Si coated	Dursan coated
MPY (mils-per-year)	389.36	16.31	1.86
Improvement Factor	---	23.90	209.8

Table 4: Sample weight loss after 24 hour exposure to 6M (18%) HCl

3.3 Wear Resistance

A final factor for consideration in selecting an appropriate surface treatment for making trace level H_2S sampling possible is the coatings' wear resistance. Down-hole sampling equipment is often subjected to significant abrasion from repeated deployment and retraction in the field. This exposure can easily remove an a-Si coating or crack PTFE/PFA coatings in only one us. To that effect, wear resistance is desired in order to avoid frequent and costly replacements. Table 5 summarizes the data obtained from wear studies conducted on both non-treated and treated surfaces. Data were generated using a pin-on-disk tribometer (Nanovea, Irvin, CA). The experiment uses a flat plate loaded onto the test rig and the indenter applies a precise force to surface. The plate is then rotated and forces are measured between the pin and the disc. Results from this experimental method can produce wear behavior and friction coefficients of the plate surface [4]. Results from this experiment demonstrate that the Dursan coatin wears less than untreated steel and an a-Si or a functionalized a-Si coated surface. The improved wear resistance of the substrate as a result of the coating leads to longer lifetimes of oil and gas down-hole sampling equipment while drastically reducing the number of required replacements.

Pin on Disc; 2.0N	316 stainless steel	Dursan coated 316 stainless steel	a-Si coated 316 stainless steel
Wear rate ($x10^{-5}mm^3/Nm$)	13.810	6.129	2
Improvement Factor over Stainless Steel	---	2 times	1/3 times

Table 5: Wear resistance of coatings

4 CONCLUSION

Today, coatings are well-accepted means for drastically improving the performance characteristics of steels and metal alloys used for oil and gas service. While the corrosion resistance of untreated alloys may suffice for traditional exploration and extraction methods, these surfaces must be enhanced in order to make them suitable for analyzing trace levels of impurities, namely hydrogen sulfide (H_2S), in crude oil and natural gas yields. Offering a combination of chemical inertness, corrosion resistance and wear resistance, these coatings have different limitations depending on the base substrate. The Dursan® coating outlined in this paper offers the required inertness for accurate quantification of H_2S and other compounds, while also demonstrating physically robust qualities that alternatives such as PTFE/PFA lack. Regardless of the surface treatment that is used, the ability to accurately evaluate the contents of oil and gas fields will continue as an increasingly specified requirement.

REFERENCES

[1] Harris, P. and Pelligrini, M., "Mass Transport in Sample Transport lines Adsorption Desorption Effects and their Influence on Process Analytical Measurements", International Society of Automation 56th Analysis Division Symposium (2011).
[2] Hashem, et al., "Low-Level Hydrogen Sulphide Detection using Wireline Formation Tester", International Petroleum Technology Conference, IPTC 11582 (2007)
[3] Biela, B., Moore, R., Benesch, R., Tabert, B., Jacksier, T., "The Do's and Don't's in the Analysis of Sulfur for Polyolefin Producers", presented as paper 081 at the Gulf Coast Conference, Galveston Island, TX, October 22, 2003.
[4] Tchon, J., "Analytical Report: Nanovea – TRB101217-30", Nanovea Corp., Internal study report, December 2010

Fabrication and Design of Carbon Nanomaterials Based Scale Monitoring Smart Sensor Systems

Hammad Younes^, Madina Jelbuldina*, Samuel O Sofela*, Lina Tizani*, Amal Al Ghaferi^ and Irfan Saadat*

^ Dept of Mechanical and Material Engineering * Dept of Electrical Engineering and Computer Science, Masdar Institute of Science and Technology, Abu Dhabi, United Arab Emirates

ABSTRACT

The deposition of scale in oil pipelines leads to unavoidable damage of the equipment parts. Therefore, suspension of oil operations becomes essential to replace the damaged parts. Such interruptions are escorted by extremely high costs. In this paper, we report the fabrication of a new type of sensor and system for complete online monitoring of scale deposition with great accuracy and reliability. The fabricated sensors are based on CNTs and graphene. The sensors are fixable that can cover large area of the pipelines. Two different methods are used to prepare the CNMs sensor; smear casting and the inkjet printing technique, in addition, the sensor is integrated into membrane structure and fabricated in cleanroom.

Keywords: scale, sensors, carbon nanotubes, graphene, inkjet printing.

1 INTRODUCTION

Carbon nanomaterials (Carbon nanotubes (CNTs), graphene etc….) have received considerable attention as a new class of nanomaterials since their discovery in by Lijima 1991[1]. The sp^2 carbon–carbon bond in the plane of the graphene lattice is among the strongest of all chemical bonds. CNTs show unique mechanical, chemical and physical properties [2, 3 4] which led to a variety of applications, i.e. nanofluids[5-8], scanning probes[9], sensors[10 11, 12], grease [13, 14], nanocomposites[15 16] and electron magnetic shielding interface [17-19]. Scale is the creation of inorganic soluble salts from aqueous brines during the oil and gas production, which deposits at the internal surface of the pipelines under the supersaturation conditions. It is often managed or mitigated by the use of scale inhibitors. The current industry practice for managing scale programs relies on the determination of scale inhibitor concentration in the scaling brine and then relates this concentration to program performance. In most cases, scale inhibitors tailored to the specific oil well are used. Often, monitoring and testing for scales requires offline sample testing which impacts safety and productivity. The deposition of the scale leads to unavoidable damage of the equipment parts. Therefore, suspension of oil operations becomes essential in order to replace the damaged parts. In petroleum industry, such interruptions are escorted by extremely high costs. There has been multiple studies and research conducted on scale monitoring, testing and optimization of inhibitors [20 21 22]. In general, these studies focused on the development of using new sensor materials that enabled monitoring and offline testing. In this paper, the sensitivity and selectivity of CNTs and graphene to detect scale in oil and gas pipelines for online monitoring is dicussed.

2 EXPIREMENTAL PART

Single wall carbon nanotubes (SWNT) were purchased from sigma Aldrich USA Inc. in. The CNTs inkjet "CNTRENE® 3015 A3-R" used for this printing was purchased from Brewer Science, Inc USA. DMP2831 Inkjet Printer was used to print the CNT inkjet and the recipe was optimized. The chemical surfactant sodium dodecyl benzene sulfonate (NaDDBS) were purchased from Sigma–Aldrich. Epoxy resins and hardener were purchased from west system Inc, Zayed port Abu Dhabi. Releasing agents, Frekote, were purchased from Logistics Company Limited, in Dubai, UAE. Sonication was performed using a Branson Digital Sonifier, model 450. Scanning electron microscopy (SEM) images were acquired using the backscattered electron detector on a Zeiss Supra40VP variable pressure system. The electrical conductivity was obtained by Elite 300 Semi-automatic Probe Station & Keithley 4200-SCS Parameter Analyzer.

Casted technique: NaDDBS surfactant was added to epoxy with a weight ratio of 4 times carbon nanotubes and mixed manually for five minutes. The mixture was then sonicated until the surfactant was dissolved completely and clear solution was obtained. 0.5 weight percentage of carbon nanotubes were added to the mixture and sonicated again. The mixture was then mixed manually for 5 min with hardener by weight ratio of 10:2.63. Glass plate was coated with a layer of releasing agent (freekote) and the mixture was casted on the glass plate by smear casting technique. The sample was cured at room temperature for 24 hours. The CNTRENE® 3015 A3-R 'the CNT ink used for this printing' was purchased from Brewer Science, Inc USA. DMP2831 Inkjet Printer was used to print the CNT inkjet and the recipe was optimized in order to have a very uniform continuous printing.

3 RESULT AND DISCUSSION

As it was stated early two different techniques were used to prepare the carbon nanomaterials used to fabricate the sensors; smear casted technique and the inkjet printing technique.

3.1 Smear Casting Technique

Figure 1 shows the dispersion of the CNTs in the epoxy matrix. As it can be seen in the figure the CNTs are dispersed very will among the matrix and form a network structure.

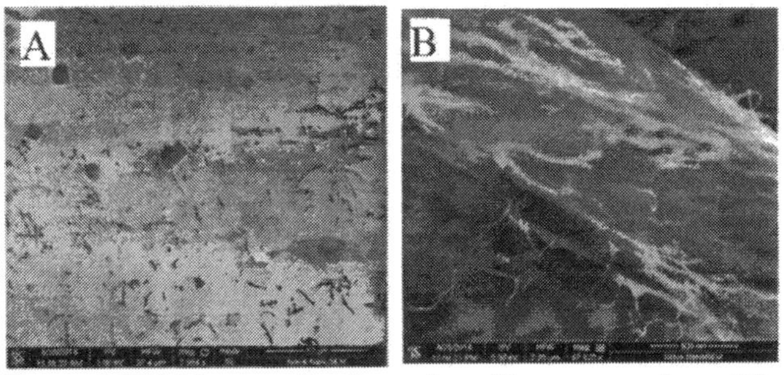

Figure 1 SEM images for A: surface, B: cross section of the 1.0 wt % of CNTs in epoxy.

Electrical characterization was conducted on the CNT-epoxy to determine the conductive path within the sample Fig. 2 shows the IV curve for a voltage sweep of 0-1 V. The resistance was found to be about 40.00 kΩ. This is because of a good dispersion of the CNTs into the epoxy resin and the network structure that the CNTs has formed as it can be seen in the SEM images figure 1. One drop of brine solution was added to the same film and the resistance was measured again and was found to be about around 90.00 kΩ. As it noticed that adding one drop of brine changed the electrical rsistance of the CNTs casted film. The film was washed with water and the electrical resistance measured again and found close to electrical resistance of the virgin sample.

After adding a drop of brine solution to the surface of the CNTs casted film, EDAS was conducted in order to detect the brine on the film. figure 3 shows SEM images for the surface of the casted CNTs. As it can be seen on the figure there are some white sopts, which was anticipated to be sodium chloride, therefore EDAS was performed. Figure 4 shows the EDAS spectrum for the white sopte marked with red cross as it can be seen in figure 3. As it can be seen in the spectrum, the sample has sodium and chlorine which confirms the presence of the brine solution.

Conductive atomic force microscopy C-AFM was carried out for virgin and with brine sample. The C-AFM of casted CNTs shows that virgin sample is more conductive than the with brine sample which matches the resistance curve for virgin sample that has less resistance. For with brine sample the CAFM shows less conductivity with higher resistance than virgin sample as it is shown in figure 5.

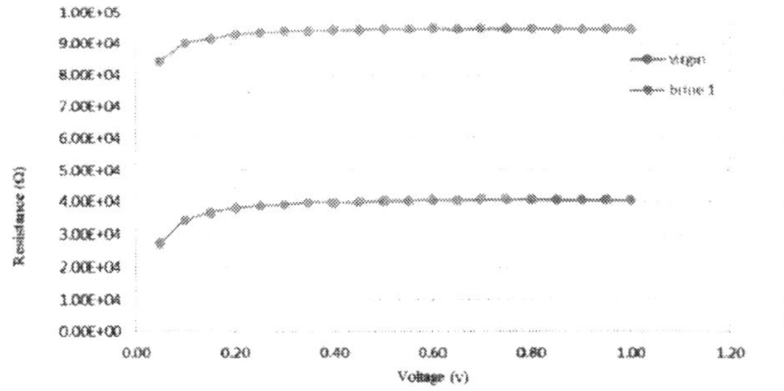

Figure 2 Four-probe voltage-resistance curve of nano-composite showing modulating resistance of the virgin sample in response to brine and DI water

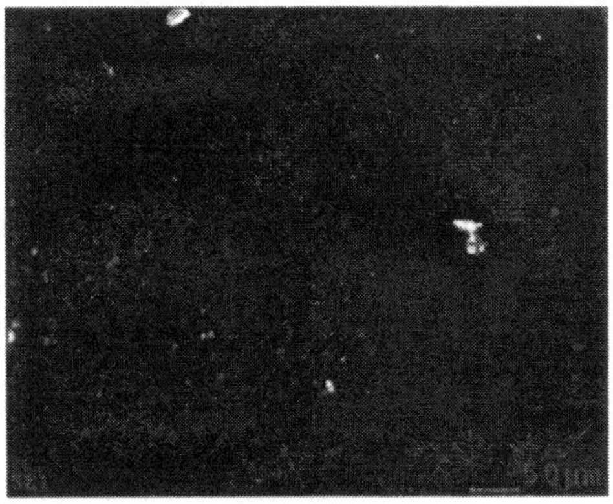

Figure 3 SEM image for the surface of smear casted CNTs shows a spot of NaCl.

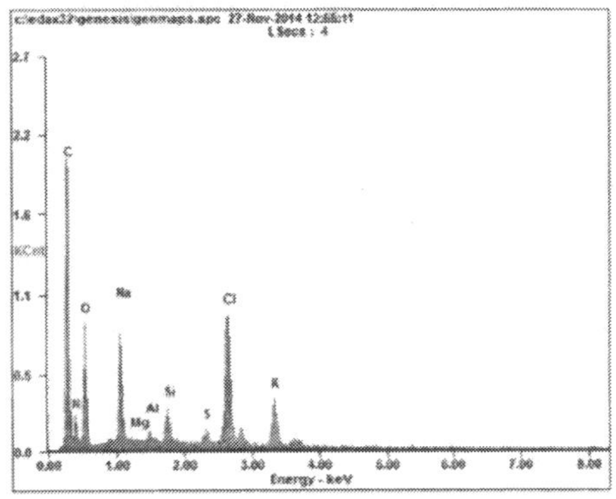

Figure 4 EDAS for casted CNTs on a glass substrate

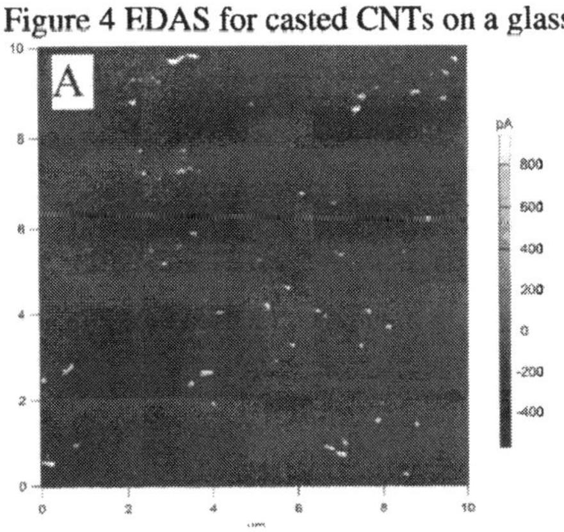

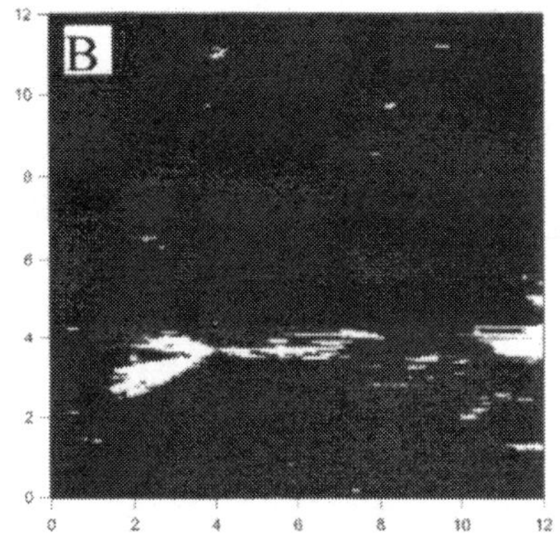

Figure 5 CAFM for current mapping of rough casted CNTs, A: Virgin sample B: with brine sample

3.2 Inkjet Printing Technique

3.2.1. Carbon nanotubes inkjet printing

The CNTs inkjet was printed on a silicon oxide substrate using DMP2831 Inkjet Printer. Fig 6 is an SEM image taken by Helios SEM shows that CNTs are bundles of CNTs that form network structure.

In general, the electrical conduction in such nanocomposites occurs either (1) tube–tube within the same bundle, (2) between neighbor bundles through their contacts. Therefore, the electrical conductivity (σ) depends on the conductivity of the nanotubes themselves and the ability of the electric carriers to tunnel between adjacent nanotubes or adjacent bundles.

The electrical resistance of the virgin CNTs inkjet printed was measured and found to be 1.00E+03 ohm. Figure 7 shows the ability of resetting film that made of CNTs inkjet printing. It was found that washing the film with DI water for one hour is essential to obtain a consistent electrical resistance. The electrical resistance for the virgin film made of CNTs inkjet printing is 1.0E+03 ohm by dipping the film for one hour in DI water the electrical resistance decreased to 5.0E+02 ohm. Adding one drop of brine increases the electrical resistance up to 4.2E+03 ohm and decreases with sweeping voltage to 3.6E+03 ohm. By immersing the film for one hour in DI water we were able to obtain an electrical resistance very close to the electrical resistance that was obtained after the DI water treatment.

3.2.1. Graphene inkjet printing

The graphene inkjet was printed on a silicon oxide substrate using DMP2831 Inkjet Printer.

The graphene inkjet printed film was subjected to two resetting cycles. As it is seen in fig 7 the virgin's electrical resistance is almost found to be 4.00E+03 ohm. Adding one drop of brine increases the electrical resistance to 9.00 E+03 ohm. The first resetting decreases the electrical resistance. The electrical resistance was found to be close to the electrical resistance of the virgin sample when the measurement was done after two days from the first reset. Adding a drop of brine for the second tine obtained the

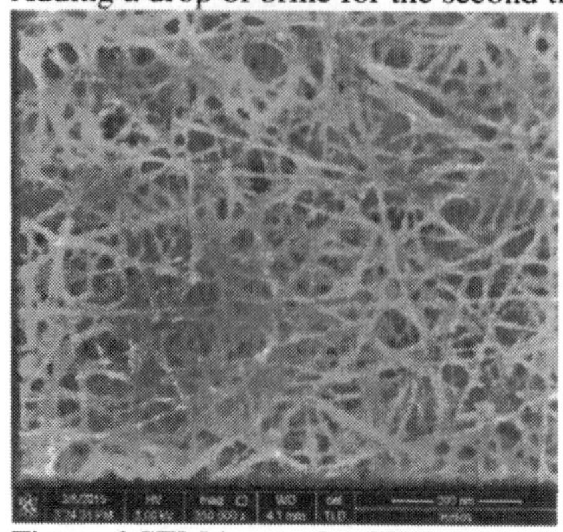

Figure 6 SEM image for CNTs inkjet printing

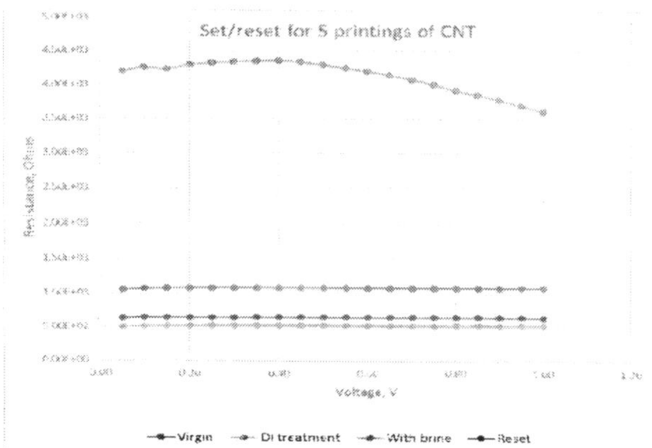

Figure 7 ability to reset the resistance values for 5 layers of CNTs printed on silicon oxide substrate.

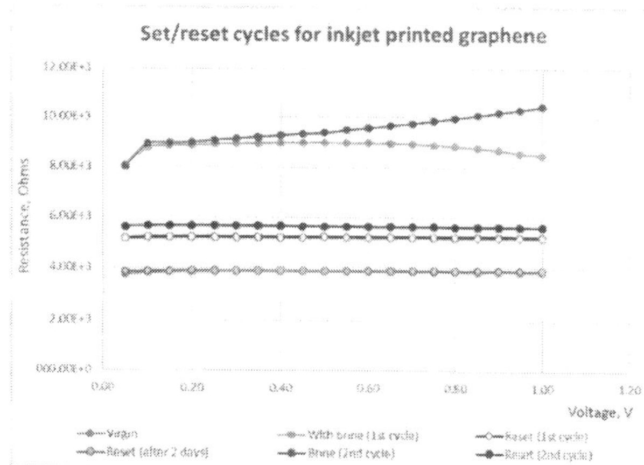

Figure 8 ability to reset the resistance values for 5 layers of graphene printed on silicon oxide substrate

same electrical resistance that was obtained for the first time that brine was added; in addition, the second resetting was close to the same values for the first resetting value.

3.3 Sensor Fabrication

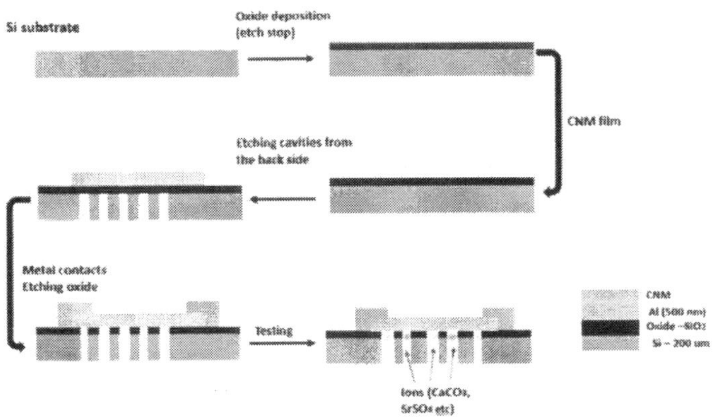

Figure 9 process flow of the membrane structure

Figure 16 shows the process flow for membrane structure. Sensor design and operation principle is based on a piezoresistive effect observed in carbon nanotubes, which undergo changes in their band structures when subjected to mechanical deformations. This phenomenon makes carbon nanotubes applicable for strain sensing applications. Fabrication steps are described in figure 16. The sensor design requires through silicon vias of different aspect ratio (with diameter in a range 20-50 μm) which can be formed using DRIE technique. In this case a layer of SiO2 is deposited on the front size to have an etch stop layer. On top of oxide layer, we apply a carbon nanomaterial film; inkjet printing technique is used for this purpose. Formation of cavities on the backside is implemented with dry etch method followed by wet etch in hydro fluoride solution to remove the oxide. Metal contacts are deposited on top of the films using thermal evaporator for further electrical testing.

4 CONCLUSION

Two methods were used to prepare the carbon nanomaterials for the sensors fabrication; casting technique and inkjet printing technique. In addition the sensors were integrated into membrane structure. The electrical resistance of both CNTs smear casted and CNTs& graphene inkjet printed were measure with and without using brine. Adding one drop of brine solution on the top of the sensor surface changed the electrical resistance by 50% 60% and 75% for the smear casted CNTs, inkjet printed CNTs and inkjet printed graphene respectively. The re-setting process was done successfully by soaking the sensors in DI-water for some time. SEM and AFM images show that the CNTs are dispersed in bundles and form network structures.

REFERENCES

[1] S. Iijima, nature **354** (6348), 56-58 (1991).
[2] A. Bassil, P. Puech, G. Landa, W. Bacsa, S. Barrau, P. Demont, C. Lacabanne, E. Perez, R. Bacsa, E. Flahaut, A. Peigney and C. Laurent, Journal of Applied Physics **97** (3), 034303 (2005).
[3] J. G. Park, J. Louis, Q. Cheng, J. Bao, J. Smithyman, R. Liang, B. Wang, C. Zhang, J. S. Brooks, L. Kramer, P. Fanchasis and D. Dorough, Nanotechnology **20** (41), 415702 (2009).
[4] T. Inoue, I. Gunjishima and A. Okamoto, Carbon **45** (11), 2164-2170 (2007).
[5] G. Christensen, H. Younes, H. Hong and G. P. Peterson, Journal of Nanofluids **2** (1), 25-28 (2013).
[6] H. Younes, G. Christensen, H. Hong and G. P. Peterson, Journal of Nanofluids **2** (1), 4-10 (2013).
[7] H. Younes, G. Christensen, M. Liu, H. Hong, Q. Yang and Z. Lin, Journal of Nanofluids **3** (1), 33-37 (2014).
[8] H. Younes, Ph.D Disertation (2014).
[9] C. V. Nguyen, Q. Ye and M. Meyyappan, Measurement Science and Technology **16** (11), 2138-2146 (2005).
[10] M. Gautam and A. H. Jayatissa, Materials Science and Engineering: C **31** (7), 1405-1411 (2011).
[11] A. Goldoni, L. Petaccia, S. Lizzit and R. Larciprete, Journal of physics. Condensed matter : an Institute of Physics journal **22** (1), 013001 (2010).
[12] M. Gautam and A. H. Jayatissa, Solid-State Electronics **78**, 159-165 (2012).
[13] H. Hong, D. Thomas, A. Waynick, W. Yu, P. Smith and W. Roy, Journal of Nanoparticle Research **12** (2), 529-535 (2009).
[14] L. Hongtao, J. Hongmin, H. Haiping and H. Younes, Industrial Lubrication and Tribology **66** (5), 579-583 (2014).
[15] S. Neupane, S. Khatiwada, C. Jaye, D. A. Fischer, H. Younes, H. Hong, S. P. Karna, S. G. Hirsch and D. Seifu, ECS Journal of Solid State Science and Technology **3** (8), M39-M44 (2014).
[16] M. M. Rahman, H. Younes, N. Subramanian and A. Al Ghaferi, Journal of Nanomaterials **2014** (2014).
[17] M. H. Al-Saleh and U. Sundararaj, Carbon **47** (7), 1738-1746 (2009).
[18] Z. Liu, G. Bai, Y. Huang, Y. Ma, F. Du, F. Li, T. Guo and Y. Chen, Carbon **45** (4), 821-827 (2007).
[19] L.-L. Wang, B.-K. Tay, K.-Y. See, Z. Sun, L.-K. Tan and D. Lua, Carbon **47** (8), 1905-1910 (2009).
[20] D. Emmons, G. Graham, S. Holt, M. Jordan and B. Locardel, presented at the SPE annual technical conference, 1999 (unpublished).
[21] D. Emmons, S. Smith, T. Weathers, M. Jordan and N. Feasey, presented at the Paper SPE 113804 presented at the SPE International Oilfield Scale Conference, Aberdeen, 2008 (unpublished).
[22] H. Al-Matar, J. K. Al-Ashhab, M. Mokhtar and S. Ridzauddin, presented at the Abu Dhabi International Petroleum Exhibition and Conference, 2006 (unpublished).

In situ Transesterification of Wet Microalgae under Acid Catalysts

Jae W. Lee*, Hanjin Im, and Bora Kim

Chemical and Biomolecular, KAIST, South Korea, *jaewlee@kaist.ac.kr

ABSTRACT

One of energy intensive processes of biodiesel production is concentrating and drying wet algal cells to facilitate the subsequent lipid extraction. The cost for this process occupies 20-30% of the total cost of biodiesel production. To this end, this work addresses a reliable *in situ* transesterification process which integrates lipid extraction from wet microalgae, and its conversion to biodiesel without having a drying process. The process enables one-step production of biodiesel by heating the mixture of wet *Nannochloropsis gaditana* with co-solvent, alcohol, and acid catalyst in one pot. Effects of reaction parameters (temperature and co-solvent volume) on the biodiesel yield will be investigated under various conditions. It will be shown that the fatty acid ethyl ester yield of 90-100% is achieved.

Keywords: biodiesel, *in-situ* transesterification, lipid extraction, wet microalgae

1 INTRODUCTION

Microalgae can take carbon dioxide from the air and grow faster than terrestrial plants. Using microalgae, high-concentration culture is possible in large quantities and even in extreme environments [1]. Producing biodiesel from microalgal lipids is feasible since microalgae can accumulate lipid levels that are higher than 50% of their dry cell weight [2]. Usually microalgae are photoautotrophic microbes, but in addition to using free sunlight as an energy source, microalgae can be cultivated using the nitrogen, phosphorus and hydrocarbons in industrial wastewater [3, 4].

After extraction of the lipids from cultivated microalgae, the extracted oils are usually converted to fatty acid ethyl ester (FAEE) using ethanol and acid catalyst in a wet transesterification reaction. Such FAEE can be used as a biodiesel after purification [5]. Because microalgae are grown in liquid media, extraction is usually classified as wet extraction or dry extraction depending on whether or not the culture broth was dried before the lipid extraction step. Most of the current extraction processes are carried out after drying the wet microalgae to increase the amount of lipid extracted from the biomass [6, 7]. However, this drying process is responsible for 59% of the total energy consumed during biodiesel production [8].

To reduce the impact of the energy consumed for the current drying process, improvement of the efficiency of wet extraction by adding other operations has recently emerged as a very important issue [9-11]. These operations include mechanical pressing, as well as ultrasonic, microwave, supercritical fluid, and solvent extraction methods, all of which have been used for lipid extraction. Mechanical pressing is easy to design, but it has the drawbacks that it consumes lots of energy and time. The ultrasonic and microwave methods need shorter extraction time than other methods, and allow improved penetration of solvent with less solvent used, but are difficult to scale up [6, 12]. When it comes to extraction using a supercritical fluid, there is the benefit of not needing to use toxic solvents. However, because of intensive energy consumption and large capital costs, it still faces hurdles for scaling up [11, 13]. Solvent extraction, except for toxicity issues, provides reasonable performance and has the advantage of low costs for scaleup [14].

The oils (triacylglycerides) recovered from microalgae can be converted into biodiesel by a transesterification reaction. Usually extraction and transesterification proceed separately, but there is a process called 'in-situ transesterification' which integrates the two processes into a single-step process as an economically feasible alternative. The in-situ process allows direct extraction of the lipid from microalgae, and its transformation to biodiesel, while reducing the processing cost [15-17]. This in-situ process delivers verified success in the dry state, but with moisture present, low yields are typical without an additional energy input such as microwave [6, 18].

In contrast, our new process presents extraordinarily high yields of FAEE using wet microalgae by combining the extraction and transesterification processes into a single step. Heating wet microalgae in a mixed solution of ethanol, solvent, and acid catalyst leads to production of FAEE with its yield of around 90 wt.%. This is a significant improvement over *in situ* transesterification of wet microalgae; without additional energy inputs. This work focuses on the effects of reaction parameters (e.g., amounts of solvent, ethanol and temperature) on conversion yields for one-pot extraction and transesterification.

2 METHOD

2.1 Chemicals and reagents

Extra pure grade sulfuric acid and reagent grade chloroform were obtained from JUNSEI Chemical (Japan).

Extra pure grade ethanol (99.5%) was obtained from Samchun (Korea). Heptadecanoic acid ethyl ester as a standard material for GC analysis from Sigma Aldrich (> 99%) was also used. The *Nannochloropsis gaditana* powder was purchased from AlgaSpring (Netherlands) and stored at a dark room.

2.2 *In situ* transesterification procedure

Numerous experiments for optimizing the reaction parameters used for biodiesel production were performed with wet *Nannochloropsis gaditana*. The algal paste used in this study was confirmed to contain 65 wt.% moisture. In-situ transesterification was started by mixing ethanol, solvent and sulfuric acid with the wet microalgae in a teflon-seal, screw-capped tube (14 ml). Ethanol and sulfuric acid were used as transesterification reactant and catalyst. Simultaneous extraction and reaction were conducted by immersing the tube in a waterbath. To minimize leakage of ethanol and chloroform from the vapor head, the tube was capped tightly. After the reaction, sodium hydroxide was added to prevent further reaction and the mixture was centrifuged at 3700 rpm for 10 min to achieve phase separation. Then, ethanol containing sulfuric acid and water formed an upper layer and chloroform containing the converted FAEE and unconverted TAG formed a lower layer. Reaction parameters (temperature and choroform solvent) were varied to find optimal conditions for maximizing the FAEE yield.

2.3 FAEE analysis and yield calculation

For the quantification of each sample, the liquid organic phase including FAEE and was injected to GC (Agilent 7890b, USA) equipped with HP-5 column. The GC analysis was performed with a flame ionization detector (FID) and helium was used as a makeup gas. At each analysis, 1 μl of the liquid sample was used. The yield of FAEE was calculated based on the ratio between the peak area of standard C17 ethyl ester and the peak area of the corresponding chemical compound. To determine maximum yield of transesterifiable lipids in *N. gaditana* cells, experiments based on Folch's method were conducted [19]. 10 mg dried cells and 2 ml 2/1 v/v mixture of chloroform and ethanol were added to the teflon-sealed, screw-capped tube and lipids were extracted by vortexing for 10 min. One ml ethanol and 300 μl sulfuric acid were added and the tubes were heated to 100°C for 20 min. After the transesterification reaction, the tube was cooled down to room temperature and left to settle about one hour for phase separation. 1 ml chloroform containing 0.5 mg heptadecanoic ethyl ester (C17:0) was added to the tube for GC analysis to determine the maximum FAEE yield. Then the FAEE yield is calculated as follows: the maximum amount of FAEE that could be derived from the lipids was determined as 12.05 mg FAEE / 100 mg dry cell. Thus, the FAEE yield could be obtained by dividing an experimentally determined FAEE amount by the maximum FAEE amount

3 RESULT AND DISCUSSION

3.1 Reaction temperature effect on the FAEE yield

The results from previous studies of *in situ* transesterification showed higher yields of FAEE when higher reaction temperatures were used [18]. In this study, the reaction temperature was varied from 65 °C to 140 °C To 0.3 g dry biomass, 0.3 ml sulfuric acid and a 2 ml, 2/1 v/v mixture of chloroform and ethanol were added. For wet condition, moisture content was maintained to 65%. Then the reaction was allowed to continue in a water bath for two hours at a specified temperature.

It is clear that the FAEE yield increases with rising temperature (Fig.1). At the dry condition, the FAEE yield was already 88 wt.% at 95 °C and reached almost 100 wt.% with flattening out after 110 °C. Less amount of FAEE was produced at the wet condition and FAEE yield more than 90% was obtained between 125 - 140 °C. Although other studies obtained an optimal reaction temperature near 80°C higher temperatures in this study helped chloroform, ethanol and sulfuric acid disrupt the cell walls as well as accelerate the transesterification reaction [6, 18]. To confirm FAEE production, the distribution of FAEE was illustrated in Fig.2. Other FAEE components were not shown because they account for only 7.3 % of the total FAEE. Each FAEE component shows a similar trend in which the yield increases along with the temperature rise. From the result, *N. gaditana*, is suitable to be selected as a biodiesel source considering that the ester products of C16-C18 has been a proper candidate for biodiesel [20].

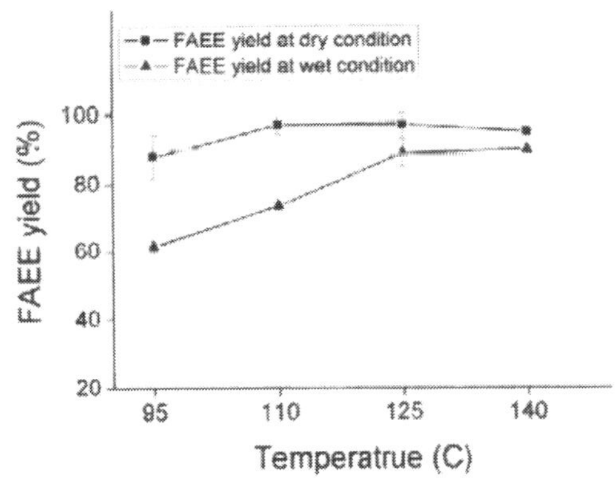

Fig. 1 FAEE yields subject to various reaction temperatures.

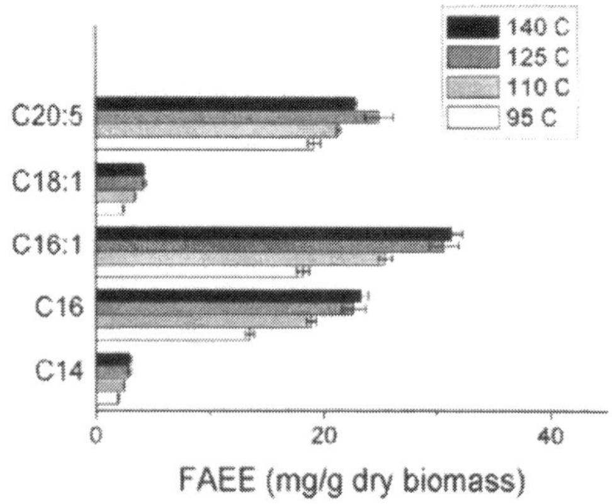

Fig. 2 FAEE product distribution at the wet condition.

3.2 Chloroform effect on the FAEE yield

Originally, chloroform was employed in order to improve efficiency of extraction and conversion of microalgal lipid as a solvent [16]. To confirm the effectiveness of chloroform for the production of FAEE, the solvent volume varied from 0 to 3 ml/ethanol in *in situ* transesterification at 125 °C while other experimental conditions remained the same as in Section 3.1.

Without chloroform, the FAEE yields was 66 wt. %, respectively, as shown in Fig. 3 . However, addition of 0.5 ml chloroform made a dramatic rise of the yields – 86 % of FAEE yield. In the reaction tube containing wet microalgae, ethanol, sulfuric acid and chloroform, ethanol and sulfuric acid with wet microalgae generate the aqueous phase and chloroform forms the organic phase. According to the molecular structure, FAEE is easily miscible with the organic phase and ethanol prefers staying in the aqueous phase. Thus, in the sample tube, the reactant (ethanol) and the product (FAEE) were stabilized in the separate phases with the enhancement of the forward direction of transesterification reaction.

Once lipids from *N. gaditana* were released in the aqueous phase, it reacts with ethanol via transesterification from which FAEE and glycerol were produced. Because FAEE has a low solubility in the aqueous phase, it remains in the organic phase. With regard to the aqueous phase, the remaining product in transesterification was only glycerol. The depletion of product FAEE from the aqueous phase was effective enough to make more FAEE by the reaction equilibrium shift.

4 CONCLUSION

The wet, in-situ transesterification process in this work has been demonstrated to achieve conversion yields as high as 90% from wet *N. gaditana* (65% moisture) using chloroform, ethanol, and sulfuric acid in the one pot. This significant improvement of FAEE yield is possible without using an energy-intensive drying process before extracting the lipid. The reaction parameters for producing biodiesel via this process have been successfully analyzed. This *in situ* approach can be used to produce biodiesel directly from harvested algae and provides a useful step forward for the next generation of renewable biodiesel.

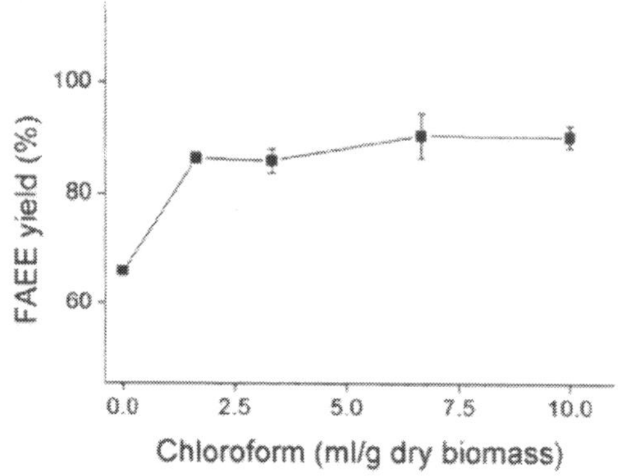

Fig. 3 FAEE yield changes depending on the amount of chloroform.

5 ACKNOWLEDGEMENT

This work was supported by the Advanced Biomass R&D Center(ABC) of Global Frontier Project funded by the Ministry of Science, ICT and Future Planning (ABC-2011-0031354)

6 REFERENCE

[1] A. Demirbas and M.F. Demirbas, Energy Conversion and Management. 52, 163, 2011.
[2] W. Xiong, X. Li, J. Xiang and Q. Wu, Applied Microbiology and Biotechnology. 78, 29, 2008.
[3] N. Mallick, Biometals. 15, 377, 2002.
[4] J.K. Pittman, A.P. Dean, and O. Osundeko, Bioresource Technology. 102, 17, 2011.
[5] X. Li, H. Xu, and Q. Wu, Biotechnology and bioengineering. 98, 764, 2007.
[6] B.D. Wahlen, R.M. Willis, and L.C. Seefeldt, Bioresource Technology. 102, 2724, 2011.
[7] I. Rodriguez-Meizoso, et al., J Pharm Biomed Anal. 51, 456, 2010.
[8] L. Yanfen, H. Zehao, and M. Xiaoqian, Energy Policy. 45, 142, 2012.
[9] L.F. Razon and R.R. Tan, Applied Energy. 88, 3507, 2011.
[10] L. Xu, et al., Bioresource Technology. 102, 5113,2011.
[11] K. de Boer, et al., Journal of Applied Phycology. 24, 1681, 2012.

[12] M. Koberg, M. Cohen, A. Ben-Amotz, and A. Gedanken, Bioresource Technology. 102, 4265, 2011.
[13] M. Herrero, J. A. Mendiola, A. Cifuentes, and E. Ibáñez, Journal of Chromatography A. 1217, 2495,2010.
[14] M. Herrero, P.J. Martin-Alvarez, F.J. Senorans, A. Cifuentes, and E. Ibáñez, Food Chemistry. 93, 417, 2005.
[15] M. J. Haas, K. M. Scott, T. A. Foglia, and W. N. Marmer, Journal of the American Oil Chemists' Society. 84, 963, 2007.
[16] H. Im, H. Lee, M. Park, J. Yang, and J. W. Lee, Bioresource Technology. 152, 534, 2014.
[17] B. Kim, H. Im, and J. W. Lee, Bioresource Technology. DOI: 10.1016/j.biortech.2015.02.092, 2015.
[18] E.A. Ehimen, Z.F. Sun, and C.G. Carrington, Fuel. 89, 677,2010.
[19] J. Folch, M. Lees, and G. Sloane-Stanley, Journal of Biological Chemistry. 226, 497, 1957.
[20] G. Knothe, Energy & Fuels. 22, 1358, 2008.

Technical and Economical Feasibility of Pyrolysis of Kraft Lignin

Sherif Farag* and Jamal Chaouki**

Research Center In Process Engineering-Biorefinery (CRIP),
Department of Chemical Engineering, Ecole Polytechnique Montreal
P.O. Box 6079, Station Centre-ville, Montreal, Quebec, H3C 3A7, Canada
*sherif.farag@polymtl.ca
**jamal.chaouki@polymtl.ca

ABSTRACT

This research project discusses the technical and economical feasibility of converting kraft lignin to value-added chemicals. Two pyrolysis techniques were implemented: conventional pyrolysis using classical heating and microwave pyrolysis using electromagnetic irradiation. The product distribution and the composition of the bio-oil from the conventional pyrolysis were compared against those from the microwave pyrolysis. Gas chromatography–mass spectrometry analysis was carried out to identify the organic compounds in the obtained oil. To estimate the minimal selling price of the lignin pyrolysis oil and the expected cumulative cash flow at the end of the life cycle of the plant, a first step economics evaluation was completed using the method "net present worth." Subsequently, different scenarios for reducing the estimated price, which would help for taking a place among the fossil fuel-based chemicals, were debated.

Keywords: pyrolysis, kraft lignin, bio-oil, economics

1 INTRODUCTION

In North America, the forest industry is one of the cornerstones of the national economy. In Canada, for example, it contributes $80 billion per year to Canada's total gross domestic product and provides more than 900,000 jobs across the country. Presently, however, this industrial sector is at a crossroads as it is facing unexpected challenges due to competition from low-cost sources of wood and a decline in demand. One of the most promising paths to ensure a sustainable future for the industry is the production of value-added products in addition to the traditional commodity. This approach leads to diversifying the range of products from the forest industry and, consequently, opens the door to new markets.

In the papermaking process, the wood chips and an aqueous solution are cooked in a digester to separate lignin from the feedstock. Thereafter, lignin is produced as weak black liquor (at 15-20 wt% of solid concentration) whereas the remaining material, which consists of wood fibres, becomes paper. The most popular technique that deals with black liquor is to increase the solid concentration and then burn it in a recovery boiler. This technique mainly produces heat energy and recycles the inorganic chemicals for reuse in the pulping process. Another technique, which is precipitating lignin from black liquor, has been established to enable expanding pulp and paper production without having to expand the capacity of the recovery boiler. Accordingly, the production of lignin is dramatically growing, which demands more attention for discovering new techniques for dealing with the massive quantity of lignin produced worldwide.

Lignin is the only renewable source of aromatics in nature and the third most abundant natural polymer after cellulose and hemicellulose [1]. The structure of lignin, which is a complex heterogeneous aromatic polymer with many functional groups, makes lignin a promising material for the production of a variety of commercial products. Chemical, biological, and physical technologies are the main methods for the production of lignin-based products. Pyrolysis, which is the thermal decomposition of chemical bonds of a target material by supplying heat energy in an oxygen-free environment, is one of the techniques applied in the chemical technology. Generally speaking, pyrolysis of lignin produces three main products: condensable gas, solid, and non-condensable gas. The condensable gas product (bio-oil) is mostly aromatics containing useful and valuable phenolic compounds, such as phenol, dimethylphenol, guaiacol, catechol, and syringol [2]. These phenols can be used in many industrial applications, for example, the production of phenol formaldehyde resins, the synthesis of pharmaceuticals, gasoline additives, antioxidants, food additives, polymerization initiators, pesticides, and colorants [2]. The solid product (bio-char) is mostly carbon and can be used in a wide range of applications: tire manufacturing processes, soil additive, activated carbon, and solid fuel. The non-condensable gas (bio-gas) is combustible, which can be burned to generate the heat energy that is required for performing the pyrolysis process. In addition, it could be used in the Fischer Tropsch process for the production of several hydrocarbon products.

Although there are few studies on the production of highly priced lignin-based products, such as conductive materials, pyrolysis is still the only technique that can deal with the huge quantity of lignin produced in pulp mills across the globe. Therefore, it is of great interest to efficiently implement pyrolysis to convert lignin into value-added chemicals that could replace petrochemicals in many applications. Such an approach does not only lead to dealing with the huge quantity of lignin produced across the world but also decreases the consumption of fossil fuel based chemicals and, therefore, addresses the problems due to the rapid increase in the price of petrochemicals.

In this research project, pyrolysis of kraft lignin was accomplished by applying two different techniques: conventional pyrolysis (CP) using traditional heating and microwave pyrolysis (MWP) using electromagnetic irradiation. Gas chromatography–mass spectrometry (GC-MS) was performed on the bio-oil to understand the composition of the obtained chemicals. Subsequently, a first step economics evaluation was completed for estimating the minimal selling price of the lignin pyrolysis bio and the expected cumulative cash flow at the end of the life cycle of the plant.

Such studies would present valuable insight for the industrial sectors to encourage them to pursue the idea or guide them in another direction to deal with lignin.

2 EXPERIMENTAL WORK

2.1 Raw Material

In this work, the raw material was softwood kraft lignin supplied by FPInnovations, Quebec, Canada. It was characterized by CHNS analysis: C=63.27%, H=5.79%, N=0.07%, and S=1.56%, and proximate analysis: fixed carbon=37%, volatiles=62%, and ash=1%. To get a better understanding for the obtained products as well as the raw material, the ^{13}C and ^{31}P NMR analyses were performed on the virgin material and reported in Farag et al. 2013, kindly refer to that reference for the full details.

Since bio-char is a strong microwave-to-heat convertor, it was mixed with the virgin material for enhancing the heating processes [3-5]. The concentration of the bio-char in the initial payload mass was chosen in a way that gives a heating rate similar to that of CP.

2.2 Experimental Setup

In the case of MWP, the experimental work was carried out using a bench scale microwave-oven, while in the case of CP a conventional heating oven was used. A quartz semi-batch reactor was used after connecting its outlet with a condensation system kept at -18 °C using a traditional freezer. In order to prevent a vapor condensation barrier to the condensation system, the line that connects the reactor with the condensers was kept at 200 °C using external-temperature heat cables. The detailed information regarding the experimental setup can be found in the references [6-8].

3 RESULTS AND DISCUSSION

3.1 Product Distribution

Figure 1 demonstrates the product distribution for the CP and MWP of kraft lignin. The presented values are the average of three repetitions performed under the same conditions, and the error bars are the standard deviations. The yield of each run was calculated based on the initial dry mass basis of the feedstock.

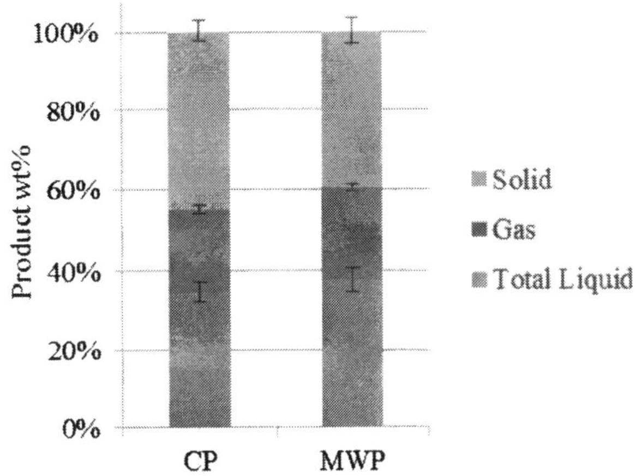

Figure 1: Product distribution of the conventional and microwave pyrolysis of kraft lignin

As one see, applying electromagnetic irradiation in pyrolysis produced a liquid yield that is slightly higher than what was obtained from the conventional pyrolysis, 38 wt% and 35 wt%, respectively. If the uncertainty of the experimental data is taken into account, the effect of microwaves on the bio-oil yield becomes insignificant. On the other hand, the solid residue obtained from MWP is 15 wt% (relative to the char yield of the CP) less than what was obtained from CP. This difference might be the consequence of the heat and/or mass transfer limitations taking place in the conventional heating processes and considered one of the main issues in such a process. Since the bio-gas yield was calculated by subtracting the liquid and solid yields from the initial lignin mass, its yield is related to the yield of the solid and liquid products. Thus, its interpretation is implicitly included in the previous explanation.

In fact, a quantitative comparison between the product distributions from the applied pyrolysis techniques is somehow misleading since it does not consider the quality aspect. Therefore, further information regarding the composition of the produced bio-oils is discussed.

3.2 Bio-oil Composition

As a first step toward a qualitative analysis, the water content in all the liquid samples (12 samples in total: 6 oil phases and 6 aqueous phases) was determined. The obtained data was processed for calculating the yield of the organics and water products in each of the liquid samples, as manifested in Figure 2. The error bars shown in the figure are the standard deviation.

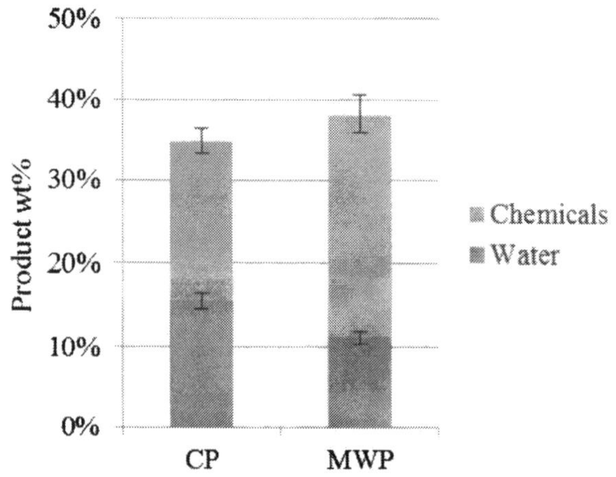

Figure 2: Yields of the chemicals and water products in the produced oils

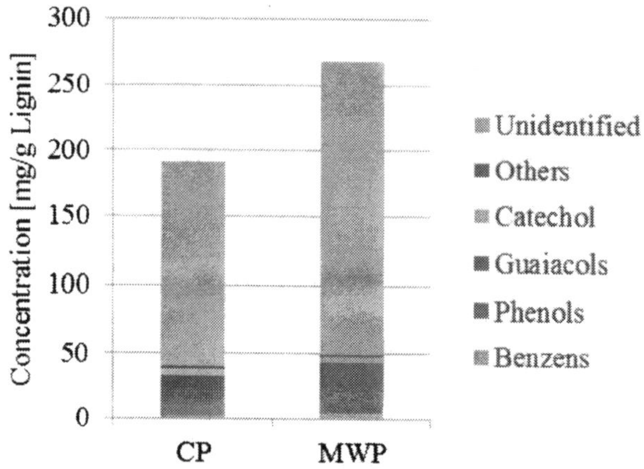

Figure 4: Identifies chemical families using a gas chromatography–mass spectrometry

It is evident that the impact of microwave irradiation on the composition of the produced oil is considerable although its influence on the liquid yield is not confirmed. The organics yield obtained from microwave-assisted pyrolysis is 42 wt% (relative to the organics yield from the CP) more than what was achieved from the classical pyrolysis. Consequently, the yield of the produced water was 31 wt% (relative to the water yield of the CP) less than that of the traditional pyrolysis.

Since conventional heating is a superficial heating process, which in most cases presents a surface hotter than the core, it would lead to further degradations for the pyrolysis vapor. This cleavage of the chemical bonds might be responsible for the decomposition of the aliphatic hydroxyl groups and, as a result, increase the formation of water during pyrolysis, kindly refer to Figure 3. Farag et al. 2014 believe that the decomposition of the OH bonds is mostly attributed to forming water and/or unsaturated sites and, accordingly, it results in a decrease of up to 90% in the concentration of the OH aliphatic side chains [7]. On the other hand, microwave heating is a volumetric energy transfer mechanism, which under controlled conditions presents a surface hotter than the core [9]. This aspect would lead to preserving the structure of the obtained products and, therefore, would produce less water yield compared to the classical pyrolysis.

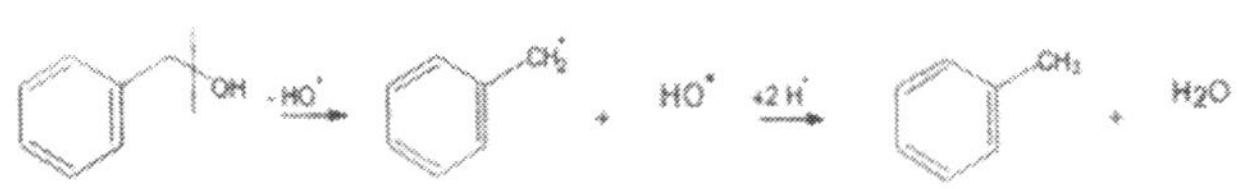

Figure 3: A suggested degradation pathway for the aliphatic hydroxyl group

To obtain further information regarding the composition of the produced bio-oils, GC-MS analysis was performed on the liquid samples. The obtained data was processed to calculate the yield of the common chemical groups, as shown in Figure 4.

It is not an exaggeration to claim that MWP produces the maximum yield of both the identified and unidentified chemical compounds. The obtained phenols' yield is more than twice that measured in the CP oil. This means that employing MWP in pyrolysis would lead to improving the quality of the obtained product. However, further investigations to confirm this statement are needed.

3.3 Estimating the Selling Price of the Lignin Pyrolysis Oil

To estimate the selling price of the obtained bio-oil, a first step economics evaluation for converting lignin to value-added chemicals was carried out. An on-site CP plant was designed, and the updated total capital investment was calculated. The main sections of the designed plant are pre-treatment, pyrolysis, condensable gas recovery, and combustion. To calculate the mass and energy of each section, Aspen PLUS^{TM} applications were used. The method "Net Present Worth," which equals the result of subtracting the present capital investments from the present worth of the cash flows, was applied [10].

The above mentioned preliminary economic assessment leads to appraising the profitability of an investment, which is the basis for selecting or rejecting the investment. It should be noted that an accuracy of 20-30% is expected in such assessments [10-12].

The hypotheses of this economics model include 25% of the applied tax, 9% of the discount rate, 30% of the minimum acceptable rate of return, 10 years of the investment period, and 7000 hours/year of the low average operating conditions. $14M was estimated for the total capital investment, considering 100 t/d for the capacity and a working capital of 15% of the total capital investment. $500/t was considered for the purchase price of the raw material. The kinetic parameters of pyrolysis were obtained from the study published by Farag et al. 2014 [13].

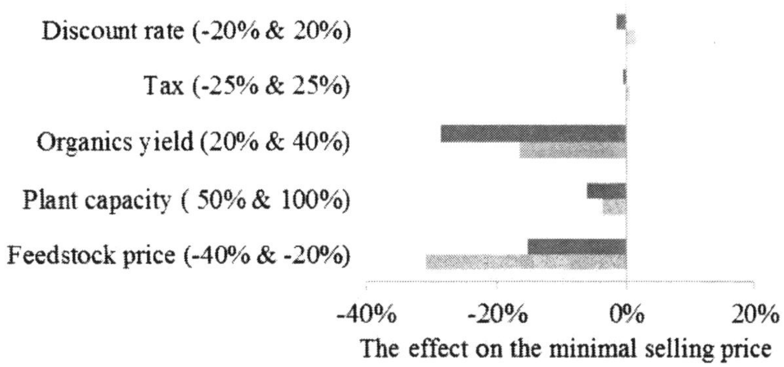

Figure 5: Effect of the crucial variables on the selling price of the kraft lignin pyrolysis oil

Based on the above hypotheses, the bio-oil from the pyrolysis of kraft lignin can be sold at 3200 \$/t. The cumulative cash position during the 10-year plant life span shows that the expected cumulative cash flow at the end of the life cycle of the plant is \$9M. As a matter of fact, this price could lead to challenging the replacement of the lignin pyrolysis oil by the fossil fuel-based chemical because the estimated selling price is about twice that of the fossil fuel-based products, such as phenols. The sensitivity analyses shown in Figure 5 explains that using a raw material at a low purchase cost in addition to enhancing the organics yield in the bio-oil will dramatically decrease the minimum selling price of lignin pyrolysis oil. In addition, the capacity of the designed plant, estimated tax, and discount rate do not affect the selling price as much as the above-mentioned variables. Therefore, further investigations for presenting lignin at a low purchase cost and to enhance the yield and composition of the lignin pyrolysis oil are essential. In addition, new applications for the produced bio-char will help for overcoming this issue.

4 CONCLUSION

The technical and economical feasibility of the pyrolysis of kraft lignin to value-added chemicals has been discussed. The key conclusion includes that the impact of the electromagnetic irradiation on the yield of the lignin pyrolysis oil could not be confirmed. Microwave pyrolysis looks like it preserves the structure of the pyrolysis vapor and, therefore, its effect on the quality aspect is considerable. According to a first step economics evaluation, the obtained bio-oil from pyrolysis of lignin might not take a competitive place among the fossil fuel based chemicals due to the uncompetitive minimum selling price and the cash flow calculated at the end of the plant life span.

REFERENCES

[1] Farag, S., DÉPARTEMENT DE GÉNIE CHIMIQUE, Philosophiae Doctor, 2013.
[2] Fu, D., et al., Bioresource Technology, 154(0), 101-108, 2014.
[3] Undri, A., et al., Fuel, 116(0), 662-671, 2014.
[4] Undri, A., et al., Journal of Analytical and Applied Pyrolysis, 105(0), 35-42, 2014.
[5] Domínguez, A., et al., Bioresource Technology, 97(10), 1185-1193, 2006.
[6] Farag, S. and J. Chaouki, Applied Thermal Engineering, (0), 2014.
[7] Farag, S., et al., Journal of Analytical and Applied Pyrolysis, 109(0), 249-257, 2014.
[8] Doucet, J., et al., Waste and Biomass Valorization, 5(1), 1-10, 2014.
[9] Farag, S., et al., Applied Thermal Engineering, 36(0), 360-369, 2012.
[10] Vanreppelen, K., et al., Chemical Engineering Journal, 172(2–3), 835-846, 2011.
[11] Max, P., T. Klaus, and W. Ronald, 2012.
[12] Zhang, Y., et al., Bioresource Technology, 127(0), 358-365, 2013.
[13] Farag, S., L. Kouisni, and J. Chaouki, Energy & Fuels, 28(2), 1406-1417, 2014.

The Scale-up of Fluidized-bed Gasifier

Zhihong LIU[*], Toshiyuki SUDA[**], Masahiro NARUKAWA[***], and Makato TAKAFUJI[****]

IHI Corporation, 1, Shin-Nakahara-Cho, Isogo, Yokohama 235-8501, Japan
[*]shikou_ryuu@ihi.co.jp
[**]toshiyuki_suda@ihi.co.jp
[***]masahiro_narukawa@ihi.co.jp
[****]makoto_takafuji@ihi.co.jp

ABSTRACT

Scale-up of fluidized-bed gasifier is important issue for the industry. It necessary to know how the efficiency of energy convergence changes when scale of gasifier changes. The residence time distribution (RTD) of lignite particles in a gasifier of fluidized bed impacts its energy conversion efficiency. The RTD of lignite in continuous bubbling fluidized bed is calculated by segregation-convection-dispersion model. The effect of bed length on the RTD of lignite is investigated by this model, and the results are verified by experiments. It is found that when the length of bed becomes longer, the residence time becomes longer and therefore the energy conversion efficiency becomes higher. Energy conversion efficiency predicted by this model is verified by experiments.

Keywords: scale-up, residence time distribution, fluidized bed, scale effect, segregation

1 INTRODUCTION

Petroleum and natural gas prices are unpredictable in recent years. The importance of utilization of coal is increasing to stabilized the price of fuel. In order to use coal with high efficiency, IGCC (Integrated coal Gasification Combined Cycle) is being developed globally. On the other hand, gasification of coal can be used to produce a synthesized chemical raw material. The key for these techniques is coal gasification.

The fluidized-bed gasifier is widely used for the gasification of low-rank coal, for example lignite. The fluidized-bed gasifier of industries is usually of large scale. Generally, the gasifier is developed from laboratory scale, to pilot scale, and then to demonstration scale, finally to the commercial scale. Generally linear scale-up is not applicable. In this paper, we describe how we scale up the gasifier.

2 TIGAR®

IHI is developing a gasifier named TIGAR® (Twin IHI GAsifieR) for gasification of lignite by steam. The schematic diagram of TIGAR® is shown in Fig.1, which comprises two beds, Bubbling Fluidized Bed (BFB) and Circulating Fluidized Bed (CFB). The solid fuel is fed into the BFB (gasifier), where it reacts with steam to produce syngas (H2, CO, etc.). The residue of solid fuel is transported to the CFB (combustion chamber) to burn and thereby heat the bed material (sand). The hot sand is returned to the BFB to supply the heat necessary for the gasification reaction. The product syngas of TIGAR® is of high calorific value, because it does not contain nitrogen. Initial costs can be reduced, because expensive material for providing a high pressure and high temperature environment is not needed. Fuels with high moisture content can be gasified by TIGAR®. It is important to predict gasifier characteristics when enlarging existing facilities. This paper deals with the scale-up of TIGAR® for lignite.

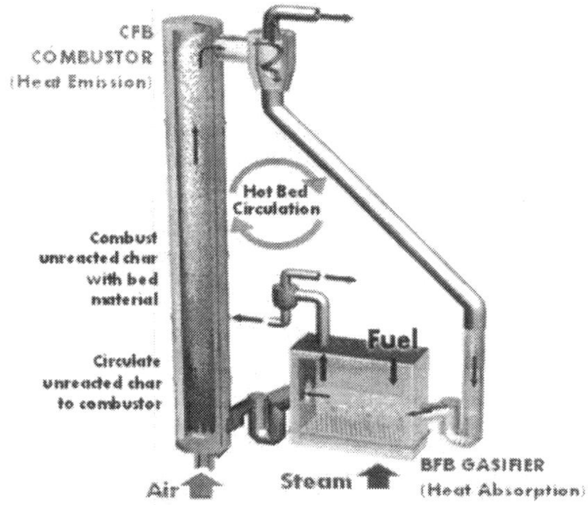

Fig. 1: Schematic diagram of TIGAR® system

3 RESIDENCE TIME DISTRIBUTION

Gasification of lignite by steam need long time, for example, Hundreds of seconds. The chemical reaction of lignite particle and its residence time influence each other. For simplicity, we split the gasification process in bubbling fluidized beds into two processes, namely the physical process (particle motion) and the chemical process (reaction of solid fuel). The particles' motion is described by RTD) of the fuel particles in the continuous fluidized bed. In our study the bed height is fixed (h=1m), assuming that the bubble behavior does not change. The dispersion coefficients of particles etc. do not change, if bed height is fixed. By this way, the scale effect is rendered simple to the effect of bed length.

3.1 Segregation-Convection-Dispersion Model

Usually, convection-dispersion model is used to calculate RTD of lignite particle in a bed of sand, because lignite particles tend to rise to the top of sand-fluidized-bed(segregation). The segregation heavily impacts the RTD of lignite. A segregation-convection-dispersion model[1] has been developed, which is outlined in Fig.2.

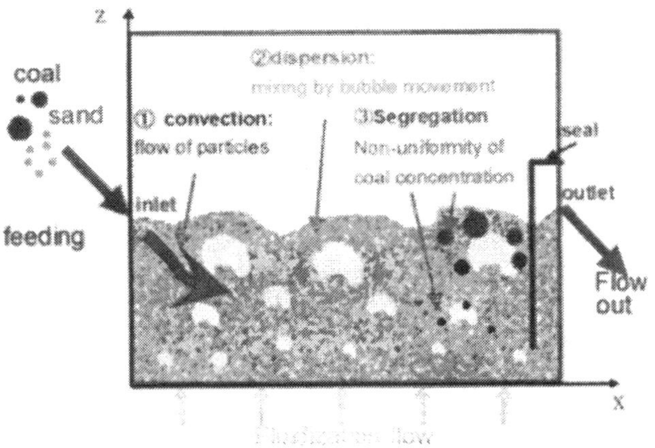

Fig.2 Segregation-convection-dispersion model

A concentration equation for lignite has been derived as Eq.(1)[1],[2],[3]

$$\frac{\partial C}{\partial t}+u\frac{\partial C}{\partial x}+wS\frac{\partial C}{\partial z}=\frac{\partial}{\partial x}(D_x\frac{\partial C}{\partial x})+\frac{\partial}{\partial z}(D_z\frac{\partial C}{\partial z}) \quad (1)$$

where S is the segregation index, which describes the bias of concentration distribution of lignite in vertical direction. C is the concentration of lignite, u,w are particle velocities in horizontal and vertical direction, Dx and Dz are dispersion coefficients in horizontal and vertical direction.

S is determined by experiment, an example is shown in Fig.3. It can be observed that the more lignite particles bias in the top of bed, the few particles are in the bottom of bed.

The concentration of lignite particle at exit of bed $C_{out}(t)$ can be predicted by solving equations of continuity, momentum and concentration Eq.(1). A residence time distribution function E(t) can be define as Eq.(2), if the feeding concentration at inlet C_{in} is pulse at time=0.

$$E(t)=C_{out}(t)/C_{in} \quad (2)$$

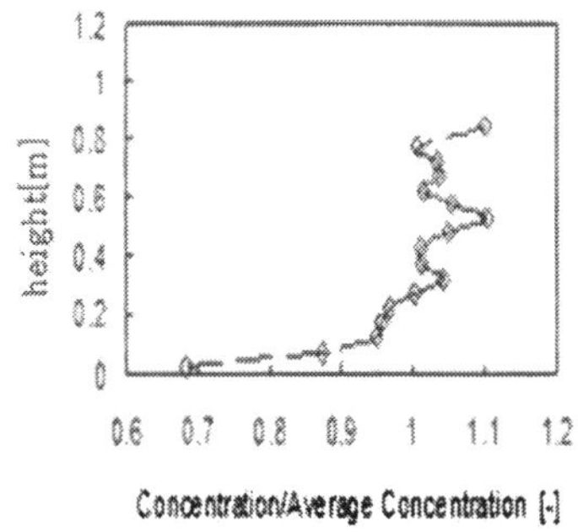

Fig.3 Segregation of fuel particle

3.2 Effect of Bed Length on RTD

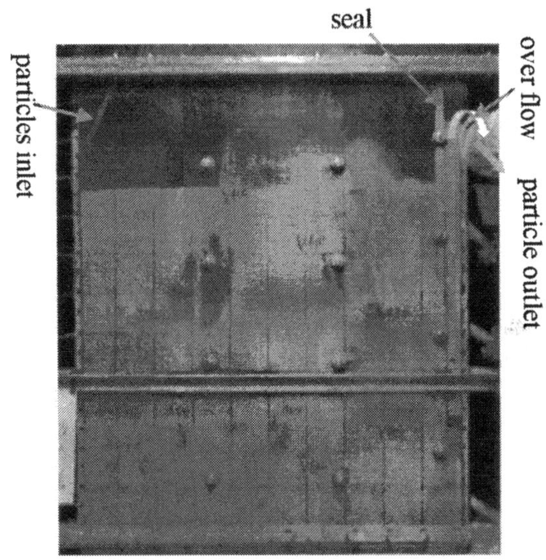

(1) experimental apparatus(L=1m)

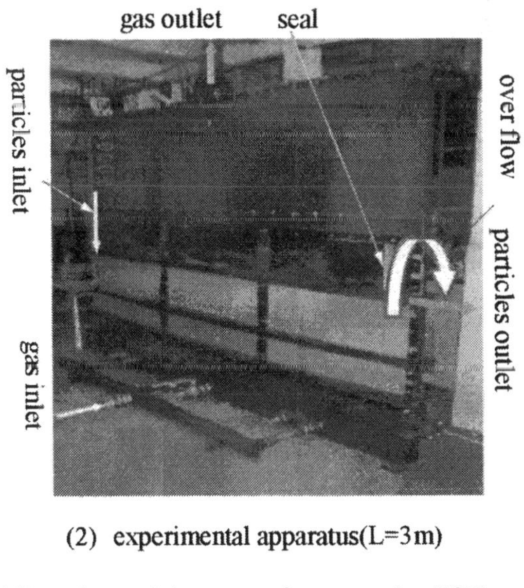

(2) experimental apparatus(L=3m)

Fig.4 Experimental Apparatus for measuring RTD

The RTDs are calculated for various bed lengths: L=1m (lab scale), L=3m (bench scale), L=9m (pilot scale) and L=18m (commercial scale), keeping the nominal residence time τ constant, where τ=W/G, with W denoting weight of the bed in [kg] and G feeding rate of particles into bed[kg/s].

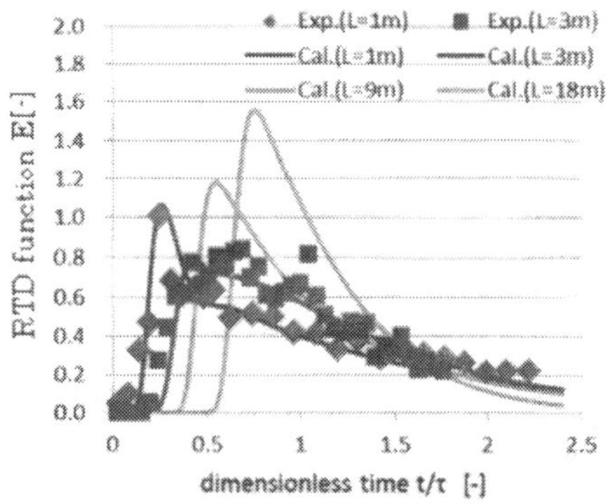

Fig.5: Experimental validation of predicted effect of bed length on RTD

Experiments have been done to validate the experiment. The experimental apparatus is shown in Fig.4. The comparisons of L=1m and L=3m are shown in Fig.5. It can be seen that the predictions agree experiments well. The predictions for L=9m and 18m are also shown in Fig.5. It can been seen that the longer the bed length(that is, the larger scale of gasifier) becomes, the more onset and peak of lignite concentration at the outlet are delayed. Therefore, the residence time of lignite is longer when they are in the larger gasifier, even though the nominal residence time is the same.

4 PREDICTION OF GASIFICATION PERFORMANCE

Experiments are made to measure the gasification performance of lignite (Fig.6). Experiments are made in a non-continuous fluidized bed with batch feeding (residence time of solid fuel is infinity) to obtain the carbon conversion X(t), which is shown in Fig.7. The carbon conversion is defined as ratio of the amount of carbon in gasified gas to that in original lignite.

The conversion is normalized by the conversion at time=τ.

Assuming that the chemical reaction and residence time of lignite do not influence each other, the carbon conversion η_c for the continuous fluidized bed can be calculated by

$$\eta_c(T) = \int_0^T E(t)X(t)dt \tag{3}$$

Results for the carbon conversion calculated according to Eq.(3) are compared with experimental data in Fig.8. The calculations agree well with the experiments.

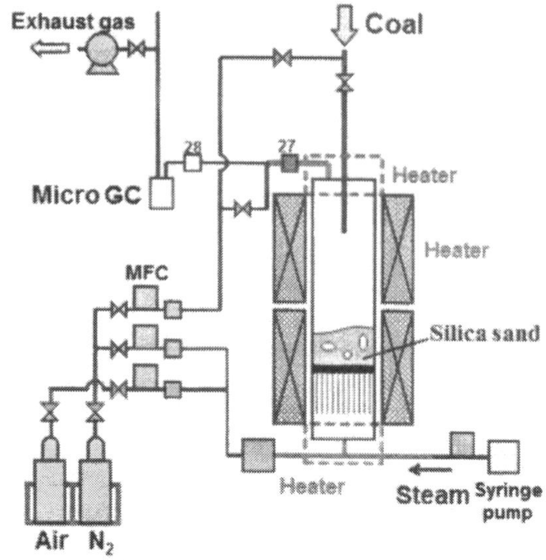

Fig.6 Scheme of Experimental Apparatus for measuing carbon conversion in non-continuous fluidized bed

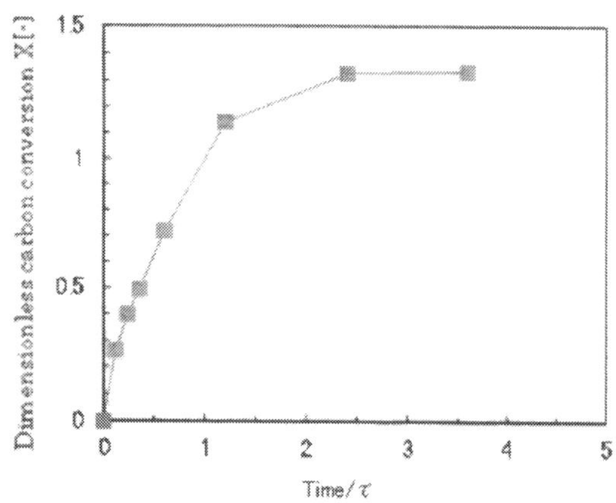

Fig.7: Carbon conversion of lignite in non-continuous fluidized bed

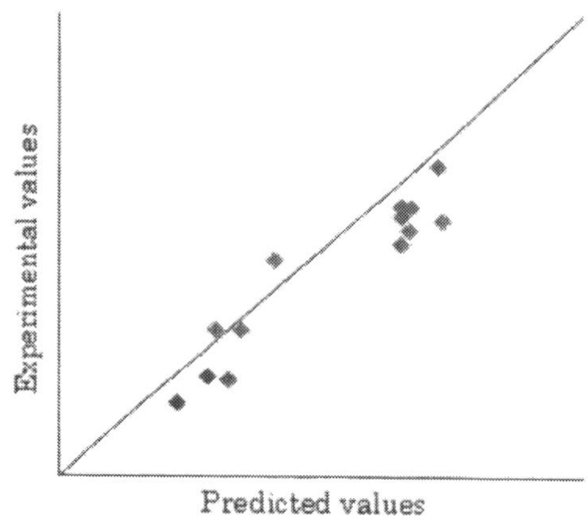

Fig. 8: Evaluation of numerical prediction for carbon conversion η_c

5 CARBON CONVERSIONS FOR GASIFIERS OF VARIOUS SCALES

The carbon conversion is calculated with the residence time distribution as defined by Eq.(3), shown in Figure 9. Gasification furnace lengths are 1,3,9 and 18m. When the bed length is longer, the carbon conversion is larger in the initial phase ($t / \tau <= 1$), while carbon conversion is reduced in the late phase ($1 < t / \tau < \infty$) is. There are no significant differences in carbon conversion at all stages, but it can be seen that the carbon conversion is higher for the longer bed.

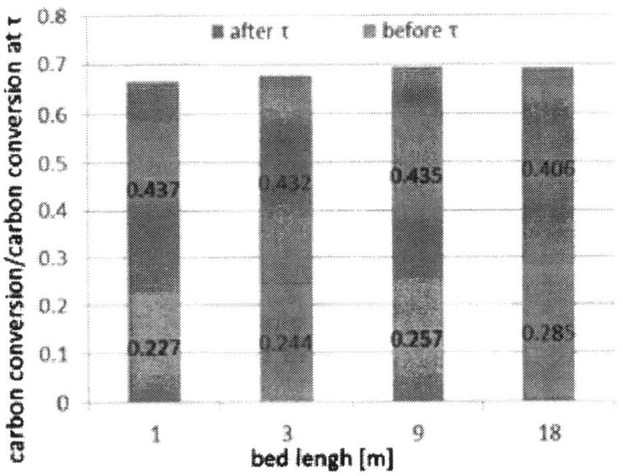

Figure 9: Carbon conversion of coal in continuous fluidized beds with various bed length.

6 CONCLUSIONS

Using segregation-convection-dispersion model, the residence time distribution of lignite particles in a sand-fluidized-bed, can be correctly predicted. The effect of bed length is predicted and validated by experiments. The carbon conversion in a continuous bed can be approximated by Eq.(3). We can correctly predict carbon conversion for an enlarged gasifier.

REFERENCES

[1] Z.Liu, M.Narukawa, M.Takafuji, T.Suda, " Prediction and verification of effect of bed length on RTD of coal in a bubble fluidized bed", Fluidization XIII, 2011

[2] T.Suda, Z.Liu, M.Takafuji, K.Hamada, H. Tani, "Gasification of Lignite Coal and Biomass Using Twin IHI Gasifier (TIGAR®)", IHI Engineering Review, Vol. Vol.52 No.1 (2012)

[3] Z.Liu, M.Narukawa, T.Suda, M.Takafuji, "The Effect of Length Scale of Fluidized Bed on Residence Time Distribution of Coal", J. of Combustion Society of Japan, Vol.56 No.177 (2014)

[4] A.B.M. Heesink, J. Klaus, W.P.M. Van Swaaij, " The influence of particle residence time distribution on the reactivity in fluidized bed reactors", Chemical Engineering Science, Vol.49, Issue 14, pp. 2243-2261, 1994

[5] S.Rodríguez-Rojo, M.J.Cocero, "Supercritical fluidized bed modeling", J. of Supercritical Fluids Vol.50 pp.54-60 (2009)

[6] Lu Huilin, ZhaoYunhua, Jianmin Ding, Dimitri Gidaspow, Li Wei, "Investigation of mixing/segregation of mixture particles in gas–solid fluidized beds", Chemical Engineering Science Vol.62 pp.301- 317 (2007)

Theoretical Synthesis of Mixed Materials for CO_2 Capture Applications

Yuhua Duan

National Energy Technology Laboratory, United States Department of Energy, Pittsburgh, PA 15236
Yuhua.duan@netl.doe.gov

ABSTRACT

Since the current technologies for capturing CO_2 are still too energy intensive, to develop new materials that can capture CO_2 reversibly with acceptable energy costs are needed. At a given CO_2 pressure, the turnover temperature (T_t) of an individual solid capture CO_2 reaction is fixed. Such T_t may be outside the operating temperature range (ΔT_o) for a practical capture technology. In order to adjust T_t to fit the practical ΔT_o, in this study, by combining thermodynamic database mining with first principles density functional theory and phonon lattice dynamics calculations, three scenarios of mixing schemes are explored. Our calculated results demonstrate that by mixing different types of solids it's possible to shift T_t to the range of practical operating temperature conditions.

Keywords: CO_2 capture, mixed solid sorbents, ab intio thermodynamics

1 INTRODUCTION

Nowadays, the burning of fossil fuels is still the main energy source for the world's economy. One consequence of the use of these fuels is the emission of huge quantities of CO_2 into the atmosphere creating environmental problems such as global climate warming.[1-5] In order to solve such problems, there is a need to reduce CO_2 emissions into atmosphere by capturing and sequestrating CO_2.[4, 6, 7] One approach to solve such environmental problems is to capture and sequester the CO_2. Current technologies for capturing CO_2 including solvent-based (amines) and CaO-based materials are still too energy intensive. Hence, there is critical need for development of new materials that can capture and release CO_2 reversibly with acceptable energy costs. In particular, solid oxide sorbent materials have been proposed for capturing CO_2 through a reversible chemical transformation leading primarily to formation of carbonate products. Solid sorbents containing alkali and alkaline earth metals have been reported in several previous studies to be promising candidates for CO_2 sorbent applications due to their high CO_2 absorption capacity at moderate working temperatures.[8-10]

During past few years we developed a theoretical methodology to identify promising solid sorbent candidates for CO_2 capture by combining thermodynamic database searching with *ab initio* thermodynamics obtained based on first-principles density functional theory (DFT) and lattice phonon dynamics.[8-15] The primary outcome of our screening scheme is a list of promising CO_2 sorbents with optimal energy usage.

For a practical CO_2 capture technology, it has optimal operating conditions, such as the absorption/desorption temperature range (ΔT_o). As a good sorbent, its CO_2 capture/release temperature should fit into such range. However at a given CO_2 pressure, the turnover temperature (T_t) of an individual solid capture CO_2 reaction is fixed. Such T_t may be outside the operating temperature range (ΔT_o) for a particularly capture technology. In order to adjust T_t to fit the practical ΔT_o, in this study, we demonstrate that by mixing different types of solids, it's possible to shift T_t to the range of practical operating conditions.

The complete description of the computational methodology together with relevant applications can be found in our previous publications.[8-16] From the calculated relationships among the Gibbs free energy, temperature and CO_2 pressure(P_{CO2}), at given P_{CO2} for pre- and post-combustion capture technologies, the turnover temperature T_t can be determined for each capture reaction.

2 RESULTS AND DISCUSSION

For a given CO_2 capture process, its optimal working conditions (CO_2 pressures of pre- and after-capture, absorption/desorption temperature range (ΔT_o), *etc.*) were fixed. However, at a given CO_2 pressure, the turnover temperature (T_t) of an individual solid capture CO_2 reaction is fixed. Such T_t may be outside the operating temperature range ΔT_o for a particularly capture technology. In order to adjust T_t to fit the practical working through reversible chemical transformations, the chemical properties of solids must be modified to change the $\Delta G(T, P)$ in Eq. (1). If we want to increase the T_t to high temperature range, the $\Delta G(T, P)$ should be more negative. To achieve it, we can either destabilize the reactants (sorbents) or stabilize the products or do both. On contrary, if we want to decrease T_t to low temperature range, the $\Delta G(T, P)$ should be less negative, which can either stabilize the reactant (sorbents) or destabilize the products or do both. In other words, mixing stabilizer/destabilizer in solid could change the thermodynamic properties of their CO_2 capture reactions to shift T_t. Some mixing examples are given in Table 1. As one can see that the mixed sorbent could be a new formed solid (e.g. lithium silicates) or just simple mixture (e.g. $MgO+Na_2CO_3$) to change the chemical properties of reactants and products. Depending on the main captor A and the direction of T_t shifts, different mixed solid B and mixing ratio could be determined. Although one can mix any number of solids to form a new CO_2 sorbent, in order to focus exploring the nature of mixtures; here we restrict ourselves with cases of two and three mixed solids.

Table 1. Mixing schemes and examples of mixed solids and their effects on CO_2 capture

Main captor A	Mixed solids B (one or more)	New formed sorbent C	Examples of CO_2 capture reaction	Effect on CO_2 capture capacity & operating T	Refs.
N_2O, N=Li, Na, K MO, M=Mg, Ca	SiO_2, TiO_2, ZrO_2, Al_2O_3, MeO (Me: transition metal) Two solids mixing with different ratios	Lithium silicates, lithium zirconates, calcium aluminum oxides	$Li_4SiO_4 + CO_2 = Li_2CO_3 + Li_2SiO_3$ $Li_2ZrO_3 + CO_2 = Li_2CO_3 + ZrO_2$ $^{1}/_{12}Ca_{12}Al_{14}O_{33} + CO_2 = CaCO_3 + ^{7}/_{12}Al_2O_3$	(1) B stabilizes A, but does not capture CO_2; (2) Compared to pure N2O, the T_t shifts to low-T range; (3) The maximum CO_2 capture capacity decreases with increasing solid B.	[9, 11, 16, 18-20]
MO, M= Mg, Ca	N_2O or N_2CO_3, N= Li, Na, K Two solids mixing with different ratios	MO+N_2O MO+N_2CO_3 MgO+$CaCO_3$	$MgO+Na_2CO_3+CO_2 = Na_2Mg(CO_3)_2$ $MgO+CaCO_3+CO_2 = MgCa(CO_3)_2$	(1) N_2O (or N_2CO_3) stabilizes the product to form double salt and does not capture CO_2 in the ΔT_o range; (2) Compared to pure MO, the T_t shifts to high-T range; (3) The maximum CO_2 capture capacity decreases with B.	[21-23]
N_2O, N= Li, Na, K	SiO_2+MeO Three solids mixing with different ratios	Li_2MeSiO_4 (Me= Fe, Co, Ni, Mg, Mn, Zn, *etc.*)	$Li_2MeSiO_4 + CO_2 = Li_2CO_3 + MeSiO_3$	(1) B stabilizes both A and product. The final effect depends on the bonding strength of MeO with A and SiO_2; (2) T_t may shift to low- and high-T range; (3) The maximum CO_2 capture capacity decreases with B;	
N_2O, N= Li, Na, K	N_2O+SiO_2 N_2O+ZrO_2 $MO+SiO_2$ Three solids mixing with different ratios	$Li_{2-x}Na_xZrO_3$ $Li_{2-x}K_xZrO_3$ $Li_{4-x}Na_xSiO_4$ They also can be treated as doped materials	$Li_{2-x}N_xZrO_3 + CO_2 = ZrO_2 + ^{(2-x)}/_2Li_2CO_3 + ^{x}/_2N_2CO_3$ $Li_{4-x}N_xSiO_4 + CO_2 = ^{(2-x)}/_2Li_2CO_3 + ^{x}/_2N_2CO_3 + Li_2SiO_3$	(1) B stabilizes A and portion of B (N_2O) also involves in CO_2 capture; (2) T_t may shift to low-T range; (3) The maximum CO_2 capture capacity decreases with weaker portion of B (SiO_2, ZrO_2, *etc.*)	[24, 25]

As shown in Table 1, as effective main CO_2 captors through chemical reactions to form carbonates, alkali and alkaline earth metal oxides are important of interests due to their easy reacting with CO_2 and low costs. The problem is that they can strongly react with CO_2 to form carbonates, but their turnover temperatures T_t are very high and only can be regenerated at very high temperatures which are unsuitable for many CO_2 capture technologies. Hence, mixing with other solids to shift their T_t becomes important for their suitability as CO_2 sorbents. Generally, when we mix two solids *A* and *B* to form a new sorbent *C*, the turnover temperature of the newly resulted system (T_C) is located between those of *A* and *B* (T_A, T_B) individuals. Here it was assumed that *A* is a strong CO_2 sorbent while *B* is a weak CO_2 sorbent and $T_A>T_B$. Also, we assumed that the desired operating temperature T_O is between T_A and T_B ($T_A>T_O>T_B$). Depending on the properties of *A* and *B*, as shown in Table 1 we have typically three scenarios to synthesize the mixing sorbent *C*:

Case I: $T_A>>T_B$ and the *A* component is the key part to capture CO_2

In this case, by mixing B into A, a new solid C is formed. Since B does not capture CO_2, it serves as a stabilizer to stabilize A to form solid C which reaction possesses lower energy than A and B individuals. Therefore, the energy state of reactant is lower than pure A. therefore, the mixed sorbent shifts T_t to lower temperature range. The amount of shifted temperature depends on the mixing ratio.

An example of this case is represented by Li_2O. Li_2O is a very strong CO_2 sorbent which forms Li_2CO_3. However, its regeneration from Li_2CO_3 only can occur at very high temperature (T_A). In order to move its T_A to lower temperatures, one can mix some weak CO_2 sorbents (such as SiO_2, ZrO_2). With different mixing ratio of Li_2O/SiO_2(or ZrO_2), different stable lithium silicates (zirconates) can be formed, such as Li_8SiO_6, Li_4SiO_4, $Li_6Si_2O_7$, Li_2SiO_3, $Li_2Si_2O_5$, $Li_2Si_3O_7$; Li_8ZrO_6, $Li_6Zr_2O_7$, Li_2ZrO_3, *etc.* Fig. 1 shows the turnover temperature and the CO_2 capture capacity of Li_2O/SiO_2 and Li_2O/ZrO_2 mixtures versus the ratio of Li_2O/SiO_2 or Li_2O/ZrO_2.[8-11, 16, 18, 19]

When the SiO_2/Li_2O or ZrO_2/Li_2O ratios are increased by adding more SiO_2 or ZrO_2 into Li_2O, as shown in Fig.1, the turnover temperatures of mixed sorbents (T_t) are shifted more to lower temperature range. Therefore, by controlling the mixing ratio, it's possible to move its CO_2 capture temperature down to the right range required by certain capture technology.

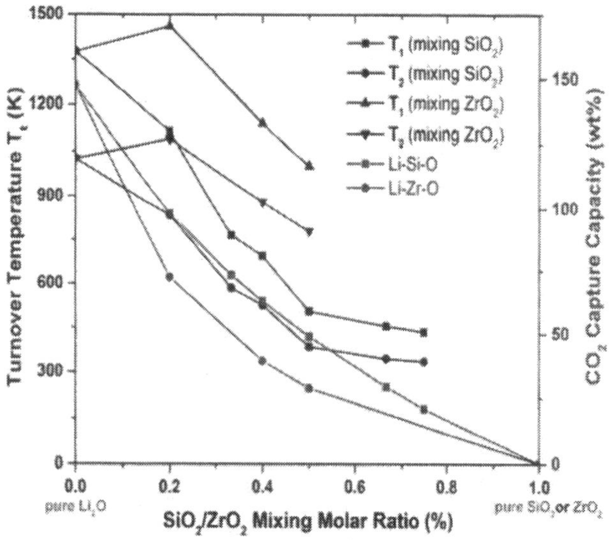

Fig. 1 The calculated turnover temperatures and CO_2 capture capacity versus molar percentage of SiO_2 or ZrO_2 in the lithium silicates or zirconates. T_1 are the turnover temperatures under pre-combustion conditions with CO_2 partial pressure at 20 bars, while T_2 are the turnover temperatures under post-combustion conditions with CO_2 partial pressure at 0.1 bar.

Case II: $T_A >> T_B$ and B component is the key part to capture CO_2

Opposite to previous case, in this case we want to increase turnover temperature (T_t). Since T_B is lower than T_O, by mixing *A* into *B* will increase the turnover temperature T_C of the *C* solid to values closer to T_o. In this way, the function of A is either to destabilize solid A or to stabilize the captured products. For example, pure MgO has a very high theoretical CO_2 capture capacity. However, its turnover temperature (250 °C) is lower than the required temperature range of 300-470 °C used in warm gas clean up technology and its practical CO_2 capacity is very low, and therefore, it cannot be used directly as a CO_2 sorbent in this technology.[22, 23, 26] Fig. 2 shows an example of mixing Na_2CO_3 (or K_2CO_3, $CaCO_3$) with MgO to shift the T_t of MgO to higher temperature range.

Since the mixed carbonates do not react with the reactant MgO directly, but react exothermically with the carbonate $MgCO_3$ to form double salt (such as $Na_2Mg(CO_3)_2$) and stabilize the product. Therefore, the turnover temperature of Na_2CO_3-promoted MgO sorbent capturing CO_2 is higher than pure MgO. Generally, as shown in Fig.2, by mixing alkali metal oxides M_2O (M=Na, K, Cs, Ca) or carbonates (M_2CO_3) into MgO, the corresponding newly formed mixing systems have higher turnover temperatures making them useful as CO_2 sorbents through the reaction MgO + CO_2 + M_2CO_3 = $M_2Mg(CO_3)_2$.[21, 22]

From Fig. 2, one can see that in the CO_2 capture reactions, the Na_2CO_3-propmoted MgO possesses larger T_t shift to higher temperature range than the K_2CO_3+MgO and $CaCO_3$+MgO do. Obviously, since the T_t of the mixture is lower than the dissociation temperature of promotors (Na_2CO_3, K_2CO_3, $CaCO_3$), during CO_2 capture sorption process, these promoters act as stabilizers to react with $MgCO_3$ to form double salt, while during the CO_2 desorption, the double salt dissociates to MgO and the promoter.

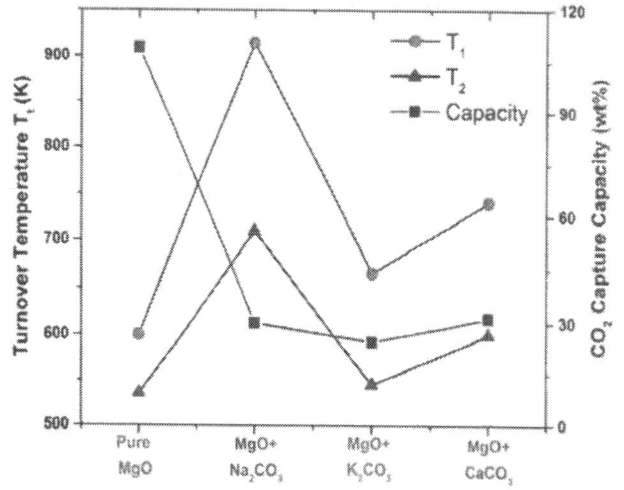

Fig. 2 The calculated turnover temperatures and CO_2 capture capacity of pure MgO and its carbonate-promoted sorbents.

Case III: Both A and B are active for CO_2 capture

In this case, we want both *A* and *B* components are active to capture CO_2, and the CO_2 capacity of the mixture is the summation of those of *A* and *B*. As we know another potential advantage of mixing solids is to increase the surface area of the solids in order to have faster reaction rates. Such a mixing scenario doesn't show too much advantage in shifting the capture temperature, but may enhance the kinetics of the capture process and eventually make the mixtures more efficient. Although up to know there is no such report in literature, we think such an attempt is worthwhile and are working on several doped systems. For example, the Li-/K- doped Na_2ZrO_3 capture CO_2 have the following reaction (b):[24, 25]

$$M_xN_{2-x}ZrO_3 + CO_2 \leftrightarrow \frac{x}{2}M_2CO_3 + \frac{2-x}{2}N_2CO_3 + ZrO_2 \qquad (M,N = Li,Na,K)$$

When the Na_2ZrO_3 doped with different molar ratio of Li or K, the thermodynamic properties of the doped systems are quite different from pure Na_2ZrO_3. The turnover temperatures versus the Li-/K- doping levels are shown in Fig.3.

Based on the calculated relationships among the Gibbs free energy change, CO_2 pressure and temperature for CO_2 capture reactions by $Na_{2-x}M_xZrO_3$ (M= Li, K, x= 0.0, 0.5, 1.0, 1.5, 2.0), compared to pure Na_2ZrO_3, overall, the Li- and K-doped mixtures $Na_{2-x}M_xZrO_3$ have lower turnover temperatures (T_t) shown in Fig.3. The calculated results show that the shift in T_t depends not only on the doping element, but also depends on the doping level. The Li-doped systems have larger T_t decreases than the K-doped systems. When increasing the Li-doping level x, the T_t of corresponding mixture $Na_{2-x}Li_xZrO_3$ decreases further to a low temperature range. However in the case of K-doped systems $Na_{2-x}K_xZrO_3$, although initial doping of K into Na_2ZrO_3 can shift its T_t to lower temperature range, further increasing the K-doping level x results in an increase in T_t. Therefore, compared to K-doing, lithium inclusion into Na_2ZrO_3 structure has a larger influence on the CO_2 capture performance.

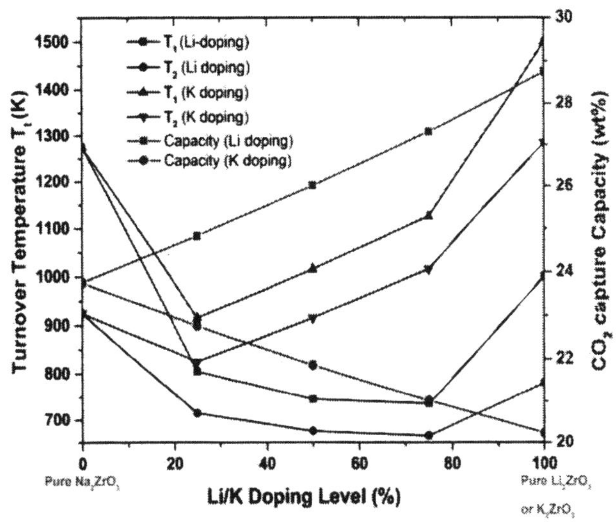

Fig. 3 The dependence of the turnover temperatures and of CO_2 capture capacity on Li/K doping molar percentage in $Na_xM_{2-x}ZrO_3$ (M=Li, K)..

3 CONCLUSIONS

The obtained results showed that by changing the mixing ratio of solid *A* and solid *B* to form mixed solid *C* it's possible to shift the turnover T_t of the newly formed solid *C* to fit the practical CO_2 capture technologies. In this study, we investigated three scenarios of mixing schemes. The obtained results can be used to provide insights for designing new CO_2 sorbents. Therefore, although one single material taken in isolation might not be an optimal CO_2 sorbent to fit the particular needs to operate at specific temperature and pressure conditions, by mixing or doping two or more materials to form a new solid, the calculated results showed that it is possible to synthesize new CO_2 sorbent formulations which can fit the industrial needs. Our results also show that computational modeling can play a decisive role for identifying materials with optimal performance.

Acknowledgements

The author wishes to his acknowledge to Drs. D. C. Sorescu, H. Pennline, K. Johnson, H. Pfeiffer, B. Y. Li, K. L. Zhang, D. L. King, X. S. Li, X. F. Wang and D. Luebke for their collaborations and fruitful discussions.

REFERENCES

[1] D. Aaron, and C. Tsouris, Separation Science and Technology **40**, 321 (2005).

[2] C. M. White, B. R. Strazisar, E. J. Granite, J. S. Hoffman, and H. W. Pennline, J. Air Waste Manag. Assoc. **53**, 645 (2003).

[3] R. S. Haszeldine, Science **325**, 1647 (2009).

[4] B. Y. Li, Y. Duan, D. Luebke, and B. Morreale, Applied Energy **102**, 1439 (2013).

[5] M. R. Allen, D. J. Frame, C. Huntingford, C. D. Jones, J. A. Lowe, M. Meinshausen, and N. Meinshausen, Nature **458**, 1163 (2009).

[6] H. Pfeiffer, and P. Bosch, Chem. Mater. **17**, 1704 (2005).

[7] E. Ochoa-Fernandez, H. K. Rusten, H. A. Jakobsen, M. Ronning, A. Holmen, and D. Chen, Catal. Today **106**, 41 (2005).

[8] Y. Duan, D. Luebke, and H. W. Pennline, Int. J. Clean Coal and Energy **1**, 1 (2012).

[9] Y. Duan, and D. C. Sorescu, Phys. Rev. B **79**, 014301 (2009).

[10] Y. Duan, and D. C. Sorescu, J. Chem. Phys. **133**, 074508 (2010).

[11] Y. Duan, and K. Parlinski, Phys. Rev. B **84**, 104113 (2011).

[12] Y. Duan, B. Zhang, D. C. Sorescu, and J. K. Johnson, J. Solid State Chem. **184**, 304 (2011).

[13] Y. Duan, D. R. Luebke, H. W. Pennline, B. Y. Li, M. J. Janik, and J. W. Halley, J. Phys. Chem. C **116**, 14461 (2012).

[14] Y. Duan, B. Zhang, D. C. Sorescu, J. K. Johnson, E. H. Majzoub, and D. R. Luebke, J. Phys.-Condensed Matter **24**, 325501 (2012).

[15] B. Zhang, Y. Duan, and J. K. Johnson, J. Chem. Phys. **136**, 064516 (2012).

[16] Y. Duan, J. Renewable Sustainable Energy **3**, 013102 (2011).

[17] M. W. J. Chase, J. Phys. Chem. Ref. Data, **9**, 1 (1998).

[18] Y. Duan, J. Renewable Sustainable Energy **4**, 013109 (2012).

[19] Y. Duan, Phys. Chem. Chem. Phys. **15**, 9752 (2013).

[20] Y. Duan, H. Pfeiffer, B. Li, I. C. Romero-Ibarra, D. C. Sorescu, D. R. Luebke, and J. W. Halley, Phys. Chem. Chem. Phys. **15**, 13538 (2013).

[21] Y. Duan, K. Zhang, X. S. Li, D. L. King, B. Li, L. Zhao, and Y. Xiao, Aerosol Air Quality Res. **14**, 470 (2014).

[22] K. L. Zhang, X. H. S. Li, Y. Duan, D. L. King, P. Singh, and L. Y. Li, Int. J. Greenh. Gas Control **12**, 351 (2013).

[23] K. L. Zhang, X. H. S. Li, W. Z. Li, A. Rohatgi, Y. Duan, P. Singh, L. Y. Li, and D. L. King, Advanced Materials Interfaces **1**, 1400030 (2014).

[24] Y. Duan, ScienceJet **3**, 56 (2014).

[25] Y. Duan, J. Lekse, X. F. Wang, B. Y. Li, B. Alcantar-Vazquez, H. Pfeiffer, and J. W. Halley, Physical Review Applied **3**, (2015)in press.

[26] J. L. Chi, L. F. Zhao, B. Wang, Z. Li, Y. H. Xiao, and Y. Duan, Int. J. Hydrogen Energy **39**, 6479 (2014).

[27] M. L. Gray, K. J. Champagne, D. Fauth, J. P. Baltrus, and H. Pennline, Int. J. Greenh. Gas Control **2**, 3 (2008).

Pelletized silica and amine composites for CO_2 capture via adsorption

Gregory P. Knowles[1,2], Christian Vogt[1,2] and Alan L. Chaffee[1,2]

[1]Cooperative Research Centre for Greenhouse Gas Technologies (CO2CRC), [2]School of Chemistry, 17 Rainforest Walk, Monash University, Vic 3800 Australia,
Email: greg.knowles@monash.edu, christian.vogt@monash.edu or alan.chaffee@monash.edu.

ABSTRACT

A broad range of mesoporous silica – amine composite powders were prepared, and readily shaped into robust forms, as are preferred for industrial applications, in accord with recently patented binderless shaping technologies. Such materials are of key interest to facilitate CO_2 capture, such as to help mitigate global warming. Shaping of the powders was otherwise found to be problematic due to product extrusion and/or texture. Select shaped products prepared in accord with the patent technologies were then subjected to CO_2 partial pressure swing adsorption (PPSA) processing. These products were found to exhibit up to ~5 wt % and ~11 wt % CO_2 working capacity, as estimated for the post combustion capture (PCC) of CO_2 from simulated flue gases via pressure swing adsorption (PSA) and thermal swing adsorption (TSA) processing respectively. The patent technologies were thus found to provide a promising opportunity to facilitate the preparation of robust sorbent options for CO_2 capture applications.

Keywords: pellet, amine, adsorption, carbon capture, CO_2

1 INTRODUCTION

Solid amine type sorbents are being developed to provide highly selective, water tolerant and high capacity sorbent options to facilitate CO_2 capture from mixed gas streams such as post combustion flue gas, natural gas and air [1]. These developments are motivated by strategic interests to establish carbon capture technologies to reduce global atmospheric greenhouse gases, and so to reduce global warming, whilst facilitating growing world energy demands [1]. Adsorption driven separation process technologies are of particular interest to provide cheaper options than traditional absorption driven technologies [2]. Such developments have typically focussed on the preparation and characterisation of powdered type sorbents, however, shaped (pelletized) sorbents are typically preferable for industrial gas separation processes [1],[3],[4].

"Molecular basket" [5] type solid amine sorbents such as those prepared via the combination of polyethyleneimine (PEI) with mesoporous silicas such as MCM-41 [5], SBA-15 [1], and mesocellular siliceous foam (MCF) [6],[7], are of particular interest for the simplicity of their preparation, and large potential CO_2 working capacity. PEI loaded MCF composite powders in particular were found to adsorb up to ~10 wt % CO_2 from both anhydrous and wet simulated post combustion flue gases (105 °C) [7]. However, shaping of such composites incorporating 80, 85 & 90 % MCF pore volume equivalents of 1200 MW branched PEI was initially found to be problematic, due to product extrusion under compression force, and stickiness [8],[9].

The patented binderless shaping technology Process A was then developed, providing a means to overcome these problems, to prepare robust dry sorbent pellets with promising potential for CO_2 capture applications [8],[9]. The following demonstrates the broader relevance of this technology to likewise facilitate the shaping of other molecular basket type amine sorbents, which incorporate various PEI loadings, different mesoporous supports and different amines, as is also now patented [10].

MCF and SBA-15 type mesoporous silicas were prepared, and Davisil 60 silica sourced, as supports. N_2 physisorption at 77 K was conducted for each to facilitate estimation of their mesopore volume, and subsequent amine loading on a per unit mesopore volume basis.

MCF-PEI composites were prepared via wet impregnation of MCF with 1200 MW PEI at 60, 70, 75 and 100 % MCF mesopore volume equivalents, to compare with those previously prepared at 80, 85 and 90 % MCF mesopore volume equivalents, to further investigate the effect of amine loading. The same PEI was separately loaded into SBA-15 and Davisil 60 mesoporous silicas at 100 % mesopore volume equivalents to investigate the effect of different support materials.

Composite sorbents were also separately prepared by loading 1st generation polypropyleneimine tetramine dendrimer DAB-Am-4 (DAB) [N,N,N',N'-Tetrakis(3-aminopropyl)-1,4-butanediamine], polyallylamine (PAA) and tetraethylenepentamine (TEPA) into MCF type mesoporous silica, and tris(2-aminoethyl)amine (TAEAH) formate into SBA-15 type mesoporous silica, at 100 % mesopore volume equivalents respectively, to investigate the effects of different amines.

PEI composite powder products were also prepared by a modified process "B" to investigate its potential to also better facilitate their shaping.

A 200 g batch of a MCF-PEI composite sorbent incorporating 80 % MCF mesopore volume equivalents of branched 1200 MW PEI was also prepared, to facilitate testing the fitness for purpose of such sorbents under more realistic process conditions.

The composite sorbent powders were then separately compressed, following initial pre-activation (drying), and in accord with the patented Process A. Visual observation of the shaping process was conducted in each case, and the form of the pellets (sticky/dry, weak/robust), and the significance of product extrusion, or not, noted. Both the amine and silica precursors were accurately pre-weighed prior to their combination, to facilitate estimation of the nominal composition of their adducts. Product characterisation was also furthered via He pycnometry.

Select sorbent powders were purged with anhydrous CO_2 at 60 °C for ~16 h, and then cooled within the same atmosphere to facilitate a preliminary assessment of their relative CO_2 capacities. CO_2 partial pressure swing adsorption processing of select sorbents was subsequently conducted in-situ within a thermogravimetric analyser (anhydrous and humidified conditions) to facilitate the assessment of their potential for PCC applications.

2 EXPERIMENTAL

2.1 Chemicals

Pluronic "P123" (EO_{20}-PO_{70}-PO_{20} – Aldrich), 1,3,5-trimethylbenzene (TMB - Fluka 99%), ammonium fluoride (Aldrich, 98 %), HCl (BDH, 32%), tetraethylorthosilicate (TEOS - Aldrich 98 %), ethanol absolute (EtOH – Merck), methanol (MeOH – BDH Analar), PEI branched (Sigma-Aldrich, 1200 Mw, 50 wt % in H_2O), TEPA (Aldrich, tech.) PAA (Polysciences, av 15,000 Mw, ~15 wt % in H_2O), DAB (Sigma-Aldrich, #460699) and Davisil grade 634 silica gel (D60 – Aldrich), or equivalents, were used as supplied. Tris(2-aminoethyl)amine (TAEAH) formate was used as supplied by Prof Douglas MacFarlane and Dr Vijay Ranganathan (Monash University, Clayton, Vic., Aust).

2.2 Preparation of MCF and SBA-15 silicas

MCF was prepared via neutral templating techniques using P123 as template, TMB as swelling agent and TEOS as silica source. **MCFa** was prepared by a method based on that reported by Han et al 2007 [11]. **MCFb** was prepared by a method adapted from that reported by Schmidt-Winkel et al 2000 [12]. Hydrothermal treatments were conducted within a glass reagent bottle, within a pressure cooker, at 105 °C as previously described [1]; the templates were then removed via calcination at 550 °C.

SBA-15 type mesoporous silicas **SBAa, SBAb and SBAc** were also prepared via a neutral templating technique, using P123 as template and TEOS as silica source, as per methods adapted from that reported by Zhao et al [13]; template removal was via calcination at 550 °C.

2.3 Preparation of Composite Sorbents

The range of MCF-PEI composites were in general prepared via infiltration of the amine from ethanol solution into ~1 g of the respective pre-dried mesoporous silica, while magnetically stirred for 24 h at ~40 °C under parafilm seal. The bulk solvents were subsequently removed via rotary evaporation under vacuum (up to 60 °C). **MCFa-PEI80**, **MCFa-PEI85** and **MCFa-PEI90** were thus prepared by combining MCFa with 80, 85 and 90 % MCFa mesopore volume equivalents of PEI as previously reported [8],[9]. **MCFa-PEI60, MCFa-PEI70 and MCFa-PEI75** were likewise prepared by combining MCFa with 60, 70 and 75 MCFa mesopore volume equivalents of PEI respectively. **MCFb-PEI100, MCFb-DAB100, MCFb-PAA100, MCFb-TEPA100** were similarly prepared by combining MCFb with a 100 % MCFb mesopore volume equivalents of PEI, DAB (synthesis in MeOH), PAA and TEPA respectively.

D60-PEI100, SBAa-PEI100 and **SBAb-TAEAH100**, were similarly prepared by combining D60, SBAa and SBAb with PEI and TAEAH respectively.

MCFa-PEI80b was similarly prepared by combining ~70 g MCFa with PEI as previously reported [10].

MCFb-PEI100-B1, MCFb-PEI100-B2 and SBAc-PEI100-B were prepared by combining MCFb and SBAc with PEI via the more recently patented modified infiltration process (Process B) at room temperature, 60 °C and 60 °C respectively [10].

Each of these composite sorbents was then activated, by degassing via vacuum at 110 °C, excepting TEPA and TAEAH products which were activated under vacuum at 25 and 60 °C respectively, prior to further processing.

The products were then compressed as activated alone, and/or in accord with the patented Process A, to form pellet (shaped) products via a Specac 15 ton manual hydraulic press. Additionally, MCFa-PEI80b was shaped via a Shanghai Chengxiang Machinery Co. TDP-6T Automated Single Punch Tablet Press (SPTP) fitted with a 4 mm standard concave quadrasect die set, to investigate the potential of the patented process technology to facilitate the shaping of such products on a larger scale.

2.4 Characterisation of Materials

N_2 physisorption at 77 K was measured via a Micromeritics Tristar 3020 gas sorption analyser, and for some samples, via a Coulter Omnisorp 360 CX gas sorption analyser. The mesopore volume of products was estimated from the amount of N_2 remaining adsorbed at p/p0 ~0.90, from desorption isotherms. MCFa, MCFb, SBAa, SBAb, SBAc and D60 exhibited volumes of 2.83, 2.06, 1.04, 1.12, 0.93 and 0.80 ml/g respectively.

He densities were measured by a Micromeritics Accupyc 1340 pycnometer.

The CO_2 sorption capacity of composite powder products was measured via a Mettler Toledo AB204s balance, by weighing them both before and after first purging them with CO_2 for 16 h at 60 °C, and then cooling them to room temperature within the CO_2 purge.

Thermogravimetric analysis of the CO_2 partial pressure swing adsorption processing of select sorbents was conducted via a custom modified [1] Setaram TAG24-16 Simultaneous Symmetrical Thermoanalyser.

3 RESULTS AND DISCUSSION

Table 1 lists the nominal amine content (wt%) and the helium density of each of the activated composite powder products. The nominal amine contents are consistent with the target loadings. The helium densities show general agreement with the nominal amine contents, assuming the density of each amine to be ~1.0 g/ml and the helium density of each silica to be ~2.2 g/ml, and so likewise support the preparation of the products as intended.

Additionally, Table 1 shows the amount of CO_2 adsorbed by the activated powder products following their prolonged exposure to a CO_2 purge at 60 °C. These results show that each of the products exhibited substantial CO_2 capacity under these conditions, excepting the PAA products, highlighting their potential for CO_2 capture applications. PAA type sorbents are knwon to more readily adsorb CO_2 at higher process temperatures [14].

Observations made for the shaping of these powders as compressed following activation alone, and as processed in accord with the patent Process A, via the manual hydraulic press, are also shown in Table 1. These results show that shaping via compression following activation alone of each of the PEI composite powders prepared via the standard wet infiltration process, with different PEI loadings and different silica supports, was in each case found to be complicated by the production of a sticky or greasy and/or weak pellet, and/or product extrusion. However, shaping of the same powder products via Process A was found to lead to dry robust pellets in each case, thus demonstrating the broader relevance of this patented shaping technology.

Similarly each of the other composite powders prepared via the standard wet infiltration process incorporating the different amines, excepting that for the SBAb-TAEAH100, were similarly complicated when shaped via compression as activated alone. Nevertheless, each of these sorbents was readily prepared into a dry robust pellet when prepared in accord with the patented Process A. Thus these results further highlight the broader relevance of this technology.

Table 1 also shows the observations made for compression of the powder products prepared via the modified infiltration Process B. These results show that each of these products formed dry robust pellets when shaped in accord with Process A, and that compression via activation alone of the MCF-PEI sorbents was complicated by product stickiness and extrusion. Nevertheless, these results show that the SBAc-PEI100-B product formed a dry robust pellet via compression following activation alone, demonstrating the potential of this Process B technology.

Table 1 also lists the helium density measure recorded for each of the products shaped via Process A. These pellet products are thus found to exhibit essentially the same helium density as their powder analogs, indicative of their like composition. Thus it is concluded that the shaping Process A does not lead to any substantial net change in the chemical composition of the sorbent.

	AC	HD powd	CO_2 powd	Char'n of shaping (*refer to key)		HD pellet
Product Id.	wt%	g/ml	wt%	Act'd	Proc A	g/ml
MCFa-PEI60	64.5	1.26	16.7	SRE	DRN	1.27
MCFa-PEI70	68.0	1.23	16.9	SRE	DRN	1.25
MCFa-PEI75	69.5	1.23	14.3	SRE	DRN	1.23
MCFa-PEI80	71.2	1.30	12.5	SWE	DRN	1.24
MCFa-PEI85	72.4	1.28	14.3	SRE	DRN	1.23
MCFa-PEI90	73.9	1.28	14.0	SRE	DRN	1.21
MCFb-PEI100	68.8	1.23	16.1	SRE	DRN	1.23
MCFb-DAB100	88.8	1.22	17.6	DRN	DRN	1.26
MCFb-PAA100	67.6	1.29	2.2	GRE	DRN	1.30
MCFb-TEPA100	67.2	1.19	23.2	GWN	DRN	1.38
SBAb-TAEAH100	58.1	1.37	12.0	DRN	DRN	1.42
SBAa-PEI100	48.9	1.50	-	SRN	DRN	1.49
D60-PEI100	46.4	1.42	6.85	DWN	DRN	1.42
MCFa-PEI80b	70.9	1.21	13.8	SRE	DRN	1.24
MCFb-PEI100-B1	68.8	1.25	16.4	SRE	DRN	1.24
MCFb-PEI100-B2	68.9	1.24	15.3	SRE	DRN	1.24
SBAc-PEI100-B	51.9	1.40	-	DRN	DRN	1.41

Table 1: Basic product and shaping characterisation: nominal amine content (AC), helium density (HD) and CO_2 capacity of powder products, observations for shaping as Activated (Act'd) alone or via Process A (*Dry, Sticky or Greasy - Robust or Weak – with notable product Extrusion or Not), & the helium density (HD) of pellets (Process A).

Figure 1 shows the gravimetric record obtained for the CO_2 partial pressure swing processing of MCFa-PEI60, MCFa-PEI70 and MCFa-PEI75 pellets as were shaped in accord with Process A via the manual hydraulic press. These records show these products to exhibit substantial CO_2 sorption capacities, as appear useful for the PCC of CO_2 from typical flue gases via pressure swing adsorption (PSA) and/or thermal swing adsorption (TSA) processing, and that the capacities of these products increase with their PEI loading, up to ~5 wt % (1.1 $mmol.g^{-1}$) and ~9 wt % (2.0 $mmol.g^{-1}$) for PSA and TSA processing respectively. However, it is noted that this MCFa-PEI75 pellet was found to exhibit slightly larger PSA capacity to analogous products prepared with higher PEI loadings [8],[9], despite slightly less TSA working capacity, suggesting a limit to the benefit of such higher loading.

MCFa-PEI80b powder was also readily shaped into dry robust pellets in accord with Process A via the SPTP as separately used in both manual and automated modes. Figure 2 shows the gravimetric records obtained for the isothermal (90, 105 and 115 °C) CO_2 partial pressure swing adsorption processing (within a humidified (~15 % H_2O) gas stream) of such a shaped MCFa-PEI80b product, including its initial 60 minutes pre-exposure to humidified (15 % H_2O) Ar, following initial activation in dry Ar. These records show that this product adsorbed negligible water under these conditions, however sorbed substantial CO_2 in response to CO_2 pressure. The CO_2 sorption capacity of the sorbent appears to decrease with temperature, however sorption kinetics improved with

TechConnect Briefs 2015, TechConnect.org, ISBN 978-1-4987-4728-8

temperature such that it would seem that this sorbent is best suited to facilitate CO_2 capture via PSA within the range 105 – 115 °C. The sorbent exhibits similar PSA and TSA capacities to the MCFa-PEI75 pellet product shown in Figure 1. Thus these results demonstrate the potential of the patent technology to facilitate a more realistic path for the shaping of such sorbents, and the promising potential of these products for the PCC of CO_2 application in particular.

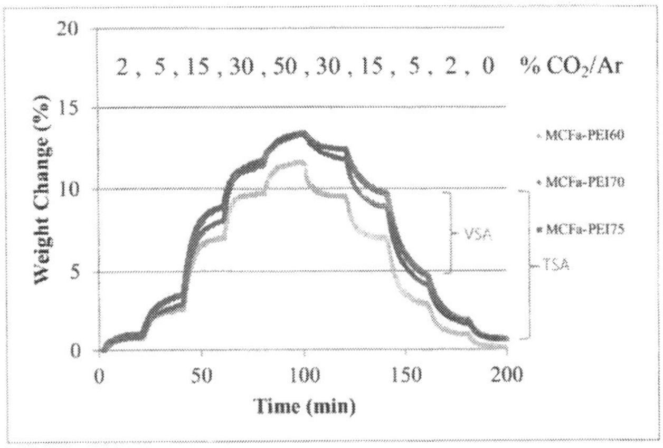

Figure 1: Gravimetric records for anhydrous isothermal (105 °C) PPSA processing of products shaped in accord with Process A via the Specac press.

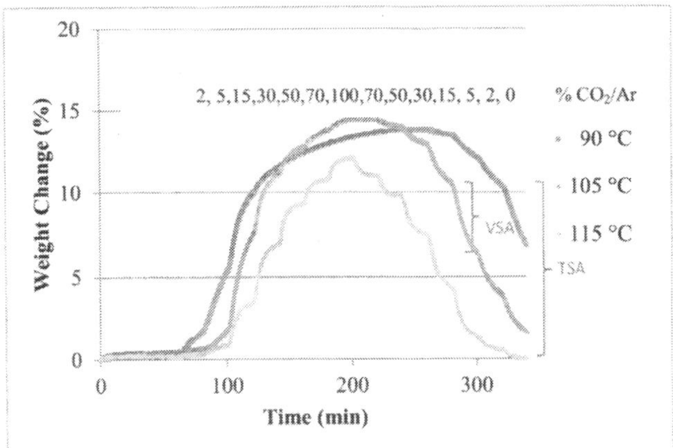

Figure 2: Gravimetric records for isothermal PPSA processing of MCFa-PEI80b (as shaped via Process A with the SPTP (manual mode)) within humidified (50 °C) CO_2/Ar atmospheres.

4 CONCLUSION

The patented processes A & B were found useful to facilitate the shaping of a broad range of powdered mesoporous silica–amine composite adsorbents into robust pellet forms such as were found to be of promising potential to facilitate CO_2 capture. Further improvements are envisaged for optimisation of traditional process parameters (e.g. compaction pressure and temperature).

Acknowledgements

The authors acknowledge funding from the Australian Government through its Cooperative Research Centres funding program. The authors also acknowledge Zhijian Liang, Jinyi Zhuang and James Bahn for the synthesis of some of the powder products.

REFERENCES

[1] Knowles, G.P., Webley, P.A., Liang, Z., Chaffee, A.L., in *Recent Advances in Post-Combustion CO_2 Capture Chemistry,* 2012, (Ed. M. Attala), Am. Chem. Soc. 1097 (Ch 9), 177-205, 2012.

[2] Davidson, R., *Post-combustion carbon capture-solid sorbents and membranes*, IEA Clean Coal Centre Reports, 2009.

[3] Cavalcante, C.L., Jr., Lat. Am. Appl. Res., **30,** 357-364, 2000.

[4] Mohanty, D. and D.J. Matty, *Crosslinked polymer-carbon sorbent for removal of heavy metals, toxic materials and carbon dioxide,* Patent WO2011103529A1, Central Michigan Uni, 2011.

[5] Xu, X., Song, C., Andresen, J.M., Miller, B.G., Scaroni, A.W., Energy & Fuels, **16**(6), 1463-1469, 2002.

[6] Subagyono, D.J.N., Liang, Z., Knowles, G.P., Chaffee, A.L., Chem. Eng. Res. Des., **89,** 1647-1657, 2011.

[7] Subagyono, D.J.N., Marshall, M., Knowles, G.P., Chaffee, A.L., Micro. Meso. Mater., **186**(0), 84-93, 2014.

[8] Knowles, G.P. and A.L. Chaffee, *Pelletized form of MCF-PEI composite and method of producing the same.,* Patent PCT/AU2013/001364, Monash University, 2013.

[9] Knowles, G.P. and A.L. Chaffee. *Pelletized silica and polyethyleneimine composites for CO2 capture via adsorption.* in CTSI-Cleantech 2014. ISBN 978-1-4822 5819-6, 170-173, 2014.

[10] Knowles, G.P. and A.L. Chaffee, *Pelletized form of a composite material and method of producing same,* Patent PCT/AU2014/050376, Monash University, 2014.

[11] Han, Y., Lee, S.S., Ying, J.Y., Stud. Surf. Sci. Catal., **165**, 829-832, 2007.

[12] Schmidt-Winkel, P., Lukens, W.W.Jr., Yang, P., Margolese, D.I., Lettow, J.S., Ying, J.Y., Stucky, G.D., Chem. Mater., 12, 686-696, 2000.

[13] Zhao, D., Huo, Q., Feng, J., Chmelka, B.F., Stucky, G.D., J.Am. Chem. Soc, **120**(24), 6024-6036, 1998.

[14] Knowles, G.P., Zhuang, J., Stenz, C., Chaffee, A.L., in Prepr. Pap. *-Am. Chem. Soc., Div. Energy & Fuels, 58(2), 372-373,* 2013.

Hollow Granules of Nitrogen/Boron co-doped Carbon for CO_2 Capture

Jinhong Kim, Jisun Han, Daegwon Ha and Shinhoo Kang*

*Department of Materials Science and Engineering, Seoul National University, Seoul, 151-742, Korea, shinkang@snu.ac.kr

ABSTRACT

Carbide-derived carbon (CDC) has been examined due to its useful conformal character. However, its variety of chemical composition is limited because functionalized atoms in its carbonaceous structure, especially with nitrogen, are unstable during high-temperature processing. In this paper, we report the enhanced nitrogen stability and an increase in surface area by boron addition. We also synthesized a hollow granule structure of nitrogen/boron co-doped porous carbon (CNB) in high yield by using a nitrogen-rich precursor without any additive template. High CO_2 capture (3.7 mmol/g) was measured at ambient temperature and 1bar with an excellent CO_2/N_2 selectivity.

Keywords: Porous carbon, CDC (carbide derived carbon), Boron Nitride, CO_2 capture, Boron, Nitrogen Doping.

1 INTRODUCTION

For negligible reactivity of pure carbon, heterogeneous atoms, such as nitrogen and boron, has been substituted in carbon structure. Nitrogen-doped porous carbon (CN) materials have maintained extensive research interest. The most popular and effective synthesis method employs templates with suitable nitrogen precursors, e.g., melamine, acetonitrile, and diaminobenzene. Boron-doped porous carbon (CB) has rarely been reported on, because the bonding between carbon and boron is unstable under moderate processing conditions.

In recent years, nitrogen-and-boron-co-doped porous carbon (CNB) has received much attention, with examples being verified as catalysts, supercapacitors, and gas sorbents. However, these materials usually consist of two separate phases, a C–N- and a B–N-rich phase owing to the C–B instability. Given the difficulty of obtaining a homogenous CNB framework, an effective way of the CNB synthesis is highly sought after.

In this paper, we discuss our breakthrough synthesis of boron-aided nitrogen-doped porous carbon to create CNBs with a robust matrix. The CNB synthesis was by a chlorination method, which is similar to the carbide-derived carbon (CDC) process [1], easily produced well-developed pores without requiring any templates. Small amounts of boron were found to act as a stabilizer for the nitrogen-doped carbon structure at a relatively high temperature of 800 °C. We denoted samples as CN-XY and CNB-XY according to whether a boron source was added; XY represents the carbon-to-nitrogen ratio of the $Ti(C_xN_y)$ precursors ($y = 1 - x$).

Furthermore, a hollow spherical CNB structure, which is generally only prepared by the hard template method[2], was achieved by using a nitrogen-rich precursor. According to our experiment, despite its moderate surface area, hollow CNB37 (synthesized from the $Ti(C_{0.3}N_{0.7})$ precursor) has a high capacity for CO_2 capture. The CO_2/N_2 selectivity, one of the key factors for post-combustion CO_2 capture, is improved significantly. This result might contribute to aid carbon capture and storage technology, and thus might be very useful, considering that there is an urgent requirement for greenhouse gas reduction.

2 RESULTS AND DISCUSSION

Figure 1 shows the N_2 sorption of CNs and CNBs at 77 K. With the exception of non-porous CN37, all samples have type I and IV pores. Boron addition causes a sharp increase in the adsorption at $P/P_0 = 0.90 - 0.99$ with an H2-type hysteresis loop, especially when precursors with high nitrogen contents are used. The results indicate that the size and shape of the pores are not fixed and that the pore size spreads from the micro- to the macro-pore regions. No significant textural differences exist between CN73 and CNB73 except for a small decrease in mesopore volume, which seems to be related to the chemical stability of the precursor.

The results are different for CN55 and CNB55. Given that the nitrogen functional groups are not sufficiently stable during high-temperature chlorination, the CN55 structure collapses partially because of a deficiency in neighboring carbons caused by the formation of CNCl and C_xN_y gas. Thus, the BET surface area decreases as pores are encroached in CN55. Pore retention is observed in CNB55 and becomes very obvious when CN37 and CNB37 are compared. Since most of the CN37 structure is lost as gaseous CNCl and C_xN_y, the yield is extremely low (1.5%) and pore formation is also retarded. However, after boron addition, the yield is increased significantly to 18.3% with pore development. The BET surface area also increases significantly from 183 m^2/g to 730 m^2/g.

TEM images of CN37 and CNB37 are shown in Figure 2. The hollow spherical shape of CNB37 (Figure 2(b)) is very different from that of CN37 (Figure 2(a)). The sphere shell thickness is estimated to be approximately 7–8 nm. Two diffusive rings were obtained from the selected-area electron diffraction (SAED) pattern: (100)/(101) and

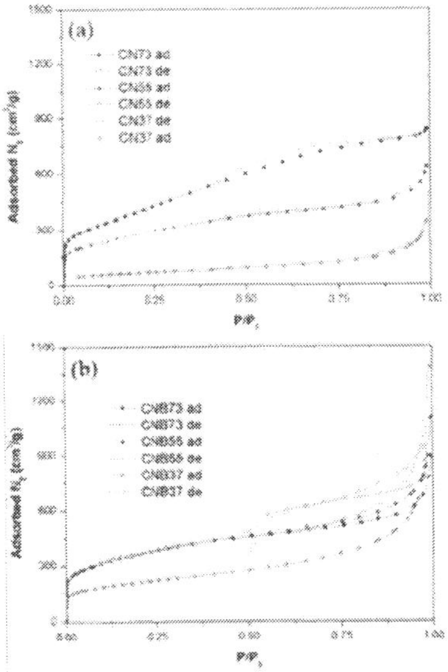

Figure 1 N_2 adsorption (ad, filled circles) and desorption (de, unfilled circles) curves of CNs (a) and CNBs (b) at 77K.

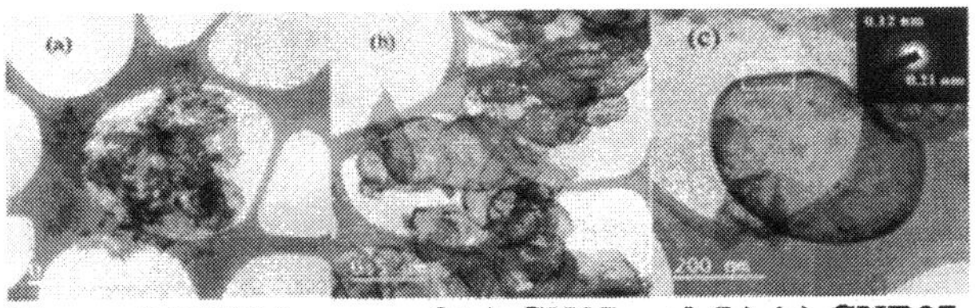

Figure 2 TEM images of (a) CN37 and (b),(c) CNB37. (d) High-resolution TEM image of CNB37. Inset in (c) shows SAED pattern at the shell area marked in (c) by the rectangular dashed line.

(110)/(112) patterns with 0.21 nm and 0.12 nm d-spacing, respectively. (Figure 2(c)). This character is also supported by two broad peaks in the X-ray diffraction spectrum. Hollow CN or CNB structures have rarely been reported. This is because C–N bonds less stable than C–C or B–N bonds, and tend to decompose easily, resulting in the collapse of the structure during synthesis.

The CNB hollow surface is stabilized not only by the direct interaction of boron with the surface layer but also by BCl_3 that is formed from the reaction between TiB_2 and Cl_2. BCl_3 is a strong Lewis acid, thus Lewis acid–base (nitrogen functional group) interaction induces a relatively lower exterior surface energy than that of the inner surface. As a result, the shell is formed and densification follows as the loosely-connected CNB shell gains part of the C–N moieties from the interior. Thus, we conclude that the presence of boron in CN stabilizes the C–N bonding and facilitates the formation of a robust CNB hollow structure.

Each peak is deconvoluted into several functional groups as N_{pyr} (398.2 eV), N_q (400.8 eV), and N–O (402.8 eV).

To determine their chemical composition and bonding nature, CNs and CNBs were analyzed by X-ray photoelectron spectroscopy (XPS) and the result is shown in Figure 3. According to B1s spectra, C–B–N and B–N

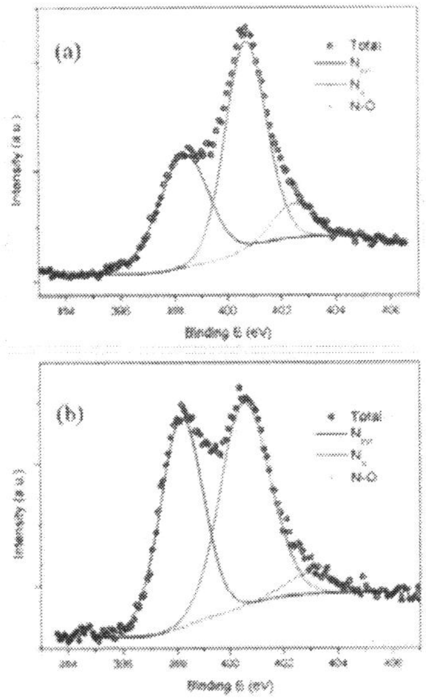

Figure 3 XPS N 1s spectra of (a) CN37 and (b) CNB37.

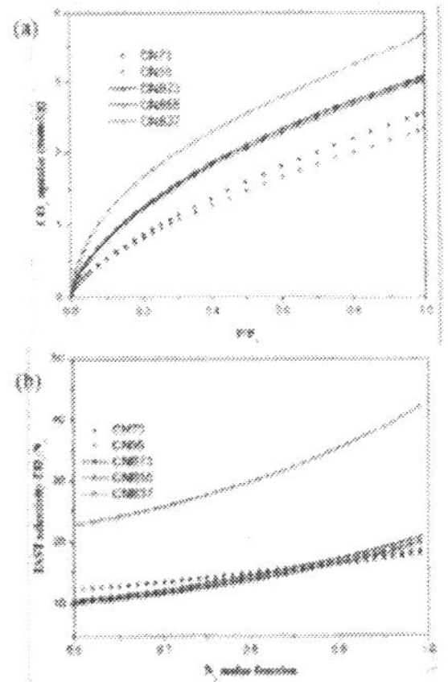

Figure 4 (a) CO_2 adsorption isotherms of CNs and CNBs measured at 298 K and (b) their CO_2/N_2 selectivity versus N_2 molar fraction. Evaluation at 1 bar total pressure by the IAST method.

bonding (190.9 eV) are the main sources of boron interaction. However, a small BC_3 contribution also appears as a sub-peak (189 eV), and this peak increases as the nitrogen content in the precursors increases. Even though BC_3 bonds are formed, they are unstable and break easily. However, once boron is bound to carbon, it stabilizes the reaction, and strengthens the C–N bonds. Given that its low boron content, the high nitrogen content of CNB37 proves that this nitrogen increment is not ascribed to that from B–N bonding. It seems that the role of BC_3 is more vigorous in strengthening the C–N bond under nitrogen-rich conditions.

CO_2 adsorption by the nitrogenous porous carbons were studied to evaluate their potential for carbon capture (Figure 4(a)). Every CNB provides a better uptake than the corresponding CN regardless of surface area. A noticeable uptake of 3.7 mmol/g is achieved by CNB37. The CO_2/N_2 selectivity increases significantly for CNB37 (Figure 4(b)).

An outstanding selectivity of 35.2 was observed at a CO_2 partial pressure of 0.1 bar and a total pressure of 1 bar. This selectivity is better than the best reported for a CN material.

N_2 molar fraction. Evaluation at 1 bar total pressure by the IAST method.

This selective sorption by a non-ordered structure is usually explained by the presence of nitrogen atoms on the pore surface, which can increase the surface–adsorbate interaction [3]. The uniqueness of the hollow CNB37 in terms of its short pore features and hierarchical structure could be the reason for its enhanced ability to capture large quantities of CO_2 and its improved selectivity.

3 CONCLUSION

In conclusion, nitrogen-containing porous carbons with a large surface area were prepared by the addition of boron during synthesis at high temperature. The addition of a small amount of boron facilitates the introduction of nitrogen into the carbon structure. The CNBs obtained from the chlorination process show a higher thermal stability and productivity (yield) than the corresponding CNs. Among the CNBs, CNB37 has a hollow spherical structure. The mechanism can be explained as the direct doping of boron at the surface of the precursor particles along with a loss of interior CN and the effect of BCl_3 in a Lewis acid–base interaction. It is worth noting that a morphological modification of CN or CNB would be possible, depending on the shape control of the nitrogen-containing precursor without any post-treatment. Although CNB37 possesses the smallest area of the fabricated structures, it shows the highest CO_2 uptake and CO_2/N_2 selectivity at ambient temperature. Its hollow morphology appears to enhance the sensitivity of the interaction. Additional work for further improvement and the development of various applications, based on a self-templating hollow CNB, are now in progress and will be discussed in our future publications.

*This is a short summary of our previous paper published in J. of Mat. Chem. A, 2, (2014) 166645-51.

REFERENCES

[1] Presser, V.; Heon, M.; Gogotsi, Y. Carbide-Derived Carbons - From Porous Networks to Nanotubes and Graphene. Adv. Funct. Mater. 21, 810-833, 2011.

[2] Feng, S. S.; Li, W.; Shi, Q.; Li, Y. H.; Chen, J. C.; Ling, Y.; Asiri, A. M.; Zhao, D. Y. Synthesis of Nitrogen-Doped Hollow Carbon Nanospheres for CO_2 Capture. Chem. Commun. 50, 329-331, 2014.

[3] Li, J. R.; Kuppler, R. J.; Zhou, H. C. Selective Gas Adsorption and Separation in Metal-Organic Frameworks. Chem. Soc. Rev. 38, 1477-1504, 2009.

H_2 Purification with PVDF/GTA Blended Membranes

Hesamoddin Rabiee, Mohammad Soltanieh*, Seyyed Abbas Mousavi, Ali Ghadimi

Department of Chemical and Petroleum Engineering, Sharif University of Technology, Azadi Avenue, P.O. Box 11155 9465, Tehran, Iran
msoltanieh@sharif.edu

ABSTRACT

In this research blended membranes with poly(vinylidene fluoride) (PVDF) and Glycerol Triacetate (GTA) were fabricated for CO_2/H_2 separation.. GTA has been added to polymeric solution in different mass ratios to study its effect on membrane permeability. Membranes cross-sectional morphology was investigated by SEM and remarkable changes in membrane structure were observed. Also, FTIR spectra were used to investigate functional groups of the membranes. Permeation results showed that CO_2 permeation almost doubled, whereas the H_2 permeation remained almost constant and thus CO_2/H_2 selectivity showed acceptable enhancement.

Key Words: H_2 Purification, Blended Membranes, Glycerol Triacetate (GTA), PVDF

INTRODUCTION

Membrane technology for gas separation purposes is widely on the increase due to low cost, interesting efficiency and acceptable processability [1]. CO_2 capture and H_2 purification are the most important fields that can use membranes for separation [2, 3]. In the past few decades many efforts have been done to in this field with various types of membranes, however, polymeric materials are quite well known to be interesting for this application [4]. Several types of modifications have been taken into consideration in order to improve membranes' separation ability and the most common methods are: incorporation of inorganic particles in membrane matrix (nanocomposite and mixed matrix membrane:MMM) [5, 6] and fabrication of blended membranes (which are mixing of polymers with low ,molecular weight additives) [7-9].
Selection of suitable inorganic particles and additives are of great importance and pre-determined criteria based on the desired application and materials properties should be applied [10]. Glycerol Triacetate (GTA) is a CO_2-phil compound and has an interesting separation factor as a liquid (in particular for CO_2/H_2 separation) [11] and its ability in membrane processes has been considered as well [7, 9, 12]. Furthermore, poly(vinylidene fluoride) (PVDF) which is an inexpensive membrane with good chemical and mechanical resistance, is an excellent choice for host polymer.
In this study, PVDF and GTA have been selected as the main blending materials in order to fabricate dense blended membranes with high separation ability for CO_2/H_2 separation. GTA as a plasticizer possesses very low vapor pressure and its ability to separate CO_2 from H_2 as a solvent is high (with CO_2/H_2 solubility selectivity of around 100) [7, 11]. It should be mentioned that polymers used commercially in gas separation applications (such as separating air into oxygen-enriched and nitrogen-enriched streams) are usually made of glassy materials [13, 14].

EXPERIMENTAL

PVDF was supplied from Kynar. 1-methyl 2-pyrrolidone (NMP) which was used as the solvent was purchased from Merck. Glycerol triacetate (GTA) was purchased from Sigma-Aldrich.

The membranes were fabricated by simple blending method and after PVDF is completely dissolved in NMP, GTA at different mass ratios to polymer (PVDF:GTA = 3:1, 2:1, 1:1, 1:1.5, 1:2) was added to solution. NMP was a good solvent for both PVDF and GTA. The procedure of membrane preparation is solvent evaporation. The homogenous solution was then casted on perti dish and it was put in oven at 70 °C for 12 hr for solvent evaporation. The fabricated films were subsequently put in vacuum oven for 6 hr in order to eliminate residual solvent. The membranes were put in desiccator before characterization and performance tests.

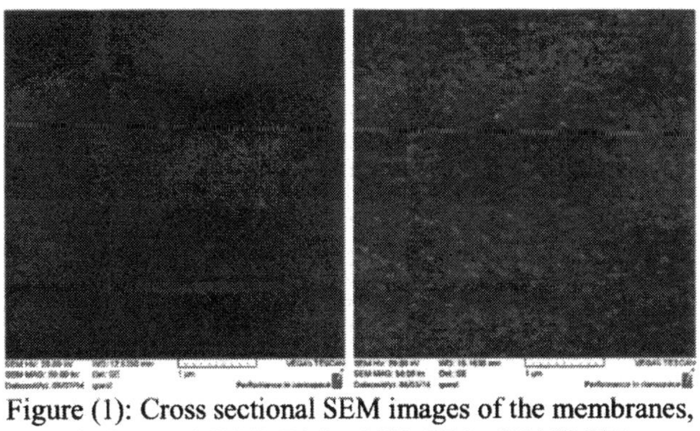

Figure (1): Cross sectional SEM images of the membranes, Left: neat PVDF, Right: 50% GTA, 50% PVDF

RESULTS AND DISCUSSION

The membranes were characterized by SEM to investigate the effect of GTA on the morphology. As it can be seen in Figure (1), remarkable changes are resulted and it seems that GTA addition has led to heterogeneous and shapeless structure which can affect permeation properties considerably [7]. Similar observations have been experienced by other blended membranes. The reason is mainly due to plasticization effect of GTA and as it locates among polymer chains, it leads to more mobility and consequently lower homogeneity. It has been reported in many other works that changes in membrane morphology and structure can significantly affect transport properties, specially diffusion coefficients [9, 15].

Structural characterization of neat PVDF and PVDF /GTA blended membrane was investigated by FTIR-ATR analysis as shown in Figure (2). The main functional groups of GTA and PVDF have been shown in their spectra. As it can be seen, with GTA addition the peak related to O–C=O loses its sharpness, as this functional group does not exist in PVDF structure. Also peaks related to symmetric and asymmetric –C–O will appear in the blended membranes which are due to GTA addition [7]. From the FTIR spectra, it is evident that the main functional groups of GTA possess highly oxygenated and carbonyl groups (C=O). It is generally well-known that materials with these properties are excellent for CO_2 capture [16].

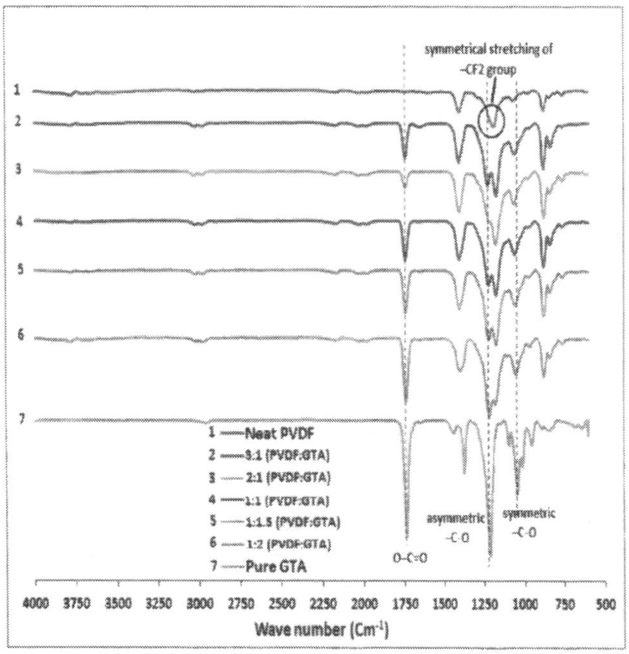

Figure (2): FTIR-ATR spectra of GTA, PVDF and blended membranes

The permeation performance of the prepared membranes was evaluated by constant volume method at 30°C and 4 bar. The membrane set-up and cell are completely described in our previous works [5, 7]. In order to measure permeability via constant volume method, the following equation has been used:

$$P = \frac{V}{A} \cdot \frac{273.15}{T} \cdot \frac{l}{p_s \Delta p} \cdot \frac{dp}{dt} \quad (1)$$

where, V is volume of permeate (cm^3), A is membrane area (cm^2), l is membrane thickness (cm), T is operating temperature (K) and p_s standard pressure (1 bar). Δp is pressure difference across the membrane (mmHg) and $\frac{dp}{dt}$ is pressure increment in permeate vessel with time (bar/sec). Eq.(1) gives permeability in Barrer unit where 1 Barrer = 10^{-10} cm^3(STP) cm cm^{-2} s^{-1} $cmHg^{-1}$.

The results are shown in Figure (3) from which it can be seen that the permeation of CO_2 and H_2 increases when GTA mass content rises. The reason for this observation is that GTA as a plasticizer possesses higher permeability in comparison with the neat PVDF membranes [12]. However, this increment is much higher for CO_2 than for H_2 and that is why CO_2/H_2 selectivity rises as GTA content increases. The main reason for this phenomenon is that GTA is highly CO_2-selective due to its functional groups and has oxygenated structure with carbonyl functional groups that are quite well known for CO_2 capture [16].

With increment in the amount of GTA in membrane structure, as the membrane morphology become more amorphous (it can be seen in SEM), molecules of gases diffuse easier, thereby for all the tested gases permeability increases.

Changes in permeation after addition of a low molecular weight additive can be discussed further by looking at the changes in solution and diffusion coefficients. Solubility of CO_2 in GTA is higher than that of H_2 and the CO_2/H_2 solubility selectivity is around 20 [11]. In addition, based on the results reported in the literature [8, 11], GTA shows more solubility coefficients for CO_2 compared to neat PVDF. Thereby, increment in CO_2 solubility with GTA is quite expected. Another coefficient which is also affected by GTA addition is the diffusion coefficient. The rate of gas molecules diffusion in membranes is directly related to their kinetic diameter and molecular size. Therefore, any changes in membranes morphology will change the ability of gases to diffuse in membrane matrix. Addition of GTA, as a plasticizer leads to mobility of polymer chains, as this additive can locate between them and gas molecules should find their way through these chains to diffuse [17]. Hence, GTA addition leads to higher values of diffusion of gases, but this growth in diffusion for CO_2 is relatively higher than that for H_2. Because the kinetic diameter of CO_2 is 3.8°A (this number for H_2 is 2.89°A), it is more dependent to changes in the space among polymer chains.

To sum up, increase in diffusion and solubility coefficients of CO_2 is more than those of H_2 and this explains more permeation increment for CO_2 with GTA addition, based on the following equation:

$$P = S.D \qquad (2)$$

where, D and S are diffusion and solubility coefficient, respectively.

Thus, it can be said that both solubility and diffusion selectivities of CO_2 over H_2, increase that lead to higher permeation selectivity. These discussions are based on the reported data about the pure components and obvious changes that can be characertized in the chemical and physical properties of the membrane. However, for understanding the exact contribution of diffusion and solubility and compare their effects, separate experiments ought to be carried out to measure their coefficients directly [7, 10, 18].

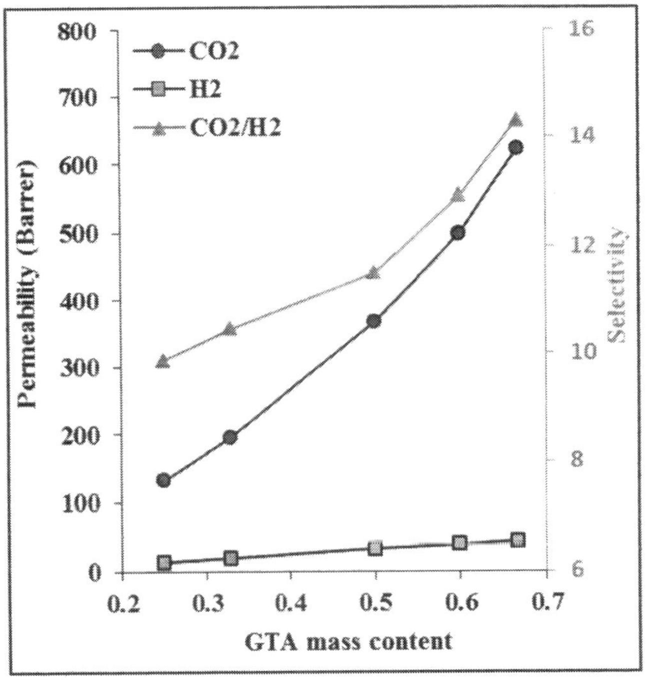

Figure (3): Permeation and selectivity through the membranes at different GTA concentrations

CONCLUSION

Membrane preparation with PVDF and GTA was studied in order to improve properties of PVDF membranes for CO_2/H_2 separation. The membranes were fabricated via commonly used solvent evaporation byusing industrially favorable solvent, NMP. The results showed enhancement in CO_2/H_2 separation by around 45% when GTA/PVDF ratio increases 0.33 to 2. The possible changes in transport properties of the membrane after GTA addition was discussed based on the published articles where much higher increment in CO_2 permeability was mentioned. This observation is quite expected due to separation ability of GTA. The membranes were also characterized by FITR and SEM to investigate changes in their molecular structure and morphology. The observed changes in membrane structure, as seen in SEM, can be considered as a way to inspect diffusion changes. The obtained enhancement in CO_2/H_2, seems significant.

REFERENCES

[1] R.W. Baker, B.T. Low, Gas Separation Membrane Materials: A Perspective, Macromolecules, 47 (2014) 6999-7013.

[2] B. Li, Y. Duan, D. Luebke, B. Morreale, Advances in CO_2 capture technology: A patent review, Applied Energy, 102 (2013) 1439-1447.

[3] K. Ramasubramanian, Y. Zhao, W.S. Winston Ho, CO_2 capture and H_2 purification: Prospects for CO_2-selective membrane processes, AIChE Journal, 59 (2013) 1033-1045.

[4] S. Alexander Stern, Polymers for gas separations: the next decade, Journal of Membrane Science, 94 (1994) 1-65.

[5] H. Rabiee, S. Meshkat Alsadat, M. Soltanieh, S.A. Mousavi, A. Ghadimi, Gas permeation and sorption properties of poly(amide-12-b-ethyleneoxide)(Pebax1074)/SAPO-34 mixed matrix membrane for CO_2/CH_4 and CO_2/N_2 separation, Journal of Industrial and Engineering Chemistry, (2015).

[6] Y. Shen, A.C. Lua, Preparation and characterization of mixed matrix membranes based on PVDF and three inorganic fillers (fumed nonporous silica, zeolite 4A and mesoporous MCM-41) for gas separation, Chemical Engineering Journal, 192 (2012) 201-210.

[7] H. Rabiee, M. Soltanieh, S.A. Mousavi, A. Ghadimi, Improvement in CO_2/H_2 separation by fabrication of poly(ether-b-amide6)/glycerol triacetate gel membranes, Journal of Membrane Science, 469 (2014) 43-58.

[8] H.Z. Chen, P. Li, T.-S. Chung, PVDF/ionic liquid polymer blends with superior separation performance for removing CO_2 from hydrogen and flue gas, International Journal of Hydrogen Energy, 37 (2012) 11796-11804.

[9] S. Feng, J. Ren, Z. Li, H. Li, K. Hua, X. Li, M. Deng, Poly(amide-12-b-ethylene oxide)/glycerol triacetate blend membranes for CO_2 separation, International Journal of Greenhouse Gas Control, 19 (2013) 41-48.

[10] H. Rabiee, A. Ghadimi, T. Mohammadi, Gas transport properties of reverse-selective poly(ether-b-amide6)/[Emim][BF4] gel membranes for CO_2/light gases separation, Journal of Membrane Science, 476 (2015) 286-302.

[11] R.C. Ernst, F.E. Miller, Separation of carbon dioxide from gaseous mixtures, in: US patent, Serial No. 762633, 1960.
[12] M.K. Barillas, R.M. Enick, M. O'Brien, R. Perry, D.R. Luebke, B.D. Morreale, The CO_2 permeability and mixed gas CO2/H2 selectivity of membranes composed of CO_2-philic polymers, Journal of Membrane Science, 372 (2011) 29-39.
[13] K. Ghosal, B.D. Freeman, Gas separation using polymer membranes: an overview, Polymers for Advanced Technologies, 5 (1994) 673-697.
[14] S. Matteucci, Y. Yampolskii, B.D. Freeman, I. Pinnau, Transport of Gases and Vapors in Glassy and Rubbery Polymers, in: Materials Science of Membranes for Gas and Vapor Separation, John Wiley & Sons, Ltd, 2006, pp. 1-47.
[15] A. Ghadimi, M. Amirilargani, T. Mohammadi, N. Kasiri, B. Sadatnia, Preparation of alloyed poly(ether block amide)/poly(ethylene glycol diacrylate) membranes for separation of CO_2/H_2 (syngas application), Journal of Membrane Science, 458 (2014) 14-26.
[16] H. Lin, E. Van Wagner, B.D. Freeman, L.G. Toy, R.P. Gupta, Plasticization-enhanced hydrogen purification using polymeric membranes, Science (New York, N.Y.), 311 (2006) 639-642.
[17] W. Yave, A. Car, K.-V. Peinemann, M.Q. Shaikh, K. Rätzke, F. Faupel, Gas permeability and free volume in poly(amide-b-ethylene oxide)/polyethylene glycol blend membranes, Journal of Membrane Science, 339 (2009) 177-183.
[18] P. Bernardo, J.C. Jansen, F. Bazzarelli, F. Tasselli, A. Fuoco, K. Friess, P. Izák, V. Jarmarová, M. Kačírková, G. Clarizia, Gas transport properties of Pebax®/room temperature ionic liquid gel membranes, Separation and Purification Technology, 97 (2012) 73-82.

Gas-Assisted Gravity Drainage - A New Process for CO_2 Sequestration and Enhanced Oil Recovery

B.D.Saikia and D.N.Rao

Louisiana State University,
Baton Rouge, LA, USA, www.pete.lsu.edu

ABSTRACT

The current commercial practice of gas-based enhanced oil recovery processes involves either continuous gas injection (CGI) or Water-Alternating-Gas (WAG) injection. Over 60 commercial projects in West Texas and other parts of the world have amply demonstrated that these CGI and WAG processes have been technically successful and commercially profitable. However, the oil recoveries from the CGI and WAG processes fall in the range of 5-15% of the remaining oil. The Gas-Assisted Gravity Drainage (GAGD) process, invented and patented at LSU, has yielded oil recoveries in the range of 65 – 95% in laboratory experiments conducted at realistic reservoir conditions. The GAGD process involves utilizing several vertical wells for injection of CO_2 in addition to drilling long horizontal wells for production. Injected CO_2 accumulates at the top of the payzone due to gravity segregation and displaces oil, which drains to the horizontal producer. This maximizes the volumetric sweep efficiency. The gravity segregation of CO_2 also helps in delaying, or even eliminating, CO_2 breakthrough to the producer as well as preventing the gas phase from competing for flow with oil. Thus GAGD enables sequestering injected CO_2 in addition to yielding much higher recoveries of trapped oil from depleted reservoirs.

Keywords: carbon dioxide sequestration, enhanced oil recovery

1 INTRODUCTION

CO_2 EOR is the fastest growing EOR process in the United States. The target for CO_2 EOR is over 400 billion barrels of oil[8] in the mature oil fields in the US alone. Water Alternating Gas (WAG) process has remained the default industry standard for CO_2 injection with almost 80-85% of commercial projects employing WAG. WAG attempts to overcome natural gravity segregation in Continuous Gas Injection (CGI) process by alternating gas injection with water and that has yielded better EOR performance in WAG floods than continuous gas injection (CGI) field projects. However, WAG is still a method to 'combat' the natural phenomenon of gravity segregation. In attempting to resolve one problem of adverse mobility, the WAG process gives rise to other problems associated with increased water saturation in the reservoir including diminished gas injectivity and increased competition to the flow of oil. The disappointing field performance of WAG floods with oil recoveries in the range of 10-15% is a clear indication of these limitations[2].

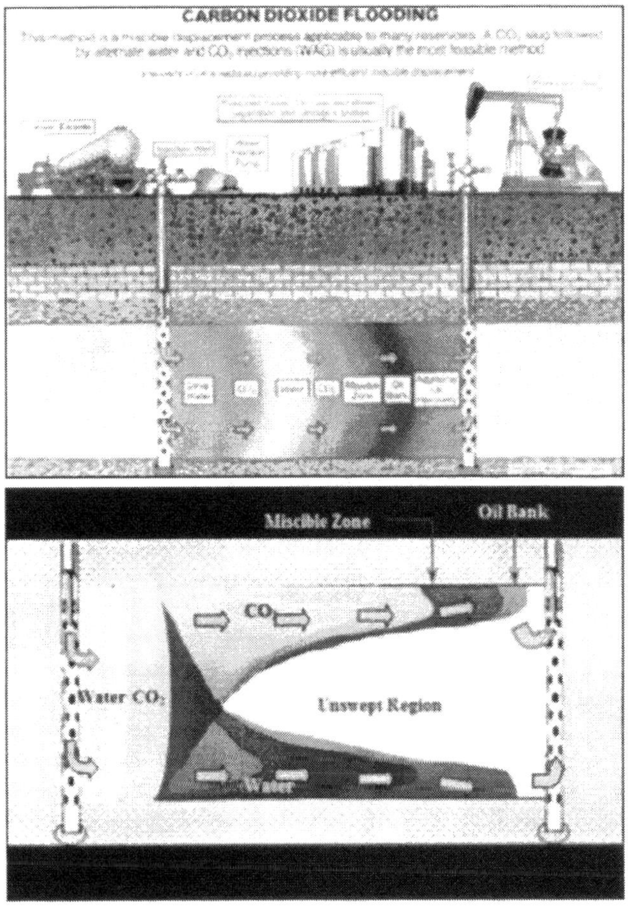

Figure 1: (top) - Conceptual view of WAG (DOE Website) (bottom) - A more realistic view of WAG showing override of injected gas (Ref: Rao et al.)

In order to find an effective alternative to WAG, Gas-Assisted Gravity Drainage (GAGD) process was developed in EOR labs at LSU. Unlike WAG, the GAGD process takes advantage of the natural segregation of injected gas from crude oil in the reservoir. Although gravity-stable gas floods have long been practiced in selected dipping reservoirs and pinnacle reefs, this project was aimed at a systematic development of a recovery process that would be widely applicable to different reservoir types in both secondary and tertiary modes. The GAGD process consists of placing a horizontal producer near the bottom of the payzone and injecting gas through new/existing vertical wells used in prior waterfloods. As the injected gas rises to the top to form a gas zone, oil and water drain down to the horizontal producer.

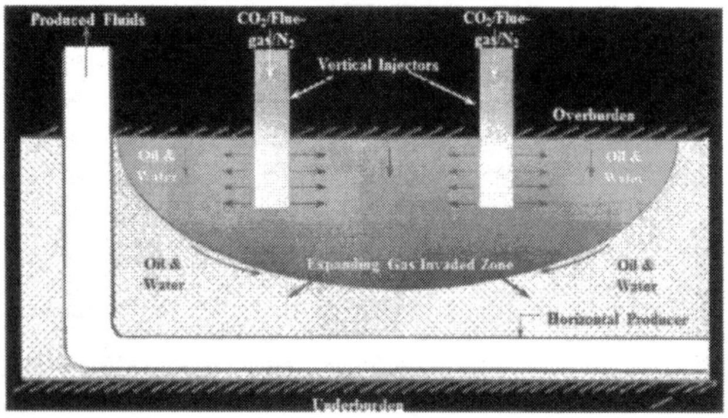

Figure 2: Conceptual view of GAGD process (Ref: Rao et al.)

2 VISUAL MODEL EXPERIMENTS OF GAGD PROCESS

A glass model was constructed and used to visualize the process and to discern the mechanisms operative in GAGD. The model experiments have shown that GAGD is a viable process for secondary and tertiary oil recovery. Oil recovery in the immiscible secondary mode was as high as 83% IOIP and the oil recovery in the immiscible tertiary mode was 54% ROIP. The visual model experiments have also demonstrated that GAGD is applicable to naturally fractured reservoirs. The presence of the vertical fracture in the physical model seemed to improve GAGD recoveries. Additionally, the GAGD process was found to be viable for higher viscosity oils as well, where secondary immiscible oil recovery was 64% IOIP[3]. These experiments have indicated three possible mechanisms responsible for high oil recoveries: Darcy-type displacement until gas breakthrough, gravity drainage after breakthrough, and film drainage in the gas invaded regions.

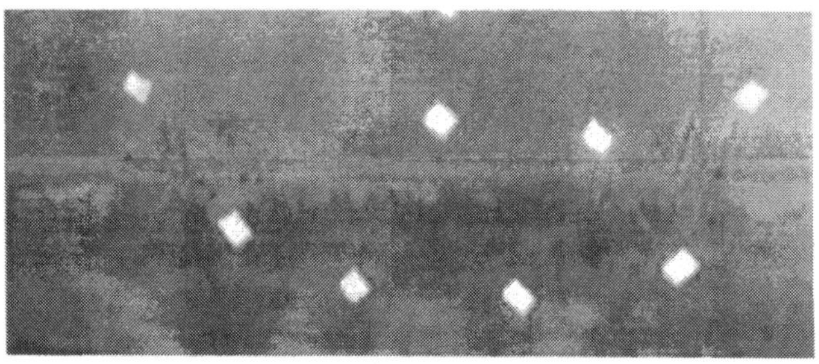

Figure 3: Development of near-horizontal flood front in visual glass GAGD model

3 PARTIALLY SCALED PHYSICAL MODEL EXPERIMENTS

This part of the research was focused on evaluating the effect of the wettability of the porous medium and the presence of a vertical fracture on GAGD performance. In the physical model experiments, a Hele Shaw type model was used along with glass beads/silica sand as the porous medium. Silanization with dimethydichlorosilane, an organosilane, was used to alter the wettability of the glass beads/silica sand from water-wet to oil-wet. The wettability alteration was confirmed visually as well as through the calculation of fractional flow curves. The effect of injection strategy on the oil recovery, i.e. secondary and tertiary mode production, was also evaluated in the experiments. The experiments showed a significant improvement of oil recovery in oil-wet experiments than in water-wet runs, both in the secondary as well as in the tertiary mode. When comparing the experiments conducted in the secondary mode to those run in the tertiary mode an improvement in the oil recovery, as high as 5 % OOIP, was also evident.

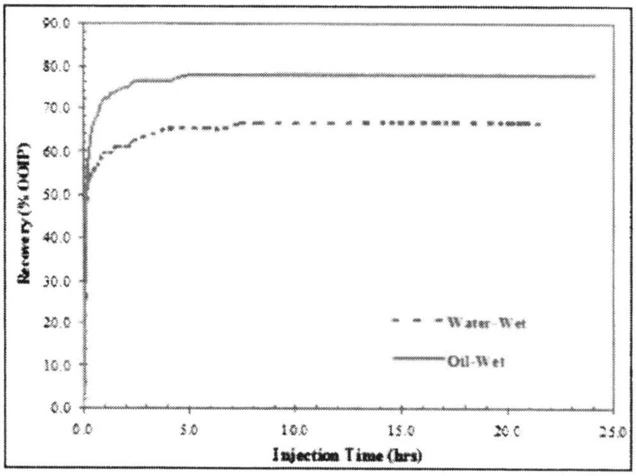

Figure 4: Effect of Wettability on GAGD recovery

Since, some of the prolific oil and gas reservoirs are carbonate reservoirs and most of the carbonate reservoirs are naturally fractured, the effect of vertical fractures on GAGD process was also investigated. The presence of vertical fractures improved oil recovery acting as conduits for oil/gas counter current flow between matrix and fracture. Fractured models yielded 8-10% higher recoveries than unfractured models. This was not the case with WAG process, wherein the presence of vertical fractures was detrimental to the process due to the adverse effect of fractures on gas-liquid segregation. The GAGD process thus provides the industry with a viable alternative to conventional EOR processes applicable in both water-wet and oil-wet reservoirs, regardless of the presence of fractures.

TechConnect Briefs 2015, TechConnect.org, ISBN 978-1-4987-4728-8

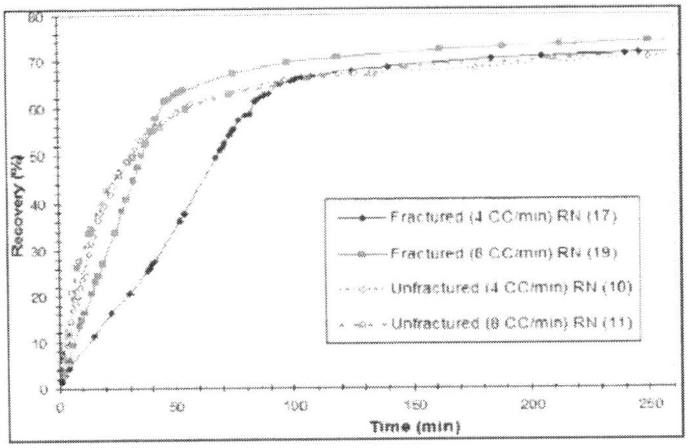

Figure 5: Effect of vertical fractures on GAGD recovery

4 COREFLOOD GAGD EXPERIMENTS AT RESERVOIR CONDITIONS

To allow for scalability of the laboratory experiments, the reproduction of the various multiphase mechanisms and fluid dynamics, which have been found to be influential in the success of the gravity stable gas floods, is crucial. Literature reviews of multiphase mechanics and fluid dynamics, suggests that dimensionless characterization of flood parameters to generate analogous field scale multiphase processes into the laboratory, is one of the most effective and preferred scaling tools. Therefore nine gravity stable gas injection field projects were used to calculate five dimensionless groups, which were then employed to optimize the laboratory experimental protocol for optimization of the GAGD process.

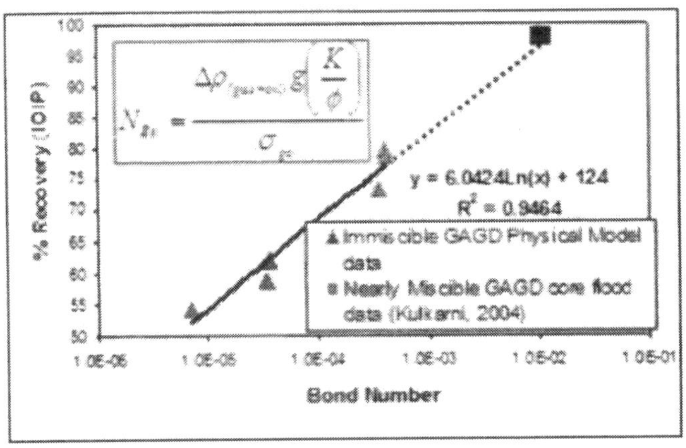

Figure 6: Effect of Bond No. on GAGD recovery[9]

Fifteen reservoir condition miscible and immiscible GAGD corefloods (with CO_2) were conducted to examine the effects of parameters such as injection strategy, rate, and fluid characteristics on GAGD performance. To provide with a common comparison basis, scaled Water-Alternating-Gas (WAG) coreflood experiments were also completed at similar conditions. Scaled experimental results clearly demonstrated that all the GAGD miscible floods, irrespective of the rock or fluid characteristics, such as presence of fractures, mode of injection, fluid systems, out-performed the WAG floods. The GAGD floods were able to substantially decrease the residual oil saturations and appear to be a useful tool for improved recovery from mature oil reservoirs. The presentation will include key results from miscible and immiscible GAGD and WAG corefloods as well as the dimensional analysis of the GAGD process for improved optimization.

5 APPLICATION OF GAGD PROCESS IN A LOUISIANA OIL FIELD

This part of the presentation highlights the results of reservoir characterization and compositional reservoir simulation studies carried out at LSU for GAGD process implementation in a Louisiana depleted oil reservoir. At first reservoir geological model was built and then three dimensional porosity and permeability distributions were generated from well log and core analysis data using geostatistics. Measured relative permeability data from the reservoir condition corefloods were used for obtaining a reasonable history match of prior production. Several strategies such as gravity segregation effects on reservoir fluids distribution with time and placing of horizontal producing wells just above the oil-water contact were adapted during the GAGD simulation. The placement and configuration of vertical injection wells as well as horizontal producing wells and the gas injection rates were optimized for maximum GAGD oil recovery. Successful reservoir flow simulation studies indicated an optimum GAGD incremental oil recovery of about 65% OOIP (≈90% ROIP), which agrees well with visual and scaled physical model experiments and reservoir condition core flow tests. Based on these results, a field test of GAGD process on a northern Louisiana field is ongoing.

6 CARBON CAPTURE, UTILIZATION AND STORAGE

Large quantities of waste gases containing carbon dioxide and nitrogen are continuously being released into the atmosphere by almost all the industrial activities involving the production or utilization of energy from fossil fuels. There is growing consensus among nations for the need to cap the emission of greenhouse gases like CO_2 to combat global warming. In the future, it is anticipated that there will be even stricter emission restrictions as the effects of climate change are felt by the global community. This has led to the push for Carbon Capture and Storage (CCS) and CO_2 sequestration. A study[6] by the Massachusetts Institute of Technology evaluated and prioritized research needs for the capture, use and storage of carbon dioxide from fossil-fuel-fired power plants. The idea behind CCS is, human-generated (anthropogenic) greenhouse gases are separated and captured at its sources such as power plants, refineries, etc. and injected into sub-surface formations such as abandoned oil and gas reservoirs, deep coal seams & aquifers etc. Though an interesting idea, the very cost of separating and capturing the greenhouse gases as well as uncertainty in the assessment of the long-term integrity of the sequestration sites has limited the widespread use of CCS. Even governmental subsidies and carbon credits are not able to tip the scale in favor of CCS. This is where CO_2 EOR promises to be the brightest beacon of hope. As stated earlier, CO_2 EOR is a proven technical and commercial success story. However, the current industrial practice of CO_2 EOR suffers from its dependence on the easy availability of CO_2 gas. Only a handful of companies own and monopolize natural sources of CO_2 and there is a need for pipeline infrastructure as well as cost involved in transporting the CO_2 from its natural sources to production fields. This is where there can be a profitable union between CO_2 emitter and CO_2 user companies. West-Ranch CO_2-EOR project[7] is a glowing example of such a union where NRG, J X Nippon and Hilcorp have joined hands. More such joint ventures will surely broaden the popularity of CO_2 EOR and CCS. At this point we must note that, such a win-win proposition is generated by the anticipated additional recoveries of only 5-15% OOIP. GAGD, with recoveries in excess of 65% OOIP, could be expected to provide scope for more such joint ventures as profit margins would greatly rise. There is a difference between CO_2 for enhanced oil recovery and CO_2 sequestration. While EOR processes seek to utilize the CO_2 efficiently, there is no requirement to ensure that the CO_2 remains permanently underground. In WAG processes, CO2 often breakthroughs to the producing well. Thus the WAG operation may not be quite CCS compliant. On the other hand in a GAGD process, the gravity segregation of CO2 helps in delaying, or even eliminating, CO2 breakthrough to the producer as well as preventing the gas phase from competing for flow with oil. Thus GAGD enables sequestering injected CO_2 much better than WAG in addition to yielding much higher recoveries of trapped oil from depleted reservoirs.

7 SUMMARY AND CONCLUSIONS

- GAGD has been successfully tested through stages of partially scaled visual models, reservoir condition corefloods and showed recoveries in the range of 65-95% OOIP.
- GAGD process works with nature and hence results are more likely to be duplicated in the field.
- This is further supported by a detailed compositional reservoir simulation of a planned field test, which yielded recovery of 65% OOIP (88% ROIP) on field scale.
- GAGD process is insensitive to reservoir heterogeneity, thus making several fractured reservoirs suitable for GAGD.
- GAGD process is amenable to reservoirs of all wettability states.
- More results are awaited with the ongoing field test in a Louisiana oil field.
- GAGD is more suited for better compliance with CCS objective as well as more likely to promote the confluence of CO2 emitter and CO2-EOR companies because of wider profit margins.

REFERENCES

[1] Rao, D.N., "Gas Injection EOR- A New Meaning in the New Millennium" Invited Article for the Distinguished Author Series, *Journal of Canadian Petroleum Technology*, Vol. 40, No.2, pp. 11-18, February 2001.

[2] Christensen, J. R., Stenby, E. H., Skauge, A., "Review of the WAG field experience", SPE 71203-PA, April, 2001, SPE Reservoir Evaluation & Engineering.

[3] Rao, D.N., Ayirala, S.C., Kulkarni, M.M., Paidin, W.R., Mahmoud, T.N.N., Sequeira, D.S., Sharma, A.P., "Development of Gas-Assisted Gravity Drainage Process for Improved Light Oil Recovery", Final Technical Report 15323R15 (October 1, 2002 – September 30, 2006) submitted to US-Department of Energy, Cooperative Agreement No. DE-FC26-02NT15323, December 2006.

[4] ARI report for US DOE, "Basin Oriented Strategies for CO_2 Enhanced Oil Recovery: Onshore Gulf Coast", February, 2006.

[5] Rao, D.N., "Gas-Assisted Gravity Drainage Process for Improved Oil Recovery", United States patent 8,215,392 B2, July 2012.

[6] Massachusetts Institute of Technology (Herzog, Drake and Adams), "CO_2 Capture, Reuse and Storage Technologies for Mitigating Global Climate Change", January, 1997.

[7] Pulsinelli, O. "NRG, JX Nippon, Hilcorp Break Ground on 'world's largest' Carbon Capture Project in Fort Bend." Houston Business Journal, 5 Sept. 2014.

[8] Kuuskraa, V.A., Godec, M. L. and Dipietro, P. "CO_2 Utilization from 'Next Generation' CO_2 Enhanced Oil Recovery Technology", Energy Procedia 37 (2013) 6854 – 6866.

[9] Kulkarni, M., (2005). "Multiphase Mechanisms and Fluid Dynamics in Gas Injection Enhanced Oil Recovery Processes", Louisiana State University, Baton Rouge, LA, USA.

Global Gas Flaring -The Energy Waste with Serious Environmental Impacts: Case Study of Iran

Mohammad Soltanieh*, Angineh Zohrabian**, Mohammad Javad Gholipour***

* Sharif University of Technology, Department of Chemical and Petroleum Engineering, Tehran 11155/9464, Iran, msoltanieh@sharif.edu.
** Sharif University of Technology, Tehran, Iran, msoltanieh@energy.sharif.ir.
*** National Iranian Oil Company, Tehran, Iran, gholipour@nioc.ir.

ABSTRACT

Gas flaring in oil production and oil and gas refining is a major environmental issue with great energy and economic loss. In this paper the global gas flaring and its adverse environmental impacts have been reviewed briefly. It is shown that emission factor, defined as the amount of gas flared per barrel of oil produced, varies widely among oil production sites. There is also great uncertainty in the composition of flare gases and the actual measurements are rather limited. Various available technological alternatives to reduce gas flaring are presented and compared. A special focus has been put in new technological options that still need to be developed. In addition, the best implemented projects in Iran as a major oil producing and the world's third largest gas flaring country are presented. Finally, techno-economical constraints for gas flaring reduction have been discussed for future planning.

Keywords: Gas flaring, emission, Environmental impact, Reduction technologies

1 INTRODUCTION

Since the inception of oil exploration, drilling and production in the world, more than one and a half century ago, gas flaring and venting has been practiced as a method of disposing the gases associated with petroleum, mostly for safety, operational and economic reasons. In 2011 the global gas flaring was around 140 billion cubic meters (Bcm) equivalent to 2.4 million barrels of oil equivalent per day with 10 billion USD lost revenue at $2.00/MMBtu, and 400 million tonnes of CO_2 per year emission [1]. In spite of huge amount of annually wasted energy resources and adverse environmental impacts of gas flaring, the awareness activities on gas flaring have been enhanced only in the past few decades partly due to the concerns on global climate change. In this manner, several financial initiatives and technological solutions have been dedicated to conserve the flared gases. This study remarks the scale of tragedy of continued gas flaring and its environmental impacts. In addition, it is aimed to assess the possible technological solutions for gas flaring reduction, by reviewing the best practiced cases of Iran, as one of the biggest oil producing and the third gas flaring country in the world.

2 ENVIRONMENTAL IMPACTS OF GAS FLARING

The light hydrocarbon gases- mostly methane - are generally accompanied by volatile organic compounds (VOCs) and several other impurities such as sulfur gases, some inorganic salts, carbon dioxide, nitrogen, polyaromatic hydrocarbons (PAHs), water, etc. The amount and composition of these gases vary significantly with time and from well to well depending upon the region of oil production and the pressure and temperature of the oil in the underground reservoir [2]. Detailed measurement of flared gas volume and composition, in most parts of the world is not applied. However, the volume of the flared gas can be estimated by using emission factors – the ratio of the amount of associated gas in cubic meters to the barrel of produced oil. The emission factor varies significantly, from around 1 to more than 50 cubic meters per barrel of oil for different countries, calculated based on data provided by NOAA satellite measurements for gas flaring and EIA data for oil production. The global 5-year average emission factor is around 4.6m^3/bbl; however, Iran's emission factor stands above the average gas flaring emission at about 7.2 cubic meters per barrel of oil (See Fig. 1).

Annually around 140 billion cubic meters gas is flared globally which is responsible for around 216 million tonnes of CO_2 in 2010 [3]. Low efficiency or poor combustion in the flare results in direct emission of methane which has 25 times higher global warming potential than CO_2 (in 100 year). Moreover, air pollutants such as sulfur compounds (e.g. hydrogen sulfide and sulfur oxides), nitrogen oxides, carbon monoxide, soot and carbon black particles, VOCs, PAHs and toxic heavy metals (e.g. mercury and nickel) and other inorganic salts (mainly as chlorate and sulfate of K, Mg, Na and Ca) are emitted to the atmosphere [4], causing severe local and regional ecological destruction and greenhouse effect. It is well known that the presence of NO_x and VOCs exacerbates the photochemical smog including ozone. In addition, gas flaring causes significant amount of local thermal and noise pollution near the flaring sites. The flared gas related pollutants under unfavorable atmospheric conditions (e.g. a temperature inversion) could origin serious diseases like asthma, blood disorder, cancer, or chronic bronchitis on human living in the surrounding areas [5,6]. One of the other impacts of gas flaring is that the combusted compounds form acid rain –especially in the humid environment of the offshore–with widespread damage to the environment, devastating the vegetation and surface water. Acid rain leaches nutrients from the soil, slows the growth of trees and makes lakes uninhabitable for fish and other wildlife [5, 7].

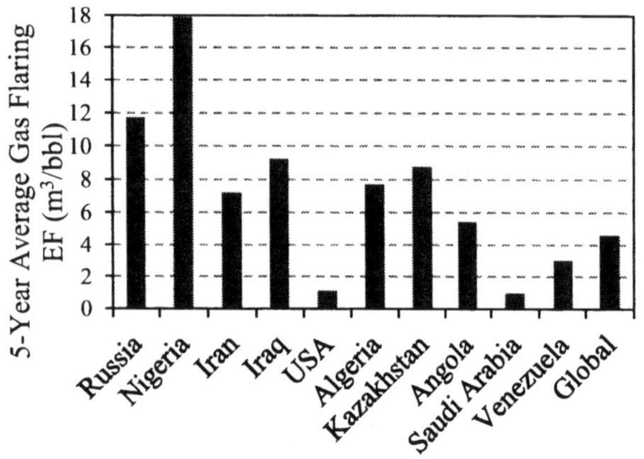

Figure 1: Five-year average gas flaring emission factor for global and the top 10 gas flaring countries in period of 2007–2011. (*Data from NOAA and EIA)*

3 GAS FLARING REDUCTION TACHNOLOGIES

3.1 Overview of Technologies

Depending on the region, proven technologies such as power generation, re-injection, gathering to add to central gas network, liquefied natural gas (LNG) and gas-to-liquids (GTL) will all have their place. These alternatives and practiced cases in Iran are summarized in Table 1.

Factors, including capital investment, technology risks, domestic market and its infrastructure, and political environment, companies' strategies, compete for decision making. A general comparison of all alternatives is difficult to make, therefore, some tips are suggested in literature.

- In offshore regions re-injection technology is often applicable in order to eliminate the need for gas transportation facilities. For offshore cases, pipeline becomes costly and challenging, with regard to the water depth and the transporting distance.
- In the case of relatively short distances to markets and low gas volumes, electricity generation or pipeline transport of the gas might be economical alternatives to flaring.
- In gas conversion to chemicals, i.e. GTL and LNG technologies, it is essential to pre-treat the gas for impurities such as sulfur and CO_2.Dong et al. [8] compared GTL and LNG and claimed that while a GTL facility is more complex, less efficient and more expensive than an LNG facility, their end-to-end supply chains are quite comparable and, thus, a decision to invest in either is challenging. Therefore, in the end, the decision to install either an LNG or GTL unit will be dependent on other factors such as local market needs, available resources, and companies' and governments' priorities, etc. [9].
- CNG is stored and transported in cylinders and can be implemented in small scale [10].

3.2 Required Technological Improvements

Although there are many available commercial technologies, it is still required to improve and develop some technologies which can be practiced in specific conditions and applications.

There are a number of advantages of CNG when compared to LNG, which have created the very strong interest in CNG that exists today. First there is no need for liquefaction or re-gasification; second the gas does not need to be cleaned to the same extent as it is necessary for LNG pre-processing [11]. CNG technology has the potential to become the preferred method of utilizing associated gas in offshore platforms where building pipeline or LNG plants are not economical and practical [12]. Since CNG is transportable, and therefore easily re-deployable, it can be used in fields with relatively short production horizons. Trans Ocean Gas is in the process of commercializing its CNG transport technology [10]. Floating LNG (FLNG) technology is also under development which is a combination of conventional LNG and floating deepwater offshore production technologies. The combined FLNG vessels will contain liquefaction facilities onboard, and can be moved to small and remote oil fields easily. Another immature associated gas utilization method is NGH or natural gas hydrate. It is crystallized form of natural gas, a chemically stable solid material in an ice state at -20°C. The advantage of this method compared to LNG is its higher stabilizing temperature. Mitsui and Mitsubishi, the BG Group, and Marathon Oil are leading the efforts to develop gas-to-solid technology [13]. GTL technology has involved high technical risks and some issues are still under study.

3.3 Best Practiced Cases

There are a number of implemented successful cases of utilization of flared gas in different countries such as Canada, Angola, Nigeria, Russia and Iran. Under the United Nations Framework Convention on Climate Change, gas flaring could be registered as a Clean Development Mechanism (CDM). The CDM program has only sparingly been used in associated gas flaring projects despite its great potential. However, since most gas flaring projects are large-scale, the total greenhouse gas reduction through these projects is considerable compared to many small-scale registered CDM projects.

Technology	Market maturity[1]	Scale	Practiced in Iran	Flared gas reduction
Re-injection	Commercial	Large and small	Darkhoin-since 2010	7 MMscmd
Gas to pipeline	Commercial	Large and small	AMAK-since 2005	6.8 MMscmd
CNG	Not commercial	Small	–	–
LNG	Commercial	Large scale	Kharg-expected 2016 Siri-since 2011[2]	18.1 MMscmd 2.9 MMscmd
GTL	Near to Commercial	Large and small	–	–
Gas to power	Commercial	Distributed and large	–	–
Gas to chemicals	Commercial	Large and small	–	–

1. Commerciality of a technology depends on many factors such as technology availability, marketing, alternatives and economical factors and may differ from case to case.
2. It has been partially implemented. Although the project is almost completed, 1.4–1.7 MMscmd is flared.

Table 1: Proven alternative technologies for gas flaring reduction and practiced cases in Iran

By the thirteen registered gas flaring reduction projects, overall 69,274 Mtonnes CO_2 emission has been cut down by the end of 2012 [14]. Iran's only registered CDM project is Soroosh & Nowrooz Early Gas Gathering and Utilization project a part of Kharg and Behregansar project. This project has been commissioned in 2010 with around 0.08 MMscmd gas transporting to Kharg Island which was increased to 0.4 MMscmd in 2011 [15]. Other commissioned projects in Iran are AMAK since 2010, Siri since 2012, Ghalenar and Nargesi since 2013.

3.4 Flare Gas Reduction Challenges

There are several technical constraints for flared gas recovery and utilization including the lack of infrastructure to collect, treat, transport and utilize the associated gases; the detachment of market from production sites; the volume and impurities of the gas and its fluctuation. Besides, the limited financial resources, lack of policy and regulatory frameworks, domestic fuel pricing and constraints to access the required technologies, in particular by developing countries like Iran, gas flaring continues at alarming rate.

4 CONCLUSION

Iran's and the world's gas flaring can be reduced significantly provided that the political, financial and technological constraints are removed. One of the most important factors for considerable gas flaring reduction is technology availability and diversity. Thereby, flared gas regardless of its volume, composition, offshore or onshore can be converted into energy or product, which may lead to several economic and sustainable environmental benefits.

REFERENCES

[1] M. F. Farina, "Flare Gas Reduction – Recent Global and Policy Considerations," GE Energy, Global Strategy and Planning, 2010.

[2] M. Johnson, L. Kostiuk and J. Spangelo, "A Characterization of Solution Gas Flaring in Alberta," *Journal of Air & Waste Management Association,* vol. 51, pp. 1167-1177, 2001.

[3] http://cdiac.esd.ornl.gov/by_new/bysubjec.html.

[4] D. Leahey, K. Preston and M. Strosher, "Theoretical and Observational Assessments of Flare Efficiencies," *Journal of the Air & Waste Management Association,* vol. 51, pp. 1610-1616, 2001.

[5] A. Hassan and R. Kouhy, "Gas flaring in Nigeria: Analysis of changes in its consequent carbon emission and reporting," *Accounting Forum,* vol. 37, p. 124– 134, 2013.

[6] S. Younessi Sinaki, A. H. Tarighaleslami and F. Jafarigol, "Study on External Costs of Flare Gases Using Asian Development Bank Method," *Chemical Engineering Transactions,* vol. 25, pp. 39-44, 2011.

[7] C. Nwankwo and D. Ogagarue, "Effects of gas flaring on surface and ground waters in Delta State Nigeria," *Journal of Geology and Mining Research,* vol. 3, pp. 131-136, 2011.

[8] L. Dong, S. Wei, S. Tan and H. Zhang, "GTL or LNG: which is the best way to monetize "stranded" natural gas?," *Petroleum Science,* vol. 5, pp. 388-394, 2008.

[9] K. Zhang and M. Pang, "The present and future of the world's LNG industry," *International Petroleum Economics,* vol. 13, pp. 55-59, 2005.

[10] B. Buzco-Guven, R. Harriss and D. Hertzmark, "Gas

flaring and venting: extent, impacts and remedies," James A. Baker III Institute for Public Policy of Rice University, 2010.

[11] P. Rynn, H. Patel and J. Gaughan, "ABS Development of a Guide for Compressed Natural Gas Carriers," in *International Society of Offshore and Polar Engineers (ISOPE)*, Seoul, 2005.

[12] C. Odumugbo, "Natural gas utilisation in Nigeria: Challenges and opportunities," *Journal of Natural Gas Science and Engineering,* vol. 2, pp. 310-316, 2010.

[13] J. Marcano and R. Cheung, "Monetizing standard natural gas," *Oil and Gas Financial Journal,* vol. 4, 2007.

[14] C. Gouvello, October 2011. [Online]. Available: http://www.worldbank.org/html/fpd/ggfrforum06/sucre/d gouvelo.ppt. [Accessed 24 May 2014].

[15] International Institute for Energy Studies (IIES), "Hydrocarbon Balance," Ministry of Petroleum, Islamic Republic of Iran, 2011.

TCAD Analysis of Silicon-Germanium (SiGe) based Back-Contact Back-Junction (BC-BJ) solar cell as an alternative for silicon based cells

R. Pandey* and R. Chaujar**

Microelectronics Research Lab, Department of Engineering Physics,
Delhi Technological University, Bawana Road, Delhi-110042, India
*Rahulpandey@dtu.ac.in
**Rishu.phy@dce.edu

ABSTRACT

In this work, an ultra-thin, Back-Contact Back-Junction (BC-BJ) Silicon-Germanium (SiGe) solar cell device structure called BC-BJ SiGe solar cell has been proposed, which shows enhanced photovoltaic properties, External Quantum Efficiency (EQE) > 76% in the spectrum range of 400-700 nm wavelength have been achieved. Results further, reveal that proposed device is less affected by carrier recombination, resulting in 13.5 % power conversion efficiency (PCE). Also we obtained the impact of material thickness on device parameters and we found that the 5μm thick SiGe cell is as efficient as 10μm thick silicon cell. The PCE of 5μm thick SiGe cell and 10μm thick silicon cell was 11.3% and 11.8%, respectively. It indicates that the proposed device significantly save the material cost over the silicon cell. All the simulations have been done using DEVEDIT and ATLAS device simulator.

Keywords: Efficiency, Recombination, Silicon-Germanim, Solar cell

1 INTRODUCTION

Crystalline silicon is the largest component of the solar photovoltaic (PV) cells. Over 80% installed solar PV in 2011 used silicon as their largest material. Silicon is widely used material for PV due to its reliability, abundance, and mature fabrication process. In Silicon PV technology, absorption coefficient of Silicon is small, to absorb enough of the solar cell spectrum. The thickness of conventional silicon cell is typically more than 100μm, in order to absorb the solar spectrum. Now since, thickness is large, in order to collect light generated carrier, minority carrier lifetime must be high enough and the recombination coefficient should be low. This requirement of large volume and high quality silicon significantly increases the module cost. In order to make photovoltaic system feasible at large scale, it should be cost effective, with module cost to be < \$0.5/W [1]. In this paper, we proposed SiGe as an alternative to silicon solar cell, due to its higher minority carrier lifetime and lower recombination coefficient. Reducing the thickness of PV materials used in solar cell production is a major goal of solar cell industry, due to its fruitful contribution to the overall cost of a PV system. However, thin silicon cells are not effective, due to severe Auger recombination and increased surface to volume ratio [2].

One promising high-efficiency solar cell concept is the back-contact back- junction solar cell [3, 4] having both, junction and the electrodes at back side of the device. Now, since the junction and contacts are at the back side, Front surface can be designed for optimum optical performance [5].

Here, we proposed novel 10μm thick crystalline SiGe back-contact back-junction solar cell, which overcomes the critical problems of thin devices: Auger and surface recombination etc. Further, we have obtained the impact of material thickness on photovoltaic parameters. In SiGe technology the electrical properties of silicon are modified with germanium. SiGe processing is simple because the physical and electrical properties of silicon and germanium are similar [6]. Also SiGe possesses higher mechanical strength, and suppresses the high intensity degradation of solar cell under illumination [7]. Optical absorption of SiGe can further be modified by varying Ge content [8, 9].

2 DEVICE STRUCTURE AND PHYSICAL MODELS

Silvaco ATLAS used to simulate the device optoelectronic characteristics. ATLAS is a device simulator capable of modeling the electrical, thermal, and optical characteristics of semiconductor devices.

In the simulation, we assume a two dimensional structure, BC-BJ solar cell as shown in Fig. 1. The $Si_{0.9}Ge_{0.1}$ was used during the simulation, very small fraction of germanium is used. Because, if we increase the germanium content the bandgap of SiGe decreases [10] The substrate was N-type, with doping density of 3×10^{15} cm^{-3}, whereas N+/P+ region has the doping density of 4×10^{20} cm^{-3}. The depth of P+ and N+ region was 3μm and 2.5μm, respectively, whereas the width was 40μm and 30μm, respectively. The Nitride (Si_3N_4) is used as anti-reflecting front surface passivation, and modeled as an optical layer with the thickness of 70 nm. The Oxide (SiO_2) is used as back surface passivation with thickness of 50nm.

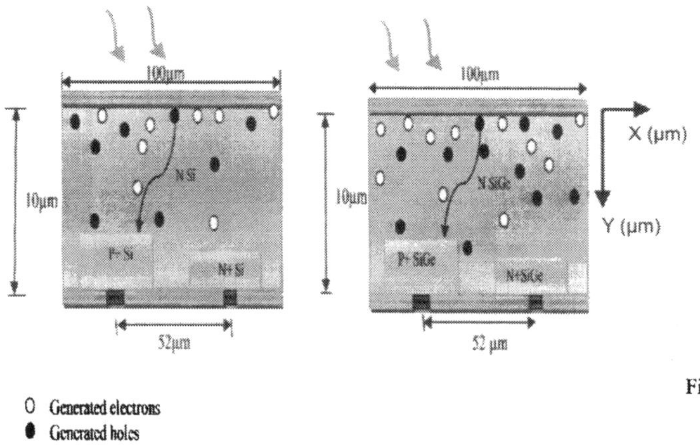

Fig.1. Simulated device: - 10μm thick BC-BJ solar cell: (a) Silicon (b) SiGe

To obtain the J-V curve under illumination, we used the standard AM1.5 solar spectrum, divided into 400 points. The models used in the simulation assume a concentration dependent mobility using default software values for electron and holes [11]. Shockley – Read-Hall and Auger recombination mechanisms were taking into account. Auger recombination was modeled with the default software values of Auger coefficient $2.8\text{x}10^{-32}$ cm^6s^{-1} for electrons and $9.9\text{x}10^{-32}$ cm^6s^{-1} for holes, in case of silicon, whereas $9.9\text{x}10^{-31}$ cm^6s^{-1} for electrons and $1.8\text{x}10^{-31}$ cm^6s^{-1} for holes were used, in case of SiGe. The minority carrier lifetime of 10^{-6}s and 10^{-5}s has been used for silicon and SiGe, respectively. The Poisson equation together with the continuity equation for electrons and holes are solved simultaneously by the software to obtain the J-V characteristics.

3 SIMULATION RESULTS

Figure 2 shows the contour plot of recombination rate. The magnitude of recombination rate in silicon cell is in the order of (10^{20} /cm^3s), whereas in proposed device it is (10^{18} /cm^3s).

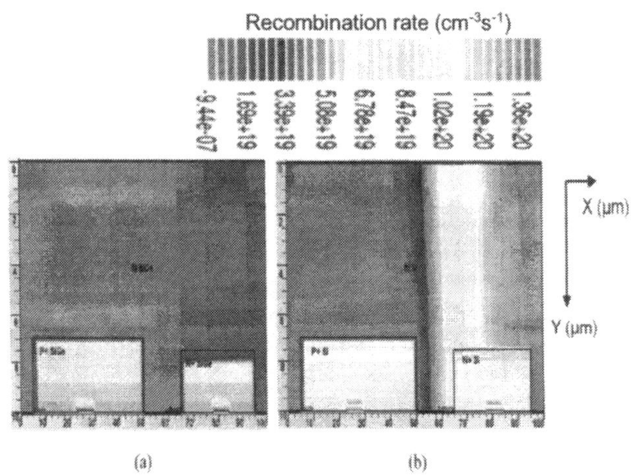

Fig. 2. Recombination rate contour (logarithmic plot): (a) 10μm thick BC-BJ SiGe solar cell (b) 10μm thick BC-BJ Si solar cell.

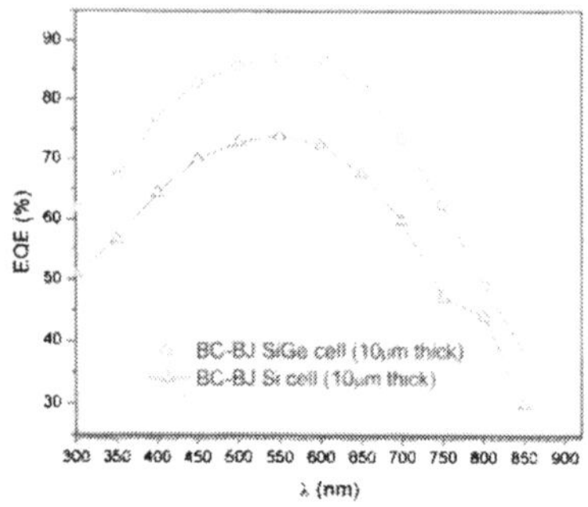

Fig. 3. Comparison of External Quantum Efficiency (EQE)

At the interface, recombination rate is higher in right side of the devices because collection probabilty of minority carriers (hole) is decreases. The lower recombination is observed in novel device because minority carrier lifetime is higher compared to silicon cell. In back contact design, most of the carriers are generated near the front surface, while the junction is far at the back surface. Lower recombination ensure higher carrier collection efficiency.

Also, the EQE of the proposed device is higher than silicon cell as shown in Fig 3. The proposed device shows EQE>76% in the spectrum range of 400-700 nm wavelength whereas in silicon cell, EQE>60% was obtained. The proposed device shows 26% higher EQE compared to silicon cell, due to the presence of Ge. Since, the no. of free electrons in the germanium is higher than the silicon at given temperature, which results in higher conductivity of Ge.

Further, we have obtained the impact of device thickness on photovoltaic parameters. We have decreased the device dimension to half the initial dimensions, discussed earlier, keeping the aspect ratio (Width/Depth) of P+ and N+ region fixed 1/15 and 1/12, respectively. The length of the pitch becomes 26μm. The electrical parameters of 10μm thick devices and 5μm thick devices are plotted in Fig 4. Fig. 4(a&d), shows the current density-volatge (J-V) curve and PCE bar for different devices, when thickness of silicon cell decreased from 10μm to 5μm, change in short circuit current density J_{SC} and PCE was 26.6% and 26.2%, respectively, whereas for SiGe it was 17.7% and 16.3%, respectively. It shows that proposed device is less affected by the thickness variation compared to the conventional silicon cell. Also Fig. 4(b&d) shows that the maximum output power density (P_M) of 10μm thick silicon cell and 5μm thick SiGe cell was approximately same, 11.8mW/cm^2 in 10μm thick silicon cell and 11.3mW/cm^2 in 5μm thick SiGe cell, respectively. Also, open circuit volatge (V_{OC}) can be obtained by taking the intercept to x- axis form the sharp minima, shown in Fig. 4c. Proposed device shows superior photovoltaic properties comparing to conventional silicon cell due to lower recombination [12]. Detailed comparisons of electrical parameters are shown in Table 1. Table 1, shows that the performance of 5μm thick SiGe was approximately equal to the 10μm thick silicon cell, indicating that the proposed device significantly saves the material cost over silicon cell.

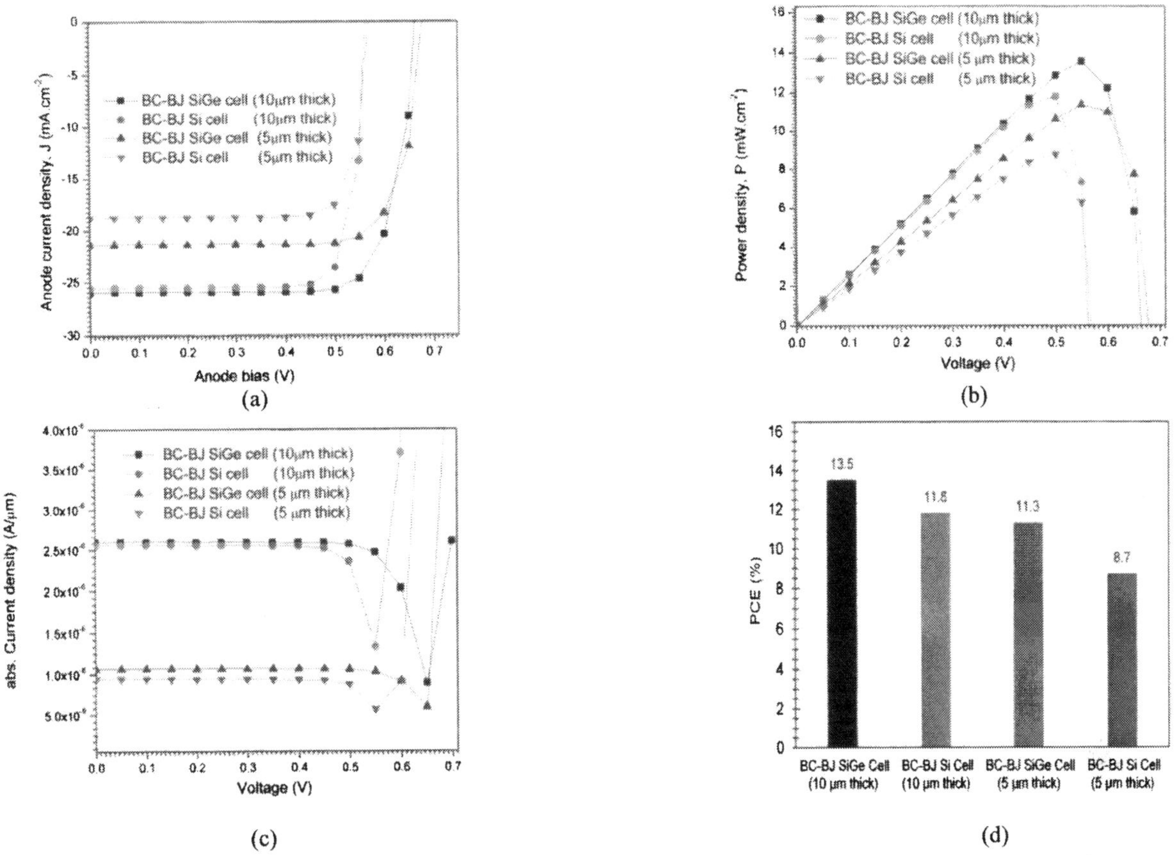

Fig.4. Comparisons of electrical characteristics of different devices: (a) J-V, (b) Power density-voltage, (c) Abs.current density-voltage (d) PCE.

Table1. Detailed comparisons of electrical parameters

Device	Voc (V)	Jsc ($mAcm^{-2}$)	FF (%)	PCE (%)
Silicon cell (10μm thick)	0.563	25.6	81.8	11.8
SiGe cell (10μm thick)	0.663	25.9	78.8	13.5
Silicon cell (5μm thick)	0.569	18.8	81.9	8.7
SiGe cell (5μm thick)	0.677	21.3	78.4	11.3
FF-fill factor, PCE- power conversion Efficiency.				

Proposed device shows superior photovoltaic parameters compared to silicon cell. Also results reveal that when the thickness of proposed device is reduced to 5μm, its PCE, 11.3% was approximately equal to PCE, 11.8% of 10μm thick silicon cell. It indicates that the proposed device significantly saves the material cost over silicon cell. Proposed device shows improved photovoltaic parameters: EQE> 76% in the spectrum range of 400-700 nm wavelength, Jsc 25.9mAcm^{-2}, Voc 663mV, fill factor (FF) 78.8% and PCE of 13.5%. Simulation result shows SiGe semiconductor as an alternative to expensive materials for thin BC-BJ solar cell design.

4 CONCLUSION

In this work, we proposed a novel 10μm thick BC-BJ SiGe Solar cell design, as a cost-effective solution for energy efficient applications. The work analyse and compares the optical and electrical characteristics of proposed design using Silvaco Atlas device simulator.

ACKNOWLEDGMENT

The authors would like to thank Microelectronics Research Lab, Department of Engineering Physics, Delhi Technological University to carry out this work. Rahul Pandey (JRF) acknowledges UGC, Govt. of India for providing fellowship.

REFERENCES

[1] S. Jeong, M. D. McGehee and Y. Cui , "All-back-contact ultra-thin silicon nanocone solar cells with 13.7% power conversion efficiency" Nature Communications 4, Article number: 2950: 1-7, 2013.

[2] R. R. King, R. A. Sinton, and R. M. Swanson, Studies of diffused phosphorus emitters: saturation current, surface recombination velocity, and quantum efficiency. IEEE Trans. Electron Dev. 37, 365–371, 1990.

[3] R. J. Schwartz and M. D. Lammert. Silicon solar cells for high concentration applications. In IEEE International Electron Devices Meeting, pages 350–352, 1975.

[4] E. V. Kerschaver and G. Beaucarne. Back-contact solar cells: A review Progress in Photovoltaics, 14(2):107–123, 2006.

[5] D.diouf, J.P. Kleider, T.desrues, P.-J. Ribeyron" Study of interdigitated back contact silicon heterojunction solar cells by two-dimensional numerical simulations", Material Science and Engineering B 159-160:291-294. 2009

[6] J.Ouellete,"Silicon-Germanium Gives Semiconductor the Edge" American institute of physics, 2002:22-25.

[7] D.Yang, X.Yu, X. Li, P.Wang, L. Wang, "Germanium-doped crystal silicon for solar cells," Solid-State and Integrated Circuit Technology (ICSICT), 2010 10th IEEE International Conference on, vol., no.,1994,1994, 1-4 Nov. 2010.

[8] C.C.Wang, D.S.Wuu, S.Y. Lien, Y.S. Lin, C.Y. Liu, C.H. Hsu, and C.F. Chen , "Characterization of Nanocrystalline SiGe Thin Film Solar Cell with Double Graded-Dead Absorption Layer " International Journal of Photoenergy Volume 2012, 6 pages

[9] H. Povolny, P. Agarwal, S. Han, and X. Deng, "Comparison Study of a-SiGe Solar Cells and Materials Deposited Using Different Hydrogen Dilution " Mat. Res. Soc. Symp. Proc. Vol. 609, 2000.

[10] G.G. Pethuraja et al., Materials Sciences and Applications, 3, pp.67-71, 2012.

[11] ATLAS User's Manual DEVICE SIMULATION SOFTWARE. May 17, 2012.324.

[12] S.M.Sze," Physics of Semiconductor Devices 2nd Edition": 811-813

Novel Applications for Flexible CIGS-based Photovoltaics for Solar Power Generation, Management and Storage

J.H. Armstrong*, S.P. Retureta*, and I. Lee*

*Ascent Solar Technologies, Inc., Thornton, CO, USA, jarmstrong@ascentsolar.com

ABSTRACT

Converting solar energy into electrical power has been an attractive alternative to fossil fuels for many years, and with the advent of lightweight, flexible, efficient photovoltaic (PV) blankets, many applications have been envisioned where off-grid power generation has become a reality. Ascent Solar's novel flexible, lightweight, monolithically-integrated CIGS on polyimide substrates offers a highly-redundant, robust power generation solution that is available in the consumer market, as well as in development for portable power solutions for today's military, onboard power generation for small unmanned aerial vehicles (sUAV), electric-powered aviation, near-space and space applications. While space vehicles have been using PV from the beginning of its existence, other non-traditional applications are just realizing the potential of on-board power generation. In this presentation, we will illustrate the inherent robust nature of our PV technology, and will be presenting the development of several key product areas using our flexible CIGS PV, including sUAV, electric powered civil aviation, and fully-integrated portable power generation, management, and storage solutions. Challenges for each area will be discussed, and the novel solutions using flexible lightweight CIGS PV will be described, and pertinent data and analyses on performance will be provided.

Keywords: flexible photovoltaics, monolithic integration, maximum peak power tracking, onboard power storage, portable power.

1 INTRODUCTION

From the first observance of the photoelectric effect in 1939 to the first demonstration of a semiconductor-based solar cell in 1904, the conversion of solar energy into electricity had been primarily a scientific endeavor or a novelty use in light-detecting sensors. However, in the mid 1950's, a practical application for solar cells, namely space exploration, pushed the development of solar cells for converting solar enertgy into useful electrical power. As these early solar cells were based on semiconductor wafer technologies, they were fragile, and in the early stages of solar energy conversion, not very efficient. However, as sunlight above the earth's atmosphere is intense and only shaded occasionally in orbit, photovoltaic power became a primary means for generating electrial power in space applications. Steady improvement of device efficiency has improved the effectiveness of PV in terms of providing the necessary power in this application, but cost has always been prohibitive. Evolution of terrestrial-based applications followed, with many off-grid applications becoming based on PV devices.

However, one common trait remains with regard to these PV technologies, namely, that the PV devices are based on crystalline semiconductor wafers. These devices are inherently fragile and thus require substantial protective packaging to prevent fracturing and otherwise damaging them in normal operation. Furthremore, as the PV devices themselves are discrete and limited in voltage, a number of devices must be interconnected in order to achieve a useful voltage for a given connected circuit.

Ascent Solar Technology, Inc. (AST) was founded to develop a PV technology based on the thin-film copper-indium-gallium-selenium (CIGS) absorber technology, with deposition upon a flexible polyimide substrate. By virtuie of the substrate being a dielectric material, unlike other CIGS manufacturs, AST has been the first to demonstrate monolithically-integrated CIGS PV modules *in production*. By its nature, monolithic integration results in a higher voltage within a smaller area than those associated with discrete cell technologies. Geometrically, monolithically-integrated cells are ideal for higher current PV technologies, such as CIGS, and are well suited for applications where higher-than-normal solar intensity is present, such as high altitude/near-space and space environments. Monolithic integration of flexible CIGS PV also has the distinct advantage over discrete cell technology (either crystalline or thin film flexible) of providing inherent redundancy in cell interconnection. Because the cells are interconnected continuously along their long side, any damage that the module withstands can be mitigated by diverting current paths around the damage (Figure 1). Thus, the monolithically-integrated module that represents a series connection of cells, will continue to operate, albeit at a reduced capacity, but more importantly, the series interconnect is not compromised.

2 MANUFACTUIRNG PROCESS BASED ON MONOLITHIC INTEGRATION

While required for thin-film PV modules on glass substrates, discrete flexible CIGS devices makes up a large portion of the industry. Ascent Solar's unique approach of monolithic integration on a flexible substrate enables larger devices, and thus, small part counts, while enabling higher voltages in smaller footprints. As shown in Figure 2, AST manufacturing process consists of three distinct steps, the first of which is the thin film deposition of all the films, followed by a patterning cell that scribes and interconnects cells in a monolithic integration process. By completely isolating deposition and monolithic integration processes, AST can provide custom modules from a common roll process to customers based their specific voltage/current requirements.

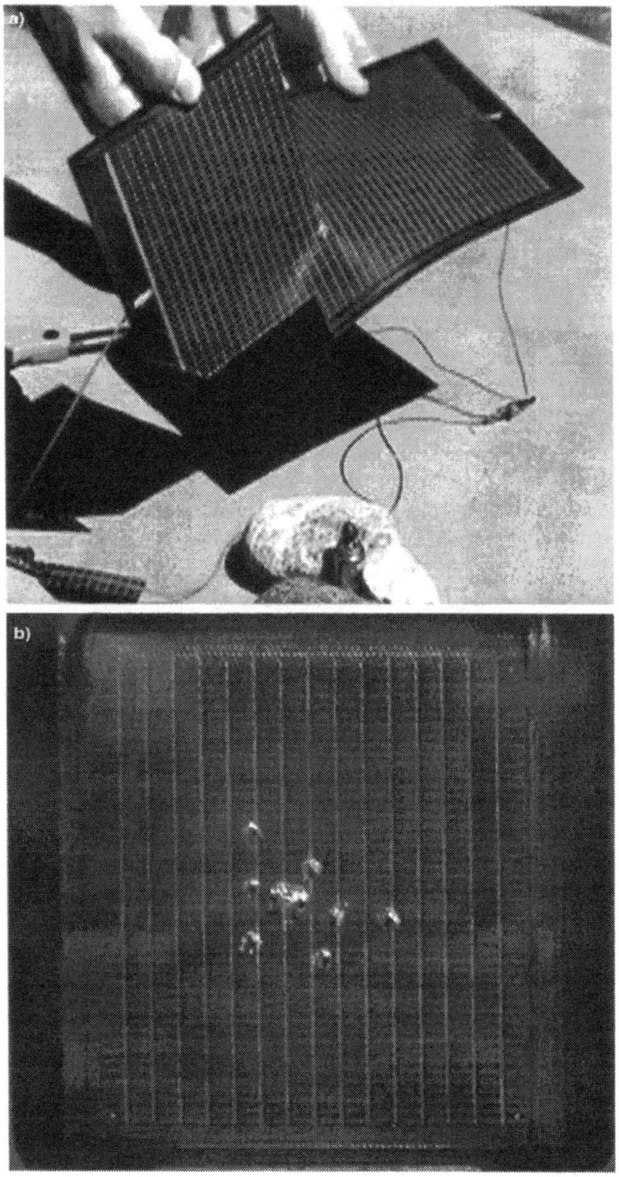

Figure 1: Demonstrated Robust Design of Flexible CIGS Module with Respect to a) Cutting or b) Penetrations.

Because monolithic integration results in narrow cells in series that are the width of the module, the distance current needs to travel to be collected per cell is much smaller than discrete cells (either crystalline or flexible thin film) and current generation per cell is thus lower than what is practical for discrete cells. Discrete cells cut this narrow would be extremely difficult to integrate into strings, while the monolithic integration process ensures that these cells are interconnected during processing. Figure 3 shows a 15cm x 15cm module that consists of sixteen (16) cells in series, producing 9VDC Voc (7.4V Vmax) at nominally 2W output. This module was designed specifically for this voltage output to power our USB charging circuitry. As a result, the inherently high current density of CIGS device chemistry can be somewhat mitigated by monolithic integration by economically integrating small area cells in series with reduced current path, and applications where the device operates under high sun intensity (e.g. solar concentration, high-altitude and space applications), monolithic integration can result in a significant reduction in module losses compared to other technologies (Figure 4).

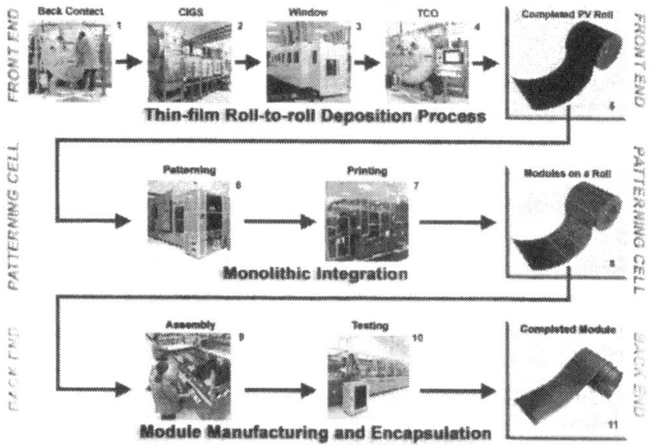

Figure 2: Schematic Representation of the AST Manufacturing Process.

Figure 3: Photograph of AST's 15x15cm 'USB' module with Sixteen (16) Cells in Series (top to bottom)

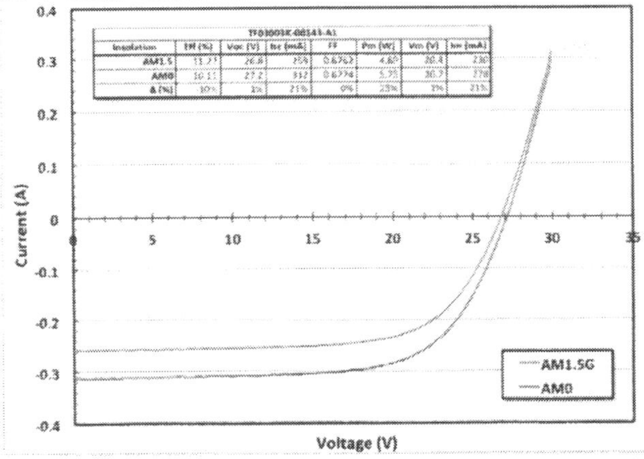

Figure 4: I-V Characteristics of a 15x30cm Module under Terrestrial (AM1.5G) and Space (AM0) Test Conditions.

3 NOVEL APPLICATIONS OF FLEXIBLE MONOLITHICALLY INTEGRATED CIGS

Because of the unique combination of lightweight flexible polymeric substrate, thin-film deposition processes, novel manufacturing process, and monolithic integration, AST's CIGS technology can be applied to a wide range of applications. Its inherent robustness makes it ideal for rugged military and off-grid portable power markets. Furthermore, its thinness and aesthetically-pleasing uniform black appearance is also ideal for consumer products; AST established EnerPlex™ as its commercial brand for portable power solutions for modern consumer electronics (Figure 5).

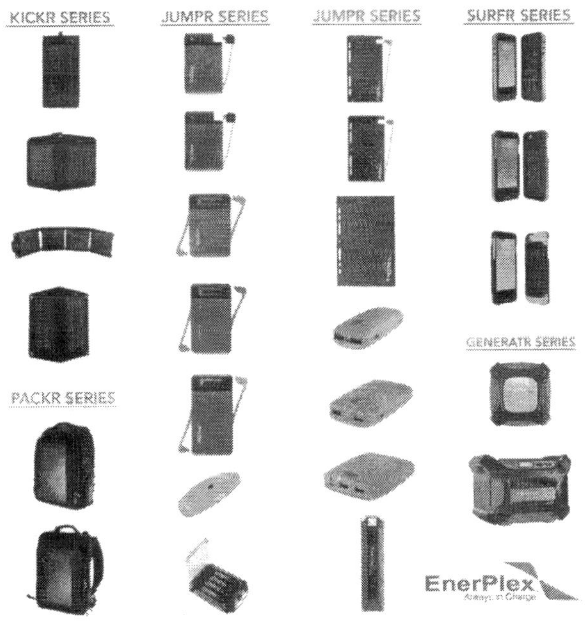

Figure 5: Photograph of EnerPlex™ Porable Power Systems that Support the Active Urban and Outdoor Lifestyles.

3.1 Portable Power Applications

On January 5th, AST announced a fully-integrated portable PV system folding PV-based power system (patent pending) at the 2015 Consumer Electronics Show (CES) that provides power generation, power management, power storage, and power distribution. Referred to as the MilPak™ family of portable PV products (Figure 6), the system contains sixteen (16) 15cmx30cm module that nominally generate 60 watts. As each module is designed to operate at the requisite peak power voltage and current, all sixteen modules are connected in PARALLEL, thus reducing the chance that damage to any module would result in a loss of the entire system. In order to facilitate maximum power conversion under a wide range of conditions (sun angle, temperature, solar intensity), a maximum peak power tracker (MPPT) maintains the output of the PV blanket by monitoring the power output and adjusting the voltage and current operating point of the PV to maintain maximum power. As solar power can vary during the day with clouds, artificial shading, etc., a lithium-ion battery pack is used as the power 'buffer' and its charge state is maintained by the MPPT as well. An internal DC power bus operates various output options, including high-current USB power charging interfaces, as well as a high-power output that ranges from 19VDC to 28VDC, depending upon the customers needs. All connectors are, at a minimum, IP-67 rated and some variants of the MilPak™ utilize military-grade connectors and switchgear throughout. MilPak™ represents a lightweight (3.7 kg) fully-integrated foldable power system (Figure 7). Specifications for this system are shown in Table 1.

Figure 6: Photograph of the MilPak™ E System.

Figure 7: Photographs of the Deployment Sequence of MilPak™ E .

Table 1: Specifications for MilPak™ E

Name:	MilPak™ E
Solar PV:	Monolithic Flexible CIGS with Maximum Peak Power Tracking
- Vmax	18.5VDC
- Imax	3.2A
- Pmax	60W
- Pnominal	52W
Storage:	• 86.5 Whr Lithium Ion with Battery Management System Protection
Output:	• Circuit Breaker Protected • 2 x 5VDC, 2.1A USB (regulated) • Either 19, 24 or 28V DC (regulated, custom voltages available) • 55W maximum (chainable)
Physical:	
- Stowed	80 x 410 x 220 mm
- Deployed	75 x 1990 x 725 mm
- Mass	3.7 kg (8.2 lbs)

3.2 Aerospace Applications

The low current/high voltage nature of AST's CIGS makes it ideal for space applications where the higher solar intensity above the ground equates to higher current per cell. By the nature of the cells in a monolithically integrated module, these devices handle higher solar intensities better than other technologies. AST has developed a lightweight flexible package that protects the PV from moisture and ultraviolet (UV) light while maintaining an overall areal density of 350-400 g/m^2. Combined with the afore-mentioned properties, the 'SuperLight' package can be used in any weight-sensitive application such as bonding onto fixed-wing surfaces to larger airship applications. Figure 8 shows the Silent Falcon™ small unmanned aerial system (sUAS) that uses AST's SuperLight construction for extending flight time. Unlike other sUAS concepts or prototypes, Silent Falcon™ is ***in production***, as are the AST flexible PV modules mounted to the wing surfaces. Flight time of Silent Falcon is designed to exceed 8 hours with contributions from the AST flexible PV blanket.

Figure 8: Silent Falcon™ sUAS Staged for Lauch with Ascent Solar Technologies SuperLight PV Integrated into the Wings for Extended Flight Time. Flight time of Silent Falcon is designed to exceed 8 hours with contributions from the AST flexible PV panels

4 SUMMARY

AST's flexible, monolithically-integrated PV product is inherently robust and is ideal for a wide range of applications where weight is a challenge, or where the PV must conform to a curved surface, such as a wing surface. Monolithic integration is also ideal for genrating a usable voltage in a small area and can be tailored to produce a specific output as required by the attached load requirements. Such an approach enables the rugged 60W military blanket with fully-integrated power storage, management, storage and distribution functions. Furthermore, the inherently small cell area makes this construction for high-intensity solar applications, such as high altitude and space applications.

Nanoscaled Microelectronics and Nanotechnology-Enabled Energy Systems For Aerospace and Robotic Applications

Sergey Edward Lyshevski[*], Mamyrbek A. Beisenbi[**], Gulzhan Uskenbayeva[**], Aliya Shukirova[**], Janar Yermekbayeva[**] and Nurlan Mukataev[**]

[*]Department of Electrical and Microelectronic Engineering, Rochester Institute of Technology, Rochester, NY 14623 USA
E-mail: Sergey.Lyshevski@mail.rit.edu URL: http://people.rit.edu/seleee

[**]Department of System Analysis and Control, L. N. Gumilyov Eurasian National University, Astana, Republic of Kazakhstan, aliya.shukirova@mail.ru

ABSTRACT

For autonomous and semi-autonomous micro air vehicles, minirobots, propulsion, remote sensing and other platforms, we study high-power and high-energy densities energy systems. These systems are designed using nanotechnology-enabled microelectronic, electronic, energy sources and energy storage components, devices and modules. Research and technology developments are performed for energy harvesting, management and storage. We apply and demonstrate key modules and components, such as: (1) Nanoscaled low-power microelectronics and sensors; (2) High-power-density semiconductor devices, circuit components and power electronics; (3) Nanotechnology-enabled solar cells; (4) Advanced energy storage devices; (5) Energy management system. The proposed solutions are substantiated by performing experimental studies. The compliance of the proposed technologies to radio-controlled mini air vehicles and robots is ensured. Proof-of-concept power systems are designed using specifications for *all-electric* autonomous aerospace, naval, robotic and security platforms.

Keywords: electronics, energy, nanotechnology

1. INTRODUCTION

Power adequacy, integrity and energy sustainability are essential to ensure functionality of propulsion, navigation, sensing, management and other systems in aerospace, land and underwater platforms. It is important to design, test and evaluate integrated self-sustained power systems for different platforms meeting application-specific requirements and specifications. Safety, affordability, energy density and other features can be ensured by using: (i) Nanotechnology-enabled components, devices and modules [1, 2]; (ii) Compliant *modular* organization; (iii) Advanced control schemes and management systems; (iv) Enabling energy harvesting and conversion solutions.

Advances in nanotechnology, microelectronics and micro-electromechanical systems (MEMS) [1, 2] result in commercialization and deployment of various minirobots, surface and air vehicles, etc. The energy module, which includes the energy source and other subsystems, is a key component. High-altitude, outer and deep space vehicles, robots as well as underwater platforms require specific modules and electronic devices which operate under extreme temperature, mechanical loads, interference and radiation. The controlled high-energy density power systems must be designed, tested and characterized.

We examine enabling inorganic and organic photovoltaic cells, nanotechnology-enabled electronic devices, front-end microelectronic components, efficient energy harvesting solutions, low-loss energy conversion and novel energy storage schemes. The controlled energy conversion, storage and distribution are accomplished by an energy management system. A coherent energy management implies consistent sensing, processing, optimization and control of energy conversion. Advanced sensing, data acquisition and processing are achieved by nanoscale electronics, optoelectronics and MEMS. The studied *modular* energy systems may operate in the range from milli-watts to hundreds of watts within continuous and pulse energy conversion and release capabilities. The proof-of-concept portable light-duty energy systems are tested and substantiated achieving a sufficient technology level.

2. SCALABLE AND *MODULAR* ENERGY SYSTEMS

With an overall objective to develop a scalable high-performance power system technology for high-energy density portable energy sources, the *modular* design is performed using nanoscale electronics and MEMS [1-4]. The power integrity, effective energy management and functionality must be guaranteed by matching compliant components and modules with the electric loads, energy harvesting and storage capabilities.

Figure 1 illustrates a *modular* power system which includes photovoltaic cells, power electronics (dc/dc converter, chargers, controllers, filters, sensors, etc.), MEMS, rechargeable battery and other modules.

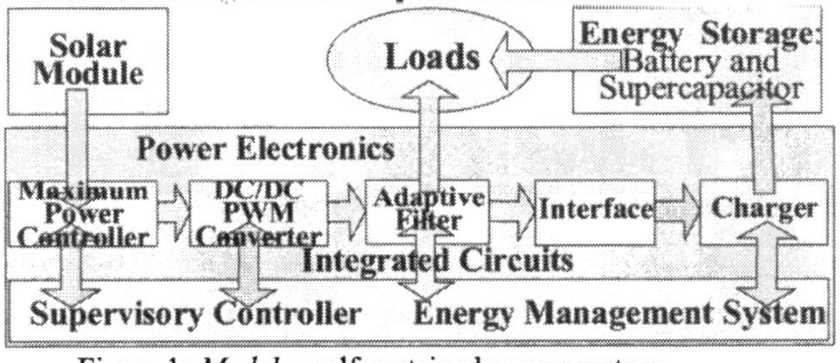

Figure 1: *Modular* self-sustained power system

One need to accommodate the application- and system-specific peak, continuous and pulse loading conditions, as well as time-varying *RLC* loads, such as electromechanical devices, antennas, communication components, electronics, etc. Therefore specific energy sources, converters and power electronic modules are used taking into account the rated and peak continuous and pulse loading conditions. Figure 2 documents the images of photovoltaic cells, super-capacitors, lithium-ion rechargeable battery, high-frequency high-efficiency dc/dc converters, etc. The varying *RLC* loads are the transmitters, receivers, permanent-magnet actuators and propulsion motors, servos, sensors and other devices [5-7].

Figure 2: Devices, components and modules of autonomous energy systems:
(a) Photovoltaic cells;
(b) Rechargeable supercapacitors and lithium ion battery;
(c) High-frequency *buck-boost* and *boost* converters;
(d) Loads: Electric motors, servos, transmitters and receivers.

3. Analysis and Control of Energy Systems

The nonlinear steady-state and dynamic analyses of energy systems component, such as photovoltaic cells and others, is reported in [3-5]. Depending on irradiation, incident angle, temperature and other factors, the output solar cell voltage $u_{cell}(t)$ varies. The output voltage, applied to the *i*th *RLC* load $u_{iRLC}(t)$, must be controlled and stabilized. The dc/dc *buck-boost* and *boost* converters are used [5,8]. The nanoscale electronics and nanotechnology-enabled components (MOSFETs, inductors and capacitors) are used. To stabilize the output voltage at the *RL* load, consider a one-quadrant *boost* converter with a filter. The converter schematics is shown in Figure 3 [5].

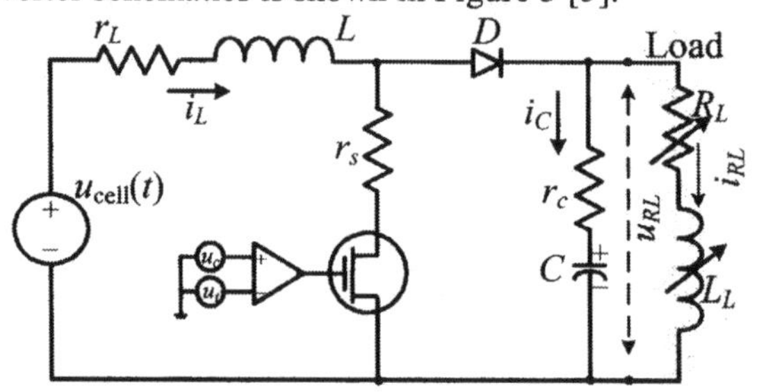

Figure 3: Controlled *boost* converter with the varying *RL* load

When the MOSFET is closed, using the currents and voltages as the state variables $\mathbf{x}=[u_C, i_L, i_{RL}]^T$, one has

$$\frac{du_C}{dt}=-\frac{1}{C}i_{RL},\quad \frac{di_L}{dt}=\frac{1}{L}\left[-(r_L+r_s)i_L+u_{cell}\right],$$

$$\frac{di_{RL}}{dt}=\frac{1}{L_L}\left[u_C-(R_L+r_c)i_{RL}\right].$$

When the MOSFET is open, the capacitor *C* is charged by the voltage source. The differential equations are

$$\frac{du_C}{dt}=\frac{1}{C}(i_L-i_{RL}),\quad \frac{di_L}{dt}=\frac{1}{L}\left[-u_C-(r_L+r_c)i_L+r_c i_{RL}+u_{cell}\right],$$

$$\frac{di_{RL}}{dt}=\frac{1}{L_L}\left[u_C+r_c i_L-(R_L+r_c)i_{RL}\right].$$

The comparator drives the MOSFET with the switching frequency *f*. Using the time when the MOSFET is *on* and *off*, one has $f=1/(t_{on}+t_{off})$. The voltage applied to the load u_{RL} is regulated by controlling the switching *on* and *off* durations t_{on} and t_{off}, respectively. The average voltage applied to the load depends on t_{on} and t_{off}. The duty cycle $d_D=\frac{t_{on}}{t_{on}+t_{off}}\in[0\ 1]$ varies between 0 and 1. Neglecting small resistances r_c and r_L, one obtains $\frac{u_{RL}}{V_d}=\frac{1}{1-d_D}$.

Using the *averaging* concept [5, 8], we have

$$\frac{du_C}{dt}=\frac{1}{C}(i_L-i_{RL}-i_L d_D),$$

$$\frac{di_L}{dt}=\frac{1}{L}\left[-u_C-(r_L+r_c)i_L+r_c i_{RL}+u_C d_D+(r_c-r_s)i_L d_D-r_c i_{RL}d_D+u_{cell}\right],$$

$$\frac{di_{RL}}{dt}=\frac{1}{L_L}\left[u_C+r_c i_L-(R_L+r_c)i_{RL}-r_c i_L d_D\right] \qquad (1)$$

The electronic components are fabricated using nanotechnologies. High-permeability and low eddy current losses ferrites are used in high-performance ferrite-core toroidal inductors. The soft, low coercivity and high-Q ferrites are the iron, zinc and manganese or nickel oxides. The high permeability soft ferrites $Mn_xZn_{(1-x)}Fe_2O_4$, high resistivity high-frequency $Ni_xZn_{(1-x)}Fe_2O_4$, and other ferrites are used. The structure and magnetic properties of $MnFe_2O_4$ ferrites depend on the preparation methods, such as the ceramic technique, combustion, co-precipitation, sol–gel and citrate. The citrate method gives the smallest lattice and ~15 nm particles, while the flash combustion results in ~40 nm. The ceramic technique results in the larger particle size, relative permeability ~2000, uniformity, etc.

In ferrite inductors, the relative permeability μ_r varies as a function of the load. For the toroidal inductors,

$$L = \mu \frac{N^2 h}{2\pi} \ln \frac{R_{out}}{R_{in}}, \tag{2}$$

where μ is the permeability, $\mu=\mu_0\mu_r=\mathbf{B}/\mathbf{H}$; μ_r is the relative permeability, $\mu_r=\mu_0^{-1}dB/dH$; B and H are the magnetic field density and intensity, $H=Ni/l$ [A/m]; l is the length; h is the thickness; R_{out} and R_{in} are the outer and inner radii.

The field intensity H and density B vary with current. The permeability μ also varies as a function of i. Using the nonlinear BH curve, one has $\mu_r(i, \frac{di}{dt}) = \mu_0^{-1} \frac{dB}{dH}$. Hence [9]

$$B(H, \tfrac{di}{dt}) = B_{max} \tanh(aH - \mathrm{sgn}(\tfrac{di}{dt}) bH_c),$$
$$L(i, \tfrac{di}{dt}) = B_{max} c\left[1 - \tanh^2\left(ci - d\,\mathrm{sgn}(\tfrac{di}{dt})\right)\right], \tag{3}$$

where B_{max}, a, b, c and d are the constants which depend on the BH curves of the nanostructured ferrites.

For a 2 mH ferrite toroidal inductor, N=60, B_{max}=0.4 T, R_{out}=14 mm, R_{in}=7 mm, a=0.001, b=0.001, c=0.005 and d=0.05. The nonlinear magnetization curve and varying inductance $L(i)=f(i,di/dt)$ are reported in Figure 4. The nonlinear model (3) is used in design. The converter with the RL load is described by nonlinear differential equations

$$\frac{du_C}{dt} = \frac{1}{C}\left(i_L - i_{RL} - i_L d_D\right),$$
$$\frac{di_L}{dt} = \frac{1}{L(i_L)}\left[-u_C - (r_L + r_c)i_L + r_c i_{RL} + u_C d_D + (r_c - r_s)i_L d_D - r_c i_{RL} d_D + u_{cell}\right],$$
$$\frac{di_{RL}}{dt} = \frac{1}{L_L}\left[u_C + r_c i_L - (R_L + r_c)i_{RL} - r_c i_L d_D\right] \tag{4}$$

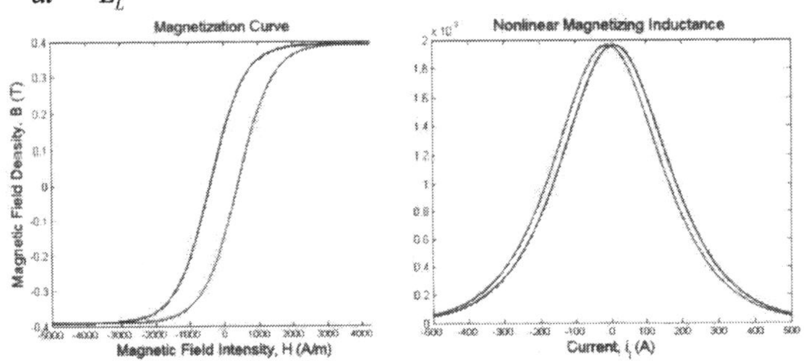

Figure 4: Inductor magnetization curve, and, nonlinear inductance $L(i, \frac{di}{dt}) = B_{max} c\left[1 - \tanh^2\left(ci - d\,\mathrm{sgn}(\frac{di}{dt})\right)\right]$ as a function of current due to the varying permeability $\mu_r(i, \frac{di}{dt}) = \mu_0^{-1} \frac{dB}{dH}$.

4. Experimental Results

We examine a high-energy density power system with high-efficiency solar cells and energy storage devices. The specific energy of super- and hybrid capacitors varies from ~1 to 30 W-h/kg, while for lithium-ion batteries, one may ensure ~200 W-h/kg. For different loads and loading conditions, depending on the spesified converter's output voltage $u_{converter}(t)$, we synthesize and verify nonlinear proportional-integral-derivative and sliding mode control laws [4, 10]. Using the tracking error

$e=u_{reference}(t) - u_{converter}(t)$,

we design and test the following control algorithms

$u(t)=k_p e + k_i\int edt + k_d de/dt$, $k_p>0$, $k_i>0$, $k_d=0$,

$u(t)=k_p e + k_{p1}e^{1/5} + k_i\int edt$, $k_p>0$, $k_{p1}>0$, $k_i>0$,

and $u=u_{max}\tanh(k_p e + k_{p1}e^{1/5} + k_i\int edt)$, $k_p>0$, $k_{p1}>0$, $k_i>0$. (5)

Nonlinear control algorithms (5) are tested in the closed-loop energy systems with the RLC loads. The experimental results are documented in Figure 5 for the time-varying R and L. The evolution of $i_{RL}(t)$ and $u_{converter}(t)$ are given by the first and second oscilloscope's channels, respectively. The steady-state values and transient dynamics are reported for different reference voltages $u_{reference}(t)$ and time-varying loads. The tracking error $e(t)$ is less than 1% under the peak loads, and, the settling time is ~1 msec. At the rated load, the converter's efficiency is ~91%. The comparison of the experimental and analytic studies is summarized in Table 1.

Table 1. Comparison of Analytic and Experimental Steady-States

Assigned $u_{reference}$ which corresponds d_D	Analytic, modeling and simulation results $u_{converter}$ [V]	Experimental results (closed-loop system) $u_{converter}$ [V]
10.6	10.54	10.6
13.4	13.35	13.4
17.4	17.35	17.4
21	20.95	21

We depart from solutions reported in [11, 12]. A dynamic maximum power tracking algorithm and efficiency optimization scheme are implemented ensuring optimal energy conversion with minimal losses. The voltage stabilization, voltage tracking and high-efficiency energy management ensure optimal energy harvesting, conversion, distribution and storage. To achieve optimal performance, we minimize the performance functional J using the power losses $P_{jLosses}$, tracking error e_j, energy, energy transfer ΔE_j and other integrands for jth components and modules. In particular, using the weighting coefficients q_j, we have

$$J = \min_{P_{jLosses}, t, e_j} \max_{\eta} \int_{t_0}^{t_{final}} \left(\sum\nolimits_j q_{P_j}\left|P_{j\,Losses}\right| + \sum\nolimits_j q_{e_j} t\left|e_j\right| + \sum\nolimits_j q_{E_j}\left|\Delta E_j\right|\right)dt. \tag{6}$$

Using (6), performance and capabilities are measured by quantitative estimates, measures and metrics. Efficiency η, energy density, robustness, stability and other measures are examined in the full operating envelope. The consistent analytic measures and estimates are consistently used. Adequate cost, durability, integrity, modularity, storage capacity and other metrics are achieved.

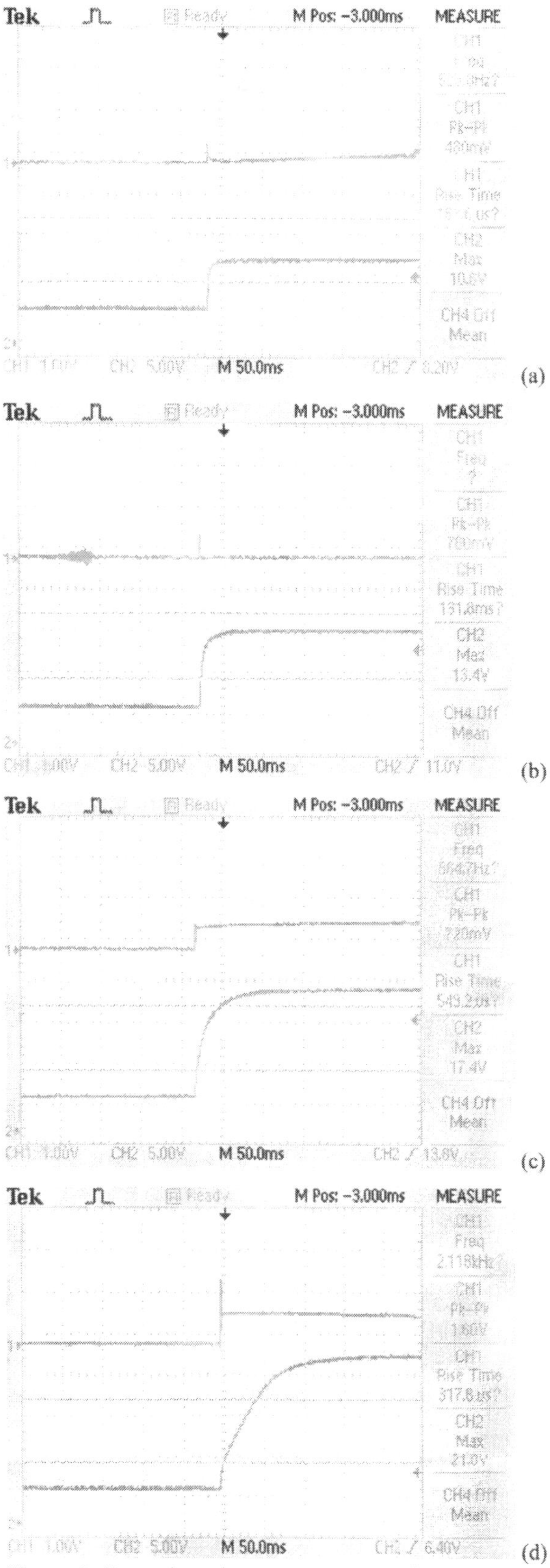

Figure 5. Dynamics of $i_{RL}(t)$ (top orange lines) and $u_{converter}(t)$ (bottom blue lines) for different reference voltages $u_{reference}(t)$ under varying *RL* loads: (a) $u_{reference}$=10.6 V; (b) $u_{reference}$=13.4 V; (c) $u_{reference}$=17.4 V; (d) $u_{reference}$=21 V.

5. CONCLUSIONS

For nanotechnology-enabled self-sustained power systems, we designed and substantiated a scalable *modular* technology. Advanced MEMS, electronic, photovoltaic and energy storage hardware solutions were used. Optimal energy conversion and management was ensured by designing practical *minimal-complexity* control laws. The proposed design ensures safety, affordability, accessibility, scalability, efficiency, effectiveness, simplicity, compliance, etc. The proposed concept meets specifications imposed in aerospace, automotive, biotechnology, consumer electronics, medical, naval, robotic, security and other applications. The effectiveness and applicability of *modular* portable energy systems were substantiated through experiments and technology transfer developments.

REFERENCES

[1] *International Technology Roadmap for Semiconductors*, 2005, 2007, 2009, 2011 and 2013 Editions, Semiconductor Industry Association, Austin, Texas, USA, 2015.

[2] *International Technology Roadmap for Semiconductors*, 2011 and 2013 Edition, *Micro-Electromechanical Systems (MEMS)*, Semiconductor Industry Association, Austin, TX, USA, 2015.

[3] S. E. Lyshevski, "High-power density mini-scale power generation and energy harvesting systems," *Energy Conversion and Management*, vol. 52, pp. 46-52, 2011.

[4] T. C. Smith and S. E. Lyshevski, "*Clean* high-energy density renewable power generation systems with soft-switching sliding mode control laws," *Proc. IEEE Conf. Decision and Control*, Orlando, FL, pp. 836-841, 2011.

[5] S. E. Lyshevski, *Electromechanical Systems and Devices*, CRC Press, Boca Raton, FL, 2008.

[6] S. E. Lyshevski, "Motion control of electromechanical servo-devices with permanent-magnet stepper motors," *Mechatronics*, vol. 7, no. 6, pp. 521-536, 1997.

[7] S. E. Lyshevski, V. A. Skormin and R. D. Colgren, "High-torque density integrated electro-mechanical flight actuators," *IEEE Trans. Aerospace and Electronic Systems*, vol. 38, no. 1, pp. 174-182, 2002.

[8] S. E. Lyshevski, "Resonant converters: Nonlinear analysis and control," *IEEE Trans. Industrial Electronics*, vol. 47, no. 4, pp. 751-758, 2000.

[9] S. E. Lyshevski, "Precision control of mechatronic systems with electromagnetically-steered moving masses," *Int. Journal of Advanced Mechatronic Systems*, vol. 5, no. 5, pp. 306-316, 2013.

[10] S. E. Lyshevski, "Robust control of nonlinear continuous-time systems with parameter uncertainties and input bounds," *Int. Journal of Systems Science*, vol. 30, no. 3, pp. 247-259, 1999.

[11] J. Colomer, J. Brufau, P. Miribel, A. Saiz-Vela, M. Puig and J. Samitier, "Novel autonomous low power VLSI system powered by ambient mechanical vibrations and solar cells for portable applications in a 0.13μ technology," *Proc. Power Electronics Conf.*, pp. 2786-2791, 2007.

[12] G. Lijun, R. A. Dougal, S. Liu and A. Jotova, "Portable solar systems using a step-up power converter with a fast-speed MPPT and a parallel-configured solar panel to address rapidly changing illumination," *Proc. Applied Power Electronics Conf.*, pp. 520-523, 2007.

Light Trapping in Solar Cells Using Nanostructured Reflectors: A Responsivity-Based Model Applied to Waveguide Mode Matching

T. Shahriad * and S.R. Shepard**
Department of Electrical Engineering
Louisiana Tech University, Ruston, Louisiana
*tsh039@latech.edu and ***sshepard@latech.edu

ABSTRACT

A responsivity-based model is used to separate the photocarrier generation events ("good absorptions") which generate usefull electricity from the material-based ("bad absorptions") which generate heat. We utilize this to study the attempt to match the spacing of the nanoparticles on the front-scatterer (at a glass/silicon interface) to the parameters of the waveguide modes in the silicon slab waveguide. We find that the perturbation of the waveguide modes by the presence of nanoparticles can prevent such designs from providing resonances at the desired wavelengths. However, if the perturbation is small enough it will primarily only shift the longitunal component of the propagation (k-vector) so that it shifts the resonance peak in a way that can be accurately calculated. This enables a mode-matching design tool which doesn't require extensive calculation of the perturbed modes.

Keywords: solar energy, photovoltaics, light trapping, nanoparticle scattering

1 INTRODUCTION

The global demand for electricity is increasing day by day in an incredible rate. To cope up with this intense thirst of electricity, scientists are coming up with different highly efficient scientific models. Light trapping in solar cells using nanostructured reflectors is one of the promising scientific models which can lead us to an astonishing solution to this problem. Light trapping refers to means by which light not converted to electricity in a single pass through the PN junction of a photodetector or a photovoltaic (PV) is reflected back into the junction. Preferably any such non-converted light is trapped within the device, e.g., by total internal reflection or via some Bragg-type of resonant mode confinement. Many techniques exist and some of the most promising include: the use of plasmonics (arrays of conductive nanoparticles) primarily to generate surface waves to redirect light; the use of photonic crystal (PC) structures typically as back-reflectors; and the use of quantum dots (nanoscale encapsulated semiconductors) which fluoresce and scatter light into different directions.

Simple bandgap models don't accurately match experimentally determined responsivities (R) and many studies focus on material absorption (rather than the photo-carrier events, which are characterized by R). At the other extreme: definitive full-model calculations, including electron/hole dynamics in a PN junction, can take hours to run on a supercomputer. Herein we describe and employ new simulation tools (utilizing experimental R data) which facilitate quick but accurate simulations for understanding and optimization. An expedited simulation methodology is utilized in which the wave equation is solved numerically but electron/hole dynamics in the PN junction are modeled through the use of the experimentally determined responsivity of the unperturbed device.

In our simulations we incorporated the material absorptions utilizing appropriate Drude models since these affect the electromagnetic field distributions of the light-trap and account for the corresponding ohmic losses. We calculated the power delivered to the silicon as a function of wavelength, in db relative to that which would be delivered without any trap and the integral of that power spectral density times the responsivity of the device times the spectral density of the input i.e., Jsc, the short circuit current of the device under such illumination (and concentration via the light-trap). Along with the db-plots we also examined the electric field magnitude distributions for particular wavelengths of interest to understand what works well and what does not.

2 MODELS AND METRICS

We created our models with C-Si and glass slabs in order to turn it into a waveguide supported by the light reflection principles. Models used in nanophotonic simulations range from those designed to analyze macroscopic properties, all the way down to those which focus on the behavior of a few molecules; and models used in the field of light trapping have a similar range of complexity. Definitive fullmodel simulations require a PN junction and a careful incorporation charge carrier dynamics; as well as a good simulation of the optics (scattering at the nanoscale). Many such simulations have been achieved but the computational requirements are more demanding than say the use of software which provides only a good simulation of the optics. Thus, one often simply omits the simulation of a PN junction and models a PV light trap by calculating how much power is trapped into (and absorbed within) a particular region, e.g., a slab of silicon [2]. This is important and useful information. To turn this metric into one more germane to the electrical power delivered from a photodetector, or a PV, a simple bandgap model could be incorporated.

Therein, one assumes the EQE (external quantum efficiency) is a step function in frequency, f ; turning off when the frequency, hence the photon energy, is too low to enable photo-electrons to jump the energy gap into the conduction band. The responsivity, R (in units of electrical amps output, per optical watt input) clearly must be related to the EQE (the number of photo-electrons generated per number of photons absorbed) via R = (q EQE)/(hf) thus, a step in EQE becomes a ramp for R (where h is Planck's constant and q is the charge of the electron). However, such simple bandgap models dont match experimentally measured responsivities; as demonstrated in Fig. 1 wherein we see that the EQE calculated from the measured responsivity of a c-Si PV is not a step function. Thus we like a means of incorporating an experimentally measured R into light-trapping simulations without having to incorporate a PN junction.

Currently, light-trapping simulations which do not incorporate a PN junction utilize material absorption as the basis of their performance metric. Again, this is important and useful information. Its important to distinguish however that these material absorptions are not the absorptions which generate photo-carrier current (which are characterized by R). This difference is quickly realized by noting that recombination in a real device would increase absorption and Rayleigh scattering dominates absorption into the UV (as $^1/_{\lambda^4}$) but neither of these processes generate relevant photo-carriers in a photodetector or PV.

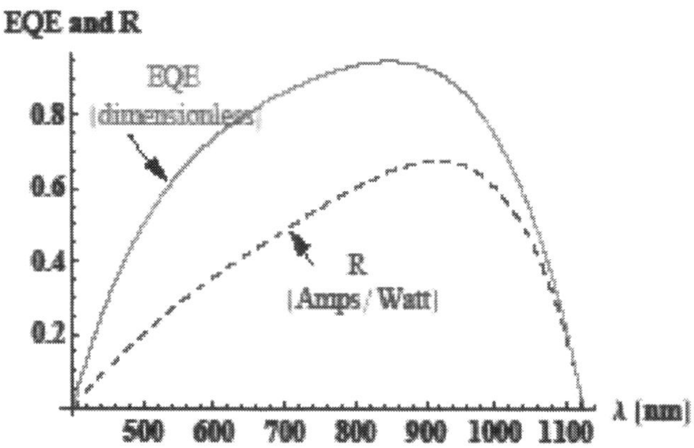

Fig 1: Responsivity of two devices.

Additionally, many papers refer to absorption and software packages (such as COMSOL which numerically solve the electromagnetic wave equation) can incorporate such as complex permittivity data or as materials models, e.g., the Drude model. It's important to distinguish however that these material absorptions are not the absorptions which generate photo-carrier current (which are characterized by R). This difference is illustrated schematically in Fig. 2 wherein the measurement of an absorption cross-section assesses the optical power transmitted through a sample, in contrast to the measurement of R which assesses the current generated at the PN junction (or within the PIN or NIP region, etc.). Recombination increases absorption and Rayleigh scattering dominates the absorption measurement into the UV (as $^1/_{\lambda^4}$) but neither of these processes generate relevant photo-carriers in a photodetector or PV.

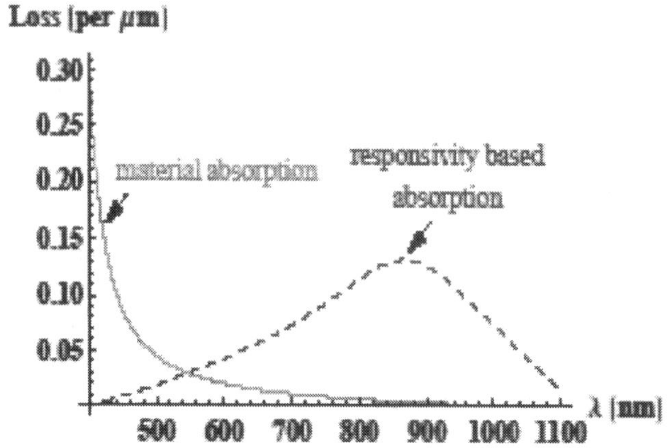

Fig 2: Schematic of the differences between aborption and photo-carrier generation.

3. MODE-MATCHED LIGHT TRAPPING

The silicon PV can function as a dielectric slab waveguide and we can incorporate a perfectly conducting electrode at the bottom of the PV via image theory [2]. Therein we simply use the anti-symmetric modes of a slab twice as thick as our actual one since they will automatically satisfy the boundary condition of zero electric field at the perfect conductor. The modes are set by the simultaneous solution of equations 1 and 2; where from boundary condions we have :

$$Y = - X \, Cot(X) \qquad \text{(eq.1)}$$

and from $k^2 = k^2_x + k^2_z$ (both inside and outside the slab) which leads to :

$$R^2 = X^2 + Y^2 \qquad \text{(eq.2)}$$

where $X = k_x \frac{d}{2}$ and $R^2 = (n_2{}^2 - n_1{}^2)\,(\frac{\pi d}{1am})^2$

and n_1, n_2 are the indecies of the glass and silicon respectively. To design a mode-matched light trap we can space the nanoparticles (by Lambda) in the front-reflector grating to match the longitudinal periodicity of the guided-mode: Lambda = 2 Pi / k_z = 285.31586537147224 nm for the wavelength for which we want enhanced trapping (we choose 900nm since that is near the peak responsivity). The material properties of silicon are accurately modeled via three Lorentzian terms [3] from which we find n_2 = Sqrt[13.20501118] = 3.6338700004265423 at 900nm. We use this in our initial design and let n_1 = 1.906271229 so that the glass cover can also funtion as an AR coating.

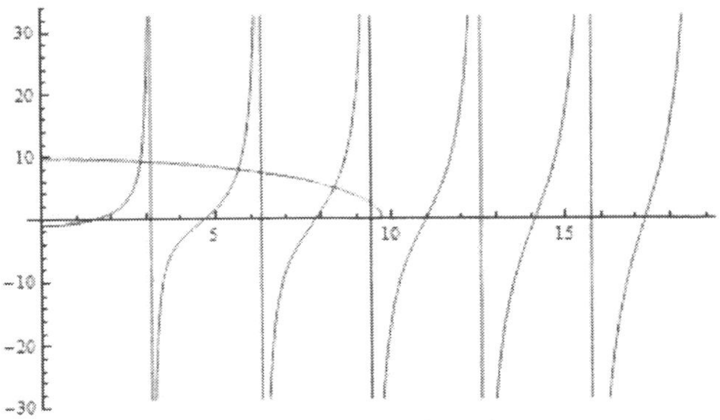

Fig 3: Mode Calculation from the formula.

Fig. 4(a) shows that for d = 400nm we have X from which we calculate k_z = 22.021857421069374 . Indeed, when we have no nano-enhanced trap then the mode pattern in Fig. 4(a) demonstrates “bright spots” (in red) spaced by 142.653 nm as predicted. When we add in nanoparticles the mode pattern is perturbed as shown in Fig.4 (b) and when the size of the nanowires increases we infer that L=(2Pi/k_z)/2 has been changed and we calculate the effective k_x and thereby the new wavelength of the perturbed mode via the simultatneous solution of eq.1 and eq.2. We obtain lam=400-1100nm and indeed Fig. 6(a) shows a strong resonance at that wavelength 567nm (of height = 4.37*10^(-07) units) whereas the initial design wavelength of 900nm only has a ;height = 1.1*10^(-10) units).

For larger perturbations, the effect goes beyond simply shifting k_z. However the above proceedure works for small perturbations and thereby provides a simple design tool for mode-matching in the presence of nanoscattering in those cases.

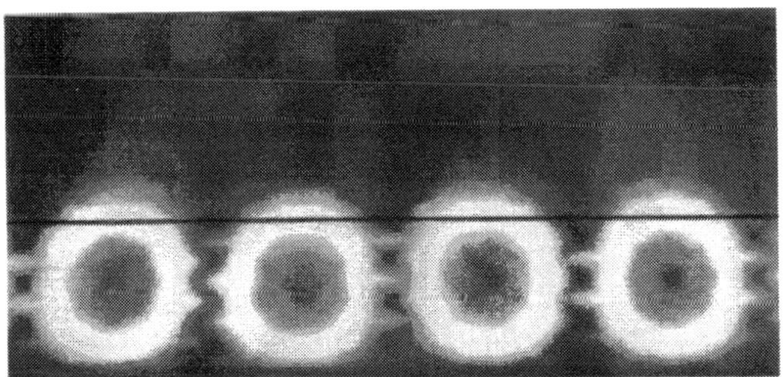

Fig 4(a): Modes created in model having no back reflectors.

We designed the models to get the resonance at 900 nm wavelength. Both Electric and Magnetic field were launched fom the top into the models to find out the efficiency of each model.

Fig 4(b): Perturbed Modes after using rectangular back reflectors.

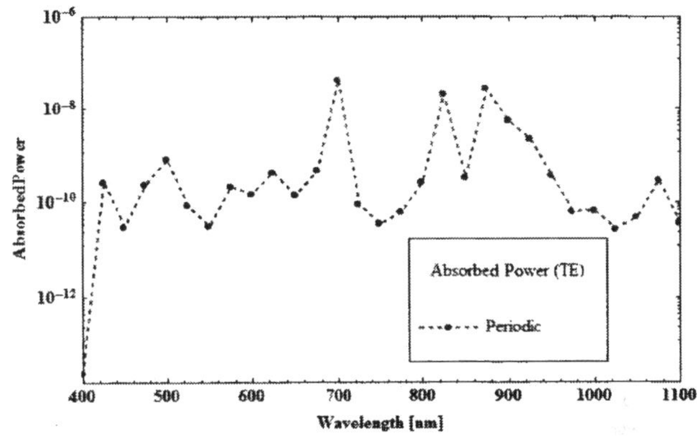

Fig 5(a): Periodic Back Reflector Enhanced model for absorbed power with respect to wavelength at TE Polarization in 25nm Resolution.

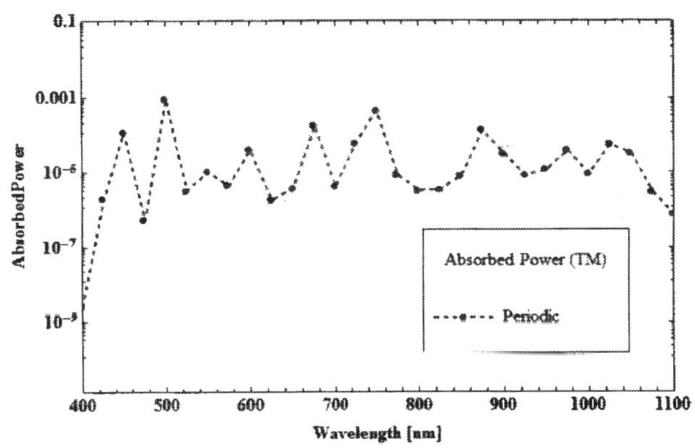

Fig 5(b): Periodic Back Reflector Enhanced model for absorbed power with respect to wavelength at TM Polarization in 25nm Resolution.

We did all the simulations in 25nm resolutions. But from the experience data we saw that the resonance frequency got shifted from 900nm to some other frequency. To find out the the new position of the resonant frequency we launced Electric and Magnetic field into our models from the left end. Due to mode purturbation we got distances between the adjacent modes for specific models at specific wavelenths for side launch. We did the back calculation from this mode perturbation distance to detect the new position of the resonant frequency.

After getting the new position of the resonant frequency we went for higher resolution (1nm) in our simulations to compare the experiment data with the theoritical data. From the experiments we got the highest responsivity points at the theoritically recalculated resonant frequency points.

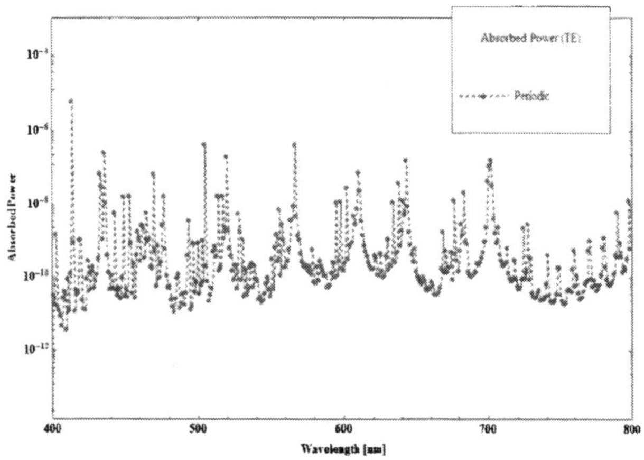

Fig 6(a): Periodic Back Reflector Enhanced model for absorbed power with respect to wavelength at TE Polarization in 1 nm Resolution from 400nm to 799nm.

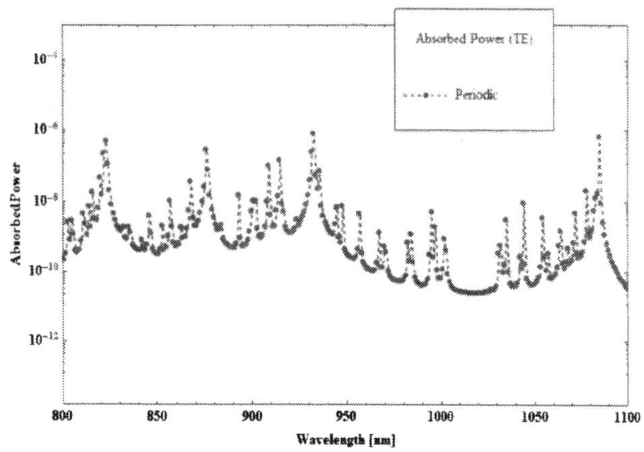

Fig 6(b): Periodic Back Reflector Enhanced model for absorbed power with respect to wavelength at TE Polarization in 1 nm Resolution from 800nm to 1100nm.

We designed our new models with 1nm resolution to get all possible data set in 400nm-1100nm range. From the high resolution data we got to see that the peaks at cetain wavelenghts were even implausibly bigger than they appeared in low resolution simulations.

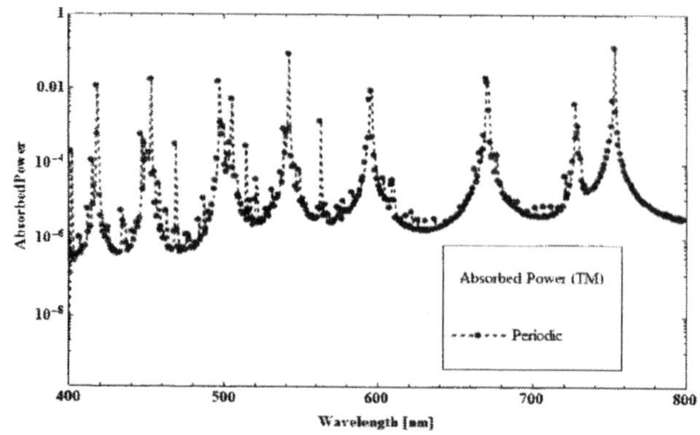

Fig 7(a): Periodic Back Reflector Enhanced model for absorbed power with respect to wavelength at TM Polarization in 1 nm Resolution from 400nm to 799nm.

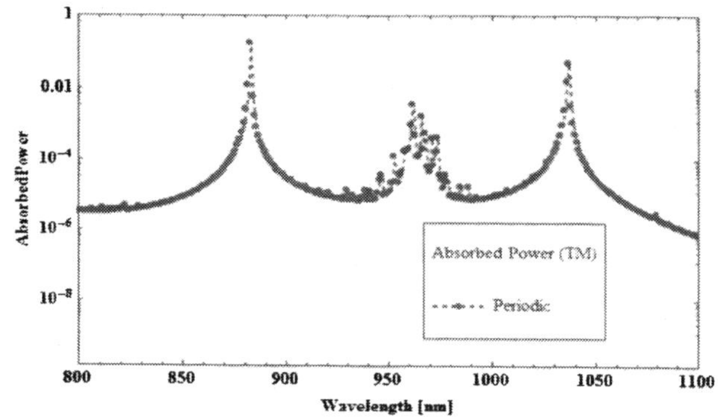

Fig 7(b): Periodic Back Reflector Enhanced model for absorbed power with respect to wavelength at TM Polarization in 1 nm Resolution from 800nm to 1100nm.

CONCLUSION

The best light trapping model can be designed by turning it into a wave guide using the calculated resonant frequency where we can get the highest level of responsivity of particular light polarization. We did the responsivity based waveguide mode maching and came up with the highly efficient model inducing peak responsivity at its resonant wavelengths.

REFERENCES

[1] S. Shepard, L. Patel, F. Rubaiat, "Periodic and Psuedo-Random Back-Reflector Nanoparticle Enhanced Light-Trapping in a Silicon Photovoltaic," Nanotech 2014 Vol. 3, 343 – 345,2014.

[2] Julius Adams Stratton, "Electromagnetic Theory," 2007.

[3] Xianqin Meng, Emmanuel Drouard, Guillaume Gomard, Romain Peretti, Alain Fave, and Christian Seassal, "Combined front and back diffraction gratings for broad band light trapping in thin film solar cell," Optics Express, Vol. 20, Issue S5, A560-A571, 2012.

Potential of Water-Responsive Materials to Harvest Energy from Evaporation

A.-H. Cavusoglu*, X. Chen** and O. Sahin***

*Department of Chemical Engineering, ac3236@columbia.edu
**Department of Biological Sciences, xc2239@columbia.edu
***Department of Biological Sciences and Physics, Columbia University, New York, NY, USA, sahin@columbia.edu

ABSTRACT

Water-responsive materials swell/shrink in response to changes in relative humidity (RH) and can potentially harvest energy from evaporation. Here, we theoretically investigate harvesting energy from naturally evaporating water due to varying weather conditions. We model the effects of energy harvesting on evaporation rate and the resulting power output and intermittency. Calculations over a range of locations across the United States predict power productions with reduction in evaporative losses. Non-steady state calculations at three test locations predict that power output is robust against daily and yearly variations in power demand & weather conditions. We find that this system can potentially surpass the energy density of wind-based approaches with lower intermittency than solar systems. These results suggest that further research into water-responsive materials could supply devices that surpass existing energy platforms.

Keywords: water, alternative energy, stimuli-responsive materials, hydrology, environment

1 INTRODUCTION

Evaporation, the flow of latent energy from a higher to lower potential, is a powerful energy flow at the Earth's surface. It is estimated that the global energy flux due to evaporation is 80 W/m^2 [1] (Figure 1). Water-responsive materials in nature, powered by evaporation, achieve a diversity of vital tasks, such as plant movement and molecular transport. Several biologically inspired materials have been developed and investigated that mimic materials observed in nature [2, 3].

With growing interest in water conservation and alternative energy, we are motivated to investigate the effect that water-responsive materials have on evaporation. These materials swell/shrink upon sorption/desorption of water and are performing work to undergo these changes in shape. It may be possible to use these materials in a system to harvest work from evaporation through a cycle of water sorption/desorption.

The power output of this system depends on the evaporation rate E and the amount of work done per unit of evaporating water w, which is determined by the load on the water-responsive material. Here, we investigate the potential of harvesting energy from naturally evaporating water due to natural weather conditions. We model the effects on evaporation rate, the resulting power output, and intermittency concerns by using a method developed by Penman [4]. Steady state calculations over a range of locations across the United States predict the average energy flux and net water savings. Non-steady state approaches at three test locations predict daily and yearly variations in power output.

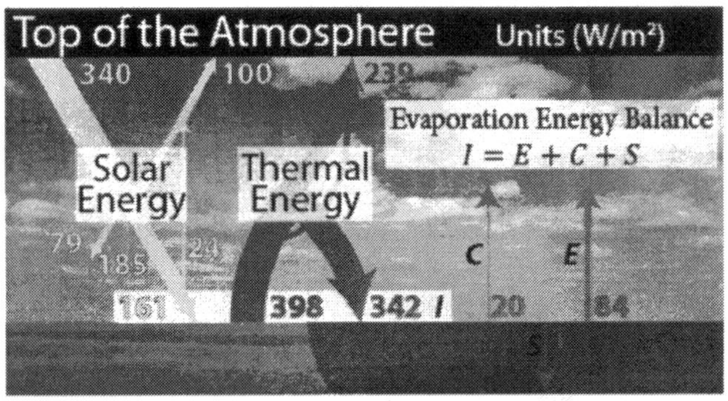

Figure 1: Energy Balance of the Atmosphere & Surface. Adapted from [1]

2 THE MODEL

The rate of evaporation is governed by the mass and heat transfer characteristics of the atmosphere and the energy balance between incoming energy from solar radiation, and heat losses due to conduction, convection, and thermal radiation. Penman's combination equation synthesizes the mass and energy balance to predict evaporation rates from meteorological data for the steady state condition. This equation has been modified to predict evaporation rates from plants [5] and soil [6] where evaporation is restricted. As a result of restriction, vapor pressure at the evaporating surface drops below saturation level. This reduction is determined empirically in the case of plants and soil.

However, in the presence of energy conversion devices, the vapor pressure is determined by the amount of work w done for a unit amount of evaporating water. This allows us to relate the work load w to the evaporation rate E. Figure 2 presents an illustration of the proposed model scheme.

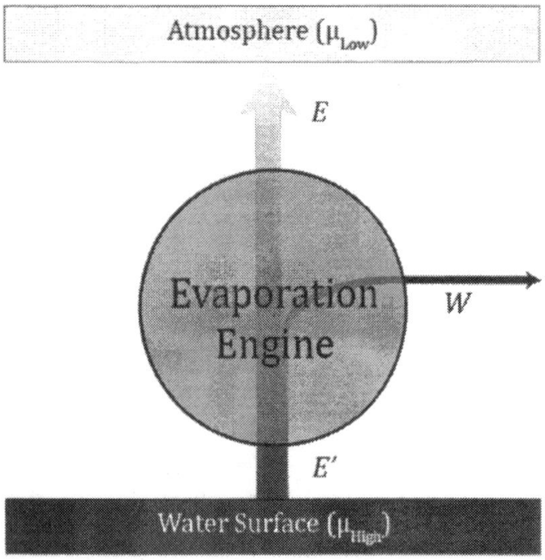

Figure 2: Scheme of an evaporative energy engine.

3 STEADY-STATE PREDICTIONS

Assuming that over sufficiently long periods of time the average temperature of an evaporating body of water remains constant, we can determine the effort that ambient weather conditions will have on the potential for evaporative energy harvesting.

3.1 Energy Availability of Typical Weather

Here, we calculated the evaporation rate, power density, and surface temperature that would result from evaporative energy harvesting as a function of the load on a potential evaporative energy harvester for mild weather conditions (Figure 3). As the work load increases, the rate of evaporation declines as the surface temperature rises. Evaporation ultimately stops at a certain work load, at which point heat is transmitted back to air entirely in the form sensible heat. Importantly, the energy flux extracted from evaporation peaks at a certain work load.

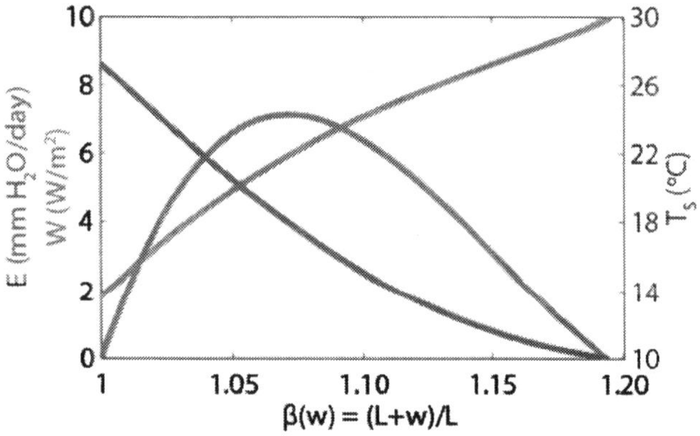

Figure 3: Predictions for Surface Temperature (red), Evaporation Rate (blue), and Power Density (green) as a function of work load for mild weather conditions.

3.2 Annual Steady State Prediction in Daggett-Barstow, CA

This calculation is expanded to give us insight into the variability in energy availability over a year. Using appropriate data for Daggett-Barstow, CA [7], we can predict the maximum power output and corresponding water savings for each day in the dataset (Figure 4). We see that during the warm, dry months of the summer, energy densities of 15 – 20 W/m^2 are available.

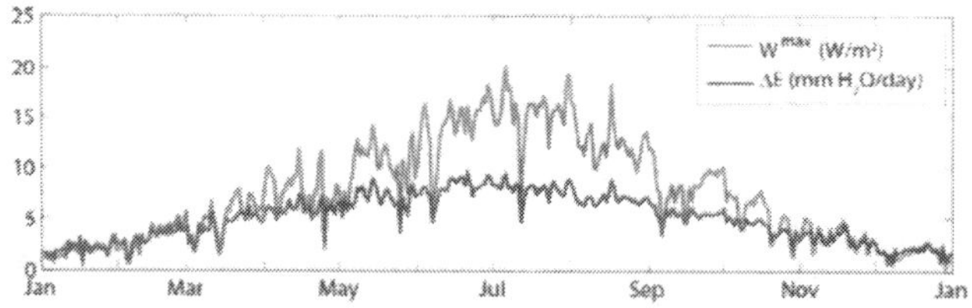

Figure 4: Maximum Daily Power Generation and Corresponding Water Savings in Daggett-Barstow, CA.

4 INTERMITTENCY & THE NON-STEADY STATE

Intermittency frequently challenges renewable energy platforms, where insufficient power is available due to external constraints such as variable demand and changing environmental conditions. To explore this challenge, we modeled a dynamically controlled non-steady state power generator subject to varying power demand and weather.

4.1 Variable Power Delivery

Each simulation model is run for three simulation years to confirm adequate stabilization and memory loss of initial conditions. Figure 5 illustrates the final year of a simulation at Daggett Barstow, CA. For clarity, the results for surface temperature and evaporation rate are excluded. In this low power density demand regime, the model matches scaled power demand over 99% of the time, exhibiting some power failure in winter days where global horizontal irradiance is low and relative humidly is high.

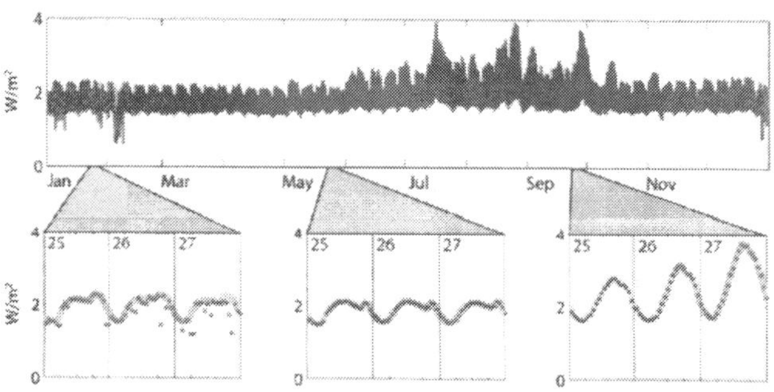

Figure 5: Power output (green) varying to match power demand (grey) during a simulation year. Matches demand over 99% of the time in Daggett Barstow, CA.

4.2 Comparison of 3 Test Locations

We observe that for each test location, the model system eventually saturates and provides no more power output with increasing power demand. A comparison of these test locations is shown in Table 1. From these results, we see that evaporative energy harvesting has the potential to provide up to 8 W/m^2 of power density along with 81% capacity with appreciable water savings.

Location	<PD> W/m^2	W^{max} W/m^2	CF @ 95% %	ΔE^{min} mmH_2O/day
CA	11791	8.08	66.9	4.62
TX	12541	5.05	80.9	3.32
NY	6292	2.46	64.8	2.05

Table 1: Comparison of Test Locations in CA, TX, & NY.

5 RENEWABLE ENERGY POTENTIAL

Another question of interest is how this model compares to wind and solar technologies. Table 2 compares this evaporative energy harvesting method to two prevalent renewable technologies, wind power and photovoltaic (PV) solar power. Data from this table is extrapolated from DOE technical reports [8-11]. The power density that can be realized by this model is squarely in the range of current renewable energy technologies, with the added benefit of increased capacity factor compared to PV solar and wind.

Clean Energy Technology	W^{peak} W/m^2	W^{gen} W/m^2	CF %
Evaporation	21.5	5.20	> 60
Wind	8.24 [8]	2.90 [9]	30 – 52 [11]
PV Solar	44.9 [8]	8.06 [10]	16 – 30 [11]

Table 2: Comparison of Evaporation Energy Harvesting Results to Wind & PV Solar Systems.

6 CONCLUSION

Evaporation across the contiguous United States can potentially provide power densities higher than current wind power plants and comparable to current solar power plants. The development of water responsive materials allows for the potential to harvest energy from evaporation. This model relates the change in evaporation rate due to work load and generates predictions based on typical weather conditions. This model predicts that harvesting energy from evaporative water flows may be possible and could provide water savings. These results suggest that further research into water responsive materials could result in novel energy harvesting devices.

REFERENCES

1. Wild, M., et al., *A New Diagram of the Global Energy Balance.* Radiation Processes in the Atmosphere and Ocean (Irs2012), 2013. **1531**: p. 628-631.
2. Chen, X., et al., *Bacillus spores as building blocks for stimuli-responsive materials and nanogenerators.* Nat Nanotechnol, 2014. **9**(2): p. 137-41.
3. Ma, M., et al., *Bio-inspired polymer composite actuator and generator driven by water gradients.* Science, 2013. **339**(6116): p. 186-9.
4. Penman, H.L., *Natural evaporation from open water, bare soil and grass.* Proc R Soc Lond A Math Phys Sci, 1948. **193**(1032): p. 120-45.
5. Monteith, J.L., *Evaporation and environment.* Symp Soc Exp Biol, 1965. **19**: p. 205-34.
6. Barton, I.J., *Parameterization of the Evaporation from Non-Saturated Surfaces.* Journal of Applied Meteorology, 1979. **18**(1): p. 43-47.
7. Wilcox, S. and W. Marion, *Users Manual for TMY3 Data Sets (Revised).* 2008. p. Medium: ED; Size: 58 pp.
8. Laboratory, N.R.E. *Land Use by System Technology.* 2013 [cited 2015 04/07/2015]; Available from: http://www.nrel.gov/analysis/tech_size.html.
9. Denholm, P., et al., *Land Use Requirements of Modern Wind Power Plants in the United States.* 2009. p. Medium: ED; Size: 46 pp.
10. Ong, S., et al., *Land-Use Requirements for Solar Power Plants in the United States.* 2013. p. Medium: ED; Size: 47 pp.
11. Laboratory, N.R.E. *Utility-Scale Energy Technology Capacity Factors.* 2013 [cited 2015 04/07/2015]; Available from: http://www.nrel.gov/analysis/tech_cap_factor.html .

A Self-Sufficient Weather Sensor Network for Microgrid Stabilization

C. Chong, Z. Dorman, K. Luong*, C. Obatake, A. Pham, R. Walser, A. Kuh**
* University of Hawaii at Manoa, HI, USA, luong97@hawaii.edu
** University of Hawaii at Manoa, HI, USA, kuh@hawaii.edu

ABSTRACT

In the Galvin Electricity Initiative, a microgrid is defined as a "small scale version of the central electricity system" [1]. Renewable energy sources, such as photovoltaic (PV), are intermittent, which raises concern when considering their integration on a large scale [2]. Through the creation of a self-sufficient, wireless weather sensor network, we present a step towards forming a method to track current weather conditions and predict future patterns. Each node in the network relays solar irradiance, barometric pressure, relative humidity and temperature to a server via a transceiver. Data is collected at approximately one sample per second and sent back to a server where it is stored into a relational database. Deployment on the University of Hawaii at Manoa campus has begun as a part of the Renewable Energy and Island Sustainability (REIS) initiative.

Keywords: Renewable energy, sensor nodes, wireless sensor network

1 INTRODUCTION

Hawaii is highly dependent on the importation of energy. In 2012, 93% of the energy Hawaii consumed was imported. In addition to this, a majority of electricity is generated from petroleum. In December 2014, 529 GWh of electricity were generated from petroleum, whereas only 102 GWh were created from renewable energy sources [3]. Due to Hawaii's heavy reliance on imported oil, it has the highest electricity prices per kilowatt in the nation. In January 2015, residents of Hawaii paid an average of 33.34 cents per kWh, which is over 20 cents above the national average [4]. In the future, our reliance on fossil fuels could be problematic when these sources become depleted.

Using alternative energy sources, such as photovoltaics (PV), have become popular in Hawaii as Hawaii now has the highest penetration of distributed solar PV in the country [9]. With large penetration of solar PV there is serious concern about the stability of the grid as PV energy generation is intermittent [2,9]. Electric companies must account for underproduction when generation of energy is inadequate and overproduction when the consumption of energy is lower than the generation of it. The outdated grid in Hawaii has an unknown tolerance to intermittent renewable energy sources resulting in an inability to integrate a large number of these sources. To address this concern, it is imperative to form a method to track current weather and environmental conditions and predict future patterns.

Through the creation of a self-sufficient weather sensor network, we present a step towards being able to predict these future patterns. The sensor nodes will be deployed on roofs of buildings around the University of Hawaii at Manoa campus. This network will monitor and relay reliable weather data for at least two years. In order to create a sensor network, a low power consumption sensor module has been designed and built. It communicates in a mesh-network configuration (relays data between sensor modules) and is self powered by a 15,600 mAh lithium ion battery pack and a 5.6 W solar panel. Each node in the wireless weather sensor network will relay solar irradiance, barometric pressure, relative humidity and temperature to a server via an XBee transceiver.

Data is collected at approximately one sample per second (this rate can be varied); a total of up to 86,400 samples per day can be collected and sent back to a server where such data is stored into a relational database (PostgreSQL). Data can then be accessed directly through the database or by means of a web application, which allows users to choose which data intervals they wish to download.

This data will be used to study and analyze the behavior of renewable energy sources, specifically solar irradiance patterns, throughout the year. The analysis and weather patterns found can be used in the future to integrate PV in optimal places on the University of Hawaii at Manoa campus.

2 MODULE OVERVIEW

The sensor node is a low cost, low power, wireless module that collects weather data, including solar irradiance, temperature, humidity, and pressure. Data from each sensor is sent back to a server for storage and analysis. Several sensor nodes will eventually be deployed on rooftops of the University of Hawaii at Manoa campus to form an environmental sensor network that monitors environmental conditions both spatially and temporally throughout the campus.

Figure 1: Primary components of current sensor node: main board and sensor board

2.1 Sensors

The pyranometer (solar irradiance sensor) used in the sensor module is the Apogee SP-110 shown in Figure 2. The pyranometer is self-powered, and has a sensitivity of 0.20 mV per Watt m^2 [5]. The current module has a 16 bit analog to digital convertor (ADC) with a 0.05 mV resolution to convert the analog solar irradiance signal to digital.

Figure 2: Apogee SP-110

The current module also has a Honeywell humidity and temperature sensor, model HIH6031, as shown in Figure 3. The sensor has a humidity accuracy of +/- 4.5% RH, and temperature accuracy of +/- 1%. The sensor has a 14 bit ADC and uses I^2C, a communication protocol, to send data to the microcontroller [6].

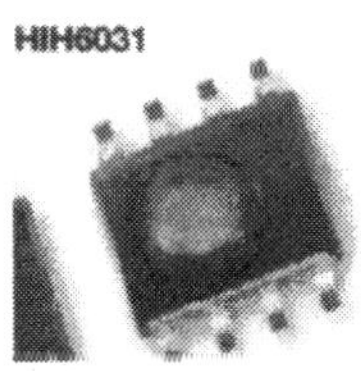

Figure 3: Humidity and temperature sensor

The barometer (pressure sensor) used in the current node design is the MPL115A2 from Freescale Semiconductor shown in Figure 4. The sensor can measure from 50 kPa to 115 kPa with +/-1 kPa accuracy [7].

Figure 4: MPL115A2 barometer

2.2 Power

In outdoor tests, the Lithium-ion battery in the previous sensor module design would discharge below the recommended 30% of energy during the rainy season. The current module has a 3.7 V 15600 mA Li-ion battery, which is twice the battery capacity of the previous module. The battery is charged by a single 6 V 5.8 W solar panel. The solar panel has dimensions of 7x8.5 in^2. A MCP7381-2CAI/ML IC is used as the charger for the battery [8].

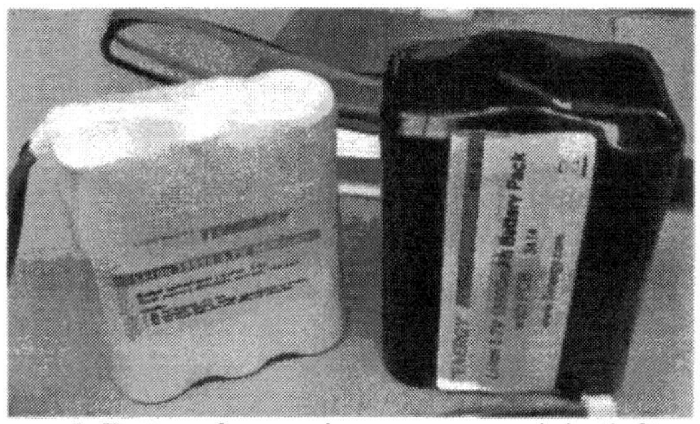

Figure 5: Battery for previous sensor module (left), and battery for current sensor module (right)

3 HOUSING

The first prototype of the housing for the current node is shown in Figure 6. The housing was fabricated using a 3D printer and Acrylonitrile Butadiene Styrene (ABS), and its dimensions are $8.5\times7\times5$ in^3. The housing consists of two pieces: the main box with the mount for the solar panel on top and the cover to seal the box and mount it to the tripod stand. Four screws on the side hold these two pieces together. There is an exposed area in the box for the sensor board. On the bottom cover for the housing, there is a custom mount joint where different types of stands can be used to mount the module. In this prototype of the housing, a camera tripod was used as the stand for the module shown in Figure 7.

Figure 6: Side and bottom view of Cranberry housing

Figure 7: Sensor module mounted to a camera tripod

4 NETWORK

Data is sent from an arduino-based unit to the laboratory server over an XBee mesh network (extended IEEE802.15). The mesh network has self-healing and re-routing capability, remaining robust through an individual network failure. This allows for larger deployments over a large area, as long as there are enough nodes available to re-route data packets in case of a failure.

The installed XBee adapters are remotely configurable through an API provided by the XBee protocol. This includes features such as power control, sampling rate adjustment, and other parameters such as 128-bit AES encryption.

A master coordinator module has access to and controls the network. Each node is given a 16-bit address, resulting in the potential number of radios in network to be 65,536. Each node, when configured as a router, can act as a bridge for any other nodes that are out of range of the coordinator.

In addition to relying on XBee transceivers to relay data back to the laboratory data store, the network can also adopt a hub-and-spoke network model, with intermediary base stations providing coverage through Ethernet or WiFi. These intermediate base stations can be easily configured using any linux-capable computing device. A successful remote deployment configuration has been accomplished using the Raspberry Pi computing platform.

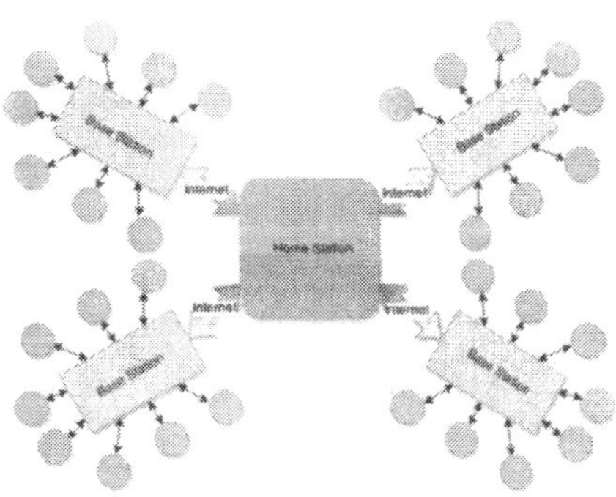

Figure 8: Hub-and-spoke network model for long-distance data relaying

5 DATABASE

Collected data is stored in a high capacity linux server running Redhat Enterprise 6.3 (Santiago) within a PostgreSQL database installation. This server allows for distributed access to be able to access data for remote analysis. Exploratory analysis can be performed by any group that has access to the server.

Because PostgreSQL is a well documented and open-source database installation, most modern programming languages have frameworks that support calls to a PostgreSQL database. Therefore, simple visualization scripts can be written in most languages within a reasonable amount of lines of code.

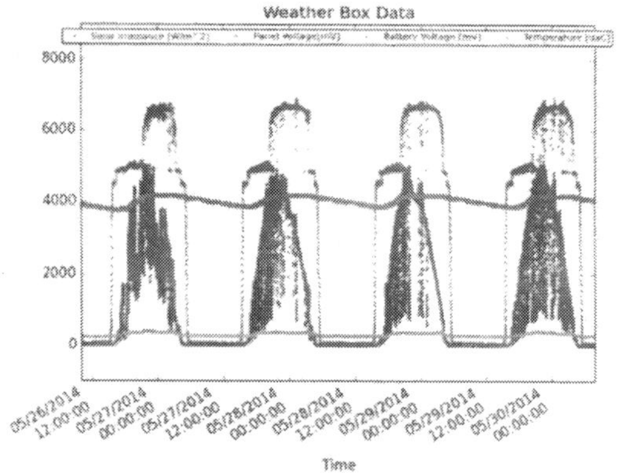

Figure 9: Simple data visualization using Python

To help facilitate data access and collaboration, a simple Representational State Transfer (REST) server was built to serve a simple data access Application Programming Interface (API). Likewise, a use-case for this REST server is a web-based on-demand visualization script written in Javascript, which is also mobile friendly.

Easy access to on-demand visualization and analysis allows for quicker iteration and verification of experimental data obtained from hardware testing. A hardware stack is shown below, describing the various technologies that are used in our database configuration.

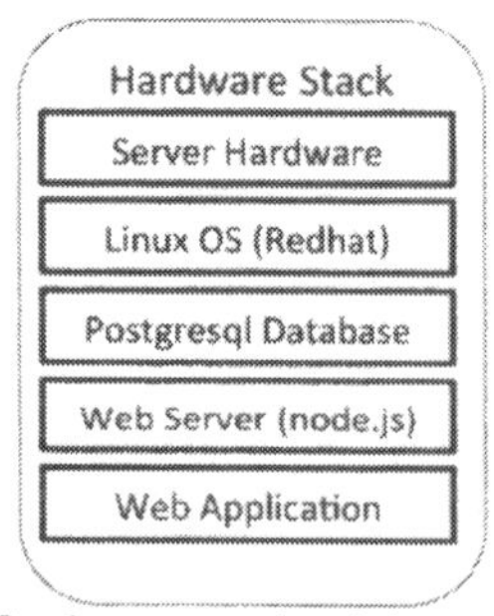

Figure 10: Hardware stack of the technologies

 TechConnect Briefs 2015, TechConnect.org, ISBN 978-1-4987-4728-8

6 DATA ANALYSIS

We used a suite of online adaptive filtering algorithms; steepest descent (SD), least mean squares (LMS), normalized least mean squares (NLMS), and recursive least squares (RLS) for solar forecasting [10]. The data collected from the sensor nodes was used to perform one step prediction of solar irradiance with a 5-tap moving average filter. Figure 11 shows the data used to test the algorithms. The algorithm uses root mean squared error (RMSE) as a performance criterion and considers the rate of convergence of the algorithms. Figure 12 shows the RMSE curve after 1000 training iterations. Another set of data was used on the same system that was obtained from the training to do cross validation. The results of cross validation RMSEs are SD 67.7, LMS 71.1, NLMS 71.6, and RLS 67.7. The rate of convergence for the algorithms ranked from highest to lowest is RLS, NLMS, LMS. The MSE of the algorithms ranked from highest to lowest is RLS, LMS, and NLMS.

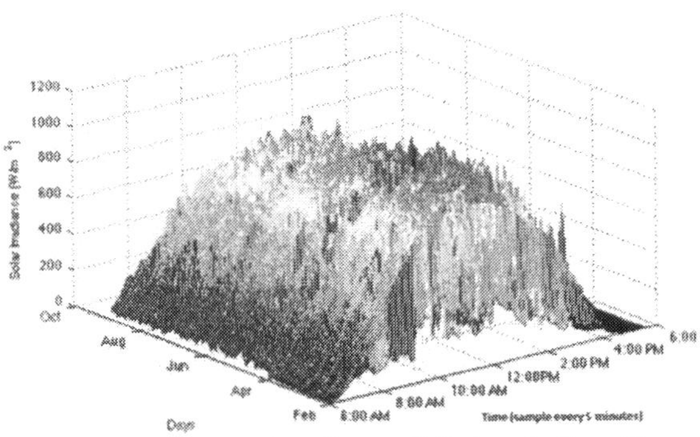

Figure 11: Solar irradiance (SI) data before filter with moving average filter, and standard normalization

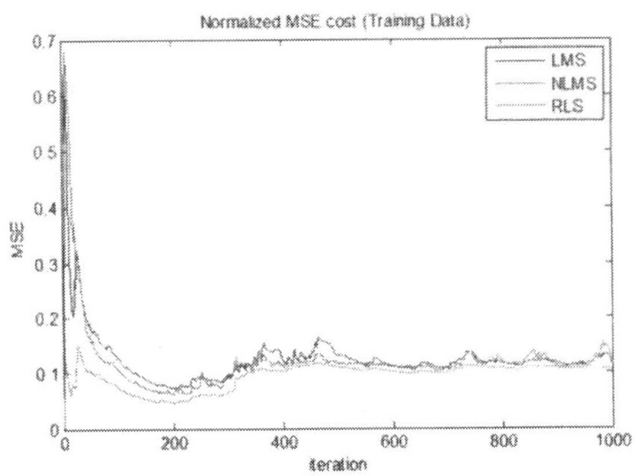

Figure 12: ***Training cost*** - LMS 68.7, NLMS 69.2, RLS 65.8

7 FUTURE DIRECTIONS

The next generation of weather sensor nodes will be an iteration of the current generation, providing lower power use, greater programming flexibility and offering a greater range of sensors.

In addition to hardware improvements, we hope to integrate additional data storage and sharing improvements into the server, lowering the cost entry for data analysis by researchers.

8 REFERENCES

[1] Galvin Electricity Initiative, 'What are smart microgrids?', http://www.galvinpower.org/microgrids/.

[2] I. P. Gyuk and S. Eckroad, 'EPRI-DOE Handbook Supplement of Energy Storage for Grid Connected Wind Generation Applications', Technical Report 1008703, EPRI and DOE, 2004.

[3] U.S. Energy Information Administration, 'Hawaii State Profile and Energy Estimates', *eia.gov*, 2014. [Online]. Available: http://www.eia.gov/state/?sid=HI#tabs-4. [Accessed: 12- Feb- 2015].

[4] U.S. Energy Information Administration, 'Electric Power Monthly', *eia.gov*, 2015. [Online]. Available: http://www.eia.gov/electricity/monthly/epm_table_grapher.cfm?t=epmt_5_6_a. [Accessed: 12- Feb- 2015].

[5] Apogee Instruments, 'Pyranometer Models SP-110 and SP-230', Pyranometer Datasheet, 2014.

[6] Honeywell, 'Honeywell HumidIcon ™ Digital Humidity/Temperature Sensors HIH6000 Series • ±4.5 %RH Accuracy', Humidity/Temperature Sensor Datasheet, 2013.

[7] Freescale Semiconductor, 'Miniature I2C Digital Barometer', Barometer Datasheet, 2013.

[8] Microchip, 'Stand-Alone System Load Sharing and Li-Ion/Li-Polymer Battery Charge Management Controller', Battery Charge Management Controller Datasheet, 2013.

[9] E. Wesoff, 'Friction at the Grid Edge: Hawaii's PUC Orders HECO to Approve Solar Rooftops', Greentech Media, Mar. 2, 2015. [Online]. Available: https://www.greentechmedia.com/articles/read/Friction-on-the-Grid-Edge-Hawaiis-PUC-Orders-HECO-to-Approve-Solar-Rooftop.

[10] S. Haykin, 'Adaptive Filter Theory', 5th Ed. Prentice Hall, 2013.

Intelligent Energy Management and Power Distribution System

Hongxia Chen,
Duli Technology LLC, hchen@dulitechnology.com

ABSTRACT

This paper presents an intelligent energy management and power distribution panel (IEMPDP) that uses a fundamentally different approach to increase energy efficiency and reduce renewable energy costs. By automating the key component of the local electric distribution system, it establishes an intelligent gateway to the power grid, local power distribution, and an energy management center. With integrated and comprehensive monitor and control, it reduces renewable energy system cost, enables local grid tied PV system to provide more stable and predictable power, allows consumers to more efficiently use energy, and helps utility power grids to be more robust and reliable.

Keywords: energy efficiency, solar energy, intelligent power electronics, intelligent energy management

1 INTRODUCTION

For decades, utility power has been connected to residential homes through a basic service panel (or a load center) where all connections are static and inflexible. This significantly limits the potential for energy efficiency. By automating the key component of the local electric distribution system, the hardwired service panel, with patent pending technologies, the IEMPDP will not only increase energy efficiency but also reduce the cost of renewable energy systems.

For example, currently most solar photovoltaic (PV) systems installed in residential homes are grid tied PV systems without energy storage due to the prohibitive costs when the energy storage system is included. The problems with grid tied PV systems without energy storage are: 1) to prevent islanding they cannot operate during a power outage; and 2) they may cause unstable and unpredictable power flow from the local electric system to the utility grid because of solar power fluctuations due to weather variations. The IEMPDP solves these problems by simplifying the structure of grid tied PV systems. This is accomplished through the integrated and comprehensive control to the local electric system and to the connection with the power grid. With the IEMPDP, the sub service panel and any additional wiring are no longer needed when installing a PV system with battery backup. It will also simplify the structure of grid tied PV systems that include energy storage. The capabilities of automatically monitoring and controlling the main and branch power lines will also enable consumers to more efficiently use energy and reduce electricity cost. These will not only reduce the PV system cost but also bring more value to consumers.

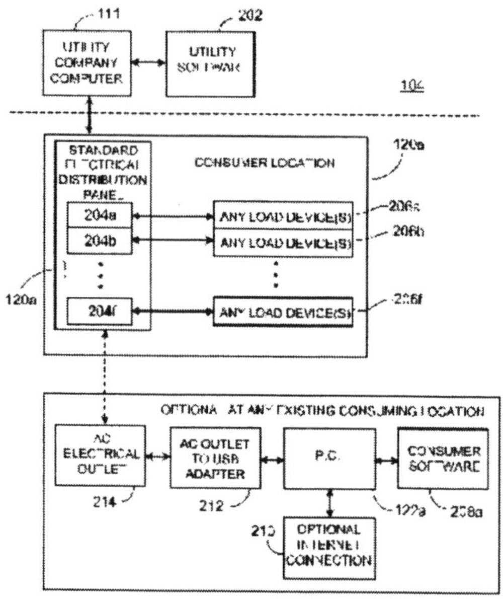

Figure 1

Why is the IEMPDP better than current power distribution panel technologies used in intelligent power distribution systems and building automation? The main difference is the approach that is used to solve these problems. Figures 1-3 illustrate how the current technologies work. Figure 1 is a block diagram of the intelligent power distribution panel (IPDP) [1]. It consists of numerous intelligent circuit breakers (ICB) that can communicate with a remote controller via power line communications or by other means. Figure 2 shows a block diagram of the ICB [1]. Each ICB includes a microprocessor or microcontroller and the means to monitor and control the connected power line as well as to communicate with a remote controller. Figure 3 shows an application environment using this technology [1]. These figures illustrate that current technologies use a top down approach to increase energy efficiency. Since this approach is effective at managing energy consumption in commercial and institutional buildings, it has been widely used in building automation [2]. This top down approach, however, is not effective in residential buildings because of several drawbacks.

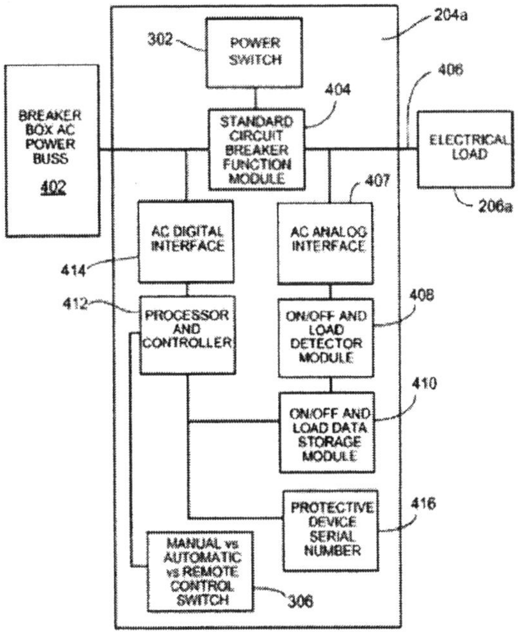

Figure 2

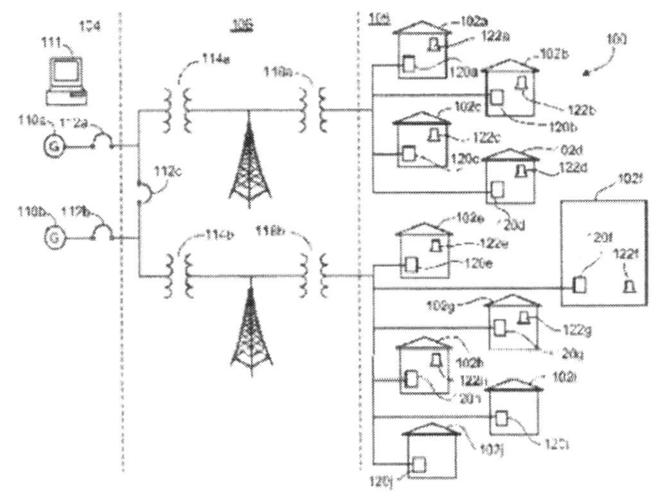

Figure 3

One drawback is the cost. Each ICB includes a microcontroller and communication interface plus a sensor, a circuit breaker, switches, etc. (see Figure 2). Each ICB costs about \$160-\$600 depending on the features that are included. This makes it very expensive for residential home owners. When adding the remote controller, the panel box, and accessories, the cost is too high given the benefits.

Another drawback is the disconnection between what the technology provides and what consumers want. Technologies that use the top down approach empower utilities to monitor and control power consumption or the usage of appliances in residential home. This enables utilities to reduce the stress on the power grid during peak hours, and therefore increase energy efficiency. But utilities cannot monitor and control the usage of the appliances inside the houses without consumers' consent, and most consumers are not comfortable giving this control to utilities and therefore do not buy into such systems.

The third drawback is that a top down approach has limited capability. The microcontroller in each ICB can only 'see' the condition on the branch power line that it is connected to, and therefore to coordinate any actions a controller is required to oversee the entire local electric system. When quick decisions are required (e.g., dynamically distributing the power according to different situations), the remote controller has to receive information and then send commands via the internet (broadband or wireless). This takes too much time and is therefore not reliable.

For these reasons electric power distribution systems in residential homes are still using decades old technology. A different approach is necessary to upgrade the local electric system to be robust, intelligent, and more reliable.

2 ADVANTAGES OF THE IEMPDP

In contrast, the IEMPDP discussed here uses a bottom up approach to increase energy efficiency [3]. Figure 4 shows a block diagram of the IEMPDP. The IEMPDP includes a central control unit (CCU), a programmatically controlled circuit breaker panel (PCCBP), and an interface unit (IU). The CCU communicates and controls each PCCB and IU through electronic signal connections. Each PCCB in the PCCBP includes a sensor, an AC switch, and control logic. The AC switch can be controlled by the sensor through control logic to provide over current protection, and by the CCU directly. PCCBs are responsible for automatically monitoring and protecting the branch power lines. Based on the information received from all of the sensors, including the sensors in each PCCB, the CCU can make informed real-time decisions. If energy storage is implemented, the CCU coordinates with the charge controller to regulate the usage of the energy storage system based on an assessment of the current dynamics of the electrical ecosystem. For example, a super capacitor and battery hybrid energy storage system (HESS) [4][5][6][7] would be monitored and controlled by the charge controller and CCU. The super capacitors are charged and discharged to absorb power fluctuations and maintain the SOC of the batteries at the optimum level. The batteries in the HESS might be discharged once a day during peak hours to help reduce the stress on the grid. When there is a problem on the grid, the local electric system will be disconnected from the power grid and the inverter will be switched from grid connected mode to standalone mode to enable the PV system to continue to operate if it is daytime. Through the IU, the CCU can communicate with the remote servers of the electric utility, smart appliances, and mobile devices.

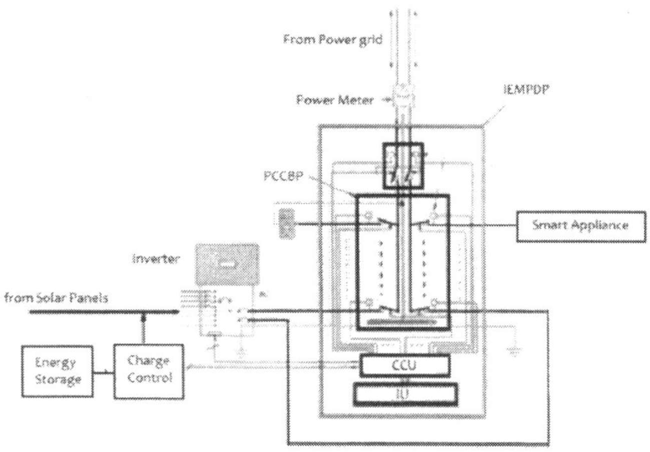

Figure 4

The IEMPDP effectively integrates the intelligent gateway to the power grid, local power distribution, and the energy management center into one system. This allows for a more comprehensive and integrated control while reducing cost by eliminating the components shared by these three systems. This also simplifies the structure of local grid tied PV systems or other renewable energy systems, reduces unnecessary power consumption, and improves the quality and reliability of the power grid.

The ICB costs \$160-\$600 compared to the PCCB used in IEMPDP that will cost only \$20-\$30. Since the CCU can monitor and control the local electric system in real-time, it will not only automatically control the thermostat and switches, but dynamically distribute the power and automatically manage the energy storage. The IEMPDP opens up great potential for energy efficiency and reduces local renewable energy system cost because it simplifies the structure of the local electric system and reduces the installation cost of local renewable energy systems.

The CCU monitors and controls residential electric systems locally, and communicates with utilities with commonly used protocols. This provides a convenient way for consumers to monitor and control the usage of their appliances according to their preferences and schedule. Since consumers control their system information they can protect any sensitive, private information, but also share data with utilities to help improve the quality and reliability of power grids.

3 FEASIBILITY

The key feature of the IEMPDP is to automatically monitor and control the main and each branch power line in real-time, reliably, at modest cost. This provides the foundation for it to intelligently interact with the power grid, to dynamically distribute power, and to coordinate with the charge controller to effectively regulate the usage of the energy storage system. The circuit topology and control algorithm are the main components that influence this key feature. The IEMPDP is feasible since it uses technologies that have been successfully used in buildings and industrial automation.

We have developed a proof of concept prototype to ensure that the IEMPDP concept works. The prototype includes one channel and is tested under 120VAC with the current limit at 12A condition. The test circuit is shown in Figure 5. The PCCB represents a proof of concept prototype. The prototype can

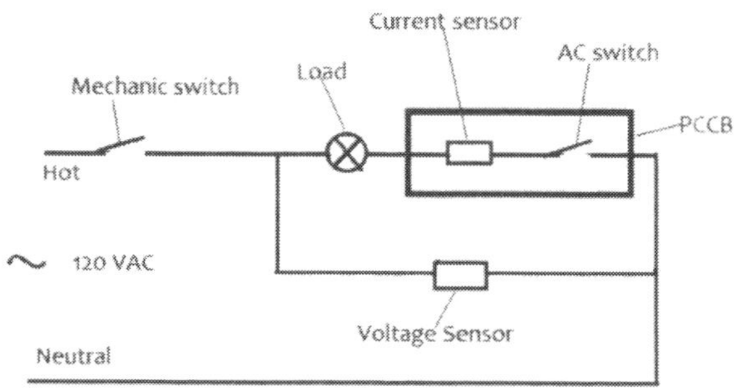

Figure 2

automatically monitor and control the power line reliably. Figure 6 displays the waveforms captured during testing. It illustrates that the prototype senses the current correctly and could be controlled automatically by a control signal. The switching, which occurs at the zero crossing point, introduced no interference into the power line.

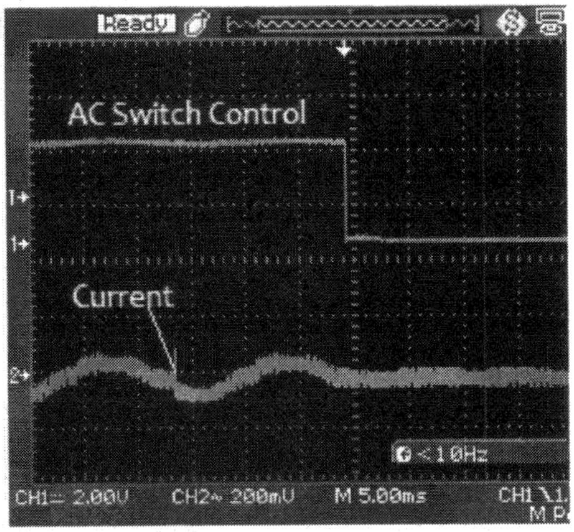

Figure 6

This confirms that main component of the IEMPDP, the PCCB works well. The next step is to develop a prototype with one main entry and several channels using the prototype PCCBs to test the control algorithm of the IEMPDP. The main function of the control algorithm is to monitor and control the power line. Once this is established many other features can be added. For example: scheduling a task to run automatically at a specified time period; communicating with an inverter and interacting with the power grid; connecting to an energy storage system (e.g., a super capacitor only, batteries, or a HESS) to coordinate with the charge controller to smooth out solar power fluctuations due to weather or store surplus energy; etc. Using the internet or mobile devices, a user can access their system easily from anywhere on the planet. The IEMPDP will unlock great potential for energy efficiency, saving people time and money, and producing a more reliable and robust power supply.

4 FUTURE DEVELOPMENT

The IEMPDP will transform the local electric system into a more robust, intelligent, reliable, and efficient system. Since the IEMPDP uses well established technologies and methods, we are confident that it will work. The main challenge is to make it robust and reliable. The IEMPDP must reliably detect any problems, robustly fix these problems, and smoothly recover from any faults. These features must be incorporated early on in the design stage. This will be challenging but very rewarding.

REFERENCES

[1] United States Patent Application 20100314940 Energy-Saving Electrical Power System

[2] Schenider Electric, Smart Panels mean visible information

[3] Patten application (PCT/US2013/060908)

[4] M.E. Glavin, W.G. Hurley, 2012, 'Optimization of a photovoltaic battery ultra capacitor hybrid energy storage system'

[5] R.Anbarasu, Dr.V.Kamatchi Kannan, Dr.N.Rengarajan, 2014, Integration of Super Capacitor and Battery for Voltage Fluctuations Mitigation in DC Hybrid System

[6] Qing Xie, Xue Lin, Yanzhi Wang, and Massoud Pedram, 2012, State of Health Aware Charge Management in Hybrid Electrical Energy Storage Systems

[7] Sathish Kumar Kollimalla, Mahesh Kumar Mishra, Lakshmi Narasamma N, 2014, Coordinated Control and Energy Management of Hybrid Energy Storage System in PV System

The Utilization of Electrochromic Materials for Smart Window Applications in Energy-Efficient Buildings

Bjørn Petter Jelle [ab]* and Tao Gao [c]

[a] SINTEF Building and Infrastructure,
Department of Materials and Structures, NO-7465 Trondheim, Norway.

[b] Norwegian University of Science and Technology (NTNU),
Department of Civil and Transport Engineering, NO-7491 Trondheim, Norway.

[c] Norwegian University of Science and Technology (NTNU),
Department of Architectural Design, History and Technology, NO-7491 Trondheim, Norway.

* Corresponding author: bjorn.petter.jelle@sintef.no (e-mail), +47-73593377 (phone).

ABSTRACT

Electrochromic materials (ECM) and windows (ECW) are able to regulate the solar radiation throughput by application of an external electrical voltage. Thus, ECWs may decrease heating, cooling, lighting and electricity loads in buildings by admitting the optimum level of solar energy and daylight at any given time, e.g. cold winter climate versus warm summer climate demands. It is crucial to be able to compare the dynamic solar radiation control for different ECWs and hence require specific ECW properties. The solar radiation control for ECWs may readily be characterized by several solar radiation glazing factors, where a comparison for various ECW configurations enables one to select the most appropriate ones for specific smart window applications in energy-efficient buildings. As an example a particular ECW based on the ECMs polyaniline, prussian blue and tungsten oxide is presented, being able to regulate as much as 60 % of the visible and 59 % of the total solar radiation.

Keywords: electrochromic, material, smart window, solar radiation, energy-efficient.

1 INTRODUCTION

Electrochromic materials (ECM) in electrochromic windows (ECW) aim at controlling the solar radiation throughput at the earth's surface, which is roughly located between 300 nm and 3000 nm. Such windows are also often denoted smart windows. The ECW solar control is achieved by application of an external voltage. The visible (VIS) light lies between 380 and 780 nm. Ultraviolet (UV) and near infrared (NIR) radiation are located below and above the VIS region, respectively. Above 3000 nm, and not part of the direct solar radiation, lies the thermal radiation called infrared (IR) radiation, which all materials radiate above absolute zero, i.e. above 0 K, peaking around 10 000 nm (10 μm) at room temperature. However, the ECWs are not aimed at controlling the IR radiation. Normally, as low as possible heat loss through windows is desired, i.e. low U-value, which is accomplished by the application of various static low emissivity coatings on the window glass panes. Miscellaneous ECMs and ECWs and their various properties are investigated in the literature [1-10]. Some commercial ECWs are also already on the market [8,9]. ECMs and ECWs may also be used together with other materials and technologies, e.g. self-cleaning glazing materials and building integrated photovoltaics (BIPV) like solar cell glazing [11-15]. In general, it is of major importance to investigate the durability of building materials and components, also newly developed ones like e.g. ECMs and ECWs, which may be performed by carrying out accelerated climate ageing in the laboratory [16]. Thus, conducting a robustness assessment of these materials and components may also be beneficial [17]. Hence, a durability and robustness evaluation of the new ECMs and ECWs should be carried out. This work presents a specific ECW and corresponding characterization by solar radiation glazing factors, i.e. ultraviolet solar transmittance (T_{uv}), visible solar transmittance (T_{vis}), solar transmittance (T_{sol}), solar material protection factor (SMPF), solar skin protection factor (SSPF), external visible solar reflectance ($R_{vis,ext}$), internal visible solar reflectance ($R_{vis,int}$), solar reflectance (R_{sol}), solar absorbance (A_{sol}), emissivity (ε), solar factor (SF) and colour rendering factor (CRF). The solar factor (SF), which is the total solar energy transmittance, is also often denoted as the solar heat gain coefficient (SHGC) and the g-value. Further details about the solar radiation glazing factors are given in the literature [10,18,19].

2 EXPERIMENTAL

Experimental details concerning the synthesis and manufacturing of ECMs and complete ECWs, and their electrochemical and spectroscopical characterization, are described in the studies by Jelle [10], Jelle and Hagen [20-

TechConnect Briefs 2015, TechConnect.org, ISBN 978-1-4987-4728-8

22] and Jelle et al. [23]. In short, the ECMs polyaniline (PANI), prussian blue (PB) and tungsten oxide (WO_3) were deposited electrochemically onto transparent conducting glass plates coated with indium-tin-oxide (ITO), i.e. indium oxide doped with tin ($In_2O_3(Sn)$), whereas solid state ECWs were then constructed by applying the solid state polymer electrolyte poly(2-acrylamido-2-methyl-propane-sulphonic acid) (PAMPS) as an ionic conductor, see Fig.1 for a schematic drawing of the ECW presented here.

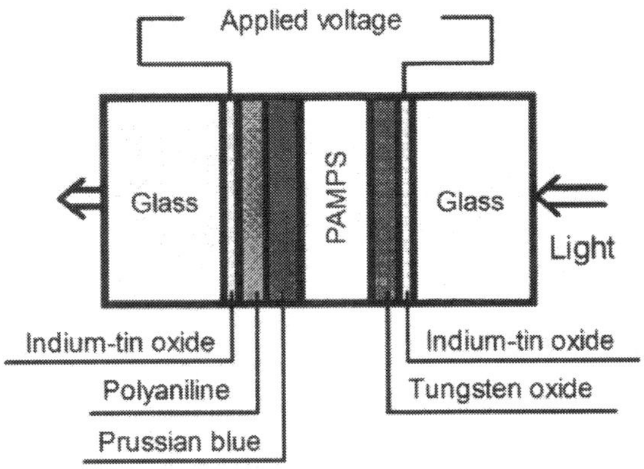

Figure 1: Schematic drawing of the ECW configuration based on the ECMs polyaniline (PANI), prussian blue (PB) and tungsten oxide (WO_3).

A Cary 5 UV-VIS-NIR spectrophotometer, with an absolute reflectance accessory (Strong-type, VW principle), was used to measure the transmittance and reflectance of the ECWs, whereas emissivity measurements were performed with a SOC-100 HDR Hemispherical Directional Reflectometer from Surface Optics Corporation connected to a Thermo Nicolet 8700 Fourier transform infrared (FTIR) Spectrometer.

3 RESULTS AND DISCUSSION

Measured transmittance spectra for an ECW at various applied electrical potentials (voltages) are given in Fig.2, with corresponding calculated solar radiation glazing factors and factor modulations in Table 1 and Table 2, respectively, for the lowest and highest electrical potential.

As to show an example of the calculation procedure for the solar radiation glazing factors, the solar transmittance (T_{sol}) is given by the following expression [9]:

$$T_{sol} = \frac{\sum_{\lambda=300nm}^{2500nm} T(\lambda) S_\lambda \Delta\lambda}{\sum_{\lambda=300nm}^{2500nm} S_\lambda \Delta\lambda} \quad (1)$$

where S_λ is the relative spectral distribution of solar radiation, $T(\lambda)$ is the spectral transmittance of the glass, λ is the wavelength, $\Delta\lambda$ is the wavelength interval, and the $S_\lambda\Delta\lambda$ values at different wavelengths are given as tabulated values [10,18].

The T_{sol} value will thus be a number between 0 and 1 (or 0 and 100 %), calculated in the main part of the solar spectrum, i.e. 300-2500 nm. A low number indicates a low transmission of solar radiation, whereas a high number represents a high solar radiation transmission. Note that the whole solar spectrum is not covered in the (standard) calculation of T_{sol}, and in future versions the wavelength range may favourably be extended to cover an even larger part of the solar radiation, e.g. from 290 nm to 3000 nm.

In similar ways the other solar radiation glazing factors may be calculated, except the ε, SF and CRF values which follow other calculation procedures [10,18]. The ECW modulation level given in Table 2 is calculated by subtracting the solar radiation glazing factors for the same ECW at the high and low potentials given in Table 1, e.g. as for ΔT_{sol}:

$$\Delta T_{sol} = T_{sol}(\text{bleached}) - T_{sol}(\text{coloured}) \quad (2)$$

where the T_{sol} values in Table 1 are calculated from Eq.1 and likewise for the other solar radiation glazing factors.

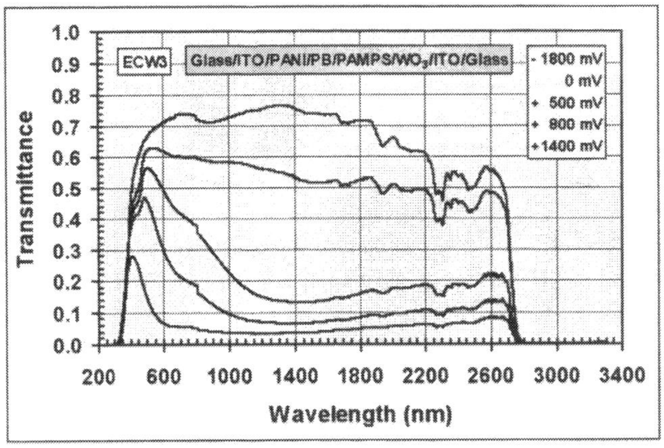

Figure 2: Transmittance vs. wavelength in the whole solar spectrum measured for an ECW at various applied potentials [10,21,23]. Highest colouration level is at +1400 mV.

It is observed from Table 1 that various solar radiation glazing factors may obtain both high and low values depending upon the applied electrical potential in the ECW, e.g. changing the T_{vis} value from 0.69 to 0.09. It is also noted that this ECW contains solar radiation absorbing electrochromic materials, i.e. not reflecting materials, as the changes with applied potential occur in the transmittance (e.g. T_{sol}) and absorbance (e.g. A_{sol}) values, and not in the reflectance (e.g. R_{sol}) values.

As expected, the highest colouration level gives the largest SMPF values, i.e. the best protection of materials is achieved with the darkest ECW, e.g. compare a SMPF value of 0.83 in the coloured state with 0.59 in the transparent (bleached) state.

The CRF value of 0.96 in the transparent (bleached) state is very high, i.e. a very good colour rendering. However, in the coloured state, the CRF value is substantially reduced, i.e. CRF = 0.59.

Table 1: Calculated solar radiation glazing factors for an ECW at two colouration levels (transparent and dark coloured), i.e. at two applied potentials (e.g. Eq.1). Corresponding transmittance spectra are given in Fig.2.

Solar radiation glazing factor	-1800 mV	+1400 mV
T_{uv}	0.08	0.12
T_{vis}	0.69	0.09
T_{sol}	0.67	0.08
SMPF	0.59	0.83
SSPF	0.97	0.97
$R_{vis,ext}$	0.09	0.09
$R_{vis,int}$	0.09	0.09
R_{sol}	0.08	0.08
A_{sol}	0.25	0.84
ε	0.836	0.836
SF	0.74	0.30
CRF	0.96	0.59

Table 2: Calculated solar radiation glazing factor modulations for an ECW. The modulation level is calculated by subtracting the solar radiation glazing factors at the high and low potentials given in Table 1 (e.g. Eq.2).

Solar radiation glazing factor modulation	Modulation between -1800 mV and +1400 mV
ΔT_{uv}	-0.04
ΔT_{vis}	0.60
ΔT_{sol}	0.59
ΔSMPF	-0.24
ΔSSPF	0.00
$\Delta R_{vis,ext}$	0.00
$\Delta R_{vis,int}$	0.00
ΔR_{sol}	0.00
ΔA_{sol}	-0.59
$\Delta\varepsilon$	-
ΔSF	0.44
ΔCRF	0.37

The ECW device has a rather large solar radiation modulation ability, e.g. $\Delta T_{vis} = 0.60$ and $\Delta T_{sol} = 0.59$, where the transmittance modulation is assumed to be due to absorbance regulation, i.e. $\Delta A_{sol} = -0.59$.

Note that reflectance values of the ECW have not been measured, but as the (absorbing) electrochromic coatings are located between two glass plates, the (low) reflectance values will be close to the values for float glass, and these are hence employed in the current calculations.

Although the solar factor modulation is lower than the solar transmission counterpart for the ECW, the solar factor modulation is still quite high, i.e. $\Delta SF = 0.44$. Hence, this ECW is able to regulate large parts of the solar radiation, and the regulation may be readily characterized by the solar radiation glazing factors. Applying the ECW into two-layer and three-layer window pane configurations reduces the total solar energy throughput modulation in the windows, which may also be seen in the ΔT_{sol} and ΔSF values, as several layers of glass and coatings will increase the total reflectance and absorbance (not depicted here, for examples it is referred to the study by Jelle [10]), i.e. less solar radiation left for the ECWs to modulate (regulate). That is, the solar radiation modulation by an ECW will decrease with the number of glass panes and low emittance coatings added to the total window configuration.

An interesting aspect is that the doping mechanisms in PANI include both redox processes and proton doping [2,24,25], and the characteristic absorbance shift from the NIR to the VIS region for PANI makes it appropriate to plot the absorbance versus both wavelength and applied electrical potential in order to enhance the visualization of the absorbance changes. Thus, a 3-dimensional graphical plot of light absorbance versus wavelength for PANI on ITO glass in an aqueous solution of 0.1 M $Na_2SO_4 \cdot 10H_2O$ and 0.001 M H_2SO_4 (pH 3.7) at different applied potentials versus Ag/AgCl (3.5 M KCl) is depicted in Fig.3 [2,10,22].

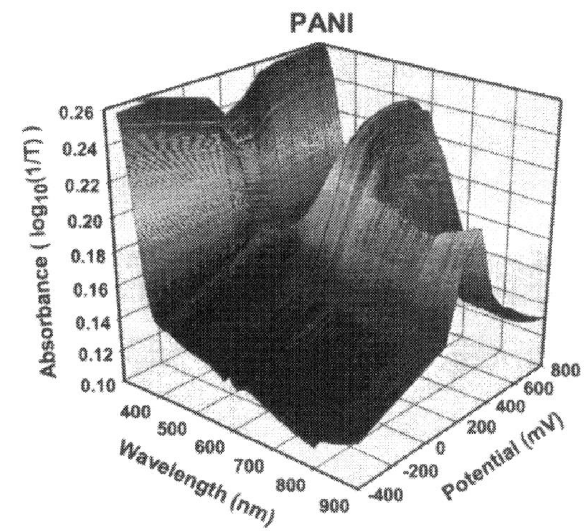

Figure 3: A 3-dimensional graphical plot of light absorbance versus wavelength for PANI on ITO glass in an aqueous solution at different applied potentials vs. Ag/AgCl (3.5 M KCl). The absorbance in the ITO glass, electrolyte and cell has been subtracted by the use of a reference cell in the double beam spectrophotometer [2,10,22].

4 CONCLUSIONS

A dynamical control of daylight and solar energy in buildings may be achieved by application of electrochromic windows (ECW). This control may be readily characterized by solar radiation glazing factors, i.e. ultraviolet solar transmittance, visible solar transmittance, solar transmittance, solar material protection factor, solar skin protection factor, external visible solar reflectance, internal visible solar reflectance, solar reflectance, solar absorbance, emissivity, solar factor and colour rendering factor. A specific ECW has been presented, demonstrating a large solar radiation modulation.

ACKNOWLEDGEMENTS

This work has been supported by the Research Council of Norway and several partners through "*The Research Centre on Zero Emission Buildings*" (ZEB).

REFERENCES

[1] B.P. Jelle, G. Hagen and S. Nødland, "Transmission spectra of an electrochromic window consisting of polyaniline, prussian blue and tungsten oxide", *Electrochimica Acta*, **38**, 1497-1500, 1993.

[2] B.P. Jelle, G. Hagen, S.M. Hesjevik and R. Ødegård, "Reduction factor for polyaniline films on ITO from cyclic voltammetry and visible absorption spectra", *Electrochimica Acta*, **38**, 1643-1647, 1993.

[3] C.G. Granqvist, "Handbook of inorganic electrochromic materials", Elsevier, Amsterdam, 1995.

[4] P.M.S. Monk, R.J. Mortimer and D.R. Rosseinsky, "Electrochromism: Fundamentals and applications", VCH, 1995.

[5] B.P. Jelle and G. Hagen, "Electrochemical multilayer deposition of polyaniline and prussian blue and their application in solid state electrochromic windows", *Journal of Applied Electrochemistry*, **28**, 1061-1065, 1998.

[6] B.P. Jelle and G. Hagen, "Correlation between light absorption and electric charge in solid state electrochromic windows", *Journal of Applied Electrochemistry*, **29**, 1103-1110, 1999.

[7] R.J. Mortimer, "Organic electrochromic materials", *Electrochimica Acta*, **44**, 2971-2981, 1999.

[8] R. Baetens, B.P. Jelle and A. Gustavsen, "Properties, requirements and possibilities of smart windows for dynamic daylight and solar energy control in buildings: A state-of-the-art review", *Solar Energy Materials and Solar Cells*, **94**, 87-105, 2010.

[9] B.P. Jelle, A. Hynd, A. Gustavsen, D. Arasteh, H. Goudey and R. Hart, "Fenestration of today and tomorrow: A state-of-the-art review and future research opportunities", *Solar Energy Materials and Solar Cells*, **96**, 1-28, 2012.

[10] B.P. Jelle, "Solar radiation glazing factors for window panes, glass structures and electrochromic windows in buildings - Measurement and calculation", *Solar Energy Materials and Solar Cells*, **116**, 291-323, 2013.

[11] S.K. Deb, S.-H. Lee, C.E. Tracy, J.R. Pitts, B.A. Gregg and H.M. Branz, "Stand-alone photovoltaic-powered electrochromic smart window", *Electrochimica Acta*, **46**, 2125-2130, 2001.

[12] C.M. Lampert, "Large-area smart glass and integrated photovoltaics", *Solar Energy Materials and Solar Cells*, **76**, 489-499, 2003.

[13] B.P. Jelle, C. Breivik and H.D. Røkenes, "Building integrated photovoltaic products: A state-of-the-art review and future research opportunities", *Solar Energy Materials and Solar Cells*, **100**, 69-96, 2012.

[14] K. Midtdal and B.P. Jelle, "Self-cleaning glazing products: A state-of-the-art review and future research pathways", *Solar Energy Materials and Solar Cells*, **109**, 126-141, 2013.

[15] B.P. Jelle, "The challenge of removing snow downfall on photovoltaic solar cell roofs in order to maximize solar energy efficiency - Research opportunities for the future", *Energy and Buildings*, **67**, 334-351, 2013.

[16] B.P. Jelle, "Accelerated climate ageing of building materials, components and structures in the laboratory", *Journal of Materials Science*, **47**, 6475-6496, 2012.

[17] B.P. Jelle, E. Sveipe, E. Wegger, A. Gustavsen, S. Grynning, J.V. Thue, B. Time and K.R. Lisø, "Robustness classification of materials, assemblies and buildings", *Journal of Building Physics*, **37**, 213-245, 2014.

[18] ISO 9050:2003(E), "Glass in building - Determination of light transmittance, solar direct transmittance, total solar energy transmittance, ultraviolet transmittance and related glazing factors", 2003.

[19] B.P. Jelle, A. Gustavsen, T.-N. Nilsen and T. Jacobsen, "Solar material protection factor (SMPF) and solar skin protection factor (SSPF) for window panes and other glass structures in buildings", *Solar Energy Materials and Solar Cells*, **91**, 342-354, 2007.

[20] B.P. Jelle and G. Hagen, "Transmission spectra of an electrochromic window based on polyaniline, prussian blue and tungsten oxide", *Journal of Electrochemical Society*, **140**, 3560-3564, 1993.

[21] B.P. Jelle and G. Hagen, "Solar modulation in an electrochromic window using polyaniline, prussian blue and tungsten oxide", in *Electrochromic Materials II*, (PV 94-2), K. C. Ho and D. A. MacArthur (eds.), pp. 324-338, *Proceedings of the Symposium on Electrochromic Materials, The 184th Meeting of The Electrochemical Society*, New Orleans, Louisiana, U.S.A., 10-15 October, 1993, The Electrochemical Society, Pennington, NJ, 1994.

[22] B.P. Jelle and G. Hagen, "Performance of an electrochromic window based on polyaniline, prussian blue and tungsten oxide", *Solar Energy Materials and Solar Cells*, **58**, 277-286, 1999.

[23] B.P. Jelle, G. Hagen and Ø. Birketveit, "Transmission properties for individual electrochromic layers in solid state devices based on polyaniline, prussian blue and tungsten oxide", *Journal of Applied Electrochemistry*, **28**, 483-489, 1998.

[24] J.-C. Chiang and A.G. MacDiarmid, "'Polyaniline': Protonic acid doping of the emeraldine form to the metallic regime", *Synthetic Metals*, **13**, 193-205, 1986.

[25] W.-S. Huang, B.D. Humphrey and A.G. MacDiarmid, "Polyaniline, a novel conducting polymer. Morphology and chemistry of its oxidation and reduction in aqueous electrolytes", *Journal of the Chemical Society, Faraday Transactions 1: Physical Chemistry in Condensed Phases*, **82**, 2385-2400, 1986.

Photocatalytic Activity of Commercial TiO_2 Oxides in the Oxidation Reaction of Nitric Oxide

E. Luévano-Hipólito*, A. Martínez-de la Cruz*, V.M. López-Gallegos**, H. Garza-del Bosque**

* CIIDIT, Facultad de Ingeniería Mecánica y Eléctrica, Universidad Autónoma de Nuevo León, Ciudad Universitaria, C.P. 66451, San Nicolás de los Garza, N. L., México.

** Pinturas Térmicas del Norte, S.A. de C.V., Aarón Sáenz 1902, Santa María, C.P. 64650 Monterrey, N.L., México

ABSTRACT

In this work, five commercial titanium oxides, R-706 from DUPONT™, P-25 from DEGUSSA™, BLR-699 from YUEJIANG™, RC-8640A and RC-8640B from RUI-CHEM™ were characterized by X-ray diffraction (XRD), scanning electron microscopy (SEM), diffuse reflectance spectroscopy (DRS) and nitrogen adsorption-desorption isotherms (BET), and its photocatalytic activity was tested in the oxidation reaction of nitric oxide (NO). The conversion degree (%) of NO reached by each oxide was associated with their physical properties such as crystalline structure, specific surface area, and particle size. In the same way, the effect of the charge of photocatalyst was also studied.

Keywords: TiO_2, photocatalysis, NO_x

1 INTRODUCTION

Nowadays, the interest by the development of sustainable materials for construction applications is increasing due to the importance to preserve the environment. Undoubtedly, one of the most interesting technologies focused in this direction is the heterogeneous photocatalysis, where the solar radiation is used by a semiconductor oxide for induce the elimination of a great variety of pollutants [1]. By far, the anatase polymorph of TiO_2 is the photocatalyst of excellence for commercial applications because of its high activity under UV irradiation, and stability to corrosion processes [2]. On the other hand, although rutile polymorph of titanium oxide has a lower activity, the combination of both polymorphs can produce a material with synergistic photocatalytic properties as the well-known TiO_2 P-25 (Degussa).

For these reasons, will be explored the potential application of five commercial titanium oxides with different crystalline structures to act as an efficient photocatalyst in the conversion oxidation reaction of NO_x gases to innocuous products as nitrites or nitrates. The generic term NO_x refers to a group of oxides formed by the chemical combination of nitrogen and oxygen in different ratios ($NO + NO_2$) and constitutes one of the most common air pollutants. This topic is of particular relevance in big cities where the accumulation of NO_x compounds in the air represents a serious problem. For example, nitrogen oxides (NO_x) are the origin of various health and environmental problems, such as irritated eyes, headache, pulmonary emphysema, acid rain, acidification of aquatic systems, ozone depletion, and photochemical smog [3]. Among the various proposed technologies to solve this problem, the heterogeneous photocatalysis has received an increasing interest in recent years [4-5].

2 EXPERIMENTAL

The commercial oxides R-706 from DUPONT™, P-25 from DEGUSSA™, BLR-699 from YUEJIANG™, RC-8640A and RC-8640B from RUI-CHEM™ were supplied by the company Pinturas Térmicas del Norte. The samples of commercial TiO_2 oxides were analyzed as were received without a further treatment. The structural characterization of TiO_2 samples was carried out by X-ray powder diffraction using a D8 Advance diffractometer with Cu Ka radiation. Diffraction experiments were performed with a step size of 0.05° and a dwell time of 0.5 s. The morphology and particle size of the samples were analyzed by scanning electron microscopy by using a FEI Nova NanoSEM 200 microscope with an accelerating voltage of 30 kV. The UV-Vis diffuse reflectance absorption spectra of the samples were obtained by using an UV-Vis spectrophotometer PerkineElmer Lambda 35 equipped with an integrating sphere. The BET surface area measurements were carried out by N_2 adsorption/desorption isotherms by means of a Bel-Japan Minisorp II surface area and pore size analyzer. The isotherms were evaluated at -196°C after a pretreatment of the samples at 150°C for 24 h.

The photocatalytic activity of TiO_2 samples was evaluated by the removal of nitric oxide from air under UV irradiation. The photocatalytic experiments were carried out in a laboratory test set-up designed according with the standard ISO 22197-1:2007 as is shown in Figure 1. The photocatalytic reactor was made of stainless steel with a volume of 0.8 L. The device has integrated a window made of tempered glass in its superior part in order to allow the pass of radiation. The photocatalyst (200 mg) dispersed previously in ethanol was deposited over an area of 0.08 m^2 in a glass substrate with the help of a small brush. As inlet gas was used a mixture 3 ppm of NO stabilized in N_2. The concentration of gas was adjusted to 1 ppm in NO by using air and the flow rate of gas was adjusted to 1 L·min^{-1}.

The source of light irradiation were two fluorescent black lamps (TecnoLite) of 20 W each one emitting between 365 and 440 nm The main contribution of the radiation source comes from the emission line at 365 nm, and, in minor proportion from two emission lines located in the visible region at 405 and 437 nm. The concentration of nitric oxide was continuously measured with a chemiluminescent NO_x analyzer (EcoPhysics CLD88p) with a sampling rate of 0.3 L·min^{-1}.

3 RESULTS AND DISCUSSIONS

Figure 2 shows the X-ray diffraction pattern of the five commercial oxides. In the samples identified as R-706 and BLR-699, the main diffraction lines were associated with the rutile structure type of TIO_2 in agreement with the JCPDS Card No. 1-1292. Additionally, the diffractogram of the sample R-706 showed the presence of ZnO in a minor proportion (JCPDS 36-1451). In the same way, TiO_2 anatase oxide was also detected in the sample BLR-699.

As is well known, the oxide usually used as reference in photocatalytic experiements (P-25), is constituted by a mixture of anatase and rutile polymorphs of TiO_2 approximately in a proportion of 70-30 % of each oxide, respectively.

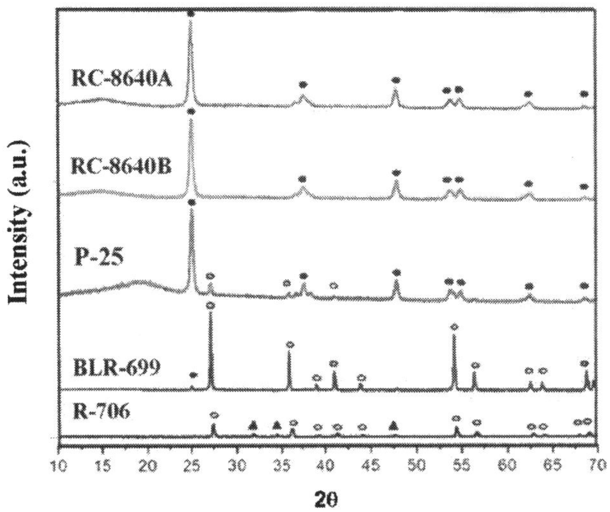

Figure 2. X-ray diffraction patterns of commercial TiO_2. (o) rutile, (•) anatase, and (▲) ZnO.

In the sample RC-8640A and RC-8640B only were detected diffraction lines associated with the anatase polymorph of TiO_2. In general, the crystallinity of the commercial oxides was decreasing in the follow order; RC-864040A > RC-8640B > P-25 > BLR-699 > R-706.

The analysis of commercial oxides by SEM revealed materials with an average of size particle of 250-300 nm (BLR-699), 350-550 nm (R-706), < 50 nm (P-25), and < 50 nm (RC-8640A and RC-8640B). As a characteristic property, large agglomerates of primary particles were observed in R-706 whose dimensions were about 3 μm.

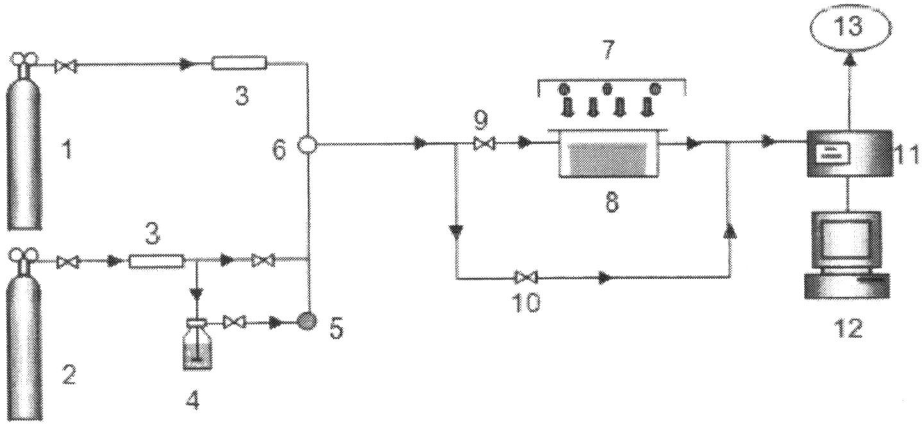

Figure 1. Schematic diagram of the photocatalytic degradation set-up. 1. NO gas supply; 2. Synthetic air; 3. Mass controller meter; 4. Humidifier; 5. Humidity controller; 6. Temperature and relative humidity sensor; 7. Light source; 8. Reactor; 9. Valve; 10. Valve; 11. NO_x analyzer; 12. Computer; 13. Vent.

The specific surface area of samples was determined throughout BET protocol. By an analyses of the profile of the N_2 adsorption/desorption isotherms was concluded that the materials were not porous and has a high energy of adsorption. Table 1 shows the specific surface area calculated for each oxide.

Sample	Surface area ($m^2 g^{-1}$)	Energy band gap (eV)
R-706	10.3	3.0
BLR-699	16.8	3.0
P-25	50.2	3.1
RC-8640A	74.5	3.1
RC-8640B	78.8	3.2

Table 1. Some physical properties of the commercial TiO_2 oxides.

As can be noticed in Table 1, TiO_2 samples present similar energy band gaps values between them. These values were ranging between 3.2 and 3.0 eV which corresponds with the typical values of the TiO_2 for anatase and rutile crystal structures.

The assessment of photocatalytic activity of the oxides was performed in a photocatalytic reactor designed according with the ISO 22197-1. Figure 3 shows the conversion degree (%) of NO reached by each oxide in the course of 60 min of lamp irradiation. For the first 25 min of reaction, the materials where the anatase polymorph was the main phase exhibited the highest photocatalytic activity, i.e. P-25, RC-8640A, and RC-8640B. In fact, the conversion degrees of NO (%) reached at this time values near to 95%. After 25 min of reaction, P-25 and RC-8640A retained almost the 100% of its photocatalytic activity. By the other hand, RC-8640B undergoes an important deactivation process which reduced the conversion degree of NO until a 60% after 60 min of reaction. In a contrary behavior, the samples in which the predominant phase was rutile (R-706 and BLR-699) showed a poor photocatalytic activity. In first instance, the materials were quickly activated but their maximums of conversion degree were reached around 12 minutes, and then a remarkably deactivation process was observed. Among rutile samples, BLR-699 was the best photocatalyst with a conversion degree value of 70% during a short time of reaction.

The crystalline structure was the main factor to condition the photocatalytic activity of the commercial oxides. It is well recognized that the crystalline structure of anatase is more active than the rutile polymorph, mainly due to a better charge transference and the existence of crystalline planes with high photocatalytic activity in the anatase polymorph [6]. Some authors have attributed the high photcatalytic activity of P-25 with a synergistic process between anatase and rutile polymorphs at the specific composition of 70/30 of each oxide.

The specific surface area is a secondary factor to taking into account in the assessment of photocatalytic activity of the commercial TiO_2 oxides. It can be the origin of the different activities observed in the rutile oxides. Due to the amount of mass of photocatalyst used in the experiments, 200 mg dispersed in an area of 0.08 m^2, the adsorption process of the NO molecule over surface of photocatalyst should have a relevant importance. By this cause, the specific surface area of oxides plays an important role in the activity observed. In relation with this physical property, the particle size can has an important influence. In both series of samples, anatase and rutile oxides, the samples

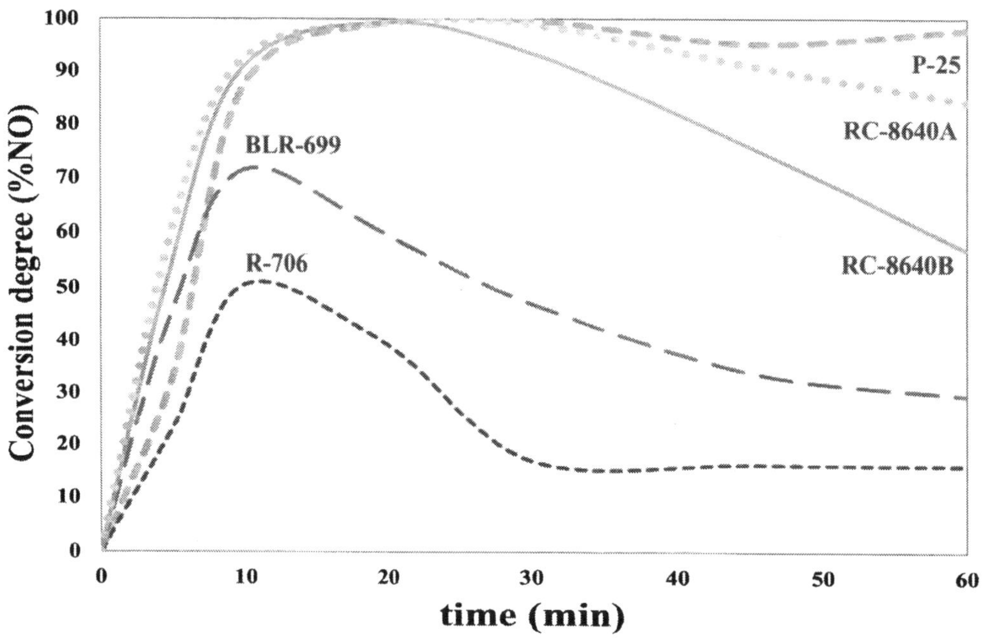

Figure 3. Photocatalytic activity of commercial TiO_2 oxides for NO conversion.

TechConnect Briefs 2015, TechConnect.org, ISBN 978-1-4987-4728-8

with particle sizes at lowest nanometric scale exhibited the highest phootcatalytic activity, i.e. P-25 and BLR-699, respectively. In this sense, a small particle size can reduce considerably the pathway to follow by the charges generated (electrons and holes) from its point of generation until the surface of the photocatalyst.

With the purpose of evaluate the activity of the best photocatalyst under more drastic experimental conditions, a series of experiments were performed where the mass of P-25 photocatalyst were reduced to 100 and 50 mg, as is shown in Figure 4. As can be seen, the performance of P-25 was similar when the mass of the oxide was reduced in a 50% (100 mg). Only when the charge of the photocatalyst was reduced in a 75% (50 mg) was appreciable a reduction in the conversion degree of NO in the stationary state, which after 60 min reached a value of 65%.

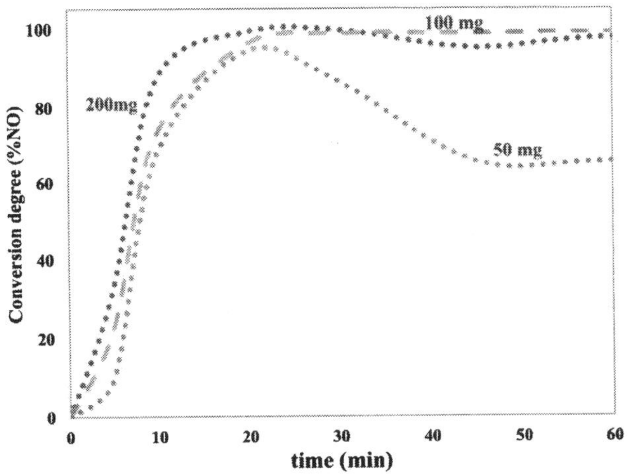

Figure 4. Conversion degrees of NO with different charge of P-25 photocatalyst.

Titanium dioxide can be incorporated as raw material in large applications such as cement, asphalt, ceramic tiles, glass, and paints. Due to these materials are usually applied over surfaces exposed to sun light irradiation, this technology can be considered to the development of new sustainable materials for the industry of construction. In this sense, an important factor to choose the commercial oxide is the cost involved. For example, the costs of R-706, RC-8640A and P-25 per kg are \$2.95, \$4.40, and \$45.0 USD. Over the basis of this parameter, and taking into account its photocatalytic activity, RC-8640 seems to be the most efficient material to be considered for a further technological application. Of course, other parameters must be considered such as the dispersion degree and interaction of TiO_2 oxide with the host matrix.

4. CONCLUSIONS

The conversion degree (%) of NO reached by each oxide was associated with their physical properties such as crystalline structure, specific surface area, and particle size, In the same way, the effect of the charge of photocatalyst was also studied.

Titanium dioxide can be incorporated as raw material in large applications such as cement, asphalt, ceramic tiles, glass, and paints. Taking into account its photocatalytic activity and cost, RC-8640 seems to be the most efficient material to be considered for a further technological application.

REFERENCES

1. F. Fresno, R. Portela, S. Suárez, J. M. Coronado, J. Mater. Chem. A 2, 2863, 2014.
2. M. Ni, M.K.H. Leung, D.Y.C. Leung, K. Sumathy, Renew. Sustain. Energy Rev. 11, 401, 2007.
3. J. Ma, H. Wu, Y. Liu, H. He, J. Phys. Chem. C 118, 7434, 2014.
4. B.Y. Lee, A.R. Jayapalan, M.H. Bergin, K.E. Kurtis, Cement and Concrete Research 60, 30, 2014.
5. C.L. Bianchi,. C. Pirola, F. Galli, G. Cerrato, S. Morandi, V. Capucci, Chem, Eng. J. 261, 76, 2015.
6. K. Nakataa, A. Fujishima, J. Photochem. Photobiol C: Photochem. Rev. 13, 169, 2012.

ACKNOWLEDGMENTS

The authors acknowledge the financial support provided by the CONACYT (PROINNOVA-2014-212775).

An inkjet printed NO_2 sensor operating under room temperature and low humidity environment

Wei Wang[1], Yu-Cheng Su[1], Wen-Yu Chuang[1], Chih-Ting Lin[1] and Wen-Jong Wu[2]

[1]Graduate Institute of Electronics Engineering, National Taiwan University, Taiwan, timlin@cc.ee.ntu.edu.tw
[2]Department of Engineering Science and Ocean Engineering, National Taiwan University, Taiwan

ABSTRACT

In this paper, a room temperature operation NO_2 sensor based on polypyrrole (PPy) /AZO (5% Al-doped ZnO)/Fe_2O_3 sensing material was developed. For the environmental monitoring, the ambient factor effect for NO_2 detection was discussed. Based on the results, oxygen plays anessential role for NO_2 sensing, which can enhance the reaction between NO_2 and PPy. The sensor shows linear response to NO_2 in range of 3-15 ppm at room temperature. It was reveal that the blending of AZO and Fe_2O_3 can improve the stability of sensor and the response to NO_2, separately. Furthermore, PPy/AZO/Fe_2O_3 based sensor performed good performance till RH 30%, which is verified the capability of environment monitoring. In summary, the blending of AZO and Fe_2O_3 nanoparticles in PPy sensing material is a promising NO_2 sensor for environmental monitoring.

Keywords: inkjet-printing, NO_2 sensor, polypyrrole

1. INTRODUCTION

Air pollution has become a serious problem in recent years. NO_2, this reddish-brown toxic gas has a characteristic sharp, biting odor and is a prominent air pollutant in large cities and the atmosphere. Therefore, many researchers developed various sensing materials for NO_2 sensing, such as SnO_2[1],Fe_2O_3[2]and ZnO[3] etc. However, most of metal oxide materials need to operate at high temperature and leads to the high power consumption sensors. To fabricate a low cost and room temperature operation sensor, polymer material based sensors were developed[4]. In traditional electrochemical sensor, the fabrication is bulky and relatively expensive, which limited the application for environment monitoring.

To resolve these problems, many research were developed a room temperature operation, and simple fabrication sensors. Sarfraz et al. [5] published printed chemical resistive of H_2S sensor on paper substrates, which can operate at room temperature. In addition, F. Molina-Lopez et al. [6] developed humidity sensor by the simple and low cost technology, inkjet-printing process. By these methods, the sensor has the advantage of low cost and low operating temperature.

In this paper, an inkjet-printed chemical sensor based on PPy/AZO/Fe_2O_3 was used for NO_2 detection. The effect of oxygen and humidity for NO_2 detection were also discussed. This sensor provides a room temperature operation and low humidity environmental monitoring capability.

2. EXPERIMENTAL

2.1.Sensor fabrication

A p-type wafer with 300 nm oxide on the top of surface was used as the device substrate. The substrate was cleaned by acetone and isopropyl alcohol. After drying the substrate by N_2 gas, the substrate was heated to remove humidity. Followed by the photolithography, the electrode was defined with W/L in the ratio of 800um/40um. Then the sensing electrode, i.e. 20nm/200nm of Cr/Au electrodes, was achieved by e-gun evaporation and lift-off process.

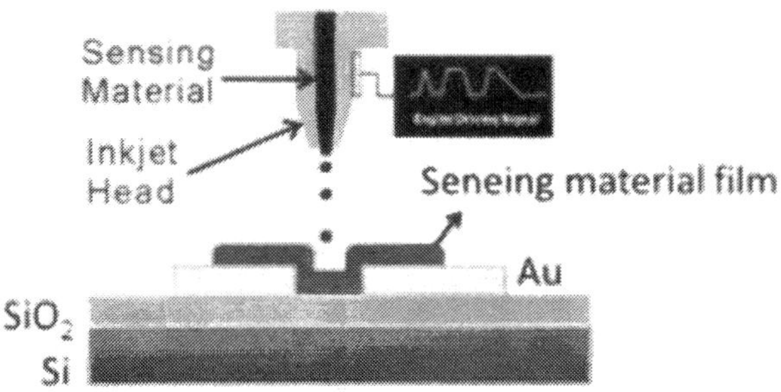

Fig. 1: the inkjet-printing process and device structure

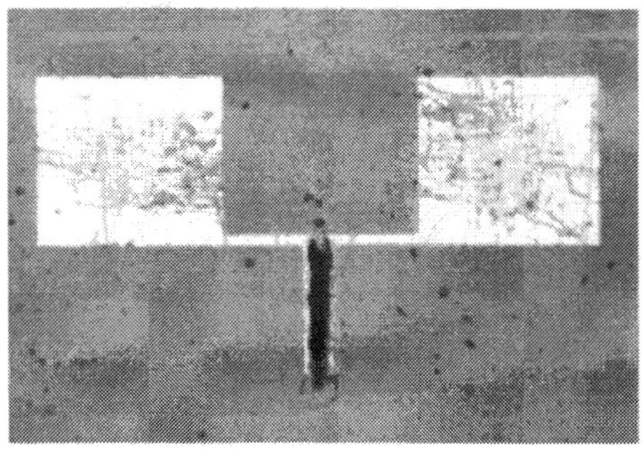

Fig. 2: the optical image of humidity sensor

This process prepared the substrate for the inkjet-printable sensing material.

Polypyrrole (PPy) was synthesized by chemical oxidative polymerization technique using monomer pyrrole. Since PPy is unsolvable in water or some common organic solvent, the sensing film takes two steps of printing process for the in-situ synthesis of PPy.

First, the oxidant solution, 11.47 wt% of $FeCl_3$ was dissolved in de-ionized (DI) water and then printed. The ratio of the $FeCl_3$/pyrrole is studied by [7]. When printing the $FeCl_3$ solution, the humidity should kept in RH20%~RH30%. Later, 4.7 wt% of pyrrole monomer was dissolved in ethanol and then blended with 0.0125 wt% AZO nanoparticle and 0.025 wt% Fe_2O_3 nanoparticle. After adding materials into solutions, the blended solution was put in a sonicator with hot water shaking for 1 hour to improve mixing. Finally, the disperse solution was printed on the oxidant and synthesized to form PPy/AZO/Fe_2O_3 film. To improve the synthesis of PPy, sensor was heated at 100°C for 5 minutes. Finally, sensor was soaked in the methanol solution for 24 hours to rinse off the excess oxidant and the solvent then dry in the ambient environment. The optical image of the fabricated humidity sensor was shown in Fig. 2. Chemical structure of thin film was examined by FTIR technique to ensure the polypyrrole was polymerized successfully.

2.2.Measurement

To measure the response of the developed sensing material, the LCR meter, Agilent E4980A, provides AC 1V to measure the impendence of sensing film with the frequency of 1 kHz. Fig. 3 shows the measure system, the gas sensing was operated in the measure chamber. The NO_2 detection in dry air and N_2 environment was measured to discuss the effect of oxygen. Before gas sensing, the chamber was vacuumed by a mechanical pump for 10 minutes to remove the ambient gases. Followed the dry air (79% N_2 and 21% O_2) or N_2 flushing as the initial condition. After removing dry air or N_2 by vacuuming for 10 minutes, NO_2 with dry air or N_2 was then flow into the chamber and wait for 5 minutes.

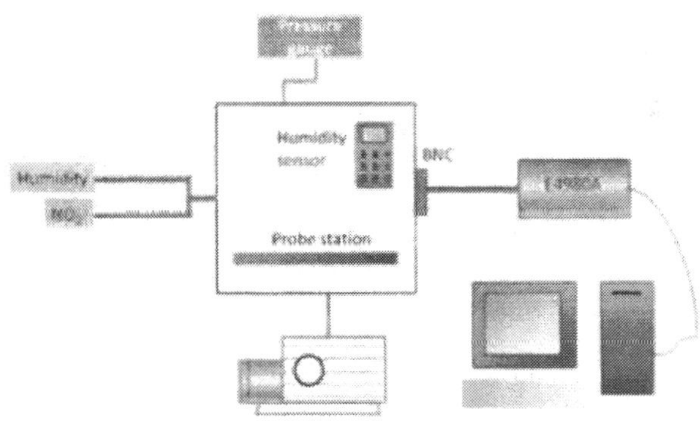

Fig.3: The schematic diagram for gas sensor measurement

Furthermore, NO_2 detected in humidity environment was also tested. For NO_2 detection with various relative humidity conditions, DI water was inkjet and flowed with air and NO_2, separately. Which is in order to make the water fully vaporize. The relative humidity concentration was detected from a commercial humidity meter. All the sensor measurements were carried out at room temperature.

3. RESULTS ANDDISCUSSION

3.1. Gas sensing properties

The sensor was tested for NO_2 and humidity. All the measurement operate at room temperature. For oxidizing gas like NO_2, the independence should decrease on exposure to NO_2. The sensitivity is defined as

$$\text{Sensitivity}\ (\%) = \frac{R_0 - R_{NO2}}{R_0} \times 100\%$$

Where R_0 represents the sensor in the initial condition. Since the resistance decreased with the rising of NO_2 concentration, the sensitivity can be defined. The reaction time is around 50 seconds, which is relative lower than other polymer sensing material. The fast respond time is due to thin film made by inkjet printing.

Fig.4 shows when the measure chamber flushed by different concentration of NO_2, the impedance decrease. The more concentration of NO_2 is flushed, the more the sensitivity is. Adeel Afzal et al. [8] reported NO_2 sensing mechanism. When PPy thin film is putting in the ambient environment, the oxygen molecules extract electron from PPy causing the formation of oxygen ions O^{2-}, O^- and O_2^- adsorbed at surface. NO_2, the strongly oxidizing gas with high electron affinity 2.28 eV, is much higher than oxygen's 0.43 eV. Hence, the redox reaction take place at the surface. PPy is a P-type material. The more NO_2 molecules interact with the surface, the more electron is absorbed, the hole is accumulated and the impedance decrease. In Fig. 5, we can find when NO_2 flushing to the chamber by air, the PPy thin film shows higher sensitivity. On the other side, NO_2 flushing to the chamber by N_2 shows no sensitivity. This result shows oxygen plays an important role in NO_2 sensing.

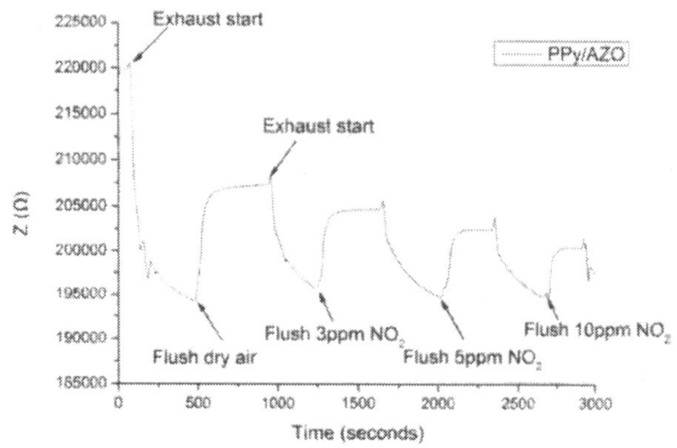

Fig.4: NO_2 response curve of PPy sensor in dry environment

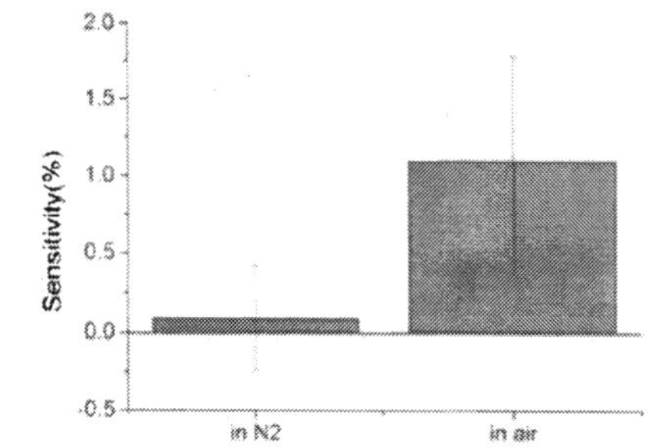

Fig. 5 : The sensitivity to a) 5ppm NO_2+ N_2 and b) 5ppm NO_2+ dry air in room temperature

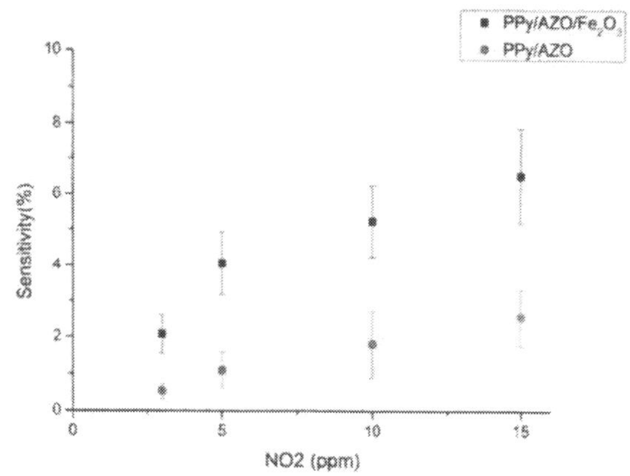

Fig. 6: Response of PPy/AZO and PPy/AZO/Fe_2O_3 film with NO_2 in dry air and room temperature

Fig. 6 shows the sensor doping Fe_2O_3 show higher response. In 3ppm NO_2, 2.07% with Fe_2O_3 relative to 0.53% without Fe_2O_3.In 5 ppm NO_2, 4.04% with Fe_2O_3 relative to 1.09%

without Fe_2O_3.

3.2. Humidity effect

Humidity plays an important role in PPy sensing. Humidity can oxidize PPy, which increase the impedance of PPy thin film [9], [10]. Also, the impedance variation by considering the dipole effect on this polar polymer [11].

As previously discuss, the sensor response to humidity and NO_2 are in opposite direction. It's difficult to distinguish these 2 target gas in high humidity. Fig. 7 shows the sensor still work in RH 20% and RH 30%. However, in high humidity (RH 50% and RH 60%), it's difficult to distinguish the NO_2 concentration. It may cause by the saturation of the PPy surface.

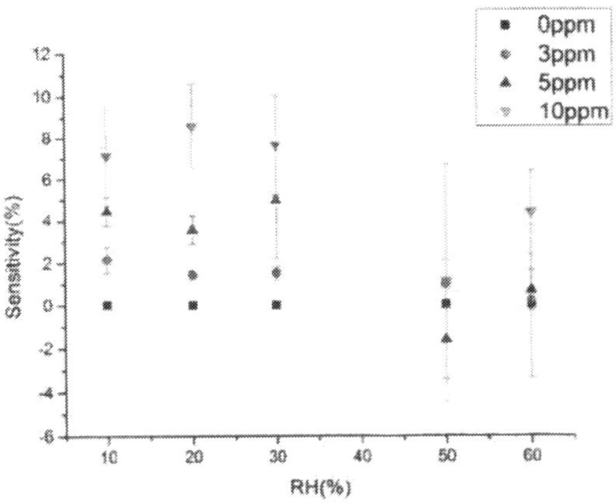

Fig. 7: PPy doped Fe_2O_3 sensor's sensitivity toward NO_2 concentration in different RH% in room temperature

4. CONCLUSION

In this paper, we made a NO_2 gas sensor by PPy. PPy thin film prepared by inkjet printing method can reduce its material waste and easy fabrication. For understanding the nanocomposite, the sensitivity between doping Fe_2O_3 or not is also studied. After adding Fe_2O_3, the sensor shows higher response. This sensor can detect down to 3ppm NO_2. In summary, the in-situ polymerized PPy sensing material by inkjet-printing technology is a promising NO_2 sensor for environmental monitoring.

ACKNOWLEDGMENT

This work is supported by National Science Council and Intel-NTU Connected Context Computing Center (NSC 101- 2628-E-002 -022 -MY3, 100-2911-I-002-001, and 101R7501). Also thanks to Material and Chemical Research Laboratories of ITRI for providing nanoparticle samples.

REFERENCES

[1]J.P. Ahn et al. Microstructure and gas-sensing properties of thick film sensor using nanophase SnO_2 powder, Sensors Actuators B 99 (1) (2004) 18

[2]S.T. Navale et al. Synthesis of Fe_2O_3 nanoparticles for nitrogen dioxide gas sensing applications, Volume 39, Issue 6, August (2013), 6453–6460

[3]ShouliBai et al. Reverse micro emulsion in situ crystallizing growth of ZnO nanorods and application for NO2 sensor, Volume 190, January (2014), 760–767

[4]Norman M. Ratcliffe et al. Polypyrrole-based sensor for hydrazine and ammonia, Analytica Chimica Acta, Volume 239, 1990, Pages 257–262

[5]J. Sarfraz et al. Printed copper acetate based H_2S sensor on paper substrate, Sensors Actuators B Chem., 173 (2012), . 868–873

[6]F. Molina-Lopez et al. All additive inkjet printed humidity sensors on plastic substrate, Sensors Actuators B Chem., 166–167 (2012), 212–222

[7]Hakam Agha et al. Polypyrrole coating films for nanoparticles, Colloids and Surfaces A: Physicochemical and Engineering Aspects 462 (2014), 217-224

[8]Adeel Afzal et al. NO_x sensors based on semiconducting metal oxide nanostructures: Progress andperspectives, Sensors and Actuators B 171– 172 (2012) 25– 42

[9]K. Suri et al. Gas and humidity sensors based on iron oxide–polypyrrolenanocomposites, Sens. Actuators B 81 (2002) 277.

[10]Ravish Garg et al. PolypyrroleMicrowires as Toxic Gas Sensors for Ammonia and Hydrogen Sulphide, Journal of Sensors and Instrumentation, (2015) Vol. 3 No. 1 pp. 1-13

[11]P.M. Borsenberger et al. Concerning the role of dipolar disorder on charge transport in molecularly doped polymers, Journal of Chemical Physics 95 (1991) 5327.

Thin-shell Concrete Structures for Low Embodied and Operating Energy Plus Durability

W. Leighty, Alaska Applied Sciences, Inc., Box 20993, Juneau, AK 99802
wleighty@earthlink.net 907-586-1426 206-719-5554 (cell)

ABSTRACT We have designed a re-usable tooling system for building thin-shell concrete dome structures of high strength and durability with low cost and embodied energy. The proof-of-concept scale model domes shown herein are ~1 cm thick, but have survived burial under 1 m sand and repeated horizontal seismic shocks of ~2.7 g. Hundreds could be made from a single forms set, on-site, maximizing use of local materials and labor. Optimum size for human utility and this method is ~ 6-7 m diameter, 5/8 sphere volume. Only quasi-spherical structures may be made of thin-shell concrete; stress concentrations at small radius must be avoided. Primary reinforcement is non-ferrous, carbon-fiber-epoxy C-Grid": http://www.chomarat.com/en/tag/c-grid/

Keywords: Thin-shell, concrete, quasi-spherical, dome, C-Grid, seismic, low-cost

This tooling and construction system and strategy was conceived for rapid construction of durable and low-cost shelter in remote places where one must minimize import of materials, tooling, tools, and expert labor, and maximize use of indigenous materials and labor:

- Replacement of failing Alaska village housing
- Disaster relief: earthquake, flood, tsunami
- Clinic, classroom, granary, water tank, storage
- Housing, office, warehouse

The forms set, shown stacked in Figure 1 and assembled in the other figures, must be transported to the construction site(s). Imported materials would include:

- Bagged premix of Portland cement, additives for workability and strength (superplasticizer, pozzolan, silica fume), for waterproofing (Xypex, for example), and plastic reinforcing fibers.
- C-Grid carbon fiber epoxy primary reinforcement, pre-cut to size for stapling to the dome form surface, as overlapping "tiles"
- Heavy fiberglass grid for embedding in the foundation ring, for structural tie-in of the dome shell during dome mortaring
- Stainless steel staples, for tiling C-Grid
- Window and door units or materials
- #3 or #4 rebar for foundation ring core

Indigenous fresh water and sand complete the concrete mix; both must be clean, perhaps washed.

Several domes, built on the same tooling, may be built adjacent to, and interconnected with, each other, to provide more space and volume in a single building than a single dome would.

At 6 m diameter, 5/8 sphere by volume, a small "loft" could be supported by a privacy partition wall, and laterally by the dome shell, for storage and for children sleeping.

DOME FORM CONSTRUCTION

The 12 "orange peel" dome segments are fiberglass covered with 60 mil EPDM rubber sheet for staple retention, with 10 mil Teflon sheet attached with 5 mil butyl rubber adhesive. The Teflon outer surface easily detaches from the cured concrete, after the spacer shims are removed to free the individual segments. Repeated use will perforate and damage the Teflon outer surface, which can be replaced in a refurbishing operation.

DOME CONSTRUCTION METHOD

See Figures.

1. A flat, compacted ground area is prepared. The foundation ring forms are assembled, with a single continuous piece of reinforcing bar, steel or other, centered in the cross-section. The ring is poured and vibrated. The vertical heavy fiberglass tie-in mesh is installed. Cure

for 1-2 days; remove and clean the foundation forms for reuse.

2. Assemble 12 dome form "orange peel" segments with inter-segment 3/8" thick shims. Attach circular top form and skylight Sonotube
3. Staple C-Grid tiles to dome form with stainless steel staples.
4. Apply rich sand-only concrete mortar, by hand trowel or spray; trowel to desired finish
5. Install or drape polyethylene sheet to exterior, for curing vapor barrier. Cure for two days.
6. Inside dome, remove all segment assembly fasteners. Remove inter-segment shims. Remove dome form "orange peel" segments, through entry door.
7. Remove lower segment shims
8. Clean form to prepare for reuse. The Teflon outer surface requires little cleaning

INSULATION AND FINISH

If required, thermal insulation is closed-cell urethane (UR) foam sprayed on interior of the concrete shell. Foam interior is then covered with a protective plaster coat and optional interior paint. The closed-cell UR foam provides a continuous vapor barrier at the structure's interior, where it is most effective.

In a full-size dome structure, plumbing, if any, should be installed only in interior partition walls, while electricity flex conduit may be glued to the concrete shell interior and buried in the sprayed interior foam insulation, if it is used.

A heating appliance chimney and a plumbing vent may be formed and mortared into the thin concrete shell.

The entry roof is also thin concrete, mortared into the dome shell in the continuous process of applying the high-strength concrete shell. Various "arctic entry" designs could be developed, perhaps approaching the classic "igloo" design.

The window frame must be carefully designed to eliminate stress-concentrating corners where shrinkage cracking will occur during concrete cure. Bare, custom-made, insulated glass units would be most economical. The skylight may be operable, to provide ventilation.

COST

Of course, cost will vary greatly, depending on many factors: degree of finish and size of structure, and with the costs of moving the forms set, tools, and materials to the construction site(s), and the number of domes built there. The C-Grid primary reinforcement will probably be the largest variable cost component for a simple, uninsulated dome built in a tropical climate, if no structural door and windows, nor plumbing and wiring, nor finished floor are needed.

EARTH SHELTERING

We have demonstrated that a 1 cm thick concrete dome may be completely buried, with 1 m of sand on top, without visible damage. This isotropic Earth load may strengthen the dome. However, we have not tested complete burial during multiple freeze-thaw cycles. Earth sheltering may be important to protect the interior from climatic extremes, and to provide some degree of protection from military attack: small arms fire and shrapnel.

SCALING UP FOR HUMAN USE

The optimum size for human utility and this construction method is probably ~ 6-7 m diameter, 5/8 sphere volume. The dome form segments may be difficult to manipulate, by hand, at larger size. A special curved ladder, supported at an apex pin at the top and on small rubber tires at the bottom, will be needed to apply mortar to dome upper areas.

Manufacturing complete forms sets, for the foundation ring, dome and entry, and window openings, will probably cost $300 – 500,000 for CAD design and tooling, from which the fiberglass form components could be made. We will invest in this step only when we confirm a market for the thin-shell concrete structures that could be built from such form sets. The sets could be sold or leased, including training to insure quality, safe, and economical construction in the field – including under difficult conditions. The business model should also include selling the bagged premix for the high-quality concrete mortar, and pre-cut pieces of C-Grid for tiling to the dome form.

Figure 1. Complete forms set, for foundation ring and dome. First prototype: proof-of concept one-third scale model, with skylight, window, chimney

Figure 2. First prototype, insulated, with insulated floor. Total weight 1,050 kg. Juneau, Alaska

Figure 3. First prototype, insulated, with floor. Total weight 1,050 kg.

Figure 4. Assembling the dome form to the foundation ring. A shim 3/8" thick is installed between pairs of form "orange peel" segments

Figure 5. Stapling "tiled" C-Grid to dome form

Figure 6. C-Grid attached to dome form segments with stainless steel staples. Note clear plastic "chairs" to space C-grid ~ 1 / 2 cm from Teflon surface.

Figure 7. Applying mortar ~ 1 cm thick

Figure 8. Applying mortar ~ 1 cm thick

Figure 9. Strength test: 1 m sand on top. No visible damage. Dial indicators inside the uninsulated concrete shell showed ~ 0.010" deflection at horizontal diameter and vertical apex.

Figure 10. Seismic test: repeated ~2.7 g shocks. Wire rope wrapped around foundation ring. No visible damage.

Silver Nanoparticles as Low-Emissivity Coating Materials for Window Glazing Applications

Tao Gao[a] and Bjørn Petter Jelle[b,c]

[a] Department of Architectural Design, History and Technology, Norwegian University of Science and Technology (NTNU), NO-7491 Trondheim, Norway, tao.gao@ntnu.no

[b] Department of Materials and Structures, SINTEF Building and Infrastructure, NO-7465 Trondheim, Norway

[c] Department of Civil and Transport Engineering, Norwegian University of Science and Technology (NTNU), NO-7491 Trondheim, Norway

ABSTRACT

The potential of silver (Ag) nanoparticles as low-emissivity (low-e) coating materials for window glazing applications has been discussed. Ag nanoparticles were prepared via a wet chemical method and applied on the surface of flat glass through spin coating. A mild heat treatment at 200°C was employed to achieve the low-e effect, which results in a total surface emissivity of about 0.015, compared to about 0.837 of the plain glass substrate. By applying such low-e coatings, the heat loss through a single-glazed window pane could be reduced by about 45% (*U*-value from 5.75 to 3.18 W/(m^{2}K)).

Keywords: Ag nanoparticle, emissivity, glazing

1 INTRODUCTION

Reducing the energy loss through the building envelope (e.g., windows, roofs, and walls) is important to increase the energy efficiency of buildings [1]. As a cost-effective technique, low-e coatings have widely been used for such purposes [2]. A low-e coating is essentially a thin film consisting of metal or metallic oxides, which is capable of emitting low levels of radiant thermal energy. Low-e coatings can reflect the radiative thermal energy back into the room in cold weather (or back to the outdoors during warm weather), thereby reducing the radiative heat transfer through the building envelopes. Low-e coatings can be applied in both the transparent (i.e. window glazings) and the opaque (i.e. roofs, walls or floors) parts of the building envelope [3]. Hence, the development of different low-e coating materials to meet the different application requirements is of both scientific and technological interest.

For window glazing applications, low-e coatings are usually prepared via sputtering or chemical vapor deposition [2]. Although these methods can be integrated with the online processes of flat-glass production, the need of high temperature, vaccum, and expensive raw materials may show certain limitations, i.e., with respect to process economics. In this regard, offline coating techniques are very interesting since they may enable more possibilities towards a cost-effective product. For example, metallic nanoparticles may represent an ideal raw material for low-e coatings [4], because they can be prepared with controllable properties [5] and be readily processed via a number of coating techniques, such as dip coating, spin coating, or inkjet printing [6]. However, to the best of our knowledge, few studies have so far been published to reveal the potential of metallic nanoparticles as low-e coating materials for window glazing applications.

We communicate in this contribution a simple method to achieve low-e coatings by using Ag nanoparticles. A preliminary test shows a total surface emissivity of about 0.015, compared to about 0.837 of the float glass.

2 EXPERIMENTAL

Ag nanoparticles were synthesized by a modified polyol process, as reported previously [7]. For a typical synthesis, 10 g polyvinylpyrrolidone (PVP) powder was added into 75 mL ethylene glycol under constant stirring. After the PVP was totally dissolved, 400 mg $AgNO_3$ was added to the PVP-ethylene glycol solution. Thereafter, the glass beaker that contained the obtained solution was immersed into a preheated oil bath and maintained at 120°C for 1 h. After the synthesis, Ag nanoparticles were separated from the reaction system by centrifugation, and washed twice with acetone and ethanol, respectively. Finally the obtained Ag nanoparticles were dispersed in ethanol to form a colloidal solution (concentration ~ 10 mg/mL).

A spin coating process was performed at 2000 rpm to prepare Ag nanoparticle films on float glass substrates. After being dried naturally in air at room temperature, the coated glass substrates were subject to an annealing treatment at 200°C for 2 hours.

Ag nanoparticles were characterized by using x-ray diffraction (XRD), scanning electron microscopy (SEM), transmission electron microscopy (TEM), and attenuated total reflectance (ATR) Fourier transform infrared (FTIR) spectroscopy. Optical transmittance spectra of the as-prepared samples were measured on a Perkin Elmer Lambda 1050 UV/Vis/NIR spectrophotometer. Reflectance spectra were measured with a 150 mm integrating sphere reflectance accessory. Reflectance spectra were also measured on a SOC-100 Hemispherical Directional

Reflectometer (HDR) in the wavelength region of 2–25 μm; Total hemispherical emissivity of the samples was then calculated by integrating the angular reflectance measurement stepwise from 10° to 80° incident angles (12 angles) and calibrated with a standard Au reference.

3 RESULTS AND DISCUSSION

In this work, Ag nanoparticles were prepared via a polyol process, as reported previously [7, 8]. The as-synthesized Ag nanoparticles are crystalline and have a face-centered cubic structure (Figure 1a). According to Scherrer's formula [9], crystallite size of the as-prepared Ag nanoparticles is about 47 nm, which also agrees with the subsequent TEM analysis. No XRD reflections related to any silver oxides such as Ag_2O were observed, indicating that Ag nanoparticles are rather stable despite of their small sizes.

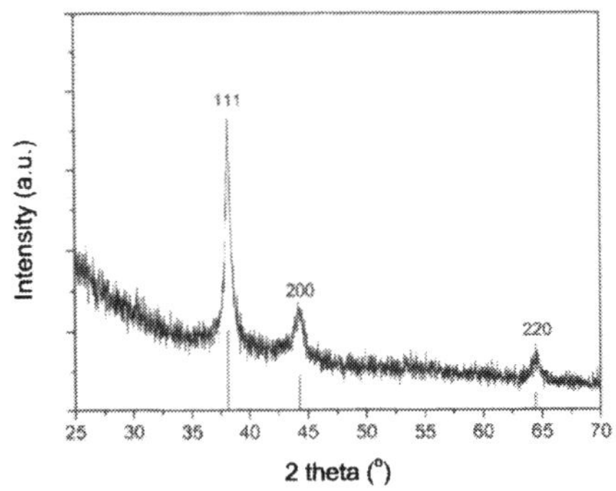

Figure 1: XRD pattern of the as-prepared Ag nanoparticles. The positions and relative intensities of the reflections are indicated at the bottom.

Figure 2a shows a typical TEM image of the as-prepared Ag nanoparticles. The particles are nearly spherical and are polydisperse with a relatively large size range from 20 to 80 nm, with an average diameter of about 50 nm, which is in agreement with the XRD results (Figure 1). Electron diffraction pattern (Figure 2b) shows typical polycrystalline features, indicating that the individual Ag nanoparticle may constitute of several primary particles.

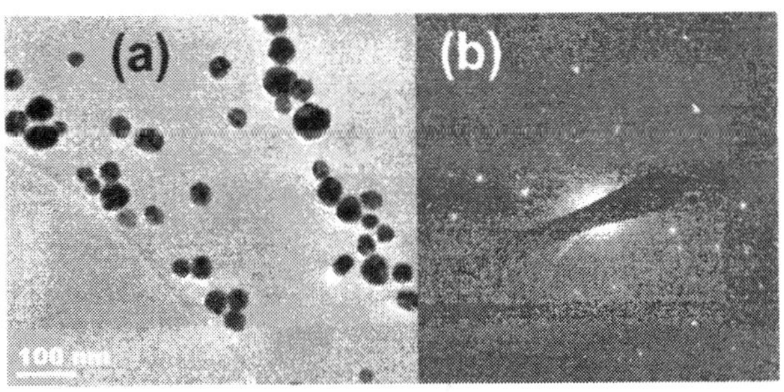

Figure 2: TEM image (a) and electron diffraction pattern (b) of the as-prepared Ag nanoparticles.

It is worth noting that, during the polyol synthesis, PVP has been added as a protective agent to prevent Ag nanoparticles from aggregation [7, 8]; therefore, the surface of the resulting Ag nanoparticles is most likely capped with PVP molecule residuals. For example, as shown in Figure 3a, the FTIR spectrum of the as-prepared Ag nanoparticles exhibit features similar to those of the PVP powders (Figure 3b), apart from the absorptions related to the glass substrate (Figure 3c). The existence of PVP on the surface of Ag nanoparticles increases not only their chemical stability against oxidation (see, e.g, Figure 1), but also workability for thin film processing. For example, the as-prepared Ag nanoparticles can be well dispersed within ethanol to form a stable colloidal suspension with different concentrations (up to 100 mg/mL), which is important for the subsequent spin coating process.

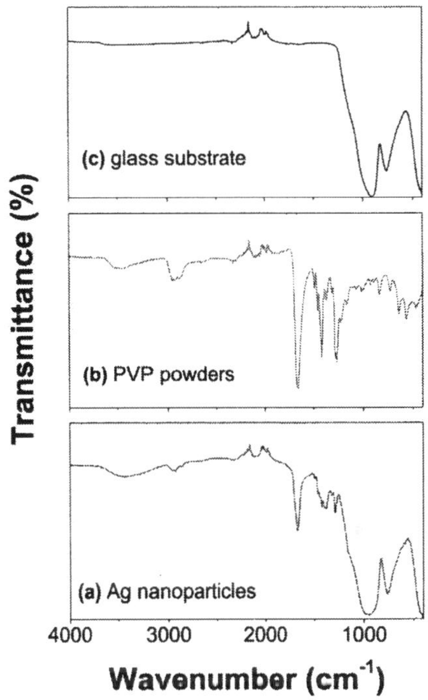

Figure 3: FTIR transmittance spectra of (a) the as-prepared Ag nanoparticles, (b) PVP powders, and (c) glass substrate. The noise around 2000 cm^{-1} is due to the ATR diamond crystal.

Figure 4a shows the SEM image of the as-synthesized Ag nanoparticle film, which is relatively dense and has a thickness of about 400 nm. After the annealing treatment at 200°C for 2 hours, a dense-porous transformation has been observed for the Ag nanoparticle film (Figure 4b), which can be attributed to the melting and growth of Ag nanoparticles. It is worth noting that small nanoparticles usually have a lower melting temperature than that of the corresponding bulk counterpart due to the size effect [10].

Figure 4: SEM images of Ag nanoparticles before (a) and after (b) the annealing treatment at 200°C.

Figure 5 shows the transmittance spectra of the obtained samples. The original Ag nanoparticle film is relative dense and shows a limited visible transparency of about 1% at 510 nm (Figure 5a). The transmission band at about 1200 nm is probably related to the absorbed PVP molecules. For example, this band disappears after the annealing treatment (Figure 5b). Moreover, the visible transparency of the Ag nanoparticle film increases to about 35% at 510 nm after the annealing treatment. This can be attributed to the dense-porous transformation (Figure 4) and the removal of surface-absorbed PVP molecules. The absorption peak at about 352 nm is related to the surface plasmon resonance (SPR) band of Ag [11, 12].

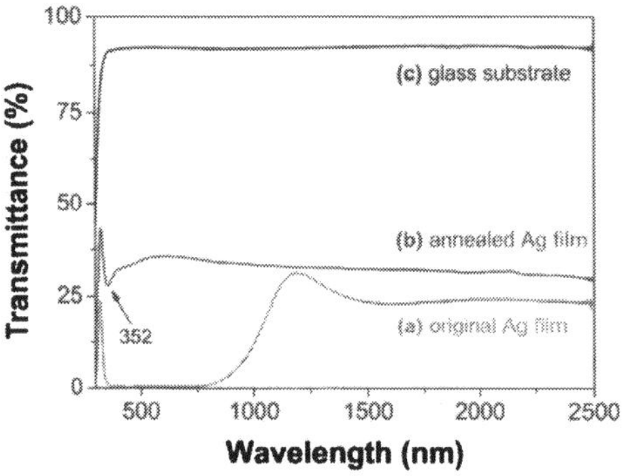

Figure 5: Transmittance of the Ag nanoparticle films before (a) and after (b) the annealing treatment at 200°C. Glass substrate (c) is also measured for comparison.

Reflectance spectra of the obtained samples were also measured, Figure 6. It is found that, compared to the plain glass, Ag nanoparticle-coated glass substrates are highly reflective. For example, the glass substrate has a reflectance of about 8% at 510 nm; in contrast, the as-prepared Ag nanoparticle film shows a higher reflectance of about 44% at 510 nm, which increases further to about 73% after the annealing treatment. Obviously, applying Ag nanoparticles on the glass surface may have the potential to achieve the low-e effect.

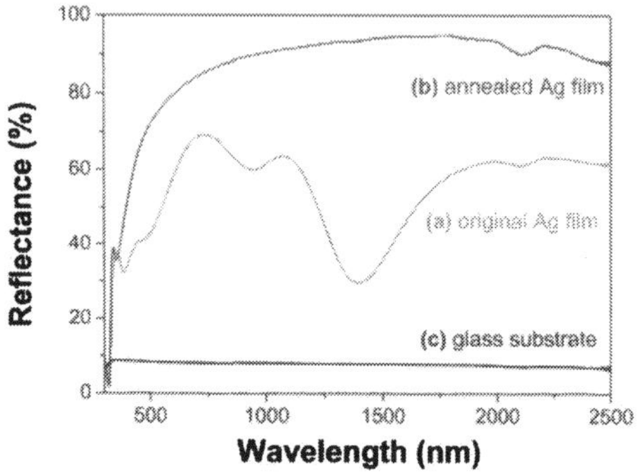

Figure 6: Reflectance of the Ag nanoparticle films before (a) and after (b) the annealing treatment at 200°C. Glass substrate (c) is also measured for comparison.

Figure 7 shows the reflectance spectra of the obtained samples, which are measured in the wavelength region of 2–25 µm in order to calculate the surface emissivity. Interestingly, compared to the plain glass substrate, the original Ag nanoparticle film does not show a significant improvement with respect to the low-e effect [7]. For example, the glass substrate has a total emissivity of about 0.837. After applying the Ag nanoparticles, the obtained sample shows a slightly lower surface emissivity of about 0.793 (Figure 7a). The low-e effect has only been achieved after the annealing treatment (Figure 7b), which results in an emissivity value of about 0.015.

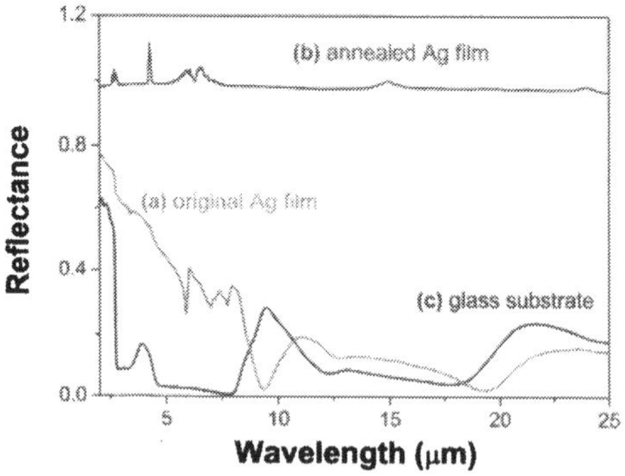

Figure 7: Reflectance of the Ag nanoparticle films before (a) and after (b) the annealing treatment at 200°C. Only the reflectance spectra measured at an angle of 10° are shown. Glass substrate (c) is also measured for comparison.

The low-e effect achieved with the Ag nanoparticle system is interesting and potentially useful. For example, by applying such a coating system, *U*-value (heat transfer coefficient) of a single-paned glazing unit can be reduced from 5.75 to 3.18 $W/(m^2K)$, indicating about 45% reduction on heat loss through a single-glazed window pane. For the most used double-glazed window, a ~ 48% reduction of the heat loss can be reached. Therefore, these low-e coatings can contribute to a large energy saving potential in the building sector [3]. Although the cost issue of Ag nanoparicles may have to be addressed [7], it seems very promising since a highly valuable property can be imparted at relatively low additive ratios and the material cost will decrease as they are produced in larger quantities [13].

Further investigations are still required for improving the Ag nanoparticle films and for a comprehensive understanding on the involved mechanisms. For example, the absence of low-e effect in the original Ag nanoparticle films (Figure 7a) is most likely related to the surface-absorbed PVP molecules, which forms an insulating layer and blocks the electrical conduction between Ag nanoparticles. An effective method for the removal of PVP molecules appears to be required to examine the proposed hypothesis; moreover, the annealing treatment may have certain limitations with respect to soft, organic substrates. In addition, for the practical application as low-e coatings in the building sector, the durability issue of Ag nanoparticle films is important to address [3].

4 CONCLUSIONS

Ag nanoparticles with typical diameters of about 50 nm have been synthesized by using a wet chemical method. These prefabricated Ag nanoparticles show a promising potential as low-e coating materials for window glazing applications. The as-prepared Ag nanoparticles can be readily deposited on flat glass through spin coating. A mild heat treatment at 200°C has been used to remove the surface absorbed organic molecules to achieve the low-e effect. A total surface emissivity of about 0.015 has been obtained, compared to about 0.837 for the plain glass substrate. By application of such low-e coatings, the heat loss through a double-glazed window could be reduced by about 48%. Further investigations are still required for an improved understanding on the involved mechanisms.

ACKNOWLEDGEMENT

This work has been supported by the Research Council of Norway and several partners through "The Research Centre on Zero Emission Buildings" (ZEB).

REFERENCES

[1] B. P. Jelle, "Traditional, state-of-the-art and future thermal building insulation materials and solutions–Properties, requirements and possibilities," Energy and Buildings, 43, 2549-2563, 2011.

[2] C. M. Lampert, "Heat mirror coatings for energy conserving windows," Solar Energy Materials, 6, 1-41, 1981.

[3] B. P. Jelle, S. E. Kalnæs and T. Gao, "Low-emissivity materials for building applications: A state-of-the-art review and future research perspectives," Energy and Buildings, 2015 (accepted).

[4] T. Gao, B. P. Jelle and A. Gustavsen, "Nanomaterials for advanced window technologies," Proceedings of CISBAT (Cleantech for Sustainable Buildings – From Nano to Urban Scale), 55-60, 2011.

[5] B, Wiley, Y. Sun and Y. Xia, "Synthesis of silver nanostructures with controlled shapes and properties," Accounts of Chemical Research, 40, 1067-1076, 2007.

[6] W. Shen, X. Zhang, Q. Huang, Q. Xu and W. Song, "Preparation of solid silver nanoparticles for inkjet printed flexible electronics with high conductivity," Nanoscale, 6, 1622-1628, 2014.

[7] T. Gao, B. P. Jelle and A. Gustavsen, "Core-shell typed $Ag@SiO_2$ nanoparticles as solar selective coating materials," Journal of Nanoparticle Research, 15, Article 1370, 2013.

[8] D. Kim, S. Jeong and J. Moon, "Synthesis of silver nanoparticles using the polyol process and the influence of precursor injection," Nanotechnology, 17, 4019-4024, 2006.

[9] J. I. Langford and A. J. C. Wilson, "Scherrer after sixty years: a survey and some new results in the determination of crystallite size," Journal of Applied Crystallography, 11, 102-113, 1978.

[10] A. Safaei, M. Attarian Shandiz, S. Sanjabi and Z. H. Barber, "Modelling the size effect on the melting temperature of nanoparticles, nanowires and nanofilms," Journal of Physics: Condensed Matter, 19, Article 216216, 2007.

[11] J. J. Mock, M. Barbic, D. R. Smith, D. A. Schultz and S. Schultz, "Shape effects in plasmon resonance of individual colloidal silver nanoparticles," Journal of Chemical Physics, 116, 6755-6759, 2002.

[12] M. K. Kinnan and G. Chumanov, "Plasmon coupling in two-dimensional arrays of silver nanoparticles: II. Effect of the particle size and interparticle distance," Journal of Physical Chemistry C, 114, 7496-7501, 2010.

[13] J. Lee, S. Mahendra and P. J. Alvarez, "Nanomaterials in the construction industry: a review of their applications and environmental health and safety considerations," ACS Nano, 4, 3580-3590, 2010.

BIOLOGICAL DECONTAMINATION OF RADIONUCLIDES IN INDUSTRIAL AND ENVIRONMENTAL WATER USING A NEW HIGHLY RADIORESISTANT MICROALGA

Corinne Rivasseau*, Diane de Gouvion Saint Cyr*, Ariane Atteia**, Marina Gromova***, A. Couté**** Emmanuel Farhi*****

*Commissariat à l'Energie Atomique et aux Energies Alternatives, CEA, Direction des Sciences du Vivant, 17 rue des Martyrs, Grenoble, France
**CNRS, Laboratoire de Bioénergétique et Ingénierie des Protéines, Marseille, France
***CEA, Direction des Sciences de la Matière, Grenoble, France
****Muséum National d'Histoire Naturelle, Paris, France
*****Institut Laue Langevin, Grenoble, France
Corinne.rivasseau@cea.fr

ABSTRACT

Nuclear energy technologies generate radioactive and chemically toxic compounds including mainly carbon-14 and radioactive metals. Treatment processes rely on physico-chemical methods, which, though efficient and robust, do not completely remove carbon-14 and are not adapted to large environmental volumes. Biological remediation technologies may constitute an interesting alternative to reduce radioactive releases in aqueous effluents or to clean-up accidentally contaminated water. We recently isolated a new microalga from a nuclear facility, *Coccomyxa actinabiotis*, which withstands huge ionizing radiation doses, up to 20 kGy. This microalga very efficiently and rapidly decontaminates most of the metallic radionuclides contained in nuclear effluents and 85% of carbon-14 in a few hours. The process feasibility was demonstrated at real-scale for *in situ* decontamination and pilote-scale for remote water treatment.

Keywords: biotechnology, microalga, nuclear industry, environmental water, radionuclide decontamination

1 INTRODUCTION

Nuclear energy technologies generate radioactive and chemically toxic compounds including tritium, carbon-14 and radioactive metals. They require processes for radionuclide decontamination inside the facilities themselves and of the liquid and gaseous effluents released into the environment. Treatment processes frequently used to remove radionuclides from liquid streams are based on physico-chemical methods including evaporation, solid/liquid separation by filtration, centrifugation or decantation possibly associated to a chemical precipitation/flocculation process, reverse osmosis, ultrafiltration, sorption, and ion exchange [1,2]. These methods are efficient and robust; however, they do not completely remove some radionuclides such as carbon-14 and they would not be adapted to the treatment of large environmental volumes. Alternative technologies are thus needed to reduce radioactive releases in aqueous effluents or to clean-up accidentally contaminated water.

Biological remediation technologies, based on living organisms or extracts from living organisms, are already used in many industrial applications and often offer efficiency, cost and environmental impact benefits against conventional technologies [3]. They may constitute an interesting alternative to clean-up radioactive streams as well. However, very few bio-based technologies have been proposed in the nuclear field. For maximal decontamination performance, such technologies would require organisms that simultaneously accumulate radionuclides while sustaining their chemical and radiological toxicity. The most radiation-resistant organisms described so far are prokaryotes, including the bacterium *Deinococcus radiodurans*. But *Deinococcus radiodurans* needs genetic engineering to acquire toxic-metal resistance and remediation capabilities [4].

We recently isolated a new green microalga from a nuclear facility which combines both properties. We characterized its radioresistance and its radionuclide accumulation properties and compared, on the basis of remediation experiments performed at lab-scale and at real-scale, its decontamination efficiency and the ultimate waste volume generated to that of the usual physico-chemical treatment. This alga is an excellent candidate for new biotechnology methods of remediation of contaminated industrial or environmental water.

2 RESULTS AND DISCUSSION

2.1. Resistance to ionizing radiation

This organism, *Coccomyxa actinabiotis*, is a unicellular green microalga of ellipsoidal shape measuring about 3.8 µm x 6.8 µm (Figure 1). It contains a big parietal chloroplast with starch granules. Cells in division can be observed in Figure 1.

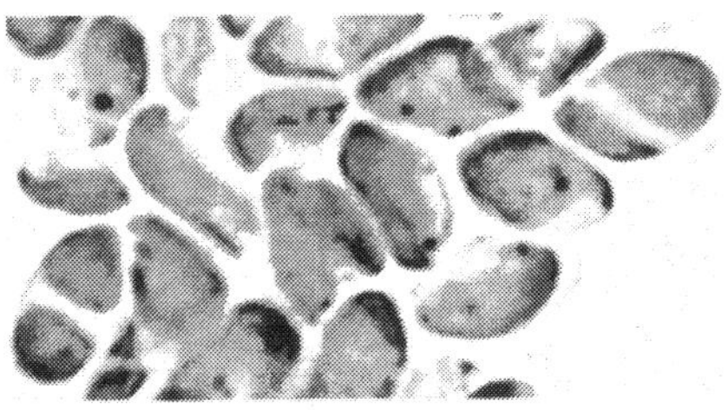

Figure 1: *Coccomyxa actinabiotis* cells.

This eukaryote was isolated from a storage pool for spent nuclear fuel elements, where it is continuously submitted to ionizing radiation. It withstands huge ionizing radiation doses, up to 20,000 Gy, when subjected to an intense γ-irradiation flux of 4,000 Gy/h [5]. The lethal dose to half the population (LD_{50}) is 10,000 Gy [6], which is comparable to that of the bacterium *Deinococcus radiodurans* and is outstanding for a eukaryote. The lethal dose corresponding to 50 % mortality after γ-irradiation for other reference species or radioresistant species is indicated in Table 1. For microalgae, this value is between 30 and 1,200 Gy [8].

Organism	Resistance to irradiation
Human	10 Gy
E. coli [7]	~ 200 Gy
Microalgae [8]	30-1200 Gy
Chroococcidiopsis [9]	2.5 kGy
Pyrococcus abyssi	3 kGy
Coccomyxa actinabiotis [6]	10 kGy
Deinococcus radiodurans [7]	6-10 kGy

Table 1: Comparison of resistance to ionizing radiation of various organisms (LD_{50}, for Human, LD_{100}).

Such high doses hardly affected the cellular metabolism of *Coccomyxa actinabiotis*. The growth after irradiation at doses of up to 6,000 Gy was not significantly different from that of the control. Nine days after irradiation, the metabolite content of the microalgae did not significantly differ compared to the non-irradiated control, whatever the irradiation dose, suggesting the complete recovery of cellular functioning [5].

2.2. Accumulation of radionuclides

Furthermore, this microalga very efficiently and rapidly accumulates most of the metallic radionuclides contained in nuclear effluents including ^{238}U, ^{137}Cs, ^{110m}Ag, ^{60}Co, ^{54}Mn and ^{65}Zn with decontamination rates above 90% [10]. It also fixes, via metabolically active processes, mineral and organic species of ^{14}C which is, together with tritium, the main radionuclide present in effluents issuing from nuclear facilities. More than 85% of $^{14}CO_2$ can be decontaminated in a few hours.

One of the most employed and efficient radionuclide decontamination method for water at nuclear facilities is ion-exchange on resins. Figures 2 and 3 show the decontamination efficiency of microalgae compared to that of the resins in suspension in the same nuclear effluent. When using 160 mg microalgae fresh weight *i.e.* 16 mg dry weight/100 ml nuclear effluent, the decontamination efficiency was similar to that obtained with 80 mg resin/100 ml effluent for the γ-emitter radionuclides ^{110m}Ag, ^{58}Co, ^{60}Co, and ^{54}Mn after a 1 h contact time (Figure 2). ^{51}Cr decontamination by microalgae was lower. But ^{65}Zn decontamination by microalgae was twice higher, probably because zinc is a physiological metal incorporated by the cells via active mechanisms.

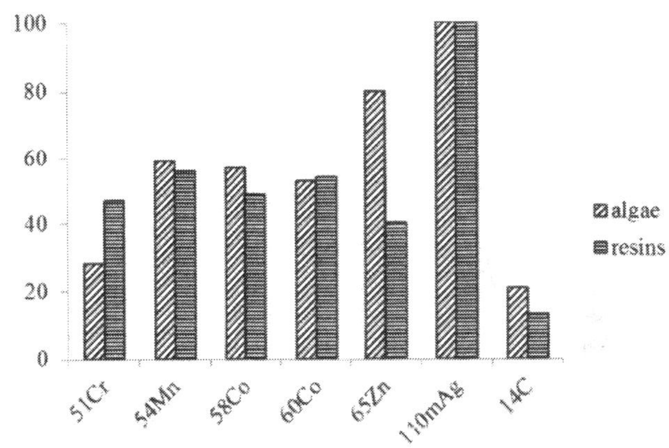

Figure 2: Comparison of decontamination efficiency of microalgae to that of ion-exchange resins after a 1 h contact time.

After a 24 h contact time, the decontamination efficiency using algae increased, though less or identical to that obtained using resins, except for ^{14}C (Figure 3).

Decontamination of ^{14}C is usually not fully achieved using ion-exchange resins and reached 27% in this experiment. Using microalgae, and in the light, ^{14}C was efficiently decontaminated owing to its incorporation inside the microalgal cells through metabolically mediated processes, particularly through photosynthesis. In that kind of application, this microalga efficiently supplements the physico-chemical decontamination methods classically employed.

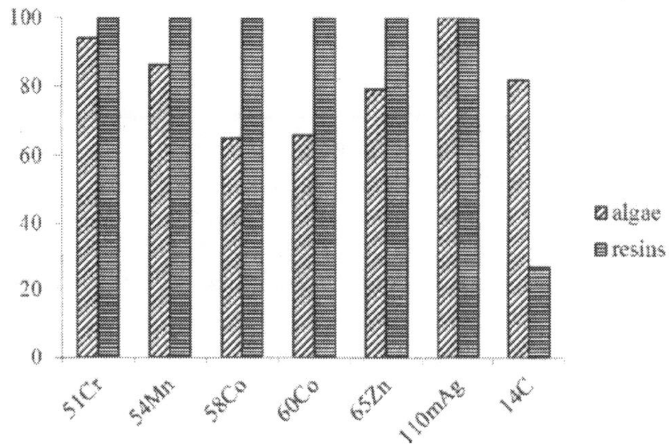

Figure 3: Comparison of decontamination efficiency of microalgae to that of ion-exchange resins after a 1 h contact time.

The microalgae volume could be reduced by 90% by drying. Once dried in the light, the algae retained the totality of the γ-emitters and 97% of the ^{14}C fixed during the decontamination step.

2.3 Real-scale decontamination test

We tested radionuclide bio-decontamination at real-scale in a 360-m^3 storage pool of radioactive components by using *Coccomyxa actinabiotis* in suspension in water and comparing its decontamination efficiency to that of the conventional method, namely ion-exchange resins (Figure 4).

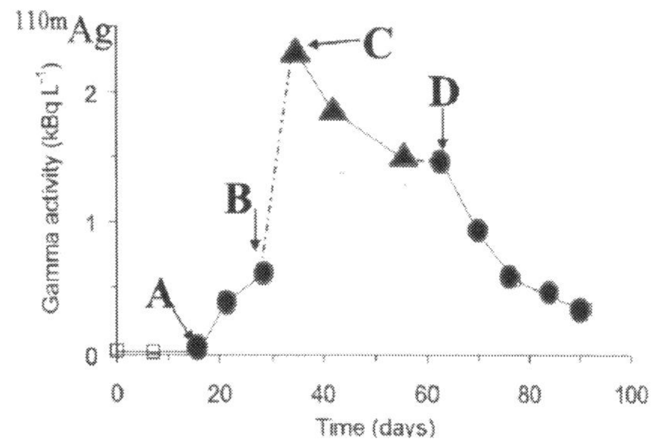

Figure 4: Decontamination of a storage pool of nuclear components using microalgae (triangles) and ion-exchange resins (circles).

Resin-based purification, used from time A to B, was stopped at time B and replaced by algae-based decontamination from time C, and for 21 days. A decrease in the main radionuclide concentration, ^{110m}Ag, was simultaneously observed, originating from its uptake by suspended algae, which were then collected onto filters. During that period, the algae removed 740 MBq of γ-emitters including 310 MBq ^{110m}Ag, 270 MBq ^{51}Cr, 90 MBq ^{124}Sb, and 30 MBq ^{60}Co from the pool. The mean activity collected by *C. actinabiotis* was 20 MBq/g dry weight. About 40 g dry weight of microalgae was used in that experiment. From time D, the resins were restarted.

3 CONCLUSIONS

In summary, we have discovered a unique microalga with an exceptional radioresistance, comparable to that of the most radioresistant bacteria. This photosynthetic microorganism can survive with only water, light, CO_2, and a few dissolved minerals. It can thrive in a radioactive environment and is also capable of capturing and concentrating, rapidly and efficiently, radionuclides in nuclear facility water. The feasibility of the bio-purification of radionuclides at real-scale was demonstrated. New means of remediation are now conceivable using this micro-alga. They may be applied in nuclear facilities where they would complement conventional methods and reduce the volume of radioactive waste, in the final cleaning of water to reduce radioactive emissions into the environment, or in decontamination of accidentally polluted water.

4 REFERENCES

[1] Efremenkov, V.M. Radioactive waste management at nuclear power plants. An overview of the types of low- and intermediate-level wastes and how they are handled. IAEA Bulletin 1989, 4, 37–42.

[2] Combined methods for liquid radioactive waste treatment; IAEA-TECDOC-1336; International Atomic Energy Agency: Vienna, Austria, 2003.

[3] I.M.M. Gillespie, J.C. Philp, Bioremediation, an environmental remediation technology for the bioeconomy, Trends in Biotechnology. 31 (2013) 329–332.

[4] Brim, H.; McFarlan, S.C.; Fredrickson, J.K.; Minton, K.W.; Zhai, M.; Wackett, L.P.; Daly, M.J. Engineering Deinococcus radiodurans for metal remediation in radioactive mixed waste environments. Nat. Biotechnol. 2000, 18, 85–90.

[5] Rivasseau C, Farhi E, Atteia A, Couté A, Gromova M, de Gouvion Saint Cyr D, Boisson AM, Feret AS, Bligny R (2013) An extremely radioresistant green eukaryote for radionuclide bio-decontamination in the nuclear industry. Energy & Environmental Science (impact factor: 9.61). 2013, 6(4), 1230-1239

[6] Rivasseau C, Farhi E, Couté A, Atteia A (2011) A new radioresistant microalga of the Coccomyxa genus WO2011/098979.

[7] Daly, M.J.; Gaidamakova, E.K.; Matrosova, V.Y.; Vasilenko, A.; Zhai, M.; Venkateswaran, A.; Hess, M.; Omelchenko, M.V.; Kostandarithes, H.M.; Makarova, K.S.;

Wackett, L.P.; Fredrickson, J.K.; Ghosal, D. Accumulation of Mn(II) in Deinococcus radiodurans Facilitates Gamma-Radiation Resistance. Science 2004, 306, 1025–1028.
[8] Effects of ionizing radiation on aquatic organisms and ecosystems. Technical reports series 172; International Atomic Energy Agency: Vienna, Austria, 1976.
[9] Billi, D.; Friedmann, E.I.; Hofer, K.G.; Grilli Ciaola, M.; Ocampo-Friedmann, R. Ionizing-Radiation Resistance in the Desiccation-Tolerant Cyanobacterium Chroococcidopsis. Appl. Env. Microbiol. 2000, 66, 1489–1492.
[10] Rivasseau C, Farhi E, Couté A, Atteia A (2010) A new radioresistant microalga of the Coccomyxa genus FR10/00578.

2-Butoxyethanol as a Draw Solution for Forward Osmotic Application

K. Sano, T. Ide and T. Imada

Corporate Research and Development Center, Toshiba Corporation
1 Komukai, Toshiba-cho, Saiwai-ku, Kawasaki 212-8582, Japan, kenji.sano@toshiba.co.jp

ABSTRACT

2-Butoxyethanol (2BE) was examined as a draw solution for forward osmotic application. The osmolality of 40wt% of 2BE aqueous solution was 2 times higher than that of 3.5wt% salt water in freezing point depression measurement. In forward osmosis experiments, 2BE induced a water flux of 0.0028m/h, 68% of that of 3.5wt% salt water when used as the draw solution under static conditions for 5 h. Under stirring at 7000 rpm, 2BE induced a water flux of 0.02 m/h, 1.2-fold that of 3.5wt% salt water in the initial 5 min. The effect of stirring is explained by breaking the concentration polarization or 2BE aggregates at the membrane surface. Aqueous 2BE solution, which is generated after forward osmotic application such as power generation and desalination, is known to have a lower critical solution temperature. Aqueous 2BE solution of the concentration from 12wt% to 58wt% separates into two phases (liquid-liquid) at elevated temperatures higher than 60 °C, allowing the 2BE-rich and water-rich phase to be recovered for the next cycle of the application. We conclude that aqueous solutions of 2BE could be a good drow solution under suitable conditions, such as with high-speed cross flows.

Keywords: forward osmosis, power generation, desalination, drow solution, phase separation,

1 INTRODUCTION

Draw solution is intensively investigated as the key material of forward osmotic applications such as power generation and desalination[1]. Although mostly investigated materials were inorganic salts, recently organic liquid such as glycol ethers were published by Lee et al[2]. The aquatic draw solutions were separated into two phases by moderate worming and water and the draw solutions were recovered for the next cycle. The solutions of this kind are known to have a lower critical solution temperature (LCST) or an upper critical solution temperature (UCST)[3]. For the draw solution, we have independently focused on organic solvents that are miscible with water and undergo phase separation in aqueous solution in some temperature region. Liquid-liquid phase separation is greatly advantageous for handling the separation process after the FO application to reclaim the draw and feed solutions. In this paper, 2-butoxyethanol is presented as an example. 2BE is well used, and investigated as an industrial material, and also a kind of glycol ether, but not reported by Lee et al.

2 EXPERIMENT

The osmolality of the 2BE aqueous solution at various concentrations was measured by freezing point depression osmometry, using Micro Osmo Master OM-815 (Vogel Inc., Germany) and by vapor pressure depression osmometry..

The suitability of 2BE as a draw solution was evaluated by the water flux in FO experiments. The experiments were carried out with a lab-built "syringe test cell" as shown in Fig. 1. The syringe test cell was made of two 1 mL polypropylene syringes. The syringes were cut and connected by rubber blocks, 1mm thick, which had a hole of 5 mm diameter and the RO membrane. The syringe test could be used as screening test of the materials. A commercially available flat-sheet RO membrane (ES20, Nitto Denko Co.) was used in the experiments, because the usual FO membrane commercially available is soluble in 2BE which has another name as cello-solve.

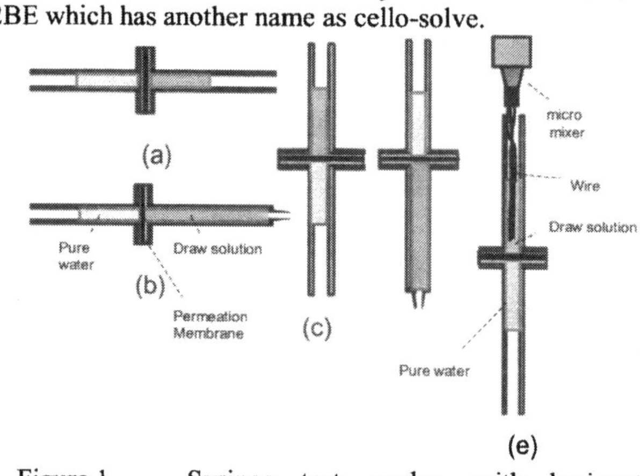

Figure.1 Syringe test modes, with horizontal arrangements in (a) and (b) and vertical arrangements in (c), (d), and (e). Arrangements (a) and (c) are used high surface tension draw solutions such as 3.5wt% salt solution. Arrangements (b) and (d) are used for low surface tension draw solutions such as 2BE. In arrangement (e), a polyvinyl chloride-coated wire was inserted as a rotator and stirred by a micro mixer.

3 RESULTS

The results of the osmolatily measurement are given in Fig. 2 in terms of osmolality against the 2BE concentration. The osmolality of 40wt% of 2BE aqueous solution was at least 2 times higher than that of 3.5wt% salt water. The osmolality measurement by freezing point depression generally has a concentration limit where the aqueous solution changes into antifreeze mixture. In this case, the limit was 40wt%. In higer concentration, larger osmolality is expected.

In forward osmosis experiments by the syringe test, 2BE induced a water flux of 0.0028m/h, 68% of that of 3.5wt% salt water when used as the draw solution under static conditions for 5 h. A series of syringe tests showed a significant decline of the water flux than expected value. However, under stirring at 7000 rpm, 2BE induced a water flux of 0.02 m/h, 1.2-fold that of 3.5wt% salt water in the initial 5 min. Considering the hitherto reports, these results indicated the concentration polarization at membrane surface or the aggregation of 2BE in aqueous solution.The experimental results are summarized in Table 1.

The fluxes of 5 h and 5 min were approximated from the slopes of linear fit with the y-intercept kept at zero. The fluxes of 5 h are defined as the average speed to compare the ability of the draw solution. The fluxes of 30 min were calculated using a variable y-intercept for comparison with the fluxes for 5 min because the observed fluxes decreased with time as shown in Fig.3. The water flux is expressed with the decreasing water in feed side of the syringe. In Table 1, the fluxes declined after the first 5 min, suggesting the accumulation of a water layer at the membrane surface or concentration polarization.

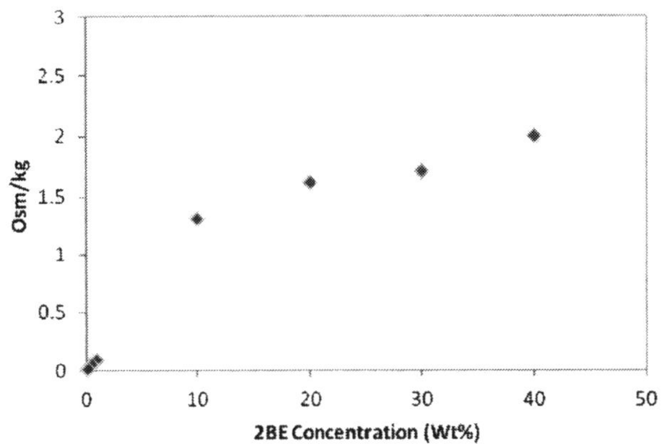

Figure.2. Osmolality of 2BE aqueous solution at various concentrations measured by freezing point depression.

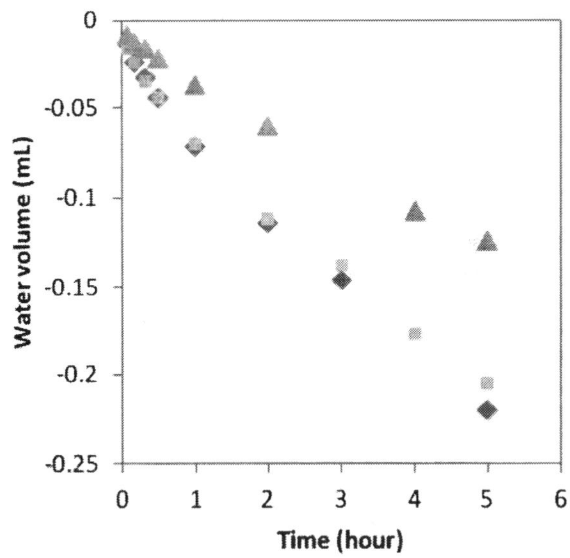

Figure. 3. Water flux in vertically arranged (draw solution in bottom) syringe tests, (◆) 3.5wt% salt water, (■)100wt%2BE, (▲)50wt%2BE.

Table 1 Water flux in the syringe tests in various conditions.

No.	Setting	Left	Right	Top	Bottom	Mode	Flux (10^{-3} m/h) 5 h	30 min	5 min
1			3.5% salt water				4.1±0.3	4.3±0.5	16±2
2	horizontal	water	100% 2BE				2.8±0.1	4.1±0.3	11±1
3			50% aq. 2BE				1.6±0.1	2.5±0.2	11±1
4				3.5% salt water			3.9±0.2	5.4±0.2	15±1
5				100% 2BE	water	static	1.2±0.1	1.7±0.2	1.2±1
6				50% aq. 2BE			1.0±0.1	1.8±0.4	3.6±1
7					3.5% salt water		2.4±0.2	3.6±0.2	8.2±1
8	vertical			water	100% 2BE		2.3±0.2	3.4±0.3	9.8±1
9					50% aq. 2BE		1.3±0.1	1.4±0.1	5.1±1
10				3.5% salt water				4.7±0.4	17±3
11				100% 2BE	water	stirring		9.1±0.5	20±1
12				50% aq. 2BE				2.7±0.3	9.4±1

In the experiments in the horizontal arrangement (No.1, 2, 3 in Table 1), the water flux of 3.5wt% salt water was higher than 2BE in all cases. The flux of 5 h was 68% of the flux of the salt water for 100wt% 2BE, and 39% of the flux of salt water for 50wt% 2BE. The difference in rate between the salt water and 2BE systems was larger in the vertical arrangement when the draw solution was in the top (No.4, 5, 6 in Table 1). The average flux of 5 h was 31% of salt water for 100wt% 2BE, and 26% of salt water for 50wt% 2BE. However, when the draw solution was in the bottom (No.7, 8, 9 in Table 1), the difference decreased to almost the same level. The flux of 5 h was 96% of salt water for 100wt% 2BE, and 54% of salt water for 50wt% 2BE.

These results can be explained by external concentration polarization at the membrane surface, as shown in Fig. 4. This is an illustration of the relation between concentration polarization and the specific gravity of the solvent. That is, in the case where 2BE is contained in the top in the syringe test, as shown on the left side of Fig.4, the transmitted water was able to remain at the surface of the membrane because the specific gravity of 2BE is 0.9 that is smaller than that of water.. This water layer acted as an external concentration polarization layer. The water flux thus decreased or could not increase anymore. In contrast, in the case with 3.5wt% salt water in the top, as shown in the right side of Fig.4, a downward flow of relatively dense salt water into the transmitted water layer at the membrane surface was able to form, accelerating the water flux rate because the specific gravity of 3.5wt% salt water is 1.02. In the case of the horizontal arrangement, the effect of the specific gravity is expected to be relatively small, and we only observed the effect of the external concentration polarization without other effects such as convection of the draw solution. In all of the above cases, the flux of 50wt% 2BE was lower than that of 100wt% 2BE.

The water flux was relatively small under static conditions and relatively large under stirred conditions (No.10, 11, 12 in Table1). Under stirring at 7000 rpm, 2BE induced a water flux of 0.02 m/h, 1.2-fold larger than that of 3.5% salt water in the initial 5 min. All the fluxes decreased to 28-46% for 30 min even in the stirring condition. The results of stirring and static experiments indicate the existence of external concentration polarization in the 2BE system, supporting the above discussion.

The results for 2BE in several modes support the effect of concentration polarization. 2BE is therefore expected to be a good draw solution in cross-flow systems which were originally developed to reduce concentration polarization.

Aqueous 2BE solution, which is formed after forward osmotic application, is known to have a lower critical solution temperature as shown in Fig.5. Aqueous 2BE solution of the concentration from 12wt% to 58wt% separates into two phases (liquid-liquid) at elevated temperatures higher than 60 °C, allowing the 2BE-rich and water-rich phase to be recovered for the next cycle of the forward osmotic application.

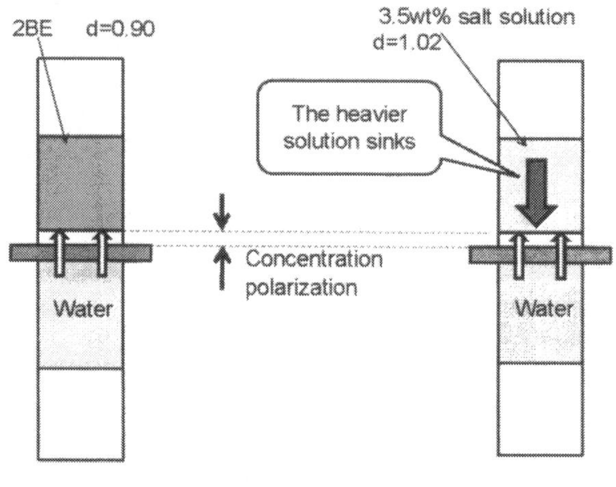

Figure.4 Illustration of the relation between concentration polarization and the specific gravity of the solvent.

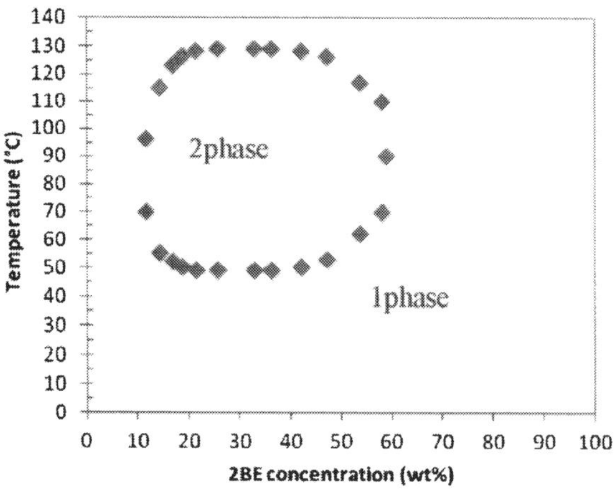

Figure.5 The phase diagram of aqueous solution of 2BE. The phase diagram was made by the data of the literature [4] consulting the literature [5] and we have confirmed the lower border line of Fig.5 by direct observation of the vial of various concentration of 2BE aqueous solution in water bath.

2BE aqueous solution was intensively studied with a number of different techniques and the detail of the solution properties was reported. Kato revealed that there exist few long-lived clusters whose component molecules diffuse together in the 2BE aqueous solution[6]. Verrall et al. reported clathrate-like structure in water rich region below 9.5wt%[7]. Siu et al. revealed repulsive 2BE-2BE interaction in the concentration region less than 10wt% , and existence of aggregation of 2BE in the region from 10 wt% to 87wt%, and existence of 2BE and H_2O cluster in the concentration region more than 87wt% in the 2BE aqueous solution at 25-35°C[8]. The second virial coefficient of 2BE aqueous solution was calculated to be -1.86 at 5°C by Scatchard et al[9]. This result is

Table 2 The 2BE concentration in each phase after phase separation.

Recovered solution	method	separation temperature (°C)	2BE concentration after phase separation (wt%) PD[a]	NMR[b]
wate-rich phase	syringe test	70	12	11.4
	φ75mm cell	60	12	11.1
2BE-rich phase	syringe test	70	58	62.8
	φ75mm cell	60	54	62.0

[a] Concentration based on the phase diagram, [b] Concentration measured by ^{1}H-NMR

corresponding to the lower osmolality of 2BE than expected in higher concentration region as shown in Fig.2. These reports indicate the existence of associated 2BE causes the decrease of osmolality in the concentration region more than 10wt%, agreeing the result of the osmolality measurement in this work, and in the concentration region more than 30wt%, the number of the aggregated 2BE is thought to have increased and the osmolality increased with it.

The separation of the solution after FO mode experiment was performed with a quartz glass cell with a membrane of diameter of 75 mm for 12 hours. 60 ml of 2BE rich solution and 8ml of water rich solution were obtained by Pasteur pipette. The collected solution was separated in a vial of 100 ml with water bath at 60°C. The separated solutions, 2BE rich and water rich, were analyzed with ^{1}H-NMR with 2,2,2-trifluoroethanol as the internal standard. The results were shown in Table 2 with the data of the syringe test samples and data expected from the phase diagram of 2BE aqueous solution. The concentration data measured by NMR was similar to the theoretical concentration based on the phase diagram. It is safe to say the recovery of the draw solution of 2BE aqueous solution could be performed quite easily.

4 CONCLUSION

2BE/water could be a good draw solution under suitable conditions, such as cross flows.

ACKNOWLEDGEMENT

We thank Dr. Masayuki Yasukawa and Professor Hideto Matsuyama at Kobe University for advice and coorporation. We thank Mr.Tomoyuki Inoue at Biomedical Science Inc. for the measurement of freezing point depression.

REFERENCES

[1] Q. Ge, M. Ling,T-S. Chung, "Draw solutions for forward osmosis process: Developments, challenges, and prospects for the future" J. Membr, Sci.,424, 225-237, 2013.

[2] D. Nakayama, Y. Mok, M. Noh, J. Park, S. Kang, and Y. Lee, Lower critical solution temperature (LCST) separation of glycol ethers for forward osmotic control, Phys. Chem. Chem. Phys., 16, 5319-5325, 2014.

[3] P. Atkins, J. de Paula, *Atkins' Physical Chemistry*,7th ed., Oxford, pp205-207, 2001. M. A. White, *Physical Properties of Materials*, 2nd ed., CRC Press, pp225-227, 2011.

[4] Y. Koga, T. Tanaka, T. Atake, P. Westh and A. Hvidt, Mixing schemes and liquid-solid phase diagram in the water-rich region of aqueous 2-butoxyethanol, Bull. Chem. Soc. Jpn., 67, 2393-2397, 1994.

[5] T. M. Bender, R. Pecora, Dynamic light scattering measurements of mutual diffusion coefficients of water-rich 2-butoxyethanol/water systems, J. Phys. Chem., 92, 1675-1677, 1988.

[6] T. Kato, Determination of velocity correlation coefficients in aqueous solutions of 2-butoxyethanol by quasi-elastic light scattering, pulsed-gradient FT-NMR, and Rayleigh-Brillouin scattering, J. Phys. Chem.89, 5750-5755, 1985.

[7] S. Kato, D. Jpbe, N.P. Rao, C. H. Ho, and R. E. Verrall, Ultrasonic relaxation studies of 2-butoxyethanol-water and 2-butoxyethanol-water-cetyltrimethylammmonium bromide solutions as a function of composition, J. Phys. Chem., 90, 4167-4174, 1986.

[8] W. Siu, and Y. Koga, Excess partial molar enthalpies of 2-butoxyethaanol and water in 2-butoxyethanol-water mixtures, Can. J. Chem., 67, 671-676,1989.

[9] G. Scatchard and G. M. Wilson, Vapor-liquid equilibrium. XIII. The system water-butyl glycol from 5 to 85°, J. Am. Chem. Soc., 86, 133-137,1964.

1 Komukai, Toshiba-cho, Saiwai-ku, Kawasaki 212-8582, Japan , kenji.sano@toshiba.co.jp, Ph :+81-44-549-2447, Fax: +81-44-520-1307

Conical, Floating Piston Isolation Valve

Bruce Farner
NASA-Stennis Space Center
Stennis Space Center, MS, USA, bruce.r.farner@nasa.gov

ABSTRACT

These pages provide the details of a recently patented valve design titled the "Conical Seat Valve". This valve concept consists of a solid piston floating in amedium to controll flowstream. The piston is designed to be axially and radially balanced within the flow stream whether the valve is in the open or closed positions. The only force imparted onto the piston is that which the operator chooses to input on to it. The advantages to this concept are the elimination of conventional actuators which are replaced by one or two simple solenoid valves; elimination of valve actuator adjustments; consistant seating force no matter the pressure drop across the valve; elimination of the valve stem and stem seals; and elimination of most flow induced thrust forces. Additionally, the valve consists of five parts with simple seals incorporated into well protected designs. This simple design allows configuration for extremely high pressures and the use of soft or harder metal seats.

Keywords: valve, balanced, actuatorless, high pressure

1 THE PROBLEM

Although the patent is titled "Conical Seat Valve" and the conical shape of the seat is a prominent feature of the valve, , especially when the concept was first developed, the key feature of the valve is the balanced floating piston that acts as both the valve plug and the actuator. The origin of this design initiated based upon the need for large extremely high pressure valves to replace the ball valves currently used as supply valves tank pressurants in rocket testing. The current valves are 12,000 psi, 8" ball valve which flow 890 lbm/sec of gas through them during testing

The ball valves currently in use have been problemmatic during operation with very short lives and expensive rebuilds and redesigns that are often required. As supplied, the valves only operate two or three cycles before failure and can not be opened with a pressure drop across them. A recent redesign has extended their life span to over twenty cycles, but they still cannot be opened with a pressure drop across them; equalization is required before opening.

Instead of looking to "improve" the current valve, the thought was to determine what was causing the failures and design that out of the valve. By assessing this, it was determined that the forces acting on the ball by the seat retainers was extremely high and the interanl pressure acting on the valve stem to eject it was also extremely high. Failures were occuring in the trunnion and in the seat itself due to the excessive force being applied by the seat retainer, which used both springs and the internal pressure of the valve to generate a very high load on the seat retainer to hold the seat onto the ball. Because of this force, a large diameter stem is required to transmit the force of a very large actuator to the ball to get it to turn. As a result, the large diameter stem is also trying to be ejected by the internal pressure of the valve, which requires a thrust takeout feature that allows turning. Rolling bearings were not desired due to damaging effects upon failure they'd have on expensive equipment, and lubrication use was limited because the work is being conducted in a very high pressure oxygen environment. These issues lead to a design that initially featured bronze sliding thrust bearings, which led to subsequent revisions, that ended up with Torlon thrust bearings. This process can be equated to trying to turn the wheels on your car while the disc brake is applied. The Torlon disc actually wears much like the pads on the disc brakes but with a much shorter life.

2 THE CONCEPT ORIGIN

In trying to eliminate or control the forces acting on the ball and stem, the the seat retainer that pressed the seat to the ball using springs plus the additional force of the pressure drop across the valve was first approached. This design, a version found in most ball valves, causes the seating force to vary greatly depending on the pressure across the valve. This is also common in other valve designs, such as globe valves. Looking at the design, it was determined that the force generated by the valves internal pressure could be removed by adding an additional set of seals, and the pressure could be used to actually work against the springs if desired to lift the seat completely off the ball, allowing for free rotation. Upon noticing that when both seats were lifted, flow would move around the ball before the ball was moved, it seemed that by increasing the retraction of the seat, or now the valve stroke, the ball could be eliminated. The ball was replaced with a fixed seat, moving the wear item over to the non-moving part and designing channels to route the flow around the seat and out the other side of the valve.

Figure 1: Original Head with flow passages

An additional chamber was added to eliminate the spring and power the valve in both the open and closed directions. The motive force for the piston could be either an external souce or the working fluid inside the valve itself. Finally, a final chamber was added to balance the force of the downstream pressure. This last innovation allowed all forces to be balanced on the floating piston no matter what position the valve is in or what the pressure drop is across the valve. This means the valve always opens or closes with the exact same force as per the design.

Figure 2: Demonstrator Valve Disassembled

The simplest way to look at the valve is as a globe valve in which all of the flow passes through the center of the plug instead of just being diverted off to the side of the plug. In this case, the plug and the seat are reversed in that the seat is a solid disc, whereas on typical globe valves the seat is a ring and all of the flow passes through the seat. The piston is designed so that it is not influenced by the pressure drop or flow inside the valve.

3 ADVANTAGES AND FEATURES

This concept has many unique features and design advantages over conventional valve concepts. First, the flow path is always axially and radially symetric, eliminating almost all of the flow induced thrust loads, even during transition from closed to open. This minimizes the need for mounting restaints and a thrust take-out structure.

Next, the flow areas where the flow is required to change direction over the seat and through the connecting passages, are designed to at least 150% of the cross sectional flow area of the inlet and outlet pipes. The inlet, outlet, and piston passages should match the pipe, exactly as ball valves are designed to do. The 150% area is estimated to reduce the velocity in the turbulent regions and minimize flow losses. Modeling indicates that this design will have an approximate 15% improvement in flow over a similarly sized standard globe valve.

The next, and largest advantage, is the elimination of the valve stem and any conventional actuator. No matter how large or how high a pressure, the valve is designed for one or two small solenoid valves that will operate the valve completely. For very high pressure valves, stem seals and bending of the valve stem are two major issues that cause valve failure. Because of the design concept of this valve, there are essentially no bending forces that have to be addressed. Since there is no valve stem, there are no stem seals to leak or to protect from gettting dirty and scaring or binding the stem.

With the elimination of the valve stem, the need for a conventional valve bonnet is eliminated as well. Instead, the body of this valve is formed by two cylindrical parts which are squeezed together with a bushing between them. The elimination of the bonnet reduces the envelop and mass of the valve by an estimated 30%.

Additionally, the features of the parts are all simple circular features, so everything can be either turned on a lathe or drilled on a simple mill. This reduces fabrication time and cost.

Because of the large diameter seat, the amount of piston movement needed to achieve full flow is greatly reduced. Typical globe valves may have a plug assembly that has to travel over three inches in order to get sufficient flow area. With this innovatve concept, travel of only about ¾" will meet the design requirement of 150% cross sectional flow area through the seat opening. This short stroke allows for reduced valve actuation times while keeping the piston velocity much lower. An issue seen with large, rapid actuation valves is that the valve stem actually stretches when the actuator tries to slow the valve, right before it slams into the seat. With the reduced velocity and increased seat contact area, slowing may not be required. If slowing is required, it will be achieved through baffling the venting process of the opening chamber and therefore, no hard pulling or stretching is required. Additionally, contact will always be with the same amount of force no matter what the pressure drop is across the valve.

Another advantage to eliminating the connection to a valve stem is the elimination of the need for valve adjustments. Conventional valves require the stroke to be adjusted as the seat wears or is coined in. Because the piston floats in this valve, it will move forward until it is stopped on the seat. Seat wear or temperature changes will not affect either the force of the closure or the contact area with the seat.

Because the piston is balanced within the flow stream and so the force required to close the valve is always the same, the amount of gas required to operate the valve is reduced by over 90% of that required for a conventional pnuematically actuated valve. Large pnuematic actuators can cost over $50,000 and be several feet in diameter and length, adding to the envelop of the valve. Although the large reduction of gas required to operate the valve may be a small cost portion of the total cost, it may allow the valves to be operated with a gas source such as a K-Bottle or a compressor, and thus eliminates the need to run high pressure gas tubing to every valve in a system.

Overall, with the fewer parts, smaller size, robust component shapes and designs, shorter stroke, and consistant seating force, the valve is expected to demonstrate increased reliability and life over comparable conventional valve designs. The cost of fabrication should also be significantly less and should require less raw material. Because of the balanced design, the valve is also inherently bi-directional. The closing force is exactly the same, even if the higher pressure fluid is on the "downstream" side of the valve.

4 OPERATION

The functional operation of the valve is very simple. The entire body and pressure boundry is formed by three parts, the main body, the bushing, and the head. In the first concept, these are all held together by bolts or studs squeezing the three parts togther with the bushing in the middle and everything sealed with o-rings.

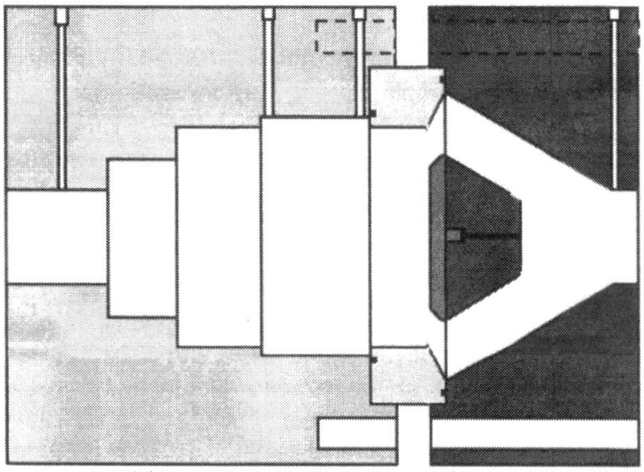

Figure 3: Pressure Boundries

The floating piston is positioned inside the body assembly and seals to the body assembly in four places with circular seals such as o-rings, or pressure energized seals, such as omni-seals. The piston and body shape combined with the four seals form three annular chambers on the outside of the piston.

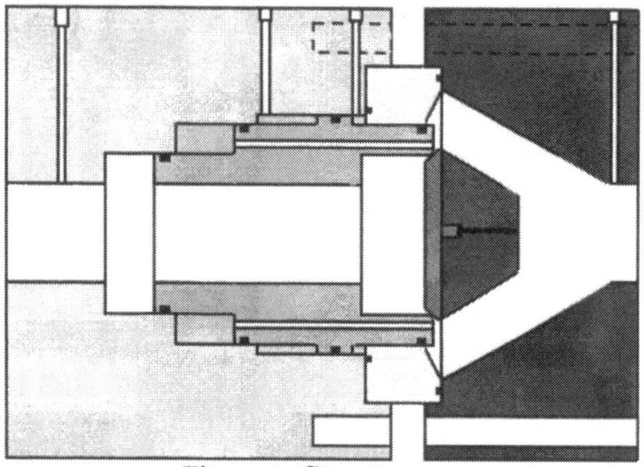

Figure 4: Chambers

The first chamber is connected to the downstream face of the piston through a series of holes drilled through the piston. This connection between the chamber and the downsteam face can be done externally through the body if desired, but in in Figure 4, it is through the piston. The annular area of the downstream face of the piston is exactly equal to the connected face area of the piston in the cavity. Because the two faces have the same area and see the same pressure, no force is generated to try to move the piston by either of these areas no matter what the pressure is inside the valve.

In similar engineering, the upstream face of the piston is equal in area to the expanding area of the flowpath inside the piston. So again, the pressure acting on both of these faces will be equal and thus the forces generated by these pressures will cancel each other out and not generate a motive force trying to move the piston.

The remaining two chambers are equal in size and connected to the outside of the valve body through a drill passage for each. These chambers are the operating chambers and provide the force to open and close the valve. The working fluid to operate the valve is up to the valve designer and user. The force required can be generated by any pressure gas by varying the outer diameter of the chamber. The working fluid can come from an external source or use pressure from the inside of the valve. The operation of the valve is simple; one chamber is pressurized and the other chamber is vented. The piston moves in the direction of the vented chamber until it is physically stopped by either the seat on closing or the main body surface on opening. The speed of the piston is easily controlled by controlling the rate at which the venting chamber bleeds off.

TechConnect Briefs 2015, TechConnect.org, ISBN 978-1-4987-4728-8

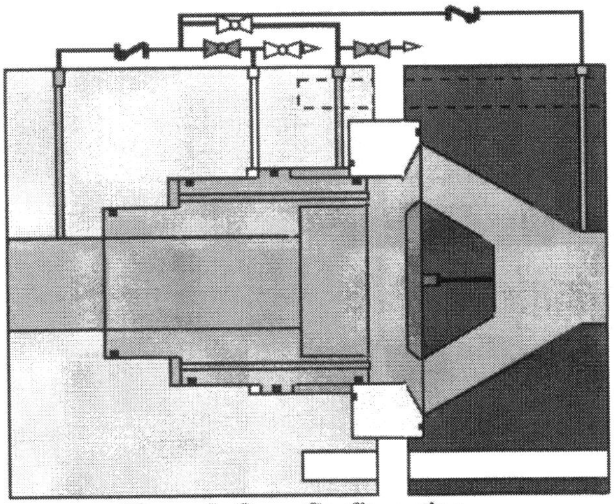

Figure 5: Open Configuration

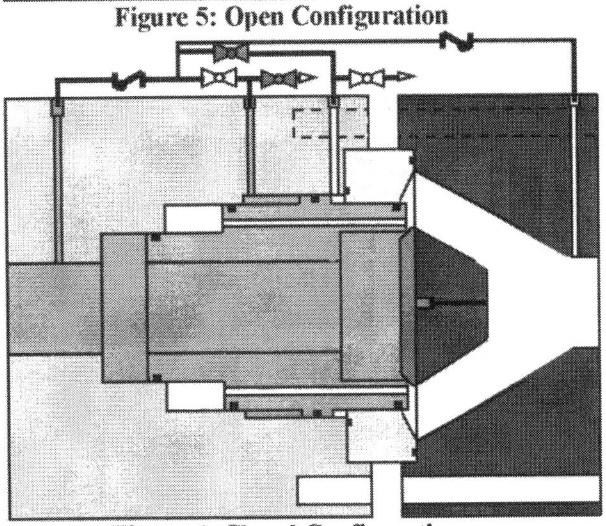

Figure 6: Closed Configuration

5 DISADVANTAGES OR CHALLENGES

The most significant challenge assocaited with this valve concept is that there is no obvious way to see the position of the piston in the valve..

Another challenge of this design is that the valve has to be removed from the line in order to be serviced. Many ball, globe, and gate valves permit the bonnet to be removed and the valve serviced while the body remains installed in the field. Designers generally design the wear parts in a way that they are accessible and replaceable in the field. That may not be possible with this valve concept, however, its fewer parts and predicted increased reliability may make that a minimal factor.

6 SUMMARY

The balanced, floating piston valve design has a wide range of potential applications of all sizes and pressure ranges. The extremely simple design and few parts makes the concept inherently reliable, simple to manufacture, and easy to work on. The valve concept works with soft or hard metal seats, and the closing force is easily adjustable so that any closing force desired can be created. The valve will consistantly close with exactly the same force as long as the driving medium is at the same pressure as designed. The fact that no adjustment is required in the design will ensure valve performance thoughout the life and operation.

The one moving part with incorporated simple seals in a well protected configuration, along with the short travel of the piston, should add to the life of the valve and reduce associated maintenance.

In addition, the reduction in commodities needed to operate the valve will reduce overall life cycle cost and enable designs that operate longer on smaller accumulators in the case of a loss of the operating medium or in power failures.

Carbon Nanotube Immobilized Membranes for Next Generation Desalination

S. Roy, S. Ragunath, K. Mitra and S. Mitra*

* New Jersey Institute of Technology
Department of Chemistry and Environmental Science, Newark, NJ, USA, mitra@njit.edu

ABSTRACT

The objective of this project is to utilize carbon nanotubes (CNTs) to create breakthrough membrane properties for the generation of pure water via membrane distillation (MD). This fundamentally new approach will reduce the cost and energy requirements for the utilization of waste and brackish water and will dramatically change the competitive landscape of conventional methods such as Reverse Osmosis (RO) and thermal evaporation. Also, MD can be operated at relatively lower temperatures (60-90°C) using low-grade heat. This makes feasible the possibility of utilizing waste heat and alternative energy sources such as solar energy. The use of electrical energy is minimal.

The novel membrane referred to as carbon nanotube immobilized membranes (CNIM) are developed by immobilizing CNTs into membrane pores where, they serve as molecular transporters and sorbents, thus providing additional pathways for solute transport. The CNTs also increase the functional surface area due to their high aspect ratio. All of these mechanisms play important roles in the transport of water vapors and lead to enhanced performance desalination

Keywords: Carbon nanotube, Membrane, Desalination, Membrane distillation

1 INTRODUCTION

Membrane Distillation (MD) is a thermal evaporative process that offers several advantages over traditional methods. Since operations in MD are at relatively low temperatures (60-90°C), it is particularly attractive for the production of high purity water. Because of lower energy requirement, MD can potentially be operated effectively using low grade heat sources that have previously been generated for example in power plants, chemical, food and other industries. It can also be run on solar energy. In essence, MD can operate as a near "zero-cost" energy technology and it will not contribute to global warming because with the exception of some electricity no additional fossil fuels are consumed in generating heat.

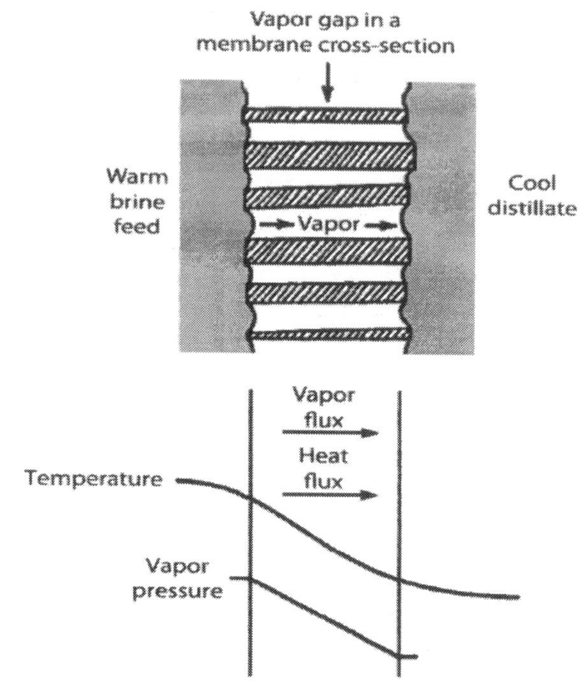

Figure 1: MD - Driven by Temperature and Water Vapor Pressure Gradients

2 CARBON NANOTUBE IMMOBILIZED MEMBRANE (CNIM)

We have utilized carbon nanotubes (CNTs) to create breakthrough membrane properties for the generation of pure water via MD. The novel membrane referred to as carbon nanotube immobilized membranes (CNIM) are developed by immobilizing CNTs into membrane pores where, they serve as molecular transporters and sorbents, thus providing additional pathways for solute transport. Immobilizing the CNTs in the pores alters the water-membrane interactions, which is one of the major physicochemical factors affecting the permeability and selectivity of the membrane [1]. Since CNTs are highly hydrophobic, they decrease the tendency of a pore to become wet with liquid, so higher transport of pure vapor can occur. The CNTs are known to have rapid sorption and desorption capacities leading to rapid mass transfer, and it is also possible that they allow the water vapor molecules to follow a surface diffusion pattern in which the solute hops

TechConnect Briefs 2015, TechConnect.org, ISBN 978-1-4987-4728-8

from one site to another by interacting with the surfaces. This increases overall vapor transport [2]. The CNTs may also provide an alternate route for fast mass transport via diffusion along their smooth surface, as well as be transported directly through the inner tubes of the CNTs, which are known to enhance vapor transport. The CNTs also increase the functional surface area due to their high aspect ratio.

It is worth mentioning that the amount of CNTs is so small enough that they do not increase the thermal conductivity of the membrane or cost.

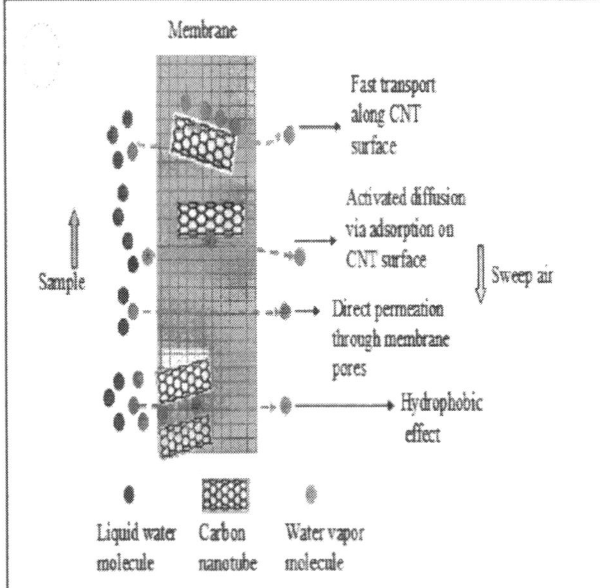

Figure 2: Mechanisms of MD in the presence of CNTs

All of these mechanisms play important roles in the transport of water vapors and lead to enhanced performance in MD. Results to date show 165% enhancement in flux [3]. The CNIM is fabricated based on a scalable approach developed in our laboratory. So far our work with sea water to produced water from hydraulic fracturing show excellent results with salt concentrations as high as seven times that of sea water.

3 ADVANTAGES OF CNIM

This fundamentally new approach will reduce the cost and energy requirements for the utilization of waste and brackish water and also treating produced water from hydraulic fracturing. This will dramatically change the competitive landscape of conventional methods such as Reverse Osmosis (RO) and thermal evaporation. MD offers several advantages over RO including the ability to handle higher salt concentrations, less stringent pretreatment requirements, less energy consumption, significantly less fouling and longer lifetime for the membranes. The method can also be effectively used to treat RO reject. Some advantages of CNIM-MD over Ro is listed below:

CNIM-MD	RO
Recovery of CNIM-MD ~90%	Recovery of the seawater RO 35 - 45%
Can handle high salt concentration including RO reject; requires significantly less feed pretreatment	Osmotic and operating pressure increases with salt concentration; high pretreatment cost
Operate at one atmospheric pressure	Typical operating pressures between 100-225 psig
Primary energy use in CNIM-DCMD is to heat the feed water (60-90°C); waste heat utilization reduces/eliminates that cost	The primary energy requirement is for operating high pressure pumps
Less fouling of membrane with high porosity	Dense nonporous membrane prone to fouling
Higher temperature significantly reduces bio-fouling	Significant bio-fouling
Longer membrane life time >5 years	Short life time, typical 5 years

Table 1: Advantages of CNIM-MD over RO:.

REFERENCES

[1] S. Roy, M. Bhadra and S. Mitra, Sep. & Purif. Technol. 136, 58, 2014.

[3] M. Bhadra, S. Roy and S. Mitra, Sep. & Purif. Technol. 120, 373, 2013.

[3] K. Gethard, O. Sae-Khow and S. Mitra, Sep. & Purif. Technol. 90, 239, 2012.

Measurement of O-phthalaldehyde (OPA) by High Pressure Liquid Chromatography (HPLC) with Post Column Derivatization

Al E. Watkins*, Danelle Dees* and Tabatha Butler*

*A2Research, John C. Stennis Space Center, Stennis, MS, USA, al.e.watkins@nasa.gov, danelle.c.dees@nasa.gov, tabatha.butler-1@nasa.gov

ABSTRACT

O-phthalaldehyde (OPA) is commonly used as a high-level disinfectant in the medical industry. On the International Space Station (ISS), waste heat loads are removed by an Internal Active Thermal Control System (IATCS) which is a water based system in conjunction with the an EATCS (External ATCS), ammonia based system. OPA is used in the IATCS water based system as a biocide to prevent negative impacts to coolant flow, heat transfer, and corrosion. Therefore the concentration of the OPA in the coolant fluid is critical to maintain optimum function of the IATCS [1]. The previously used method for validating OPA concentration had been proprietary and is no longer available to NASA. To address the critical need for ISS, a simple and inexpensive process was developed to determine OPA concentration in water by reversing an established method used to determine glyphosate concentrations in water by HPLC using post column derivatization.

Keywords: O-phthalaldehyde, HPLC, NASA, derivatization, ISS

1 BACKGROUND

On the International Space Station (ISS), systems produce waste heat which needs to be transferred from the ISS to space to achieve thermal control and maintain components at acceptable temperatures. The Internal Active Thermal Control System (IATCS) is a critical system used to facilitate heat transfer aboard the (ISS). The IATCS aboard the ISS is primarily responsible for the removal of heat loads from payload and system racks. This also includes, heat loads from crew members, experiments, electrical equipment and other systems generate heat inside the modules of ISS that also has to be rejected. To accomplish this, the IATCS consists of loops that circulate water through the interior to collect the excess heat. The IATCS is a water based system which works in conjunction with the EATCS (External ATCS), an ammonia based system, to facilitate heat transfer aboard ISS. The purpose of the Thermal Control System of ISS is to keep all internal and external equipment and payloads at temperatures within their specific thermal requirements. The IATCS circulates water through two independent loops inside the modules, a Low Temperature Loop and a Moderate Temperature Loop in order to segregate the heat loads, to simplify heat load management and to add redundancy in case of equipment failures. Water is used because it is an efficient heat transfer fluid and presents no danger to crew members in the event of an internal leak.

Since 2001, growth and proliferation of microorganisms in the IATCS on ISS has been of significant concern. NASA and Boeing condcuted testing and assessments of a variety of biocides, to determine bacterial disinfection capability, material compatibility, stability (rate of oxidative degradation and identification of degradation products), solubility, application methodology, impact on coolant toxicity hazard level, and impact on environmental control and life support systems to identify a prioritized list of acceptable biocidal agents. O-phthalaldehyde, OPA, an aromatic dialdehyde compound with the formula $C_6H_4(CHO)_2$ was selected as the optimum antimicrobial. OPA is commonly used as a high-level disinfectant (for example, for sterilization of heat sensitive medical instruments) that demonstrates effective microbicidal activity against a range of microorganisms including mycobacteria, gram-negative bacteria and spores.

Upon selection for use, OPA was thoroughly evaluated and proven effective for use in the water based portion of the coolant system as the preffered microbial growth inhibitor for use in the ISS IATCS, replacing silver as the choice antimicrobial, and was approved for use at concentrations $<$109 ppm.

The concentration of the OPA in the coolant fluid is critical to the process and previously used methods for developing, analyzing and validating the OPA concentration. The previously used method for developing, analyzing and validating the OPA concentration had been proprietary process which no longer was available to NASA. A subsequent literature search could only provide a technical method that was both labor intensive and involved using hazardous materials (hydrazine dissolved in concentrated sulfuric acid).

Therefore, NASA Environmental Laboratory located at the John C. Stennis Space Center (SSC), worked closely with NASA Kennedy Space Center (KSC) to develop a simple, inexpensive, and reliable methodology for analyzing low- to mid- level concentrations of OPA for use in IATCS coolant water.

2 INTRODUCTION

Accumulation of microorganisms on surfaces inside IATCS could result in material degradation. Should the IATCS become damaged, crew health and safety are at risk. OPA is a disinfectant that effectively inhibits the growth and recovery of viable microorganism. An OPA water solution is required and used as a high level disinfectant to effectively inhibit the growth and recovery of viable microorganisms in the IATCS coolant to prevent microbial negative impacts to coolant flow, heat transfer, and corrosion. The concentration of the OPA in the coolant fluid is critical for maintaining proper functionality of IATCS. To address this acute need for ISS, an analytical process was developed which could verify a known concentration of the disinfectant OPA.

A National Institute of Standards and Technology (NIST) traceable EPA method 547 for quantitating glyphosate concentration in water, which uses High Pressure Liquid Chromatography (HPLC) method that uses OPA to convert glycine (not glyphosate) in a post-column reactor into a compound that can be more readily detected, was reversed engineered. By reversing the premise of this process, glycine is reacted in excess with the OPA solution along with 2-mercapto-ethanol. This developed and validated method to analyze OPA offers a new effective, low cost HPLC method that substantially reduces the future process waste stream and avoids potential health hazard effects from previously identified processes.

This simple inexpensive, NIST traceable process to analyze OPA that does not require highly hazardous chemicals nor does it have a hazardous waste stream that is costly to discard.

3 TECHNOLOGY

NASA SSC operates a full scale Environmental laboratory, and utilizes Environmental Protection Agency (EPA) methodologies for most analytical processes. EPA Method 547, entitled *Determination Of Glyphosate In Drinking Water By Direct-Aqueous injection Hplc, Post-Column Derivatization, And Fluorescence Detection,* is a widely accepted method. Method 547 effectively identifies and determines Glyphosate (N-phosphonomethyl glycine) concentrations in drinking water matrices. Once Glyphosate is eluted from the column, it is oxidized using hypochorite solution to form glycine. OPA, used in conjunction with either N,N-dimethyl-2-mercaptoethylamine hydrochloride or thioflur, derivitize the glycine to form a highly fluorescnet isoindole compound [1] [2].

NASA SSC's Environmental Laboratory calibrated, using known concentrations of OPA solutions, derivitized and performed post column reaction of the OPA solutions with glycine and 2-mercapto-ethanol. The resultant reaction complex was used to determine the OPA concentration with accuracy and precision. Calibrations were established, proficiency studies (IDOPs) and a method detection limit (MDL) study were analyzed to demonstrate the precision and accuracy. This developed and validated method to analyze OPA offers a new effective, low cost HPLC method that substantially reduces process waste. Additionally, the final product of the reaction is environmentally friendly. Glycine is an active chemical neutralizer used in the medical industry for disposal of expired, unused high-level disinfectants, such as OPA. Because Glycine, a primary amino acid which is a building block for protein, OPA readily reacts with it and "attacks" the protein, and quickly becomes suspended/neutralized [3] [4].

4 DEMONSTRATION OF PROFICIENCY

4.1 Calibration and Linearity

The calibration range for this method was established specifically for the optimum concentration of OPA in IATCS solution. A five point calibration range was used with the low standard concentration set at 100 mg/L, and a high standard of 500 mg/L. The coorelation coefficient for the calibration was determined to be 0.99979. Second source calibration verification standards were used to verify the calibration. A 300 mg/L second source standard was analyzed and reported at 293.3 mg/L, a 97.8% recovery.

4.2 Precision and Accuracy Qualification

Upon establishing and verifiying calibration procedures for OPA concentration determination, precision and accuracy studies were performed over three non-consecutive days, using a 250 mg/L standard, four replicates were analyzed per sample run for a total of 12 data points. The average recovery for this study was reported to be was analyzed four times each . Over the course of three separate runs, the average result reported for the 250 mg/l standard was 255 mg/L, or 102% recovery. The relative standard deviation was 8.02 for the three data sets.

4.3 Establishment of Method Detection Limit

Method detection limit (MDL) studies were performed over the course of three months utilizing the lowest standard on the calibration curve. A 100 mg/L standard was analyzed in three separate analytical runs to establish an MDL of 7.4 mg/L. MDLs were established following guidelines set for in the *US Code of Federal Regulations* Title 40 Chapter 1 [5].

5 CONCLUSIONS

OPA has demonstrated effective inhibitory growth and recovery of viable microorganism in the IATCS coolant when the concnetration of OPA is greater than the minimum inhibitory concentration of 100 ppm. Levels of up to 300 mg-OPA/L are also used in various loops of the IATCS aboard the ISS to keep microbiological populations in check.. Standard validation practices for OPA concentration have to be accessible to ensure effectiveness of implmentation.

The implementaion of OPA in the IATCS has culminated in a safe, compatible, effective, and sustinable method to control microbioal growth in the ISS IATCS to prevent microbial impacts to the coolant flow, heat transfer, corrosion and crew health and safety. This method enables a cost-effect apporach for OPA qualification, which could potentially used future space-based and medical application.

REFERENCES

[1] American Institute of Aeronautics and Astronautics, International Space Station (ISS) Internal Active Thermal Control System (IATCS) New Biocid Selection, Qualification, and Implemetation. Mark E. Wilson, Boeing, Harold E. Cole, Boeing, Tony Rector, John Steele, and Jerry Varsik, Hamilton Sundstrand Space Systems International, Inc., Windsor Locks, CT

[2] METHOD 547-"Determination of Glyphosate in Drinking Water by Direct-aqueousinjection HPLC, Post-Column Derivatization, and Fluorescence Detection" (applicable to glyphosate), which is incoporated by reference in accorance with 5 U.S.C. 552(2) and 1 CFR part 51, July 1990 T.W. Winfield W.J. Bashe (Technology Applications Inc.) T.V. Baker (Technology Applications Inc.), Environmental Monitoring Systems Laboratory Office of Research and Development, US Environmental Protection Agency, Cincinnati, OH

[3] Rutala, W.A. and D.J. Weber, "New Disinfection and Steriization Methods", Emerging Infectious Diseases, Vol 7, No. 2, 348-353, 2001.

[4] Rutala, W. A., and D. J. Weber, D. J. "Sterilization, high-level disinfection, and environmental cleaning". *Infectious disease clinics of North America,* 25(1), 45-76, 2011.

[5] *U.S. Code of Federal Regulations*, Title 40: Protection of the Environment, Volume 22, Chapter 1, Subpart D-Reporting and Record Keeping, Part 126, Appendix B (7-1-09 Edition) .

Designing Ultrathin Film Composite Membranes: Importance of a Gutter Layer

Moon Kattula[1], Koushik Ponnuru[1], Ioannis Karampelas, Lingxiang Zhu[1], Weiguang Jia[1], Ting Wang[1] Haiqing Lin[1] and Edward P. Furlani[1, 2]

Dept. of Chemical & Biological Engineering[1], Dept. of Electrical Engineering[2], University at Buffalo (SUNY), Buffalo, NY 14260-4200, efurlani@buffalo.edu

ABSTRACT

The performance of composite membranes that utilize an ultra-thin selective layer (<100 nm) can be improved with the introduction of an intermediate gutter layer (<100 nm) positioned between the selective layer and porous support. The properties of the gutter layer can be carefully chosen to enhance overall membrane performance, i.e., high permeance and high selectivity. However, the experimental determination of optimum gutter layer properties is very challenging, if not impossible, and modeling is needed to guide the selection process. In this presentation we address this need using a computational model to determine the effects of the gutter layer thickness and permeability on membrane performance. We show that a layer thickness of 1-2 times of the pore radius of the porous support yields the maximum improvement in permeance without significantly decreasing selectivity.

Keywords: thin film membrane, membrane design, gutter layer, permeance, selectivity, computational model.

1 INTRODUCTION

Membrane technology has been widely used for water purification, gas separation and has recently emerged as the leading technology for seawater desalination, nitrogen enrichment from air and CO_2 removal from natural gas.

Figure 1a shows a schematic of a thin film composite membrane for gas separation [1]. The thin, dense polymer layer (< 100 nm) performs molecular separation and the bulk porous support (150-200 μm) provides mechanical strength with negligible mass transport resistance. The porous support has small pores (< 100 nm) on the surface providing a smooth surface for the deposition of the selective layer [1,3,7]. However, as shown in Fig. 1b, this imposes a geometric restriction and increases the effective diffusion length (red lines) for penetrants, which decreases membrane performance and leads to a non-linear concentration profile of penetrants [4]. The effect of the support surface morphology on penetrant permeance has been studied using analytical models [5] as well as numerical models that precisely describe the concentration profile and flux within the selective layer for a given pore geometry.

To mitigate the geometric restriction derived from the porous support, a gutter layer with higher permeance than the selective layer can be used as an intermediate layer in practical membranes, as shown in Fig. 1b [5]. The gutter layer is made of extremely high permeability but low selectivity material (PTMSP & PDMS). Due to the high permeance, the gutter layer channels the permeate into the surface pores, reducing the geometric restriction without adding significant transport resistance [5]. In this presentation we use a computational model to elucidate the effect of the membrane nanofeatures on its separation performance.

2 THEORY AND MODELING

The support layer is assumed to contain a 2D array of uniformly spaced cylindrical pores. We can exploit the symmetry of this ordered pore structure and reduce the analysis to a unit cell of the membrane as shown in Fig. 2. Symmetry boundary conditions are applied on the sides of the unit cell to account for the surrounding membrane structure. The penetrant transport in the selective and gutter layers is diffusive and driven by the gradient in concentration following the solution-diffusion model. The equation that governs the steady-state concentration (CA) of a penetrant A in the membrane is [2],

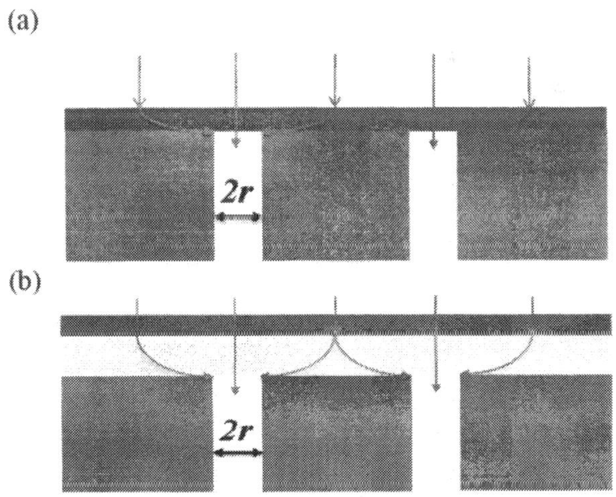

Fig.1 Schematics of thin film composite membrane: (a) a conventional two-layer thin film membrane comprised of a selective layer (with a thickness of l_s) on top of a porous support; (b) a three-layer composite membrane with an intermediate gutter layer of a thickness of l_g. The support has a pore radius of r.

$$\frac{\partial^2 C_A}{x^2} + \frac{\partial^2 C_A}{y^2} + \frac{\partial^2 C_A}{z^2} = 0 \tag{1}$$

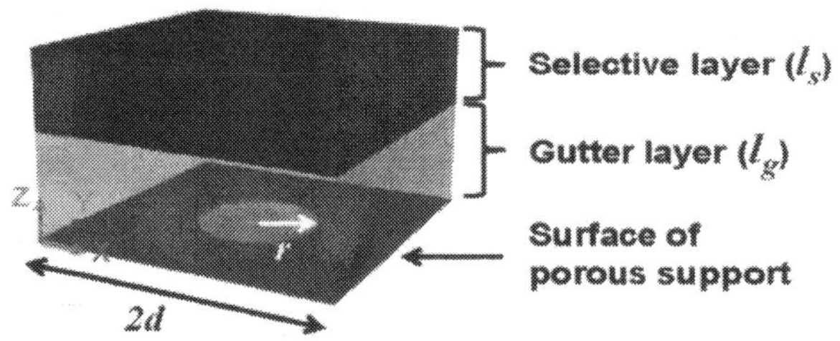

Fig 2. Schematic of a cubic unit cell: *2d*: unit cell length, *r* pore radius of the porous support.

The penetrant concentration and its flux are assumed to be continuous at the interface between the selective and gutter layers [2].

$$D_{A,g} \left. \frac{\partial C_{A,g}}{\partial z} \right|_{z=l_g} = D_{A,s} \left. \frac{\partial C_{A,s}}{\partial z} \right|_{z=l_g} \tag{2}$$

where D_A is the diffusion coefficient for penetrant A, and the subscripts of *s* and *g* indicate the selective and gutter layers, respectively. The nonporous region of the porous support is assumed to be impermeable and there is also zero-flux of the penetrant in the *x* and *y* direction through the sides of the unit cell because symmetry conditions are imposed at these boundaries.

The numerical model was implemented in the COMSOL multiphysics program (www.comsol.com), which solves the governing differential equations using the finite element method (FEM). An adaptive mesh function was used to ensure adequate refining of the mesh in the regions where the boundary condition transitions from a constant concentration to a no-flux condition. This has previously been shown to cause significant errors in the computation; therefore, refinement was continued until the calculated solution became mesh-independent.

The computational model predicts the concentration profile in the selective and gutter layers, and the resulting flux across the composite membrane. The geometric restriction of the membrane nanostructures on the observed permeance of penetrant A can be characterized as membrane permeance efficiency, β_A:

$$\beta_A = \frac{(P_A/l)_{Apparent}}{(P_A/l)_{Ideal}} \tag{3}$$

where $(P_A/l)_{Apparent}$ is the membrane permeance modeled, and $(P_A/l)_{Ideal}$ is the ideal permeance of the selective layer without the influence of the gutter layer and porous support (*i.e.*, $l_g = 0$ and $\phi = 1$). Lower β_A values indicate greater deviation from the ideal permeance, reflecting more severe effect from the porous support and/or the gutter layer.

3 RESULTS AND DISCUSSION

The simulation results depend on the porosity $\phi = \pi r^2/(4d^2)$ and the scaled selective layer thickness, $S = l_s/r$, and scaled gutter layer thickness, $G = l_g/r$, but not the absolute value of l_s or r [2,4]. In addition to the typical porosity values ranging from 0.01 to 0.1 [2,3] a porosity as high as 0.2 is also investigated in this work to explore the maximal benefits attainable by increasing the porosity. As shown in Fig. 3a, increasing porosity increases the membrane permeance efficiency. At a support porosity of 0.05 and a scaled selective layer thickness of 2, the permeance efficiency is as low as 0.17, indicating a significant flux restriction imposed by the porous support. These results are also consistent with those simulated for water transport by Ramon and coauthors, as represented as circles in Fig. 3a [2].

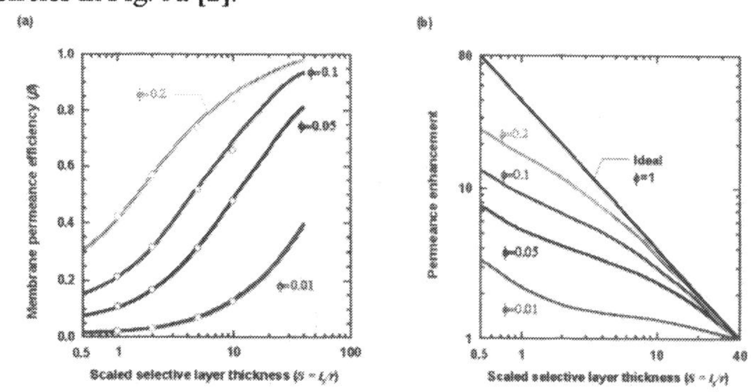

Fig. 3. Effect of the porous support porosity and the scaled selective layer thickness on (a) the membrane permeance efficiency (circles represent validation data taken from reference [2]) and (b) permeance enhancement.

Decreasing the selective layer thickness increases the ideal permeance, which, however, also decreases the permeance efficiency. Fig. 3b illustrates the compromised benefits of decreasing the selective layer thickness as a function of porosity (ϕ). However, due to the geometric restriction of the porous support, the permeance enhancement at S = 0.5 (ratio of apparent permeance at S = 0.5 to that at S = 40) is only 3.6, 7.6, and 13 at $\phi = 0.01, 0.05$, and 0.1, respectively. Even for a hypothetical porous support with $\phi = 0.2$, the expected 80 times increase in the ideal permeance yields only 25 times increase in the apparent permeance.

As shown in Fig. 1b, a gutter layer with negligible mass transfer resistance can channel the gas flow and thus mitigate geometric restriction by the porous support in the ultrathin film composite membranes. Figures 4a, 4b show that the streamlines along with the concentration profile which represent diffusive paths through the selective and gutter layers. The pores increase the path length of the molecular transport and hence decrease the permeance. The increase in the path length is diminished with increasing porosity. Since the restriction is near the pore regions, it is expected that increasing the thickness of the selective layer and/or the gutter layer would minimize the pore restriction, and increase the concentration distribution in these layers.

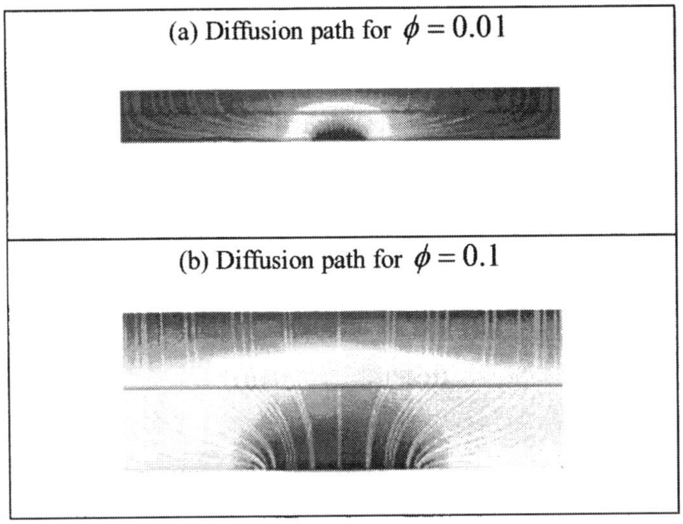

Fig. 4 Cutaway view of the 3D model unit cell showing the streamlines along with the concentration profile which represent diffusive paths through the selective and gutter layers for (a) $\phi = 0.01$ and (b) $\phi = 0.1$. In these calculations, $S = G = 1$; $P_g/P_s = 10$.

The gutter layer is assumed to have 10 times of permeability of the selective layer. The introduction of a gutter layer significantly increases permeance efficiency, though adding the gutter layer is expected to increase the mass transport resistance. As shown in the curves for membrane permeance efficiency (β) in Fig. 5a, introducing a thin gutter layer can increase the β value from 0.022 to as high as 0.13 (~ 6-fold increase) at ϕ= 0.01, and from 0.13 to 0.59 (~ 4.5-fold increase) at ϕ = 0.1. Figure 5a also shows the membrane permeance efficiency for overall selective and gutter layers (β_A'), which is defined as the apparent permeance to the ideal permeance of the combined selective and gutter layer.

$$\beta_A' = \frac{(P_A/l)_{Apparent}}{(P_A/l)_{Ideal,s+g}} = (P_A/l)_{Apparent}\left[\left(\frac{l}{P_A}\right)_s + \left(\frac{l}{P_A}\right)_g\right] = \beta_A + \frac{(P_A/l)_{Apparent}}{(P_A/l)_g} \quad (4)$$

where the subscripts of s, g, $s+g$ indicate the permeances for the selective layer only, gutter layer only, and the combined selective and gutter layer, respectively. Clearly, increasing the gutter layer thickness reduces the geometric restriction and thereby increases the β_A' values. On the other hand, further increase in the gutter layer thickness increases mass transport resistance in the gutter layer and thus decreasing the membrane permeance. Therefore, the benefit of reduced geometric restriction by the thicker gutter layer can be diminished by the increased mass transport resistance.

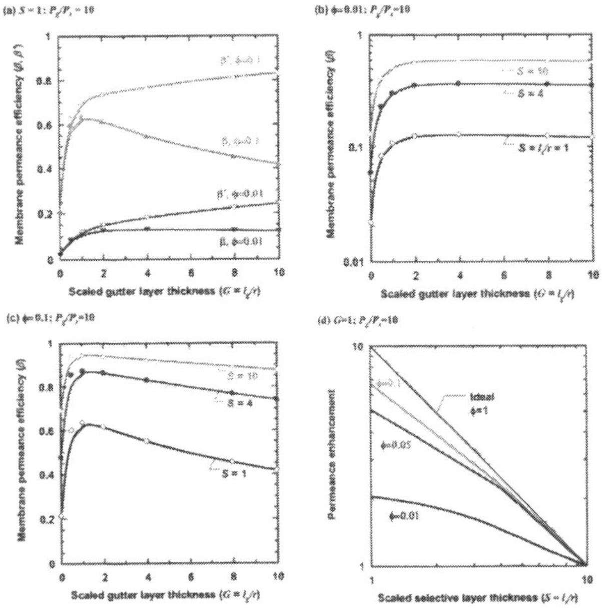

Fig. 5 Effect of the scaled gutter layer thickness G on (a) membrane permeance efficiency β_A and β_A' (defined as the ratio of the apparent permeance to the combined selective and gutter layer); (b) ϕ=0.01; (c) ϕ=0.1, and (d) permeance enhancement by decreasing selective layer thickness at G = 1 and various porosities. In all simulations, the permeability of the gutter layer is ten times that of the selective layer.

Figures 5b,c shows the quantitative effect of the gutter layer on membrane permeance efficiency at the scaled selective layer thicknesses ranging from 1 to 10, and porosity ϕ of 0.01 and 0.1. For these conditions, there seems to be an optimal gutter layer thickness of G between 1 and 2 to achieve the highest increase in permeance efficiency for various S values. Increasing the selective layer thickness increases the permeance efficiency, which is consistent to the study of two-layer composite membranes, as shown in Fig. 3. The most benefit of introducing the gutter layer is observed for the membranes with low porosity and thin selective layer, where the geometric restriction is most severe. Figure 5 also illustrates the importance of the porous supports. At ϕ =0.01, the best membrane permeance efficiency that can be achieved is 0.12 for S = 1, and 0.58 for S = 10. On the other hand, at $\phi = 0.1$, the greatest permeance efficiency achievable is 0.64 for S = 1, and 0.94 for S = 10. Therefore, porous supports with finer pores (lower r values and hence higher S values) and higher porosity are preferred. However, higher porosity may lead to the pore penetration of the coating solution during the membrane preparation, which blocks the pores and increases the support transport resistance, and thus decrease the membrane permeance [5]. Consequently, most commercial porous supports may have porosity less than 0.1. There remains a great challenge in producing the porous support with the balanced characteristics needed for membrane formation.

Figure 5d illustrates the benefit of decreasing the selective layer thickness at different porosities. The condition of $G = 1$ is chosen here for illustration, because it provides one of the greatest permeance efficiency values. While decreasing the selective layer thickness from $S = 10$ to $S = 1$ increases the ideal permeance by 10 times, the enhancement in apparent permeance is only 2.0 at $\phi = 0.01$, and 6.7 at $\phi = 0.1$. To put this on perspective, without the gutter layer, the permeance enhancement is only 1.5 at $\phi = 0.01$, and 5.1 at $\phi = 0.1$, though the permeance efficiency is much lower for the two-layer membranes compared to the three-layer membranes with the gutter layer, as shown in Fig. 5a. The results strongly indicate that the porous support and gutter layer are critical to designing high flux ultrathin membranes.

4 CONCLUSION

We have introduced a 3D computational model for predicting the performance of ultra-thin membranes and have used it to study the effect of an intermediate gutter layer on separation performance. The analysis shows that a gutter layer can be used to achieve high permeance, but it can decrease the selectivity when the porous support has low porosity. A gutter layer thickness of 1-2 times the pore radius of the porous support yields the maximum improvement in the membrane permeance without significantly decreasing selectivity. The model is readily implemented in commercially avaiable software and enable the rational design of ultrathin composite membranes.

REFERENCES

[1] R.W. Baker, Membrane Technology and Applications, 3rd ed., John Wiley and Sons, Ltd., Chichester, UK, 2012.

[2] G.Z. Ramon, M.C.Y. Wong, E.M.V. Hoek, Transport through composite membrane, part 1: Is there an optimal support membrane?, J. Membr. Sci., 415 (2012) 298-305.

[3] X. Lu, L. H. Arias Chavez, S. Romero-Vargas Castrillón, J. Ma and M. Elimelech, Environ. Sci. Techn., 2015, In press.

[4] H. K. Lonsdale, R. L. Riley, C. R. Lyons and D. P. Carosella, in Membrane Processes in Industry and Biomedicine, ed. M. Bier, Plenum Press, New York, 1971, pp. 101-122.

[5] K.A. Lundy, I. Cabasso, Analysis and construction of multilayer composite membranes for the separation of gas mixtures, Ind. Eng. Chem. Res., 28 (1989) 742-756.

[6] T. C. Merkel, H. Lin, X. Wei and R. W. Baker, J. Membr. Sci., 2010, 359, 126-139.

[7] H. Lin, M. Zhou, J. Ly, J. Vu, J. G. Wijmans, T. C. Merkel, J. Jin, A. Haldeman, E. H. Wagener and D. Rue, Ind. Eng. Chem. Res., 2013, 52, 10820-10834.

Low-Cost Membranes for Nanofiltration and Reverse Osmosis Water Purification

B. Berland[1], K. Huntley[2] and M. Misra[3]

[1]ITN Energy Systems, Inc., Littleton, CO, USA, bberland@itnes.com
[2]ITN Energy Systems, Inc. Littleton, CO, USA, kimjhuntley@aol.com
[3]ITN Energy Systems, Inc., Littleton, CO, USA, mmisra@itnes.com

ABSTRACT

ITN has developed a proprietary nanocomposite water filtration membrane that outperforms state of the art membranes with the following attributes:

- Excellent selectivity (passing water while rejecting contaminants)
- High water flux (gallons of clean water per minute produced)
- Chlorine Tolerance
- Anti-Fouling (extended clog-free operating times)
- Durabilty (long lifespan without leaking, breaking, or losing functionality)
- Low cost

ITN's membrane design includes a selective layer made from an inexpensive polymer fabricated onto a robust support to provide good water flow at relatively low pressure, which eliminates the energy and equipment costs of high-pressure pumps. Energy savings can be up to 30% [1]. In addition, the advanced chemistry of ITN's membrane ensures significantly longer membrane life without the fouling (clogging) that occurs with state of the art membranes.

Keywords: water purification, membranes, reverse osmosis, nanofiltration, nanocomposite membrane

1 WATER PURIFICATION

Worldwide demand for water purification capability has skyrocketed due to the following factors:

- Fresh water scarcity (climate change, drought)
- Population growth
- Industrial expansion
- Contamination of surface water (pollution)
- Government regulation (tougher clean water standards)

Water can be purified with various impurity removal and disinfection methods, including:

- Chemical additives to coagulate and precipitate impurities
- Ionic chemicals, ozone, and ultraviolet treatments to kill viruses and bacteria
- Membrane filtration to selectively pass water while blocking contaminants

Processes can be effective at removing certain impurities, but not others, i.e. they may kill bacteria but not remove salts or arsenic, or they may detain certain inorganic particulates without removing microorganisms. Of these methods, only membrane filtration can provide a complete purification solution.

Membrane treatment is the most reliable and proven technology for providing safe and high quality water to the growing global population and for industrial applications. Membrane systems offer operating savings and reduced water and energy footprints. New treatment systems are predominately moving to a sustainable all-membrane approach. [2]

In addition to waste water processing, salinity management (removing salt) is expected to become critical to more regions. Membranes are seen as the preferred way to address desalination issues.

Membrane filtration systems can be designed for high water selectivity, which means that contaminants are blocked, while water is allowed to pass through. Membrane systems are available in different grades, from ultra-pure water filtration to home water softeners to less selective systems designed to reduce pollutants for agricultural use, etc. Advanced water filtration technologies yield a much higher quality of water when compared to the existing conventional water filtration technologies.

Figure 1—Membrane Filtration Applications

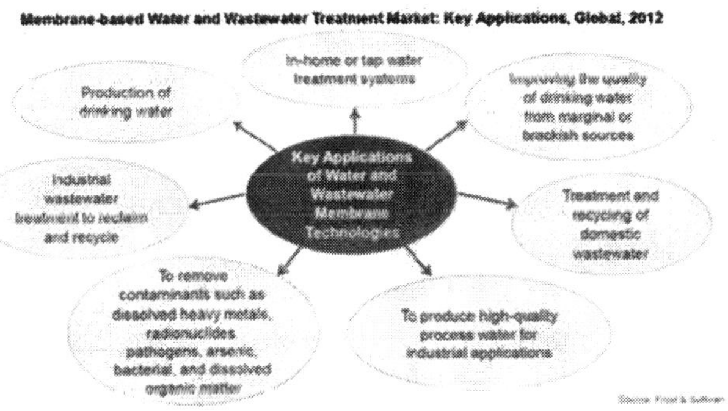

1.1 Water Filtration Technologies

Membrane filtration technologies are classified into 2 categories: 1) Nanofiltration (NF), and 2) Reverse Osmosis (RO). Both are pressure filtration techniques, and both use size sieving as the separation mechanism. RO incorporates finer pore sizes for more stringent filtration, and therefore requires higher pressure (Figure 2).

Figure 2—Nanofiltration vs Reverse Osmosis

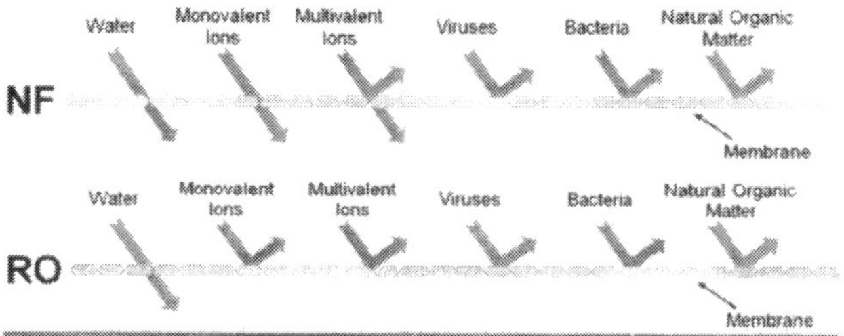

	NF	RO
Membrane pore structure	loose	tight
Operating Pressures	75 to 150 psi	250 to 800
Operating Costs	Low	High

1.2 The Opportunity

Current membrane filtration systems are expensive.

- They require high pressures to process large volumes of water, leading to expensive infrastructure and high energy costs for pumps, etc.
- Membranes are prone to frequent fouling (clogging with organic materials that degrade water flow)
- Membranes are susceptible to chemical degradation (primarily from chlorine disinfectants), which causes decreased flow, impaired function, and breakage.

Membrane filtration has benefits for removing contaminants from water.

- Unlike water treatment via coagulation, which uses significant amounts of chemicals, membrane filtration eliminates need for these chemicals.
- Similarly, while water quality of coagulation and activated carbon technologies depend on the feed water quality, membrane technology can offer consistent water quality for a variety of sourcing.
- Membrane systems require less space to produce the same quantity of water than other filtration systems.
- Membrane processes are modular to meet a wide range of market demands.
- Membrane-based systems are highly automated which enables low-cost operations and minimizes need for highly skilled labor force and/or native chemical/mechanical supply chains.

1.3 ITN's Filtration Innovations

ITN's membrane design includes an ultrathin selective layer made from an inexpensive polymer fabricated onto a robust support (Figure 3). Since water flux is inversely proportional to the membrane thickness, ITN's submicron selective layer enables high water flow at relatively low pressure, which eliminates the energy and equipment costs of high-pressure pumps. Similar approaches have been employed with thin film polyamide membranes. However, the interfacial polymerization process leads to expensive membranes and difficulty achieving uniform performance over large area in manufacturing. ITN's membrane avoids these problems as the polymers are already formed in the slurry used to cast the selective layer.

Operation of ITN's membranes at reduced pressure significantly reduces the system's long-term cost of ownership. The membrane support employs a highly porous material that doesn't significantly impeded water flow while still providing the needed mechanical strength to the composite membrane.

Figure 3—ITN's Asymmetric Nanocomposite Membrane for Water Purification.

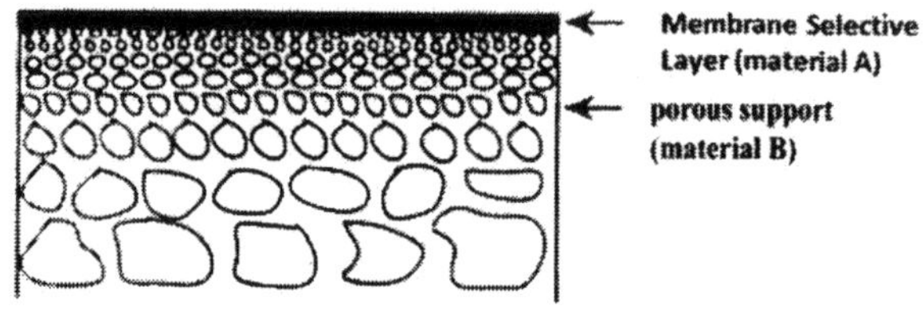

2 TECHNICAL DISCUSSION

2.1 ITN's Membrane Materials

Table 1 shows a comparison of ITN's NF membranes relative to commercially available NF membranes. Typically, one has to choose either high cost membranes with high water flux and poor fouling and chlorine tolerance or lower cost membranes with low water flux and good fouling and chlorine tolerance. ITN's membranes offer a unique combination of a cost-effective membrane with high water flux AND fouling and chlorine tolerance.

We have demonstrated the feasibility of fabricating our polymer-based thin film composite membranes for applications in NF water purification. For the same solute rejection, the expected water flux of ITN membranes in the preliminary experiments was found to be about 2.75 times higher than that of industry leading commercial NF membranes cellulose acetate (CA) membranes. ITN also has encouraging preliminary results and a roadmap to transfer this technology to RO membranes.

Figure 4—ITN's NF membrane performance compares favorably with commercial CA membranes

Large-scale testing (140 cm^2 membrane sheets)		
Performance Characteristics	Cellulose Acetate commercial NF Membrane	ITN's NF Membrane
Water flux (l/m^2-hr-bar)	2.5	4.7
$MgSO_4$ rejection (%)	98.5	99.0
Testing conditions: 200 psi, 2000 ppm $MgSO_4$ soln.		

Table 1—ITN's composite NF membrane provides a unique combination of a cost-effective membrane with high flux at low pressure AND antifouling with chlorine tolerance.

Characteristics	ITN's Composite Membranes	Cellulose Acetate Membranes (Commercial)	Polyamide Composite Membranes (Commercial)
Membrane configuration	Thin film composite membrane	Asymmetric membrane	Thin film composite membrane
Water flux	High	Low	High
Salt or contaminant rejection	>99%	97%	>99%
Free chlorine tolerance	≥ 1.0 ppm	Up to 1.0 ppm	< 0.1 ppm
Membrane fouling tendency	low	low	high

2.2 Competitive Advantages

ITN's membrane provides a unique combination of properties that result in the best overall solution for water filtration. These advantages include:

(i) Scalable, large area membrane fabrication and low cost: In the case of polyamide TFC membranes, the polymer is made in-situ on the microporous substrate to deposit a thin film. Thin films of ITN's polymer can be deposited onto a microporous support by either simple dip coating or spray coating, so ITN's membrane fabrication process is scalable to large area membranes with more uniform performance and lower cost.

(ii) Higher water flux from thinner selective layer: Since the thickness of the selective layer is submicron, the ITN membrane has higher water flux than cellulose acetate asymmetric membranes with similar solute and salt rejection properties. Higher flux results in lower energy costs from elimination of high-pressure pumps, which makes the NF technology economically viable.

(iii) Lower surface fouling: ITN's NF membrane has significantly lower fouling tendency than other thin film composites due to lower surface roughness (Figure 5).

Figure 5—ITN's NF membrane, with a smooth surface, provide greater resistance to fouling.

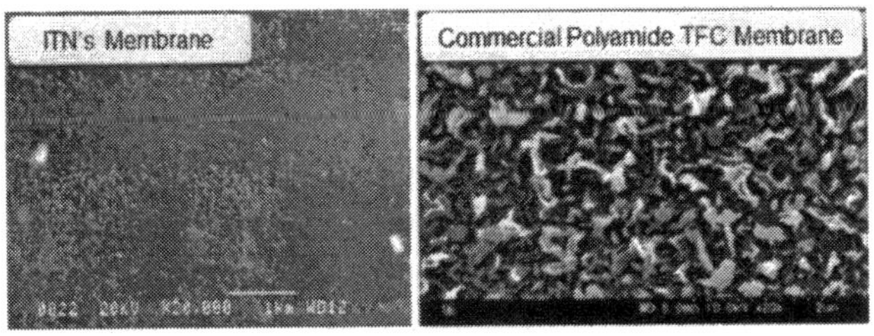

(iv) Chlorine tolerance: Many applications employ disinfectants such as chlorine to kill micro-organisms that are dangerous to humans and degrade membrane performance. However, chlorine also attacks polyamide membranes. Figure 6 shows stable operation of ITN's NF membrane with 1 ppm chlorine.

Figure 6—ITN's membranes have demonstrated stable operation in 1 ppm chlorine.

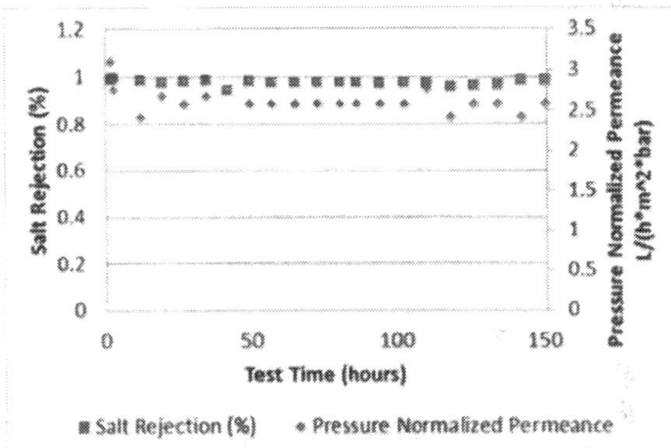

(v) Tunable membrane properties: ITN's membranes have already been demonstrated to have promising properties for NF applications. Further, ITN has promising initial formulations and established a roadmap for to meet market demands for RO membranes.

(vi) Greater durability when dry: While commercial cellulose acetate membranes tend to crack when dried out, ITN's NF membrane remains pliable. (Figure 7)

Figure 7—ITN's membrane remains intact after drying, while competing membranes crack and wrinkle when dried.

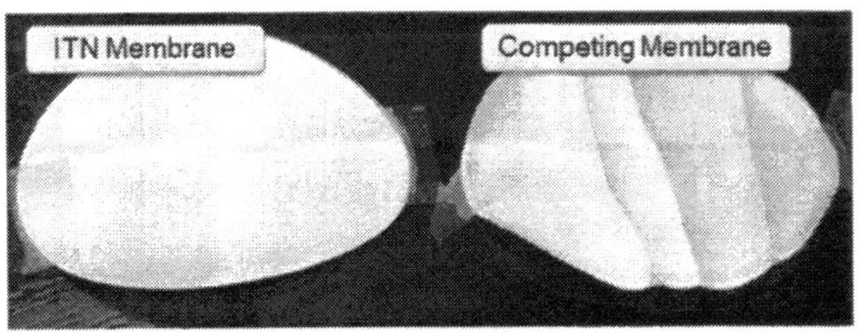

3 COMMERCIALIZATION

3.1 Market Strategy

ITN's filtration membrane technology can be used for numerous water filtration and purification applications (Figure 8).

Figure 8—Applications for Water Filtration Membranes

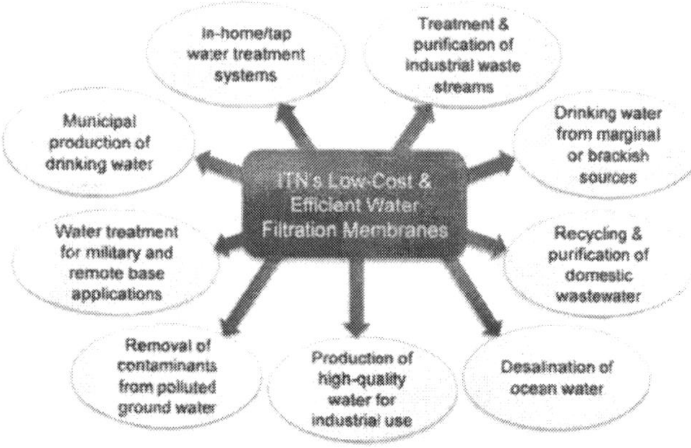

The biggest market for filtration membranes is large scale water supply systems for communities that rely on surface water purification to meet their daily needs. Municipal water use in the U.S alone is approximately 45 billion gallons of water per day (Bgal/day). Eighty percent of this (approximately 36 Bgal/day) is from surface water. In 2012, global revenue from membrane-based municipal water treatment reached almost $1.2 billion and is projected to exceed $2.5 billion by 2020.

ITN's technology will gain market penetration because of the competitive advantages described in 2.2, and because of ITN's efforts to work with industry partners with existing market presence. We expect to generate revenue through sales to multiple customer classes. First, the membranes will be sold directly to private and public operators of large scale NF water purification facilities as a direct replacement of current filtration systems. Secondly, they will be sold to filtration system manufacturers and suppliers. Companies such as Dime Water, DAICO, and Pure Aqua can use ITN's technology in new water filtration systems as well as retrofitting more expensive, less efficient water filtration systems.

ITN anticipates pilot scale production, testing, pilot production, and scale-up to mass production within five years. ITN expects to be able to capture 5% market share of the membrane water purification market within the first five years of mass production and commercialization.

Figure 9—ITN's spiral membrane modules was the first step on the roadmap to large commercial production.

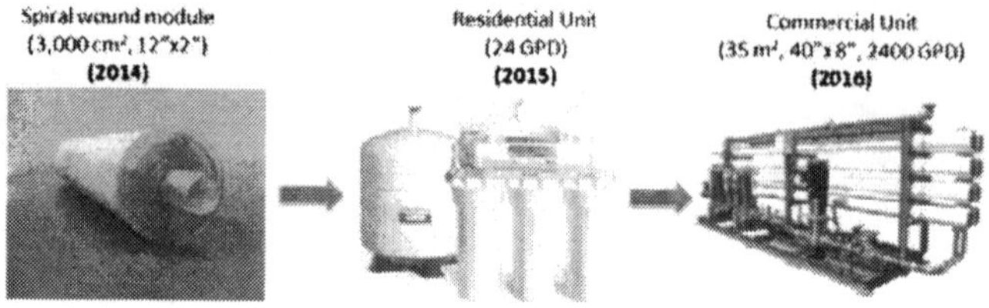

REFERENCES

[1] A. Sagle and B. Freeman, "Fundamentals of Membranes for Water Treatment" http://texaswater.tamu.edu/readings/desal/membranetechnology.pdf

[2] North American Industrial Water and Wastewater Treatment Chemicals Markets Cost-effective Solutions will Drive Growth in the Market, Frost & Sullivan report, NAE2-39, May 2012

Preparation and characterization of porous polyethersulfone nanofiltration membranes via combined phase separation method for water desalination

H. Jahanbakhsh Asl[*], M. Soltanieh[**]

[*] Department of Chemical and Petroleum Engineering Sharif University of Technology, Tehran, Iran, h_jahanbakhsh@che.sharif.ir, Tel: +989381170352

[**] Department of Chemical and Petroleum Engineering, Sharif University of Technology, P.O. Box 11155/9465, Tehran, Iran, msoltanieh@sharif.edu, Tel: +98 21 6616 6417 (Direct), +98 21 6600 5819 (Office), Fax: +98 21 6602 2853

ABSTRACT

Porous polyethersulfone (PES) membranes were prepared by combined phase separation method in which both non-solvent induced phase separation (NIPS) and thermally induced phase separation (TIPS) occured in the membrane-formation procedure. The membrane-forming solution consisted of PES, phenylbenzoate as diluent and N-Methyl-2-pyrrolidone (NMP) as solvent. The morphology of the prepared membranes was investigated by scanning electron microscopy (SEM). Pure water permeation flux, salt (NaCl) rejection rate, pure water contact angle, and thermal gravimetric analysis (TGA) were also investigated. SEM images revealed that the membrane finger-like structure was converted to a continuous porous structure by increasing mass ratio of phenyl benzoate to NMP. All membranes immersed in coagulation bath at 45°C were more hydrophobic than other prepared samples. The highest pure water permeation flux of 766.2 $Lm^{-2}h^{-1}$ (0.2 MPa) and the highest salt rejection of 73.3% were obtained via combined thermally and non-solvent phase separation methods described in this work at feed concentration of 5000 ppm.

Keywords: porous membrane, combined method, NIPS, TIPS

1 INTRODUCTION

Membrane technology requires no chemical additives, thermal inputs and regeneration of spent media. Because of these characteristics, this technology can be favored to answer the global water challenges consisted of population increase [1], overall water shortage [2] and poor water quality [3]. Pressure-driven processes are the most widely used membrane technologies in water treatment applications [4]. A desirable membrane used in water treatment should have high productivity, selectivity, fouling resistance, stability and lower cost. These favored features may be possible not only by using appropriate materials but also by using suitable formation method. Polymeric membranes have been made by phase separation [5, 6], interfacial polymerization [7] and evaporative casting [8]. Among these methods, the phase separation is the most commonly used method for the preparation of polymeric membranes and classified in two main methods known non-solvent induced phase separation (NIPS) and thermally induced phase separation (TIPS). Asymmetrical structure and finger-like pores resulted from NIPS method are unfavorable because of poor mechanical properties, creation of additional resistance against mass transfer and low flux [9]. In contrast, TIPS method leads to membranes with higher porosity and good mechanical properties [10]. However, there is some limitation in using TIPS method commercially. In this study, the combined phase separation process was employed to prepare the desired membrane structure by mixing PES, phenyl benzoate as diluent and NMP as the solvent at relative lower temperature compared to conventional TIPS method. The performance of the prepared membranes was evaluated by measuring the pure water permeation and salt rejection.

2 METHODS AND MATERIALS

2.1 Materials

Polyethersulfone (PES) (Ultarason E6020, η=82 $mL.g^{-1}$, Mw=51000, ρ=1.37 $g.cm^{-3}$) was purchased from BASF Chemical Co. Ltd. (Germany).The polymer was dried for 8 h at 100°C and 0.7 bar. N-Methy-2-pyrrolidone (NMP) and ethanol (bp=78.4°C , ρ=0.709 $g.cm^{-3}$) were purchased from Merck Co. phenyl benzoate (bp=66 °C) and sodium choloride (NaCl) were purchased from Sigma Aldrich Co. NaCl solution was used to characterize desalination performance of the prepared membranes. Also, all of the water used was distilled water.

2.2 Preparation of membranes

Mixture of solvent (NMP) and diluent (phenyl benzoate) were mixed and then an amount of dried PES were dissolved in the mixture of solvent and diluent at 140 °C. After being mixed by mechanical mixer for 2 h, a homogenous hot solution was degassed at 140 °C for 1 h to remove air bubbles. The solution was casted onto glass preheated at 100 °C with 15×15 cm^2 surface using a mechanical casting machine with a gap of 250 μm and immersed in a water bath colder than solution for 1 h to form membrane. For extraction of phenyl benzoate, the prepared flat sheet membrane was immersed into ethanol bath for 48 h.

The extracting ethanol was evaporated in open air for 1 day at room temperature. Studied parameters and conditions are listed in Table 1.

Table 1. Membrane preparing conditions

Name	Bath temperature (° C)	Mass ratio of phenybenzoat/NMP	Polymer concentration
S_1	0	1.5	27.5
S_2	0	2.25	27.5
S_3	0	3	27.5
S_4	22.5	1.5	27.5
S_5	22.5	2.25	27.5
S_6	22.5	3	27.5
S_7	45	1.5	27.5
S_8	45	2.25	27.5
S_9	45	3	27.5

2.3 Characterization methods

The fractured membranes in liquid nitrogen were deposited on holders. After sputtering with gold in vacuum, the morphology of cross section and surface of membranes were observed by scanning electron microscopy (SEM, Tescan VEGA II XMU, USA). The permeation flux of the membranes was measured using a home-made cross-flow nanofiltration cell. All experiments were conducted at room temperature and under a feed pressure of 0.2 MPa for 1 min after achieving equilibrium condition. Salt (NaCl) rejection was measured at feed concentration of 5000 ppm.

The hydrophobicity of the membranes was studied by measuring the water contact angle.

Heat resistance of the prepared membranes was investigated by TGA test (Mettler Toledo DSC/TGA 1). A small piece of flat sheet membranes was placed into aluminum pans to seal. The scanning range was 25-600 °C with rate of heating of 10 °C/min in argon atmosphere.

3 RESULTS AND DISCUSSION

In the process used for preparing membranes, the PES was dissolved in a mixed solvent at relatively elevated temperature (140 °C); the homogeneous solution was casted and immersed in water bath at a temperature lower than casted solution. Phase separation due to gradient of temperature and gradient of concentration between coagulation bath and casted solution occurred to produce a membrane structure.

The cross-sectional structure of all samples is shown in Fig.1. As shown in Fig.1 (a), typical instantaneous demixing occurred and finger-like pores were formed in the bottom section of the membrane. Porous structure was observed as dominant structure. Lower bath temperature is associated with lower exchange rate between bath and casted solution and higher viscosity. Thus, this phenomenon led to slow precipitation rate, and the delayed demixing process accrued which resulted in a porous structure in the cross section of the prepared membranes. Cellular structure happened during both L-L TIPS [11] and delayed dimixing NIPS. Therefore, it was expected that by increase of the mass ratio of diluent to solvent, from 1.5 to 2.25 and 3, the contribution of the TIPS phenomenon against the NIPS increased. Thus, finger-like pores got smaller and porous structure got more dominant as shown in Fig.1 (b) & Fig.1(c).

The cross section of immersed samples into coagulation bath of 22.5 °C are shown in Fig.1 (d), Fig.1 (e), and Fig.1 (f). Typical finger-like pores were disappeared but continues grooves from top layer to below layer (throughout the cross section) were formed. Decreasing the viscosity of a casting solution due to increase of the bath temperature will increase the tendency of macrovoid formation [12]. The same phenomenon was observed when bath temperature reached to 45°C but the formed grooves were deeper (Fig.1 (g), Fig.1 (h), Fig.1 (i)). By increase of the bath temperature, release of the solvent increased and grooves got larger and deeper under the same casting condition. Comparing structural morphology of the S_1, S_4, and S_7 membrane samples indicates this fact.

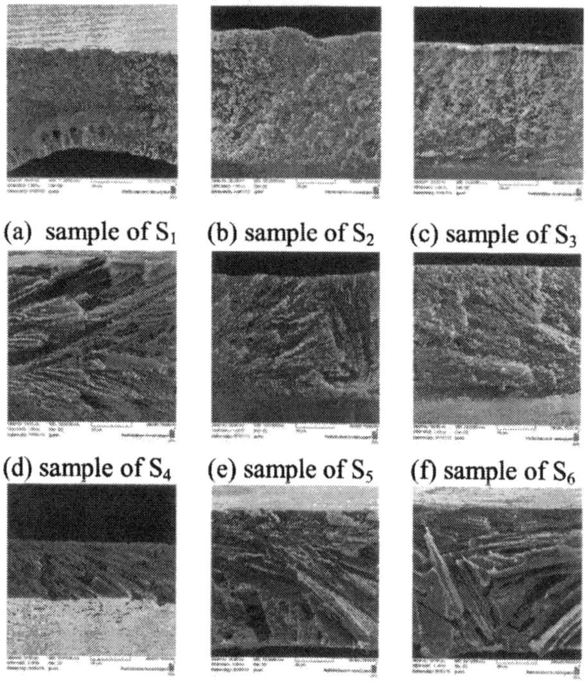

(a) sample of S_1 (b) sample of S_2 (c) sample of S_3

(d) sample of S_4 (e) sample of S_5 (f) sample of S_6

(g) sample of S_7 (h) sample of S_8 (i) sample of S_9

Fig. 1 SEM of cross section of prepared samples

In Table.2 the water permeation flux (Jw) of the porous membranes is presented.

Table 2. Pure water permeation flux of prepared membranes

	Bath. Tem (°C)	Phenylbenzoate/NMP	$Lm^{-2}h^{-}$) J_w (^{1}

S_1	0	1.5	156.7
S_2	0	2.25	110.6
S_3	0	3	35.8
S_4	22.5	1.5	356.3
S_5	22.5	2.25	296.6
S_6	22.5	3	216.4
S_7	45	1.5	774.1
S_8	45	2.25	766.2
S_9	45	3	565.4

All membranes prepared at bath temperature of 0 °C had the lowest flux compared with other membranes immersed at hotter coagulation baths. By increase of diluent to solvent ratio, pure water flux decreased in the following descending order: S_1, S_2 and S_3.

As shown in Table 2, when the ratio of phenyl benzoate to NMP was kept at 1.5, decrease of the bath temperature resulted in lower pure water flux. At the lower ratios of the diluent to solvent, NIPS process had comparable contribution to formation of the membrane. Therefore, decrease of the coagulation bath temperature resulted delayed dimixing, smaller pore sizes and lower water flux consequently. The maximum water permeation flux 766.197 $Lm^{-2}h^{-1}$ was measured for the S_8 polymeric membrane sample.

Salt rejection percent of the prepared membranes at 0.2 Mpa pressure and salt concentration of 5000 ppm is presented in Table.3.

Table 3. Salt rejection percent of samples at feed concentration of 5000ppm at 0.2 MPa

	Bath. Tem (°C)	Phenylbenzoate/NMP	Salt rej. (%)
S_1	0	1.5	60.4
S_2	0	2.25	62.1
S_3	0	3	73.3
S_4	22.5	1.5	52
S_5	22.5	2.25	54.2
S_6	22.5	3	56.8
S_7	45	1.5	46.4
S_8	45	2.25	41
S_9	45	3	47.6

Comparing of the results from Table.3 shows that as the mass ratio of the diluent to solvent increased, salt rejection increased too. Salt rejection is related to morphology of membranes directly. The effect of delayed dimixing and contribution of the TIPS enhanced the performance, which led to cellular pores and continuous structure in the cross-section of the prepared membranes. Therefore, porosity and mean pore size of the membranes decreased and rejection percent increased.

The effect of casting conditions on contact angle which is related to the pure water flux is reported in Table 4.

Table 4. Pure water contact angle of samples

	Bath. Tem (°C)	Phenylbenzoate/NMP	Contact angle
S_1	0	1.5	72.26
S_2	0	2.25	80.7
S_3	0	3	71.12
S_4	22.5	1.5	80.55
S_5	22.5	2.25	79.1
S_6	22.5	3	75.7
S_7	45	1.5	89.55
S_8	45	2.25	79.45
S_9	45	3	90.58

It was observed that the increase in the mass ratio of the diluent to solvent from 1.5 to 2.25 for S_1 and S_7 samples resulted to more hydrophobic surface. However, further increase in the ratio resulted in lower contact angles.

In addition, for membranes containing diluent to solvent ratios of 1.5 and 3 times, increase in bath temperature enhanced the hydrophobicity.

The weight loss curve of the S_4 sample is shown in Fig.2. Also TGA data of the prepared samples are presented in Table 5.

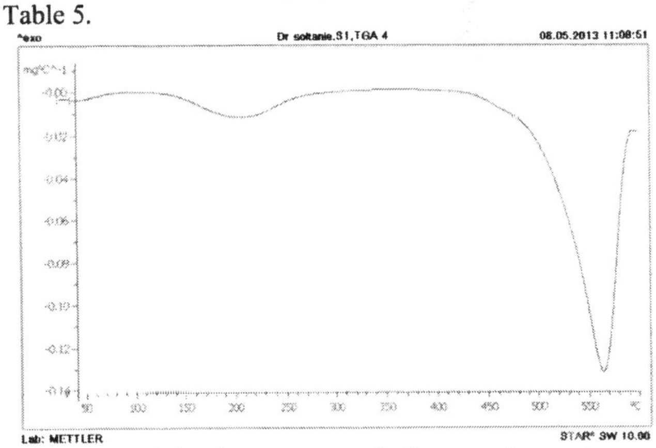

Fig. 2 Weight loss curve of S_4 sample against temperature

According to Fig.2, it can be concluded that there are two weight loss stages starting from around 200 °C and 560 °C. First stage is ascribed to the evaporation of residual phenylbenzoate or NMP and second stage is related to decomposition of PES main chains. Reported data in the Table 5 indicates good thermal resistance of the prepared membranes.

Table 5. TGA test results of S_2, S_4, S_5, S_6, and S_8 samples

	Bath. Tem (°C)	Phenylbenzoate/NMP	TGA test results	
			Temp (°C)	Percent of weight loss (%)
S_4	22.5	1.5	209.04	7.34
			564.34	49.25
S_5	22.5	2.25	208.54	6.17
			567.16	52.09
S_6	22.5	3	190.85	7.27
			566.89	50.68
S_2	0	2.25	202.40	6.81
			567.02	50.22
S_8	45	2.25	209.61	8.46
			566.64	49.83

4 CONCLUSIONS

In this work, we used combined thermally (TIPS) and non-solvent induced phase separation (NIPS) methods for preparing polymeric nanofiltration membranes for water desalination. The coagulation bath temperature and mass ratio of the diluent to solvent were two important parameters to achieve the desired structure and a balanced contribution of each phase separation method.

As water bath temperature increased, the cross-sectional morphology of the produced membranes changed from cellular structure with some macrovoids to exhibited groove-like pores with continuous cellular pores among them. The separation performance of the prepared membranes was related to the phase separation process and the resulted morphology. Because of the application of both phase separations methods in forming membrane, the resulted structure strongly differed from one casting condition to another. As the mass ratio of the diluent to solvent increased and bath temperature decreased, porosity, mean pore size of the membranes and pure water permeation flux decreased and salt rejection increased. The highest pure water permeation flux of 766.2 $Lm^{-2}h^{-1}$ (at 0.2 MPa) and the highest salt rejection of 73.3% were obtained

It is concluded that the combined method has the advantage of producing a symmetrical porous structure while at relatively lower temperatures compared to those used in the conventional TIPS process. Also, the continuous structure obtained from the combined method is more similar to structure resulted from TIPS alone.

REFERENCES

[1] Human Development Report 2006, United Nations Development Programme (UNDP), Chapter 4, 2006.

[2] Coping with water scarcity. Challenge of the twenty first century, United Nations (UN) Water, Food and Agricultural Association (FAO), 2007.

[3] Progress on Sanitation and Drinking-Water, World Health Organization (WHO)/UnitedNations International Children's Fund (UNICEF) Joint Monitoring Programme for Water Supply and Sanitation, 2010.

[4] M. Ulbricht, Advanced functional polymer membranes, Polymer, 47, 2217-2262, 2006.

[5] I.M. Wienk, R.M. Boom, M.A.M. Beerlage, A.M.W. Bulte, C.A. Smolders, and H. Strathmann, Recent advances in the formation of phase inversion membranes made from amorphous or semi-crystalline polymers, Journal of Membrane Science, 113, 361-371, 1996.

[6] I. Kim, K. Lee, and T. Tak, Preparation and characterization of integrally skinned uncharged polyetherimide asymmetric nanofiltration membrane, Journal of Membrane Science, 183, 235-247, 2001.

[7] P.W. Morgan, Interfacial Polymerization, in: J.I. Kroschwitz (Ed.), Encyclopedia of Polymer Science and Engineering, 8, 221-237, 1987.

[8] B. Kunst and S. Sourirajan, Effect of casting conditions on the performance of porous cellulose acetate membranes in reverse osmosis, Journal of Applied Polymer Science, 14, 723-733, 1970.

[9] Z. L. Xu, F. Alsalhy Qusay, Polyethersulfone (PES) hollow fiber ultrafiltration membranes prepared by PES/non-solvent/NMP solution, Journal of Membrane Science, 233, 101–111, 2004.

[10] M. Liu, Y-M. Wei, Z-L. Xu, R-Q. Guo, L-B. Zhao, Preparation and characterization of polyethersulfone microporous membrane via thermally induced phase separation with low critical solution temperature system, Journal of Membrane Science, 437, 169–178, 2013.

[11] C.B Funk, Microporous Mixed Matrix (ZeoTIPS) Membranes, PhD thesis, The University of Texas at Austin, 2008.

[12] C. S. Feng, B. L. Shi, G. M. Li, Y. L. Wu, Preparation and properties of micro-porous membrane from poly (vinylidene fluoride-co-tetra fluoro ethylene F2.4) for membrane distillation, Journal of Membrane Science, 237, 15–24, 2004.

Effect of nitrogen doping and tunable pore structures on arsenic removal

Jisun Han, Daegwon Ha and Shinhoo Kang*

*Department of Materials Science and Engineering, Seoul National University, Seoul, 151-742, Korea

ABSTRACT

Nitrogen doped carbons were synthesized by a carbo-nitride derived carbon (CNDC) at various chlorination temperatures. CN73s that have both small micro- and meso-pores with nitrogen functional group, such as pyrrolic-N at the surface of pores, show excellent adsorption ability for the arsenic complex ions. Experimental and theoretical calculation results suggested that pore structure and the types of nitrogen functional groups making the CN73s highly charged significantly influenced the adsorption of arsenic. The small micro-pores(0.7 nm) that pyrrolic-N is placed at the edge of pores are active sites for capturing arsenic complex ions. The small meso-pores provide the diffusion path for arsenic ions and enhance the adsorption kinetics.

Keywords: arsenic adsorption, nitrogen doped porous carbon, carbo-nitride derived carbon (CNDC), pore size distribution, nitrogen functional groups

1 INTRODUCTION

The arsenic and its compound are highly toxic and some of them are carcinogen. They could spread into ground water through geothermal, mining and industrial activity, and contaminate ground water. A large amount of water resources around the world contain signification concentration of arsenic ranged from 0.5 to 5000 μg/L. To secure drinking water, removal of arsenic is vitally required. Therefore, developing adsorption material and elimination technology became a big global issue and numerous studies are currently underway.

The porous carbons, especially carbide derived carbon (CDC), synthesized by a chlorination of various carbides is one of the typical adsorbent materials. CDC has well developed pore structure not only on its surface but also in bulk [1]. Thus, it is widely applied as the substrate for the gas adsorption and the electrochemistry application [2, 3]. Particularly, certain size of micro-pores and small meso-pores are uniformly formed in CDC via controlling experimental factors [4]. It is suitable for small sized gas adsorption such as hydrogen, carbon dioxide. Because micro-pores, which could catch gas molecules via overlapping of interaction force between the pore walls, act as adsorption sites and meso-pores provide the kinetic path to gas molecules enhancing diffusion of gas molecules within the bulk of CDC [5].

Meanwhile, pure porous carbon itself has high stability and is electrically neutralized. Therefore, it has low van der Waals interaction (4 kJ/mol) with gas molecules or electrolytes. To enhance the interaction energy, hetero atom doping is extensively implemented notably the nitrogen. A number of research about nitrogen doped carbons are currently in progress. Thus nitrogen doped carbons are widely applied as adsorbent, electrode for energy storage device and catalyst for electrochemical applications [6, 7].

In this paper, N-doped porous carbon was synthesized by a carbo-nitride derived carbon (CNDC), which is similar to the CDC process. It has both micro-pores and meso-pores which is the advantage of CDC. Furthermore, the nitrogen is successfully doped on edges of CNDC along with pore development. Therefore, aforementioned N-doped porous carbon has the positively charged surface that could enhance the adsorption ability for arsenic ions in water. This material shows outstanding adsorption performance for arsenic compared with other carbon materials. Also, its performance is comparable to metal oxide adsorbents, such as iron oxide, aluminium oxide, commonly used in arsenic removal. We denoted entire samples as CN73s and each sample as CN-X; In CN73s, 73 represents the carbon-to-nitrogen ratio of the $Ti(C_7N_3)$ precursors. X signifies the chlorination temperature.

2 RESULTS AND DISCUSSION

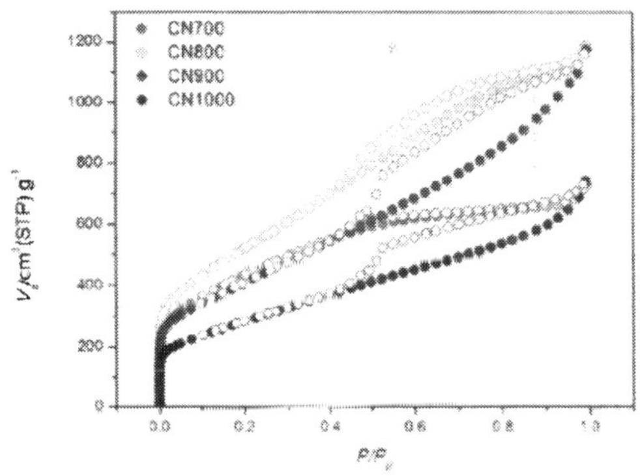

Figure 1: N_2 sorption isotherms of CN73s (solid – adsorption, open – desorption)

The N_2 sorption isotherms, shown in fig. 1, exhibit the pore structure of CN73s. CN700 have type I isotherm with mild slope of P/P_0 = 0.1~0.5, which represents the development of micro-pores wider than those of pure micro-porous materials like typical CDC. The sorption isotherms of pure CDC usually have sharp hill at low relative pressure and plateau over wide ranges. TiC-CDC, as a typical CDC, have negligible meso-pores but above

1200°C, hysteresis loop begins to appear representing meso-pores [8]. However porous CN73s, chlorinated from nitrogen substituted TiC, well develops meso-region pores from synthesis temperature of 800°C. Its sorption isotherms change into Type H3 and H4 hysteresis loops between 800~1000°C, which represent aggregations of narrow meso-pores [9]. CN73s, with similar pore forming agent like CDC, developed pore structure well with more than 1000 m^2/g surface area calculated by Brunauer-Emmett-Teller (BET) model. CN800 and CN900 have large amount of N_2 adsorption especially for meso-region whereas CN1000 decreases its N_2 adsorption by partial collapse of the pore network with longer graphitic layers growing. According to pore characteristics, CN800 showed highest values in micro- and meso- pore volumes among the CN73s. Surface area of CN1000 significantly decreased as its micro- and meso-pore volumes decrease, and it is caused by growth of pore width, which is influenced by curved graphitic layers. The textural properties are summarized in Table 1.

The pore size distribution (PSD) of CN73s, calculated by non-local density functional theory (NLDFT), are broadened by increasing the chlorination temperature displayed in fig. 3. Pores of CN700 are centered at 0.7nm, which is a characteristics of TiC-derived carbon , 1.5nm and 3nm, which is believed to be affected by CN_x desorption [10]. It has uniform tri-modal distributions, verified as parallel line (figure.1) at $P/P_0 = 0.5{\sim}0.9$ [11]. As synthesis temperature increase, 0.7nm and 1.5nm pores collapsed due to releasing a highly stressed and curved graphitic layers, and it caused the irregularity. Large pores with 20~100nm are well developed in CN1000 while total pore volumes decreased.

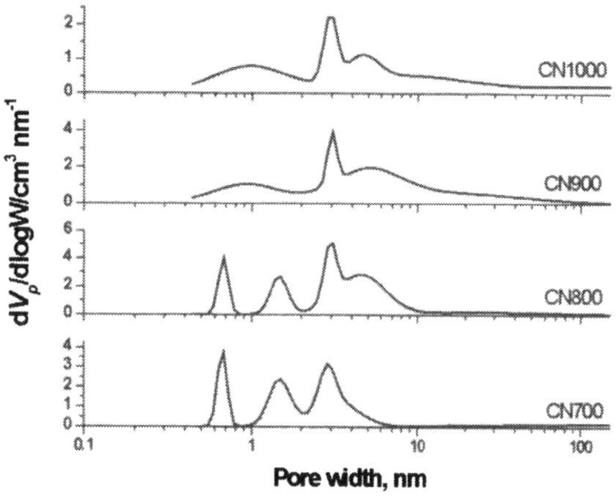

Figure 2: Pore size distributions (PSD) of CN73s calculated by NLDFT (cylinder pore model).

X-ray photoelectron spectroscopy (XPS) spectra (figure. 4a) was employed to compare chemical composition and bonding nature of CN73s with different chlorination temperature and summarized in table 2. The CN700,

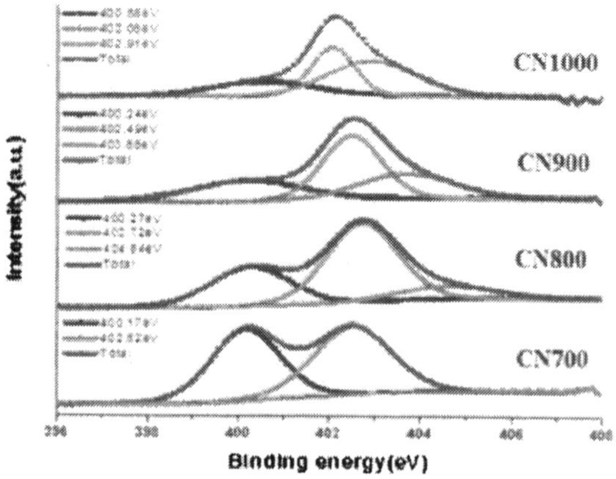

Figure 3: XPS of N1s spectra of CN37s

CN800, CN900, and CN1000 possess 3.14, 3.72, 2.67 and 2.17 wt.% nitrogen doping contents, respectively. Nitrogen doping can be divided into two types of configuration as pyrrolic-N (400.37$\pm$0.2eV) and oxidized-N (402~404eV), especially N-oxides of pyridinic-N (402.06$\pm$0.3eV). Since these functional groups are invariably located at the edges sites [12], the edges caused by pore development can be advantages of such selective doping site [13]. Consequentially, Edges along the pore growth of the whole seem to be a desired sites for an edged-nitrogen doping, given the report which proposes that pores in CDC are developed not only at the surface and throughout the bulk [1]. Micro-sized pores are formed in pure carbide systems but micro- and meso-pores are well developed in carbo-nitride systems. Pores grow over by decomposing unstable C-N bonding, such as pyrrolic-N (5-membered ring) while stable nitrogen, such as pyridinic- or graphitic-N (6-membered ring), remain or unstable one converted to stable one on the pores [14]. Pyrrolic-N is well developed in CN700 but it decreases dramatically with temperature. Pyridinic-N of all CN73s transforms into N- oxidizes of pyridinic-N on the surface [12].

Adsorption isotherm of arsenic is shown by figure 4. CN900 and CN1000 is incapable of adsorbing arsenic significantly, whereas CN700 and CN800 show strong absorption behavior. The maximum adsorption values(q_m) of aresenic are 31.08 and 30.34 mg/g, respectively. The adsorption isotherm for CN73s fits the Langmuir equation which is marked on (Eq. (1)).

$$q = \frac{q_m K_a C}{1 + K_a C} \qquad (1)$$

q_m is the maximum mass adsorbed at saturation conditions per mas unit of adsorbent (mg/g). C is equilibrium aqueous concentration (mg/L) and K_a is the empirical constant with units of inverse of concentration C (/mg)

Through the above experimental results, it is estimated that the optimized pore-structure and nitrogen functional groups located on the surface of pores are the key factors

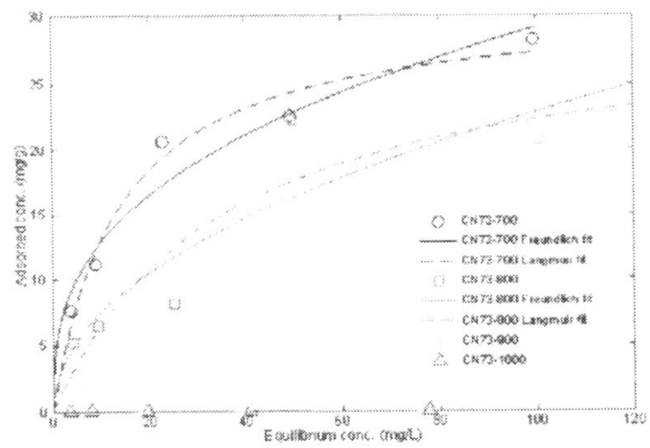

Figure 4: Adsorption isotherm of As on CN73s (700, 800, 900 and 1000 °C)

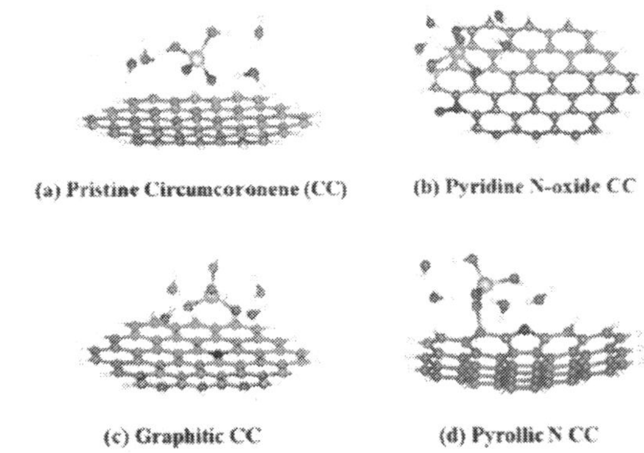

Figure 5: Optimized adsorbed structure of As complex, H_2AsO_4, and four water molecule on four adsorbents.

for adsorption of arsenic. Based on the PSD of CN73s and the arsenic adsorption isotherm, small micro-pores (0.7 nm) provide the active sites for adsorbing the arsenic ions. CN900 and CN1000 that small micro-pores are barely developed could not adsorb the arsenic. On the other hand, CN700 and CN800 have tri-modal pore distribution showing small micro-pore(0.7 nm), big micro-pore(1.5 nm) and meso -pore(3.0nm) are harmoniously developed. Thus, througth the small micro-pores, they could effectively capture the arsenic and meso-pores could be the kinetic path that enhances the diffusion of the arsenic ions to interior of CN700 and CN800. The advantage of bio-modal pore size distribution that both micro- and meso-pores are simultaneously developed is verified and widely applied to gas adsorption and electrochemistry field [5]. Not only the pore structure but also the nitrogen functional group, especially pyridinic-N could be important for arsenic adsorption. Based on the XPS results, as mentioned above, CN700 and CN800 contain considerable amount of pyrrolic-N and N-oxides of pyridinic-N. In contrast, pyrrolic-N nearly vanished and only N-oxides of pyridinic-N are dominant in the case of CN900 and 1000. Therefore, it is estimated that optimized pore-structure that both micro-pores and meso-pores are well developed and pyrrolic-N located on the surface of pores are the key factors for adsorption of arsenic.

To identify the interaction that works on between arsenic ions and nitrogen doped carbons, we calculate optimized structures and binding strengths of As complex on adsorption sites which are chosen based on the XPS spectrums in figure 3. Four water molecules surrounding As complex are included to mimic aqueous environment. Since adsorbate consists of As complex, H_2AsO_4, and four water molecule, zero-point energy and entropy effects are also included. Therefore, binding strength is evaluated from the difference of gibbs free energy before and after reaction, which is defined as $\Delta G = G_{adsorbent+adsorbate} - (G_{adsorbate} + G_{adsorbent})$. Each term represent gibbs free energies of corresponding systems and T*S (temperature * entropy) term in gibbs free energies are obtained at temperature 25℃. Four N-containing structures (pyridine N-Oxide, graphitic N, and pyrollic N) and N-free (Pristine) structure shown in figure s1k are examined. The change of gibbs free energies for adsorption on four structures are summarized in table 1k and optimized structures after adsorption are given in figure 1k. It is shown that optimized structures that As complex can be adsorbed on graphitic N and pyrollic N structure through ionic and covalent bonding, respectively. This result accurately matches with our experimental consequence as we mentioned above. Therefore, the most crucial factors for aresnic adsorption are both pore-structure and nitrogen funtional group located on the edge of pores. The micro-pores that possessed the nitrogen functional goups, such as pyrrolic-N and graphitic-N could effectively captured the arsenics ion because their surface was negaively charged. The small mese-pores(2-10 nm) well developed through the bulk act as the diffusion path for arsenic ion and improve the aresenic adsorption behavior of CN73s.

Table 1: Pore characteristics of synthesized CN73s at different temperature.

Samples	SSA (m^2/g)	Total pore Volume (cm^3/g)	Micro pore Volume (cm^3/g)	Meso pore Volume (cm^3/g)	Mean pore diameter (nm)
CN700	1572.3	1.1137	0.5594	0.4739	2.8332
CN800	1922.8	1.8309	0.6642	1.0734	3.8088
CN900	1496.2	1.7923	0.5284	1.1032	4.7916
CN1000	1037.3	1.1349	0.3677	0.6284	4.3762

Table 2: The changes of gibbs free energy when As complex surrounded with four water molecules on each adsorbent (Pristine CC, Pyridine N-Oxide CC, Graphitic N CC, and Pyrollic N CC). Arsenic complex with four water molecules can bind to selective N-containing sites.

Adsorbent	ΔG (eV)
Pristine CC	0.36
Pyridine N-Oxide CC	2.71
Graphitic N	-1.72
Pyrollic N	-1.95

3 CONCLUSION

In summary, nitrogen-doped carbons (CN73s) were synthesized through the chlorination of the carbo-nitride material. Both micro- and meso-pores were well developed in CN73s because unstable C-N bond, such as pyrrolic-N contained in CN73s induced pores growing. Among the CN73s, CN700 and CN800 have an extraordinary capacity for arsenic complex adsorption compared to other carbon adsorbents. The maximum arsenics adsorption values(q_m) for those materials are measured as 31.08 and 30.34 mg/g, respectively. It is worth noting that the most crucial factors for aresnic adsorption are both pore-structure and nitrogen funtional group located on the edge of pores. It is shown through the experiment and the theoretical calculation that the micro-pores which possessed the nitrogen functional goups, especially pyrrolic-N and graphitic-N, could act as acitve sites for capturing arenic ions. In addtion, the meso-pores devloped through the whole bulk of CN73s offered the kinetic path for arsenic ions and enhanced the accessibility to micro-pores located on the inner side of CN73s. To undertand the role of pore-structures and nitrogen functional group for arsenic adsorption mechanism in nitrogen-doped carbons more deeply, additional study is highly necessary.

REFERENCES

[1] GOGOTSI, Y., NIKITIN, A., YE, H., ZHOU, W., FISCHER, J. E., YI, B., FOLEY, H. C. and BARSOUM, M. W. "Nanoporous carbide-derived carbon with tunable pore size". Nat Mater, 2, 591-4. 2003

[2] PRESSER, V., MCDONOUGH, J., YEON, S.-H. and GOGOTSI, Y. "Effect of pore size on carbon dioxide sorption by carbide derived carbon". Energy & Environmental Science, 4, 3059. 2011

[3] CHMIOLA, J., YUSHIN, G., GOGOTSI, Y., PORTET, C., SIMON, P. and TABERNA, P.-L. "Anomalous increase in carbon capacitance at pore sizes less than 1 nanometer". Science, 313, 1760-1763. 2006

[4] YUSHIN, G., DASH, R., JAGIELLO, J., FISCHER, J. E. & GOGOTSI, Y. "Carbide Derived Carbons: Effect of Pore Size on Hydrogen Uptake and Heat of Adsorption". Advanced Functional Materials, 16, 2288-2293. 2006

[5] ROSE, M., KORENBLIT, Y., KOCKRICK, E., BORCHARDT, L., OSCHATZ, M., KASKEL, S. and YUSHIN, G. "Hierarchical micro- and meso-porous carbide-derived carbon as a high perfor-mance electrode material in supercapacitors". Small, 7, 1108-17. 2011

[6] QIE, L., CHEN, W. M., WANG, Z. H., SHAO, Q. G., LI, X., YUAN, L. X., HU, X. L., ZHANG, W. X. and HUANG, Y. H. "Nitrogen-doped porous carbon nanofiber webs as anodes for lithium ion batteries with a superhigh capacity and rate capability". Adv Mater, 24, 2047-50. 2012

[7] WEI, J., ZHOU, D., SUN, Z., DENG, Y., XIA, Y. and ZHAO, D. "A Controllable Synthesis of Rich Nitrogen-Doped Ordered Mesoporous Carbon for CO2Capture and Supercapacitors". Advanced Functional Materials, 23, 2322-2328. 2013

[8] SILVESTRE-ALBERO, A., RICO-FRANC S, S., RODR GUEZ-REINOSO, F., KERN, A. M., KLUMPP, M., ETZOLD, B. J. M. & SILVESTRE-ALBERO, J. "High selectivity of TiC-CDC for CO2/N2 separation". Carbon, 59, 221-228. 2013

[9] SING, K. S. "Reporting physisorption data for gas/solid systems with special reference to the determination of surface area and porosity (Recommendations 1984)". Pure and applied chemistry, 57, 603-619. 1985

[10] SEO, M. S., KIM, J. H., KIM, J. M., HAN, J. S., KANG, S., IHM, J. S. and KIM, D. O. "Tunable and selective formation of micropores and mesopores in carbide-derived carbon". Carbon, 60, 299-306. 2013

[11] INAGAKI, S., GUAN, S., FUKUSHIMA, Y., OHSUNA, T. and TERASAKI, O. "Novel mesoporous materials with a uniform distribution of organic groups and inorganic oxide in their frameworks". Journal of the American Chemical Society, 121, 9611-9614. 1999

[12] LAI, L., POTTS, J. R., ZHAN, D., WANG, L., POH, C. K., TANG, C., GONG, H., SHEN, Z., LIN, J. and RUOFF, R. S. "Exploration of the active center structure of nitrogen-doped graphene-based catalysts for oxygen reduction reaction". Energy & Environmental Science, 5, 7936. 2012

[13] PALANISELVAM, T., AIYAPPA, H. B. and KURUNGOT, S. "An efficient oxygen reduction electrocatalyst from graphene by simultaneously generating pores and nitrogen doped active sites". Journal of Materials Chemistry, 22, 23799. 2012

[14] PELS, J., KAPTEIJN, F., MOULIJN, J., ZHU,Q and THOMAS, K. "Evolution of nitrogen functionalities in carbonaceous materials during pyrolysis". Carbon, 33, 1641-1653. 1995

* Address correspondence to shinkang@snu.ac.kr

HydroFlame – A Low Carbon Intensity Process for Low-Mobility Oil Recovery

D.N.Rao and D.Deshini
HFT Resources, Inc.
8000 Innovation Park Drive, Bldg 3000
Baton Rouge, LA, USA
www.hydroflametech.com

ABSTRACT

HydroFlame is a new concept of burning a fuel directly inside a rotating stream of water. Because of the direct contact between the flame and water, very high heat transfer coefficients are achieved which render the HydroFlame devices as quite compact and highly efficient heat transfer systems. This process has several engineering applications, including hot water heaters, and compact Surface/Downhole steam generators for thermal heavy oil recovery. Steam generators developed using this unique "flame-in-water" concept can be applied either on surface or downhole at any depth, enabling the delivery of high quality steam along with exhaust gases, containing CO_2 and N_2, into the reservoir reducing greenhouse gas emissions. The injected gases rise to the top of the reservoir, forming a gas blanket that prevents heat loss to the overburden, thus improving the economics of thermal enhanced oil recovery (EOR).

Keywords: hydroflame, downhole steam generator, heavy oil, greenhouse gas, enhanced oil recovery

1 INTRODUCTION

With declining light oil discoveries (light oils are those with API gravity greater than 25°API), the vast reserves of heavy oil (oils with API gravity between 10°API and 20°API inclusive and viscosity greater than 100 cP) and bitumen (oils with API gravity less than 10°API and viscosity greater than 10,000 cP) are becoming increasingly important as a future energy source. According to the United States Geological Survey, approximately 2 trillion barrels of heavy oil and bitumen are located in reservoirs deeper than 2500 feet [1]. Both heavy oil and bitumen are too viscous to flow in their naturally occurring state. Steam injection into these unconventional reservoirs is widely applied to reduce oil viscosity in order for it to flow to a producing well.

Once-through surface steam generators (OTSGs) are currently the state-of-the art for steam generation. In OTSGs, steam is generated on the surface and injected downhole into the formation through thermal tubing/casing. Some of the limitations of OTSGs include:

1. Heat losses through transmission pipelines, long wellbores and vent flue gases can add up to 50% of total heat generated.
2. Surface generators cannot be applied economically to reservoirs deeper than 2500 ft due to heat losses during delivery and the loss of steam quality at the sandface (physical interface between the oil formation and the wellbore) [2]. Use of vacuum insulated tubing (VIT) is not a viable option to many producers due to the high costs (15times that of conventional tubing) and reliability issues [3]. Furthermore, heat losses from the couplings that connect the individual VIT joints results in disappointing performance [3].
3. At oil/steam ratios ranging from 0.3 to 0.5, surface steam generators produce 80 to 140 lbm of carbon dioxide for every barrel of heavy oil produced [4].Cosequently, the heavy oil producers are faced with difficulties in obtaining government permits to produce these heavy oil reserves.

In order to overcome these limitations, it is necessary to develop reliable, long-lasting steam generators that can operate on surface (for depths < 2500ft) as well as in downhole (for depths >2500ft). As opposed to surface-operated OTSGs, HydroFlame Surface/Downhole Steam generators can be lowered into the oil well close to sandface (the oil bearing zone) eliminating heat losses through pipelines, wellbores and vent flue gases. The flue gases, comprising mainly of carbon dioxide and nitrogen, are injected along with steam into the oil formation thereby not only reducing greenhouse gas emissions but also repressurizing the reservoir to enable longer production life.

2 HYDROFLAME PROCESS

Although the concept of Downhole Steam Generator (DHSG) has been and is still being explored by many, the real issue at the center, namely effective heat management to control the combustor wall temperature, remains to be addressed.

The past as well as present designs address the cooling of burner walls by running feed water in a sleeve or liner placed around the combustion chamber. Examples of some DHSGs that apply this principle include Sandia National Laboratory's DHSG designed and field tested for project

Deep Steam in the 1980s [5], Precision Combustion's downhole catalytic combustor steam generator [6], as well as other patented designs [7][8][9][10].

This method of external cooling results in extreme thermal stresses due to high temperature differences across the combustor walls that eventually lead to thermal degradation. One consequence of this is blocking of water passage by deformed wall preventing cooling to some areas of the combustor, and even melting of the walls. Another could be cracks in the combustor walls that results in water leaking into the combustion chamber [5]. These issues were observed during the Deep Steam project. Thus, a DHSG based on this pattern of outside cooling does not perform reliably over extended periods of operation.

The novel HydroFlame process in which flame is in direct contact with the feed water provides an innovative solution to this major problem. This is achieved by creating rotating films of the feed water so as to cover the combustor inner walls and igniting flame within the air core of the rotating body of water. This concept is illustrated in Figure 1 below. The combustor walls are exposed to only the water temperature, thus, eliminating thermal stresses that result in combustor failure.

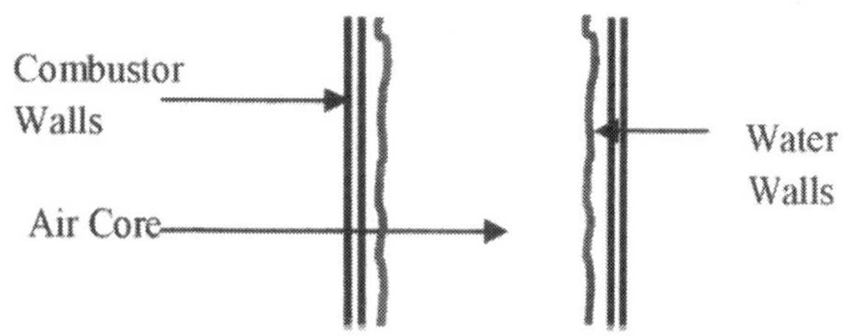

Figure 1: Illustration of air core surrounded by water walls inside combustor

The air core is formed by injecting water tangentially into the lower interior of the combustion chamber. A high intensity flame is generated in the air core by injecting air and fuel at the top of the combustor and igniting them by means of glow plugs positioned at the top of the combustor. The fuel and air mix within the combustor giving rise to a diffusion flame that is safe to operate and also is self-sustaining. The films of rotating water protect the combustor walls from the flame in the middle of the combustion chamber.

This "flame-in-water" concept of HydroFlame is the fundamental change in the way combustion has been traditionally carried out in various devices designed for generating heat energy. This innovative concept provides opportunities to develop compact DHSGs required in heavy oil and bitumen recovery operations.

3 TECHNICAL PROGRESS

As part of research &development, a 1-MMBtu/h 100 psi working prototype of the steam generator has been built and successfully tested at HydroFlame facility in Baton Rouge, Louisiana. Figure 2 shows a natural gas flame surrounded by water films in a 1-MMBtu/h steam generator. A polycarbonate tube forms the walls of the combustion chamber and is unharmed by the high intensity flame due to protection by water walls thus depicting the significance of HydroFlame process.

Figure 2: 1-MMBtu/h -100 psi steam generator in operation at HydroFlame facility in Baton Rouge

This was followed immediately with the design, development and successful testing of two 5-MMBtu/h - 500 psi steam generators. These are fully automated, skid mounted trailer units that are applicable for the surface or downhole operations. A trailer-mounted compact HydroFlame steam generator is shown in Figure 3 below. Although designed for 750 psi with 500 psi operational pressure, it was successfully operated at 600 psi in the field and has turndown ratios of 2-6 MMBtu/h depending on pressure.

Figure 3: 5-MMBtu/h – 500 psi HydroFlame steam generator on a skid-mounted trailer – (a) Control room (b) Control valves (c) Flow meters (d) 5-MMBtu/h -500psi HydroFlame Tool

4 FIELD OPERATIONS & RESULTS

The first surface field test of 5-MMBtu/h steam generator was conducted in a heavy oil field in Louisiana during February – April, 2012. Reservoir depth and oil gravity were 1100 feet, and 16.8° API, respectively. The Test well last produced in 1985 at 5 BOPD that declined to 2 BOPD and it was shut-in since then. Production attempts were made in 2006 which failed to produce any oil. When steam and flue gas were injected from the surface using the 5-MMBtu/h HydroFlame steam generator in a Cyclic Steam Simulation (CSS) process, a total of 1153 bbls of oil were produced over a few months when the well was put on production after a week of steam injection/soaking. In April – May, 2014, a second cycle of steam injection was started on the same well for one week. The well tested at 14 BOPD when placed on production on second week of May, 2014. It accumulated 1698 bbls of oil as of October, 2014 and, as per the last report from the Producing Company, it was still continuing to produce at 9 BOPD.

Figure 4: 5-MMBtu/h – 500 psi HydroFlame steam generator in operation at an oil field in Louisiana

With success of these two surface field tests, preparation for a downhole test, and research and development of higher capacity units have begun.

5 ADVANTAGES OF HYDROFLAME

The impacts and benefits of HydroFlame process steam generators compared to other conventional steam generators are:

- *Reduction in greenhouse gas emissions* - All exhaust gases from the combustion reaction are injected into the reservoir along with the steam, resulting in natural sequestration of CO_2.
- *Pressure maintenance* - Reservoir pressure declines as fluids are produced from the reservoir. As a result, the production rate declines and eventually stops when the reservoir pressure is depleted. At this point, mechanisms to improve production such as artificial lift (pumps), water injection, or gas injection are implemented. In the application of the HydroFlame DHSG, flue gases (carbon dioxide and nitrogen) are injected into the reservoir along with the steam. These flue gases rise to the top of the reservoir and form a blanket. The reduction in reservoir pressure when fluids are withdrawn is offset by the increase in pressure when these gases are injected thereby maintaining the reservoir pressure or significantly lowering the pressure decline rate. This enables higher production rates, longer production times, and improved recoveries – without the need for additional pumping equipment.
- *Smaller footprint* – Since flame is in direct contact with water, very high heat transfer coefficients are achieved which render the HydroFlame devices as quite compact and highly efficient heat transfer systems. The compact HydroFlame DHSG has a significantly smaller footprint than the presently used oil-field steam generators.
- *Reduction in heat losses and increased thermal efficiency* - Heat losses of up to 50% that occur in conventional surface steam generators in deep heavy oil applications are eliminated by the use of HydroFlame DHSG. These heat losses arise from stack gases, pipelines and wellbore losses. With surface steam generators, large insulated pipelines from the location of the steam generator to the well head are required to reduce pipeline heat losses. Building insulated pipelines adds to the capital cost of an already expensive steamproject, in addition to leaving a large footprint. Even if the HydroFlame DHSG were operated on the surface, it could be operated in close proximity to the well head, eliminating both pipeline losses and stack gas losses.

- *Production of heavy oil reserves from deep reservoirs and offshore* - The HydroFlame DHSG appears to have no depth limitation. Hence, it can be applied to reservoirs that are too deep for surface steam injection (typically, reservoirs located at depths greater than 2500ft). In addition, conventional steam generators cannot be applied offshore due to lack of space for equipment. However, this limitation would not be an issue for the HydroFlame DHSG, due to its compact design and high efficiency and low space requirement for compression equipment.
- *Reduced water treatment* - Because the transfer of heat between the hot combustion gases and the water is direct in the HydroFlame DHSG, and because no heat conduction materials are required, water of lower quality can be used, compared to that used in surface steam generators. In situations where scaling ions (carbonates and sulfates) are present in the produced water in high concentrations, scale inhibitors can be injected with the water to prevent scaling. Less stringent water quality requirements mitigate the need for expensive water treatment.

REFERENCES

[1] R.F. Meyer, E.D. Attanasi and P.A. Freeman, "Heavy Oil and Natural Bitumen Resources in Geological Basins of the World", U.S. Geological Survey, Reston, Virginia. Open File-Report 2007-1084.

[2] Steam Injection Techniques for Deep Reservoirs", SPE California Regional Meeting, 18 – 20 April 1979, Ventura, California

[3] L. Capper, M. Kuhlman, G. Vassilellis, M.J. Schneider, N. Fitzpatrick, "Advancing Thermal and Carbon Dioxide Recovery Methods beyond their Conventional Limits: Downhole Innovation", SPE Heavy Oil Conference and Exhibition, Kuwait City, Kuwait, 12-14, December 2011.

[4] T. Kovscek, "Heavy Oil, Overview", JPT Online, Vol. 58, No. 4, April 2006.

[5] B. Marshall, "Field Test of Two High-Pressure, Direct-Contact Downhole Steam Generators", Volume I. Air/Diesel System. Enhanced Oil Recovery Division 9755. Albuquerque, New Mexico: Sandia National Laboratories, May 1983.

[6] Precision Combustion Inc. "Downhole Oxyfuel Steam/CO_2 Generator for Production of Gas from Hydrates", NETL, http://www.netl.doe.gov/technologies/oil-gas/FutureSupply/MethaneHydrates/projects/DOEProjects/ER85773_PrecisionCombustion.html

[7] A.G. Castrogiovanni, R.T. Voland, C.H. Ware, B.A. Folsom and M.C. Johnson, "A Downhole Steam Generator and Method of Use", Patent international application number: PCT/US2011/027398, 7 March 2011.

[8] B.A. Donaldson and D.E.Hoke,"Downhole Steam Generator with Improved Preheating/Cooling Features", Patent 4,411,618, 25 October, 1983.

[9] S. Eisenhawer, A.J. Mulac, B.A. Donaldson and R.L. Fox, "Steam Generator Having a High Pressure Combustor with Controlled Thermal and Mechanical Stresses and Utilizing Pyrophoric Ignition", Patent 4,648,835, 10 March, 1987.

[10] W.R. Wagner, D.E. Wright and R.L Binsley, "Direct Firing Downhole Steam Generator", Patent 4,336,839, 29 June, 1982.

Advanced oxidation processes for water purification: compact vs nanoporous anodic TiO_2

V. Piazza[1], M.V. Diamanti[1,*], A. Mazare[2], MP. Pedeferri[1], P. Schmuki[2]

[1]Politecnico di Milano, Department of Chemistry, Materials and Chemical Engineering "Giulio Natta", Via Mancinelli 7, 20131 Milan, Italy

[2] Department of Materials Science WW4-LKO, University of Erlangen-Nuremberg, Martensstrasse 7, 91058 Erlangen, Germany

* Corresponding author. Tel: +39223993137, Fax: +39223993180, E-mail: mariavittoria.diamanti@polimi.it

ABSTRACT

The work herein presented is dedicated to a comparison of photodegradation properties of anodic TiO_2 films, involving the study of both oxides with nanotubular geometry and compact layers. The comparison has been performed both through characterization techniques (SEM, XRD) in order to analyze the influence of anodic oxidation parameters on the oxide structures, morphology and crystal structure, and by evaluating photocatalytic degradation of a dye used in textile industry, known as Acid Orange 7 (AO7).

Different electrolytes and process parameters were used for the fabrication of both compact oxides and nanotubular layers in order to point out the effect of specific anodic parameters on the photodegradation rate.

The characterization of these structures aims to a comparative analysis – between different structures and between similar structures with different features – of photoactivity and related phenomena, which are strongly dependent on the electrolyte used during anodizing, as it influences both morphological features and crystallinity degree.

Keywords: *Anodizing, TiO_2, Nanotubes, Compact Layers, Characterization, Photodegradation*

1 INTRODUCTION

In the last years, research on the engineering of titanium dioxide has been oriented to the obtaining of compact thin films as well as of more complex nanostructures and architectures, for applications ranging from corrosion resistance to bioengineering, from optical devices to bipolar resistive switching memories, from electrochromic devices to environmental and energetic applications, such as advanced oxidation processes for water purification and dye-sensitized solar cells [1,2]. In this scenario, anodic oxidation can be a very effective technique in the controlled production of oxide films with high bond strength on the metal, tuning growth kinetics and oxide properties as a function of process parameters. In fact, it is possible to relate the controlling electrochemical parameters (cell voltage, chemical composition of the electrolytic solution, process time) to the thickness, morphology, chemical and structural composition of the growing oxide layer [1,3-5].

In all applications based on exploiting its photoactivated properties, two main features drive titanium dioxide photoefficiency: crystallinity and surface area. The former characteristic is generally achieved either by high voltage anodizing (anodic spark deposition, ASD) where TiO_2 nanocrystals with anatase structure are generated in an otherwise amorphous oxide on account of the onset of localized microplasma surface states; on the other hand, also completely amorphous oxides can be partially converted into crystalline ones by annealing. This is very important when considering the second feature, that is surface area: nanotubular oxides have enhanced surface area but amorphous structure, and annealing enables their successful use in photoactivity-related applications [1,6,7].

Still, few works address a direct comparison between compact oxides of nanoscale thickness, and nanotubular ones. Here we present a comparison among differently obtained anodic titanium oxides, showing either compact or nanoporous morphology, in order to define the most effective and robust treatments for the production of photoactive layers.

Photoactivity of the layers was investigated in terms of photodegradation of an organic dye, (AO7), an azo compound largely used in textile industry. Research in this field has increased in time in order to reduce the content of these compounds in wastewaters, which represent the main source of pollution in this industry. [8,9]. TiO_2 has been identified as an ideal photoactivated catalyst, due to its ability to decompose –N=N– bonds in azo compounds by formation of OH• species or by direct transfer of holes to the adsorbed molecules [10]. Different strategies were developed to improve photocatalytic activity of anodic TiO_2 layers [11,12]; however, a comparison of different nanostructures can help detecting oxide features to be controlled in order to improve photocatalytic performances.

2 EXPERIMENTAL

Several TiO_2 samples were produced during this work in order to investigate the behavior under illumination of oxides with different features. Anodic compact oxides and nanotubular layers were grown on ADVENT Ti foil, (purity 99.6%, max. impurities content: O= 0.15%, N=0.012%, H=0.005%, Fe= 0.15%, C= 0.02%. Foil thickness = 0.1 mm). All the samples were cut in squares with 1.5x1.5 cm^2 lateral dimensions. After that, dust and surface contaminants were removed by two steps of ultrasonic action, respectively in acetone and in ethanol, rinsing with de-ionized water and then the metallic substrate was dried in N_2 stream. For the production of compact oxides, anodizing was carried out at different voltages in potentiodynamic mode with a sweeping rate of 100 mV/s up to the final voltage, which was maintained, potentiostatically, for 15 minutes. A potentiostat PGU 100V-2A, with an output voltage of 140V, was used. As final voltages 20 V, 50 V and 90 V were investigated. A platinum electrode was used as a cathode while the titanium substrate was anodically polarized by means of a Cu electrode connected to the positive plug of the potentiostat. The chosen electrolyte was 1 M H_2SO_4. After anodic process, samples were rinsed with water and dried in N_2 flow. For the production of nanotubular layers, anodic oxidation was carried out potentiostatically once again with a two-electrode setup where a Pt electrode acted as counter-electrode. The titanium foil was anodized in Ethylene Glycol (EG) +0.1 M NH_4F +1 M H_2O at 35 V for 120 minutes. Then the sample was treated with ultrasonic steps in pure H_2O in order to remove the formed nanotubular layer, followed by cleaning in acetone, ethanol. After being rinsed with water and dried in N_2 flow, the pretreated substrate was used to grow the final oxide layer. For the last step, several electrolytes were used, and depending on the electrolyte, the applied voltage and anodizing time was modified. In Tab.1, the samples investigated in this work are described.

As for morphological and microstructural characterization, a field emission scanning electron microscope (FE-SEM, Hitachi S4800) and a Philips X'Pert MPD PW3040 XRD instrument were used.

Table 1 List of investigated samples

Name	Electrolyte	Sweeping rate	Applied Voltage	Anodizing time
S1	1 M H_2SO_4	100 mV/s	20 V	15 min.
S2			50 V	15 min.
S3			90 V	15 min.
EG-NH_4F	EG+0.1 M NH_4F+1 M H_2O	-	60 V	10 min.
EG-HF 8M	EG+0.2 M HF + 8 M H_2O	-	58 V	120 min.
EG-HF 12M	EG+0.2 M HF + 12 M H_2O	-	48 V	120 min.

Photodegradation tests were performed in a solution of 2.5 x 10^{-5} M of AO7 ($C_{16}H_{11}N_2O_4SNa$, Acros Organics, Belgium) with a measured pH of 5. Samples were immersed in 5 ml of the solution in a quartz cuvette and illuminated by 200 mW He-Cd LASER at 325 nm, Kimmon Electric IK3552R-G with a measured power of 60 $mWcm^{-2}$.and an expanded beam to 1 cm^2. The solution was stirred at 300 RPM during the test. Compact oxides were kept in the solution under illumination for 120 minutes, while the nanotubular layers for 160 minutes. In both cases, the concentration was measured every 20 minutes. The variation in concentration was performed by an UV/Vis machine Perkin Elmer Lambda XLS+ and evaluated by the variation of the orange peak at 485 nm in time. Kinetics of degradation was evaluated by plotting the results as $\ln(C/C_0)$ vs time.

3 RESULTS AND DISCUSSION

As shown in Fig.1, representing top morphology of compact oxides S1, S2 and S3 reported in Tab.1 and obtained by anodic oxidation at 20 V, 50 V and 90 V, respectively, SEM images of compact oxides showed occurrence of dielectric breakdown when oxides were produced in 1 M H_2SO_4 at 90 V. This breakdown onset can be observed in the formation of craters in the otherwise smooth oxide, due to sparking which is typically associated to breakdown. No relevant morphological features were noticed on the low voltage compact layers.

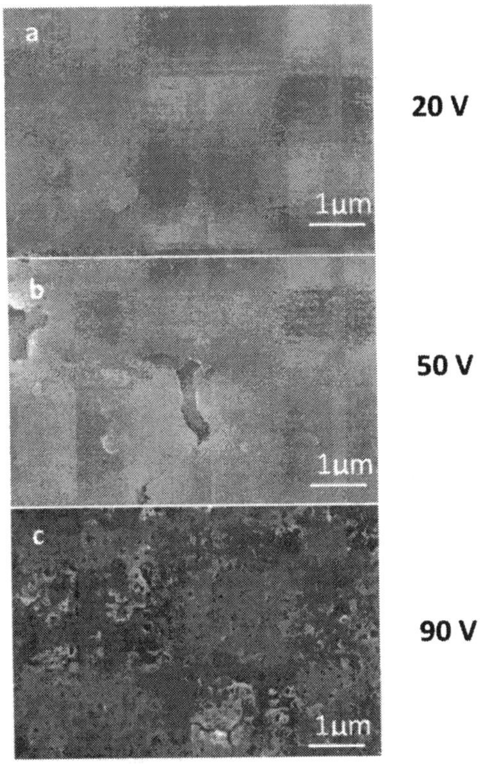

Figure 1: Top morphology of oxides grown in 1 M H_2SO_4. Final applied voltages are a) 20 V, b) 50 V, c) 90 V.

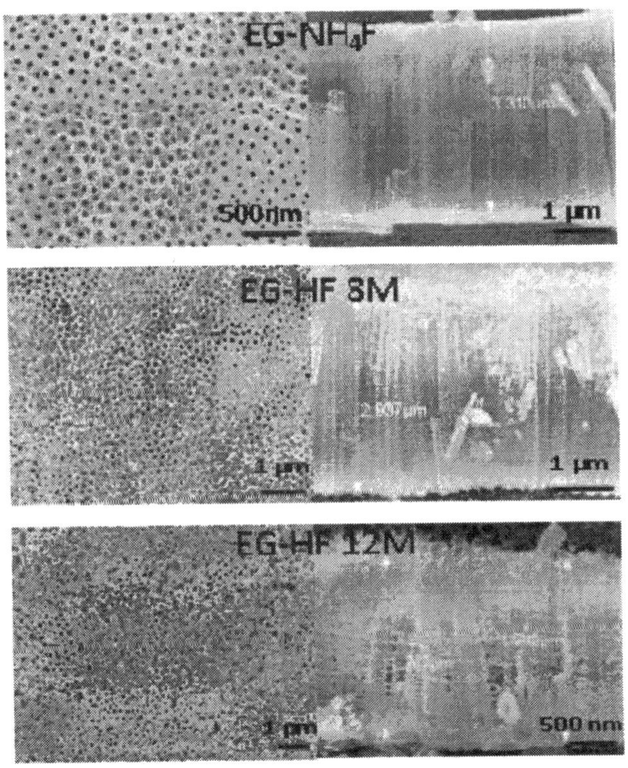

Figure 2: SEM images of nanotubular layers.

Nanotubular layers obtained in the described condition showed different morphologies, dependent on the imposed conditions. Nanotubes grown in NH_4F containing electrolyte are characterized by an initiation layer on top, conversely to oxides layers grown in HF containing electrolytes which show an open tube morphology. As it can be noticed in Fig. 2, the layer thickness ranges between 1.7 and 3.3 µm for the different nanotubular layers. All nanotubes have diameters of ≈95nm, however EG-NH_4F nanotubes are covered with an initiation layer that limits the openings at the top to ≈50nm.

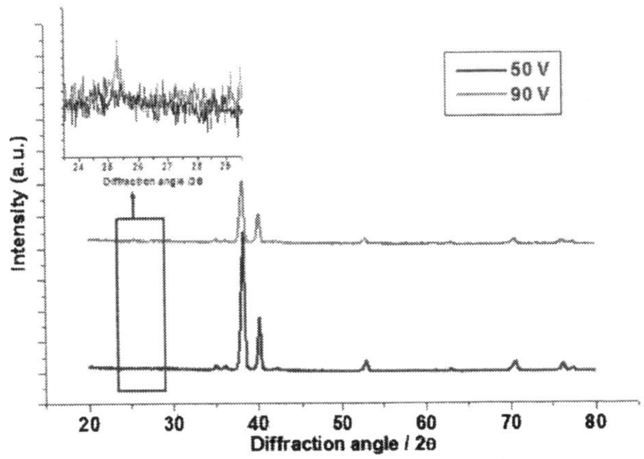

Figure 3: XRD spectra of compact oxides, with a magnification of the diffraction angle range between 23.5° and 29.5° in the inset.

While for nanotubular layers XRD measurements showed no crystalline peaks (amorphous), for compact oxides anatase peaks were recorded for the sample where breakdown occurs (90 V) (Fig 3, in the inset a magnification on the main anatase peak at 25.4°).

Fig. 4 reports the observed values of dye absorbance with irradiation time in the presence of different TiO_2 photocatalysts. In all cases, the dye concentration – which is related to its absorbance by Beer Lambert relationship – decreases under UV irradiation, to different extents depending on the oxide morphology and crystal structure.

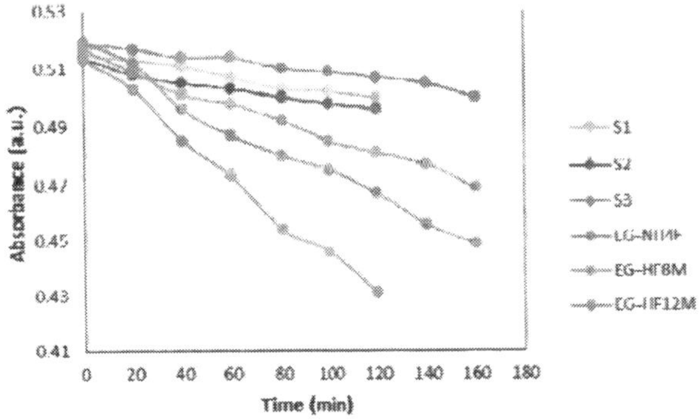

Figure 4: Absorbance of AO7 dye vs irradiation time in presence of different anodic oxides.

Furthermore, a comparison of the kinetics of degradation trends in the photodegradation tests gives interesting results. From Fig.5, in fact, it is possible to notice that concerning compact oxides, even the presence of small amounts of anatase crystals has a fundamental role in improving photocatalytic properties, enhancing photodegradation rate up to almost 10 times with respect to amorphous compact oxides.

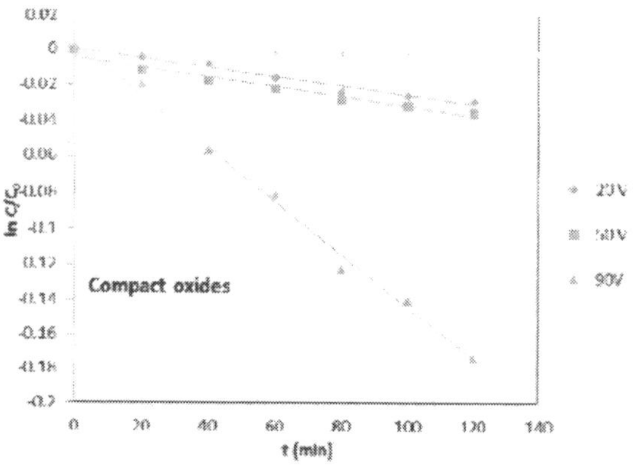

Figure 5: Photodegradation kinetics of AO7 observed in presence of compact titanium oxides.

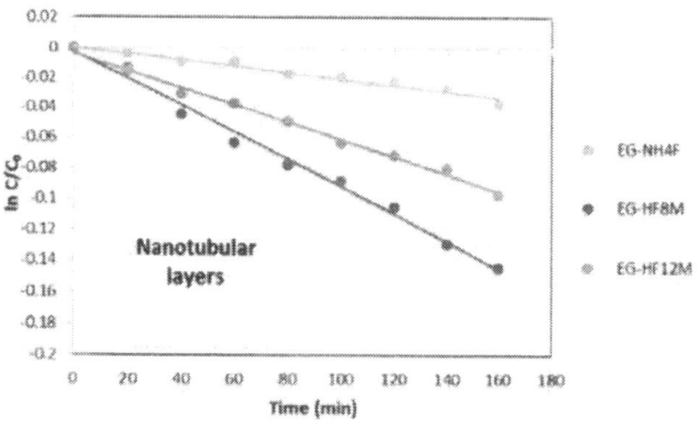

Figure 6: Photodegradation kinetics of AO7 observed in presence of TiO_2 nanotubular layers.

Dye decomposition tests on TiO_2 nanotubes showed that another fundamental factor to be optimized for enhancing photocatalytic activity is the high aspect ratio morphology. In fact, it is possible to notice that nanotubular layers characterized by the presence of an initiation layer present a lower photodegradation rate compared to structures with open tube top morphology, whose efficiency improves with nanotube length, and therefore with available surface area.

From numerical evaluation of photodegradation rates, reported in Tab.2, it can be noticed that sample EG-NH_4F has an efficiency comparable to amorphous compact oxides, while longer nanotubes, even at an amorphous state, allow a faster disappearance of the dye, approaching the values presented by the partially crystalline compact oxide.

Table 2 Photodegradation rates.

Name	Photodegradation rate (K) [h^{-1}]
S1	0.016
S2	0.017
S3	0.089
EG-NH_4F	0.013
EG-HF 8M	0.054
EG-HF 12M	0.035

4 CONCLUSIONS

Results presented here can be considered as a first step in the analysis of factors able to enhance TiO_2 photocatalytic activity. Data clearly show the positive effect of dielectric breakdown in improving performances of low aspect ratio structures, related to the presence of anatase crystals, which wasn't detected in any other oxide film. Moreover, nanotubular layers can present different photodegradation rates, not only depending on the layer thickness: more generally, performances are related to oxide morphology. It is clear that the presence of an initiation layer creates a physical barrier for solution intercalation, and thus photodegradation rate is very similar to that measured on amorphous compact oxides. On the other hand, if solutions can penetrate inside nanotubes, an increase in efficiency can be noticed dependent on the nanotubes length.

REFERENCES

[1] Lee, K.; Mazare, A.; Schmuki, P, *Chem. Rev.* **2014**, *114*, 9385-9454

[2] Grätzel, M., *Nature* **2001**, *414*, 338-344.

[3] Diamanti, M.V.; Del Curto, B.; Pedeferri, MP., *J. Appl. Biomater. Biomech.* **2011**, *9*, 55-69.

[4] Diamanti, M.V. and Pedeferri, MP, *Corrosion Science*, **2007**, 49, 939-948.

[5] Matykina, E.; Montuori, M.; Gough, J.; Monfort, F.; Berkani, A.; Skeldon, P.; Thompson, G.E.; Habazaki, H., *Trans. Inst. Met. Finishing*, **2006**, *84*, 125-133.

[6] Diamanti, M.V.; Ormellese, M.; Marin, E.; Lanzutti, A.; Mele, A.; Pedeferri, M.P., *J. Hazard. Mater.* **2011**, *186*, 2103-2109.

[7] Ohtsu, N.; Komiya, S.; Kodama, K., *Thin Solid Films* **2013**, *534*, 70-75.

[8] Chen Y., Wang K. and Lou L., *J. Photochem. and Photobio. A: Chem*, **2004**,163, 281-287.

[9] Tanaka K., Padermole K. and Hisanaga T., *Water Research*, **2000**, 34, 1, 327-33.

[10] Kostantinou I. K. and Albanis T. A., *Applied Catalysis B: Environmental*, **2004**, 49, 1-14

[11] Macak J. M., Zlamal M., Krysa J. and Schmuki P., *Small*, **2007**, 3, 2, 300-304.

[12] Paramasivam I., Macak J. and Schmuki P., *Electrochem. Comm.*, **2008**, 10, 71-75.

River, Lake and Ocean Water Cooling

J. Peterson*

*Hewlett Packard
6712 Allegheny Ave., Takoma Park, MD, USA, jpeterson@hp.com

ABSTRACT

Using water directly from rivers, lakes, and oceans is becoming an option for end users to save energy, avoid ozone depleting refrigerants, and save space. The use of water from rivers, lakes, and oceans is not a new concept, but the lower energy use, system simplicity, and lower capital and operational costs have proven to be much more viable than in the past. Already industrial facilities and power plants have been harnessing water cooling from these sources and now many more, smaller projects can also achieve these benefits. In the past many groups would attempt to work together to have aggregated cooling loads large enough for the installation of a cooling system. However, with the advances of pumping, improvements in reliability and experience in the industry many more existing and new facilities have the option to compare river, lake or ocean water cooling.

The first system in the U.S. at Ithica, NY, which serves about 51 megawatts of cooling starting in 2000, will be presented for lake water use. A data center in Hamina, Finland will be reviewed for using seawater. The potential efficiencies, comparisons of savings of equipment and other advantages and disadvantages are also illustrated. For data centers and other critical facilities, the additional requirement to meet higher reliability standards will also be reviewed.

Keywords: river, lake, ocean, efficiency, cooling

1 WATER SOURCES

For most heat transfer applications water is the best medium for moving and rejecting heat. Water can move more heat per unit (gallon or liter) than another other liquid, and it is far more efficient than air systems; because of this chilled water is often used to pull heat from airstreams and water-cooled equipment. Water is most dense at 3.98 degrees C, which is the temperature most often found for deep lakes. Since about 3% of the water on earth is fresh water, systems are being designed to operate with salt water and other non-potable water sources.

Rivers have long been utilized to support power plant operations. The water temperature of rivers tends to vary the most and they are also more closely regulated upstream, downstream, and across borders of nations.

Lakes have a more constant temperature according to the depth available, as the deep water within a lake is less effected by weather and other events.

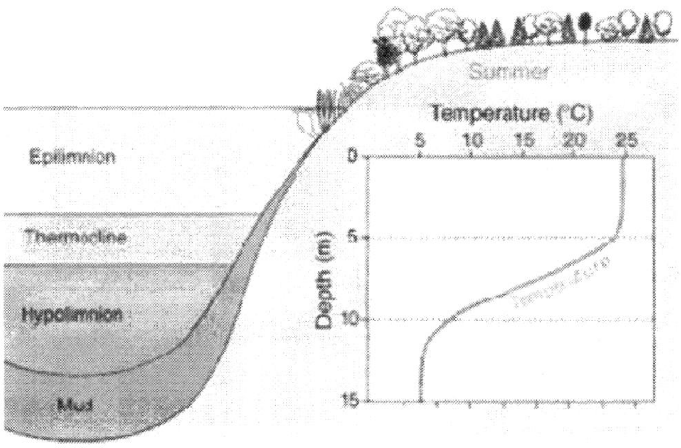

Figure 1: Thermoclines for a typical lake in North America with summer gradients.

Ocean water, like lake water, has wide fluctuations of temperatures at the surface. However, just like a lake, the ocean water temperature approaches a more constant year-round temperature that can be used for cooling. Even in the most tropical locations ocean water can be reliable to achieve 15 degrees C at 500 meters below the surface and about 5 degrees C at about 1,000 meters.

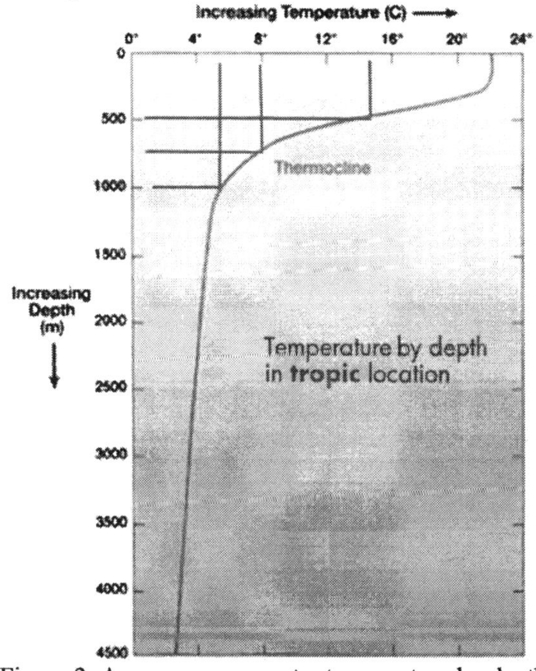

Figure 2: Average ocean water temperature by depth for most tropical locations.

2 COOLING SYSTEMS COMPARISONS

2.1 Typical Cooling Systems

A typical cooling plant is composed of several pieces of expensive and sometimes sensitive equipment. The chilled water removes the heat from the load then uses a chiller to move the heat from chilled water system to the condenser water system. A cooling tower then rejects the heat to the atmosphere. Each water system requires pumps to move the water.

These systems require specialized care and at times complex operational and maintenance procedures. Based on the type of cooling refrigerants, there are hazards related to the personnel and the environment as well as standards and regulations on the minimum equipment performance, the type of accepable refrigerant per country, and other safety considerations for having and operating the equipment. Most of these existing systems also rely on evaporative cooling as the method to reject heat to the atmosphere, and the chemicals and water use has come under more scrutiny in the last few years.

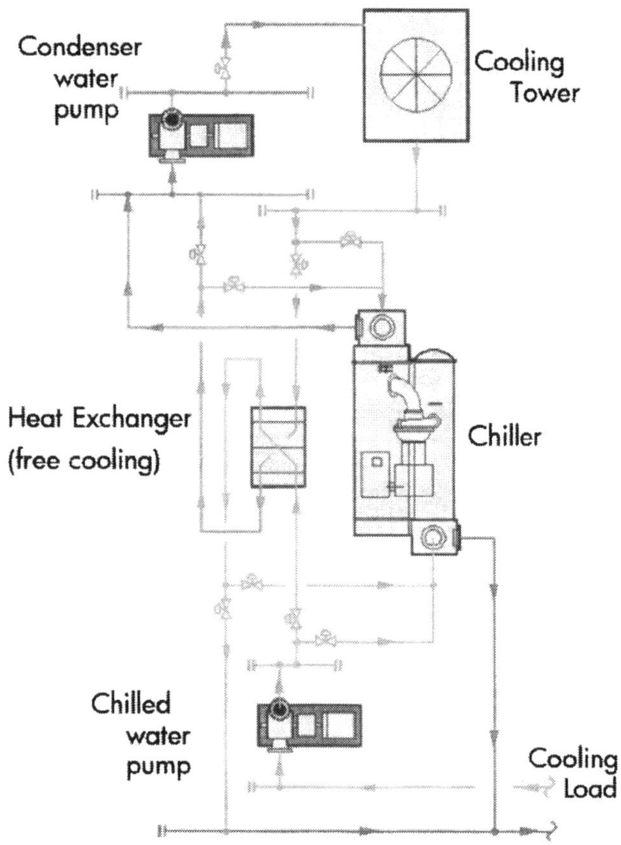

Figure 3: Typical chilled water plant with heat rejection to the atmosphere after several steps of moving heat.

Redundant components are installed to prevent loss of cooling during equipment failures. Since there are many operational components and complex modes of operation, loss of cooling may occur more frequently with inexperienced personnel or poorly maintained equipment.

2.2 River, Lake and Ocean Cooling System

A system utilizing a river, lake or ocean for heat rejection can be simplified by reducing or removing the refrigeration equipment.

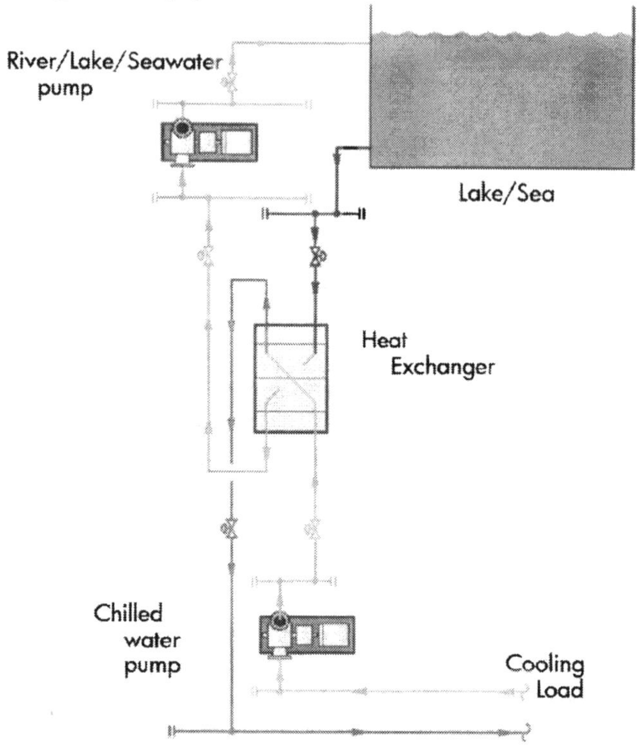

Figure 4: Cooling plant using River/Lake/Seawater for heat rejection.

Combinations can also be used, with the heat sink replacing the cooling tower only or by providing a reduction in size or number of chillers and/or cooling towers.

2.3 Efficiency Comparisons

A high-efficiency, optimized chilled water cooling plant with all-variable speed drives and other energy saving measures will have an operational coeffecient of performance of approximately 6.0, which equates to about 0.6 kW per ton of cooling. This optimized cooling plant might have chillers operating at 0.4 kW per ton and the cooling towers at around 0.1, with the remaining energy of 0.1 kW per ton needed for the pumps. By comparison, the river/lake/seawater cooling system will not need the chillers or cooling towers. However, there may be more pump energy required as factors of location, height, water depth, and more should be considered. Even if the pump energy were doubled, the energy use compared to an excellent chiller plant is about one third. For a megawatt (MW) of cooling required the amount of savings would be about 0.135 MW; over the course of a year this would be over 1 million kWh. Additionally, since the river/lake/seawater system does not require potable water for heat rejection, over a million gallons of fresh water per MW of cooling load would be saved.

Efficiency Comparison Summary		
Equipment	**Traditional Chilled Water kW/ton**	**River, Lake, Ocean Cooling kW/ton**
Chiller	0.40 - 0.54	-
Chilled Water Pumps	0.05 - 0.08	0.05 - 0.08
Cooling Tower	0.08 - 0.12	-
Condenser Water Pumps	0.07 - 0.11	-
River/ Lake/ Ocean Pumps	-	0.08 - 0.13
Total	**0.6 - 0.85**	**0.13 - 0.21**
Minimum Difference	**0.39**	

Table 1: Comparison of traditional chilled water systems and river, lake and ocean water cooling

2.4 Relability Comparisons

The risk for river, lake and ocean water cooling systems from storms, watercraft, and biological sources (fish, shellfish, etc.) should be noted and explored based on the location. Although pumps can be placed in parallel, the piping for a river, lake or ocean water cooling is often not redundant. Due to this, the piping at the shore line can be buried or otherwise hidden since this is the most vulnerable section.

While typical mechanical plants are more flexible for location, they also have vulnerabilities due to having more equipment that can fail without a rigorous maintenance program. For high-availability needs, this can be an issue if personnel are inexperienced or not trained in the details of the system such as refrigerant types, condenser water treatment chemicals,

3 EXISTING SYSTEMS

There are many projects around the globe where river, lake and ocean water is used for purposes beyond industrial, and many of these locations are where electricity, fuel and water are not abundant. Among the best locations are islands, and such projects are either in design or underway, such as Mauritius and Honolulu, HI, USA, or complete, such as the HSBC Building, Hong Kong and the more detailed examples below.

3.1 Cornell Lake Source Cooling Project

Cornell University in Ithica, NY, underwent the first lake source cooling project in the USA. The system began operation in 2000 and supports 50 MW of cooling load for the university as well as other buildings in the city. The cost was about US$59 million and offset the cost of 6 new chiller plants on the campus. By comparison with the old chiller plant systems the average energy savings is about 86%.

The lake source cooling system draws from a depth of 76m to get water at 3.8 to 5.0 degrees C. This water is then cycled through a heat exchanger near the shore of the lake then back out to the lake at a depth where the outlet temperature matches the average stratified lake temperature.

The cooling water system for the university is a closed loop and therefore needs less water treatment. Pumps supporting the water flow on each side of the heat exchanger operate as needed to ensure adequate cooling is available for the university year-round.

Before, during, and after the lake source cooling project Cornell University studied the environmental impact of the project as well as the impact of two other nearby waste water treatment plants. The annual reports for over 14 years show that the weather and tributaries had the most impact on the temperature, phosphorus, fluorescence and other measured items. In addition, the waste water treatment plants had larger attributed impacts to the lake than the lake source cooling project.

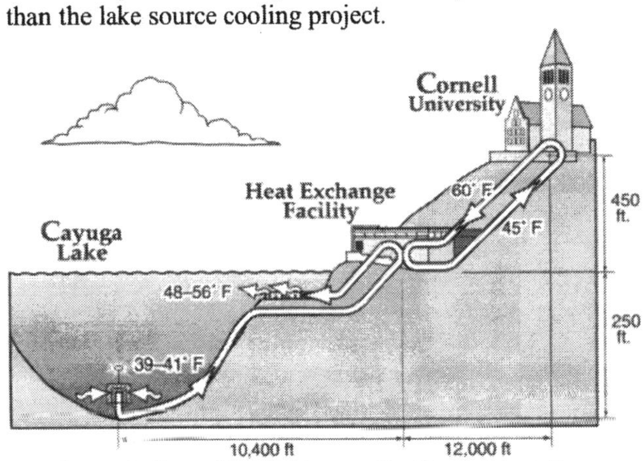

Figure 5: Cornell Lake Source Cooling project diagram

3.2 Ocean Cooled Data Center

In 2009 Google purchased a former paper plant in Hamina, Finland originally built in 1954 and converted it into a data center. The original building used a sea water tunnel 450 meters long for cooling and this was reused for the data center. Since the location is on the Gulf of Finland, the water temperatures are consistently cold at 1 to 3 degrees C; the tunnel is far enough below the surface to avoid any issues of freezing. The building went operational in May, 2010 and has been touted as one of their most energy efficient data centers.

In addition to the data center being able to use a low temperature, the discharge back to the Gulf of Finland is tempered with fresh seawater to discharge water back into the gulf at temperature closer to the the intake to minimize the environmental impact.

Figure 6: Inside the data center cooling plant rows of pumps flank their associated heat exchanger.

4 WATER SOURCE RISKS

Rivers, lakes and oceans all have changes that happen over time. While a traditional chiller plant has equipment failures, the piping for a river, lake or ocean cooling sytem may endure stress in the future as the supports beneath change with the underwater landscape.

In each country government agencies exist to oversee the installation and operation of the projects. In many ways these agencies can be similar to that of a utility such as electricity or potable water supplies. Since there are a large number of industrial sites throughout the world the agencies have a wide range of experience and will work with the design. Since a river, lake or ocean water project is only rejecting heat and not altering the water quality most authorities will allow a project to move forward.

For rivers, risks of sandbars, century-low river flows and other dangers should be examined to determine whether mitigation strategies would be warranted. Although they occur less often than earthquakes, tornadoes or other disasters, the hazards should be examined thoroughly.

Lakes pose less risk than rivers since the water flow, temperature, and shifting from flooding is much less likely.

Ocean water cooling systems have a few other issues to condend with, such as continental shelf and slope shifting on downward or lateral directions. Also underground tunneling may be considered to avoid wave action and disturbance by others.

Lastly, the risk of pipe failure, although similar in possibility to failures of a traditional piping systems, may require additional time and resources to location and fix or replace a failure. However, if the failure occurs underwater, the river, lake or ocean water system may continue to operate as normal and wait for a planned outage for repairs. When operating a data center, other means of redundancy may be required.

5 CONCLUSION

As depicted above, the efficiency gains of river, lake and ocean cooling systems are significantly large enough to examine the possibility replacing traditional cooling systems. There are many details to be examined, but with projected electrical price increases and grid reliability concerns, leveraging the local resources to save energy, water and cost is becoming an easily viable option for most new and upgrade projects.

REFERENCES

[1] ASHRAE Technical Committee 9.09, "Thermal Guidelines for Liquid Cooled Data Processing Environments," ASHRAE, 2011.

[2] ASHRAE Technical Committee 9.09, "Liquid Cooling Datacom Equipment Centers," ASHRAE, 2011.

[3] Environmental Protection Agency, "Water Use Classifications and Water Quality Standards for Surface Waters,"

[4] Environmental Protection Agency, "Clean Water Act, section 316(b): Cooling Water Intake Structures," EPA, 2014

[5] Enwave Energy Corporation, "District Cooling System," Enwave Services, 2013

Author, "Title (optional)," Journal, Volume, Pages, Year.

Investigating Environmental Weathering of CNF/CNT/ Thermoplastic Polymer Synthesized using Additive Manufacturing

E. Sahle-Demessie*, Amy Zhao[1], Chady Stephen**, Heidi Grecsek**, Nicholas A. Gagliardi, Nicholas***

*U.S. Environmental Protection Agency, Office of Research and Development, National Risk Management Laboratory, Cincinnati, OH 45268
**Perkin Elmer, Inc. Woodbridge, Ontario, Canada
*** University of Dayton Research Institute, UD, Dayton, OH

ABSTRACT

The stability of polymeric nanocomposites during the use phase and at the end-of-their useful life is not fully understood. The potential release of nanomaterials from polymer matrices becomes a concern as a high volume application of additive manufacturing is coming real. The degradation polymeric nanocomposites depends on the properties of the polymer and aging conditions. It is critically important to understand the relationship between the inherent characteristics of nanocomposite based consumer products and the likelihood of the release of nanomaterials throughout the life-cycle of the product. This research work is focused on weathering of additive manufacturing nanocomposites that contain different weight percentages of carbon nanofiber and carbon nanotubes. The weathering study was conducted using composites containing multi-walled carbon nanotubes (0 to 5 wt%) and carbon nanofiber (0 to 5 wt%). Various analytical techniques have been used for the characterization of structural changes and the potential for release of nanomaterials (NMs) from the composite matrix. Structural changes at selected times of the aging process were measured using thermogravimatric analysis and differential scanning calorimetry. These results are compared with weathered-nanocomposites that were manufactured by the conventional melt mixing techniques. Weathered materials were also characterized using FTIR, SEM and TEM and XRD, XPS, and AFM, to measure changes with increase aging. The release of CNT/CNF was analyzed using Single Particle- ICP-MS to detect metal catalysts releases that were used for the synthesis of CNF and CNT. The results of this study are essential to develop environmental risk assessment of additive manufactured materials over their life cycles.

Keywords: polymer weathering, additive manufacturing, nanorelease,

INTRODUCTION

Over the past decades the production and use of nanocomposite materials has been increasing rapidly. Many of these composites are used for outdoor applications, for example in automobiles, greenhouses, and for ropes. To meet these requirements, new polymers can be developed, or the current polymers can be modified to improve their properties such as mechanical stability, heat stability, barrier for water and air, fire retardant, optical and electrical properties. The incorporation of nano-fillers into crystalline or glassy state of the bulk samples will lead to significant change in the structure as well as morphology of the samples. At the same time, the mobility of the nano-fillers is much higher than that of the conventional composite fillers because of their comparable time scale of motion with the polymer chains. All of these factors contributed to a different energy dissipation mechanism of the bulk nanocomposites. However, it is found that the mobility of the polymer is a precondition for this mechanism to be effective and the increase in toughness in the presence of nano-particles scales with the increase in the mobility of the polymer chains. It is also why lots of samples functioning at their viscoelastic region usually would have more significant boost in properties.

Nanocomposites are materials made from a phases where one of more phases are at nanoscale. Unlike most engineering structure materials, the mechanical properties of micro-composites could vary as the direction along which measure changes that is they are isotropic. In order to change in chemical and physical properties of nanocomposites thermal analysis serves as a gross material characterization technique to determine chemical and structural changes of nano-composites.

Additive manufacturing (*aka*, 3D printing) is an emerging technology for making a new class of multifunctional nanocomposites for many niche items such as medical implants, and a wide range of industrial designed components. Additive manufacturing technology allows printing complex 3D objects layer-by-layer enabling nanomaterials in new ways toward greater control over material properties across three dimensions and could further extend capabilities of nanocomposites

Energy, material use, and emissions during the manufacturing phase of composite materials are much higher than the amounts from the conventional materials. However, nanocomposite materials lighter and require much lower energy during the use phase. Life cycle risk assessment of nano composites includes the potential release of nanomaterials during manufacturing, use and end-of-life management (Figure 1). Polymers and nano-composites are increasingly used for structural applications such as boat and air craft construction, sporting goods and agricultural uses. There is often

The rate of deterioration of materials subjected to weathering depends on the nature of the materials. The concern regarding the durability of nano composites is partly because their relative newness but also because the changes during their useful lifetime and the potential impact of at the of their use. The objective of this study whether the deterioration of nanocompostite could release imbedded nanomaterials from the polymer matrix. Although there is a large volume of work on the hazard of nanomaterials, there are very limited studies on potential release of nanomaterials from composite matrix.

EXPERIMENTAL

Materials

The base polymer for this study is a ULTEM™ 1000, which is an amorphous, transparent polyetherimide (PEI) plastic offering a glass transition temperature (Tg) of 217°C. The high Tg makes the polymer to have high stability of physical and mechanical properties at elevated temperatures. PEI resins are used in medical and chemical instrumentation due to their heat resistance, solvent resistance and flame resistance. The composite samples were prepared using additive manufacturing (or 3D printing) technique with different amounts of CNT and CNF (Table 1).

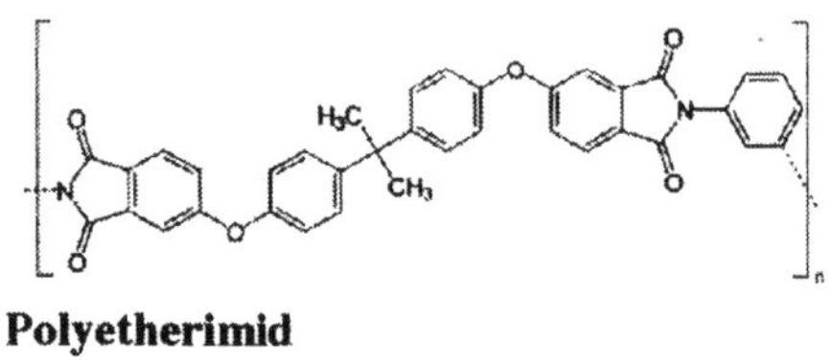

Polyetherimid

Type	Wt CNF %	Wt CNT %	Wt Polymer %
A	5	1	94
B	5	2	93
C	5	3	92
D	5	4	91
E	5	5	90
F	0	10	90
E	10	0	90

Table 1: Types of Polyetherimid CNF/CNT composite tested

Accelerated aging of nano-composite

The flow of O_2 from the pressurized tank was controlled using mass flow controller (Sierra Instruments). The ozone generator was fed with up to 1L/min pure oxygen, and it can generate up to 2.5 g/hr of ozone in the oxygen stream. A tube furnace was provided to control the temperature of the reactor. The concentration of ozone was measured following the ozone generator with an ozone analyzer (Ozone Solutions). The concentrations of O_3 was analyzer were determined by the manufacturer's calibration for the ozone sensor. The gas flow to an ozone generator that has a capacity of up to 5 wt% conversion. Ozone-O2 flow was bubbled through water before flowing into the tube furnace (Carbolite).

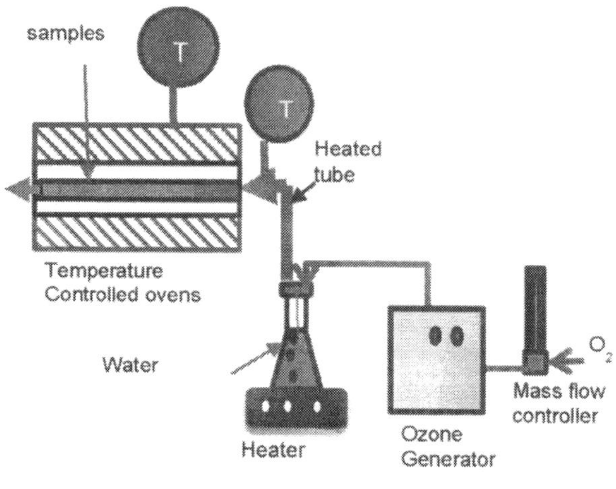

Figure 1: Schematic diagram of aging experimental setup

RESULTS

Microscopic observation

Microscopic observation of Ultem 1000/CNF/CNT composite before and after aging were compared. There were no surface cracks or other visible physical changes observed (Figure 2).

Type	Before aging	After 4 m aging
A		
B		
C		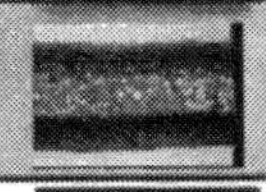
D		

Table 2: Microscopic images of CNF/CNT composites

Thermogravimetrical (TGA) Analysis of Weathered nanocomposite

TGA (Pyrus 1, PerkinElmer Cor.) is used in polymers for the determination of the thermal stability of polymers, compositional analysis, and identification of polymers from their decomposition pattern. The thermogram shown in Figure 2 illustrates weight losses of neat nanocomposite and ozone aged nanocomposites. Thernal analysis is a very simple technique for quantitatively analyzing for filler content of a polymer compound (e.g.CNT). The thermogram overlaps with the temperature range of hydrocarbon polymer degradation. The curves cannot be reliably deconvoluted since the actual decomposition range of a polymer in a polymer blend can be affected by the sample morphology. Results of thermogravimetric analysis weathered polymer samples (D) measured using PerkinElmer TGA7. TGA makes a continuous weighing of a small sample in a controlled atmosphere as the temperature is increased at linear.

The TGA data for sample D shows that PEI/CNT/CNF composite is stable up to 500 °C, above this temperature the polymer starts to degrade. There are three ways in which a polymer degrades: random scission, systematic chain scission or a combination of two. Random scission along the chain produces radicals or other reactive species. These reactive species my continue to break down into progressively smaller species, which become volatile and are lost or they may attack other polymer chains leading to cross-linked polymers, which are less prone to degradation and ultimately lead to high temperature residue referred to as a char. The onset thermal decomposition temperatures were 550- 600°C.The thermal stability of the PEI/CNF/CNT decreased with oxidative aging. The stability of PEI composite decreased slightly from 576 °C to 547 °C for the extended exposure to ozone (Figure 2).

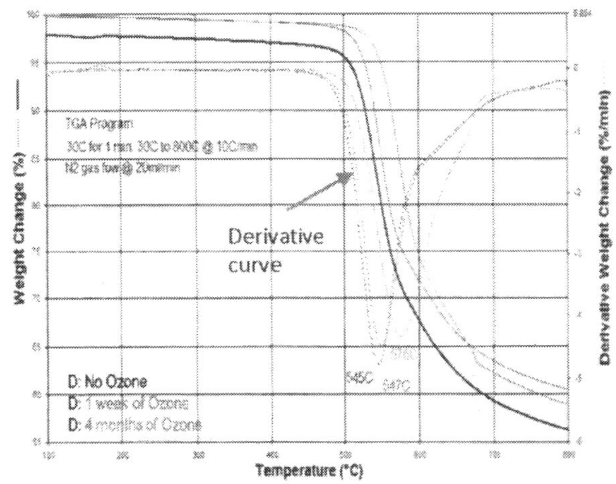

Figure 2: TGA of fresh and aged CNT/CNF polymer Composites

Fourier Transfer Infrared Spectroscopy Analysis

FTIR spectrum recorded using Cary 660 FTIR from Agilent Technologies Spectrophotometer. FTIR spectra for all samples were recorded by averaging 20 scans with 4 cm^{-1} resolution. The FT-IR spectrum of the PEI is shown in Figure 1. It exhibits characteristic imide group absorptions at 1780 and 1720 cm-1 (typical of imide carbonyl asymmetrical and symmetrical stretch), at 1355 and 743cm-1 (C-N stretching and bending), and at 1234 cm-1 (aromatic ether C-O-C) (Figure 3). The structural changes of weathered composites samples were evaluated by recording. Unfortunately no real evidence for amide and/or ester groups, no N-H clear-cut absorption band, could be detected on FTIR-ATR spectra because the neat and ozone exposed PEI/CNT/CNF spectra appear only differ slightly. This difficulty can be related to the similarity of the functional groups in both samples as well as to the FTIR-ATR technique itself. From the FTIR spectrum of weathered composites shows peaks around 2300 cm-1 (Conjugated nitrile bond) gradually disappeared. This might be an indication of breaking of the polymer crosslinking breaking. t

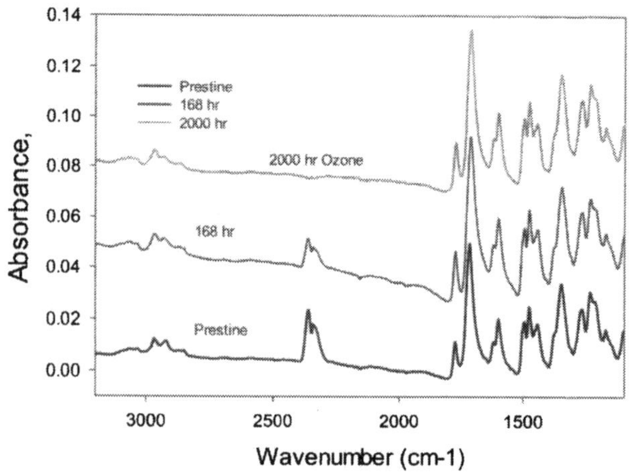

Figure 3. FT-IR spectrum of neat and ozone aged PEI/CNT/CNF

REFERENCES

[1] Bernd Nowack, Raymond M. David, Heinz Fissan, Howard Morris[d], Jo Anne Shatkin, Michael Stintz, Richard Zepp, Derk Brouwer, Potential release scenarios for carbon nanotubes used in composites, Environment International, 59(2013) 1-11.

[2] Asatulu E., Twomey J., Overcash M., Life cycle and nanoproducts: end-of life assessment, J. Nanopart Res 2012;14.

[3] Christopher Kingston, et al., Release characteristics of selected carbon nanotube polymer composites, Carbon, 68 (2014) 33-57

[4] Chen, B.K, Su, C-T., Tseng, M-C, Tsay, S-Y, Polymer Preparation of polyetherimid nanocomposites with improved thermal, mechanical and dielectric properties, Bulletin 57, 671-681 (2006).

[5] Natural weathering studies of polypropylene, J. Polymer Environ (2009) 17:191-202

The Safe & Electrical Transmission for Energy Markets and Grid Technology

Z. Strong

Copper Development Association, 260 Madison Avenue, Suite 1600, New York. New York 10016, zolaikha.strong@copperalliance.us

ABSTRACT

Energy storage in grid applications ranges from distributed community energy storage (CES) to large, centralized, bulk storage. This paper focuses on four application areas: ancillary services, transmission services, CES and other distributed storage.

Ancillary services, renewable energy integration and transmission support are promising for near term growth while CES is most promising for long-term growth in grid energy storage. Near-term growth will come from past investments as well as the emergence of new policies to promote the market. The long-term market for storage will depend on the ability of suppliers to reduce costs and policies to help formalize application markets.

Industry analysts forecast that the global market for energy storage could increase to 300 GW with a value from $200-600 billion over the next 10 to 20 years. In the short term, the size of the U.S. grid energy storage market is forecast to reach between two to four gigawatts by 2017 [1]. According to a 2014 roadmap from the International Energy Agency, there is already more than two gigawatts of installed thermal storage capacity in the country [2].

Keywords: grid, energy, storage, copper, forecast

1 INTRODUCTION

The term "energy storage" could refer to many types of storage technologies, including electrochemical, mechanical, or thermal. For specifics on energy storage initiates and detailed analyses of the copper content projections, the reader is referred to the original DNV GL (formerly KEMA) study, which is currently available online from the Copper Development Association [1].

Energy storage technology is typically classified according to rated power and discharge duration. Together, these characteristics describe how much power a storage unit can provide (in MW) and for how long (in hours). Even within a given technology type, the range of ratings can be large and there is a great deal of overlap in technology ratings.

Initial investment, operating costs and service life all are factors in evaluating the cost of energy storage. These costs are related to performance, which is measured in terms of capacity, efficiency and cycle life. Cycle life refers to the number of charge and discharge cycles that a storage device can provide before efficiency decreases so as to make it no longer capable of suitably performing the functions it needs to in an application.

A report by Sandia National Laboratories identifies 19 different energy storage grid services [3]. Grid storage is connected along the electric grid system at various locations. It can range from distributed energy storage at the community scale, which is on the order of kilowatts (kW) in capacity; to large-scale battery energy storage, which is on the order of one or two megawatts. The latter can be aggregated up to 50 MW in size. Bulk grid storage can be on the order of tens of megawatts.

The growth potential for grid energy storage remains tremendous as predicted by an earlier study [1]. Aspirations for energy independence coupled with urgent needs to stabilize peak load times is driving the demand for energy storage systems. Growth will be fueled by investments as well as governmental policies and copper will play a significant role in this growth.

Copper contributes to safe, reliable and efficient transmission of electricity in all energy markets and grid technologies. Its superior electrical conductivity is essential for many electrical products and systems, including wires and cables as well as motors, generators, transformers and protective devices. Copper provides efficient transmission and distribution of electricity. It is ideal for electrical applications because it conducts electicity better than other commercial metals, i.e., it is 65 percent more conductive than aluminum. Copper keeps the power grid working efficiently because its electric and thermal properties minimize load loss.

2. GRID TECHNOLOGY

The applications having the largest near-term market growth include community energy services, ancillary services, renewables energy integration, transmission services and other distributed storage.

Community energy services (CES) consists of a small, distributed energy storage unit connected to secondary transformers serving a few houses or small commercial loads. As the name implies, local communities are the primary beneficiaries of an energy storage device. CES enhances reliability, reduces the required capital investment by flattening peak loads, compensates for the variability of distributed renewable resources, such as roof top solar photovoltaics (PV), and provides a source of back-up power during grid events for residential and commercial and industrial customers. As such, CES combine multiple applications to serve grid needs at the end of the distribution system.

"Ancillary services" are tools used by grid operators to help maintain a continued balance between electricity production and demand. Fast-response storage devices have the ability to provide frequency regulation and spinning reserve grid services, a subset of the full set of ancillary services. Independent system operators and regional transmission organizations (ISOs/RTOs) are currently examining the potential for fast-response storage devices to act as an alternative to traditional generation technologies to help balance system supply and demand through regulation and spinning reserves services.

"Transmission services" include the deferral of transmission and distribution equipment (also known as T&D deferral), transmission support, and transmission congestion. T&D deferral is the use of energy storage to defer the installation or upgrade of transmission and distribution equipment, which can often be difficult to site. Transmission support refers to the use of energy storage to address electrical anomalies and disturbances on the grid. Transmission congestion relief refers to the use of storage to avoid the need to transmit power during periods of high system demand; for example, energy stored by devices could be used during on-peak hours when the transmission systems are congested.

Increased implementation of variable renewable generation resources, especially wind and solar, may increase grid volatility, requiring an increased need for frequency regulation services. Advanced energy storage technologies with fast-response capabilities are a promising approach to managing the volatility of renewables.

The renewable energy integration application is directly linked to intermittent renewable implementation on the grid (e.g., wind and solar) and the amount that occurs on a percentage basis. Energy storage can address three "buffering" challenges that variable energy resources introduce on the grid system: 1) *Capacity firming*, as previously discussed under ancillary service/regulation; 2) *Smoothing or ramp control*, a function to help reduce the adverse impacts of a very fast change in renewable generation level or output; and 3) *Time-shifting*, to match typically off-peak renewable energy supply with on-peak demand.

The integration of renewable resources will be helped by the ancillary services market. The intermittency of renewable resources could be addressed in the markets or outside of them. Obligations to firm wind capacity—that is to maintain the power output at a committed level for a reasonable time—vary by region. For the most part, the obligation lies with the ISO/RTO. However, there has been movement toward requiring generators of intermittent or variable energy resources to firm their power to minimum requirements before placing their power on the grid. The management of localized voltage issues could also be addressed by energy storage outside of market-based services.

Additional distributed grid storage applications besides CES are under development. For example, thermal storage currently constitutes a large portion of the storage market today. Here, storage can be used to help shift cooling or heating loads to off-peak hours. More recently, thermal storage associated with water heaters has demonstrated the ability to participate in the ancillary services markets. In addition, microgrid demonstrations are underway that integrate storage into the portfolio of onsite resources, such as distributed generation or demand response capability. Furthermore, storage can play a role in assisting with integrating intermittent renewables on a customer site, and with providing temporary back-up power services.

3. COPPER INTENSITY

Based on market assessments and copper intensities, grid energy storage could result in a sizeable demand for copper [1]. Between 900 and 3,000 tons of copper could be used in grid energy storage in the U.S.

The copper content of grid energy storage installations can also be assessed on a per megawatt basis. Copper intensity ranges from zero to more than four tons per megawatt (MW), depending the electrical equipment type, the energy storage type and the installation configuration.

On a per megawatt basis, CES installations have the largest copper intensity, with a mid-range estimate of more than four tons per megawatt. Other distributed storage could offer high copper intensity but the majority of these installations, such as thermal energy storage, offer the lowest copper intensity, e.g., 0.04 tons per megawatt. Applications for renewable energy integration, ancillary services, and transmission services can range from 0.3 to 3 tons per megawatt.

Copper intensities vary significantly by installation and storage technology. With many storage applications in the demonstration phase, energy storage installations have yet to conform to standard configurations. Distributed applications, however, generally have higher copper intensities because they use more lower-voltage equipment, which typically has higher copper intensity. Lower voltage components are more likely to use copper due its higher conductivity and lower maintenance requirements.

At the unit level, energy storage devices range from zero to nearly 0.3 tons of copper per megawatt. Copper can play a role in the fundamental design of storage units by contributing toward internal connections and current collectors in battery technologies or motors for pumped storage or CAES devices.

Lithium-ion, flow, and sodium batteries as well as flywheels, CAES, and pumped hydropower are strong users of copper at the unit level, ranging from over 0.10 to nearly 0.3 tons per MW.

Thermal storage, lead-acid batteries, and super capacitors exhibit the lowest copper intensities, ranging from 0 to 0.03 tons per MW.

With many storage applications in the demonstration phase, energy storage installations have yet to conform to standard configurations. Energy storage installations come in a wide array of sizes and configurations and few best practices have evolved with regard to defining storage configuration. Nonetheless, common pieces of equipment will likely be needed to interconnect energy storage to the electric grid. Such equipment includes transformers; interrupting devices, such as breakers and switches; protection and communication systems; monitoring and control systems; and inter- and intra-system wiring.

Estimates of the magnitude of copper demand associated with installed applications of energy storage are based on representative configurations. Various types of equipment can be used in a variety of applications, ranging from low-voltage to high-voltage installations, and best practices are yet to develop; this range of electrical equipment associated with storage installations is accounted for by using a range of copper intensities associated with these applications.

The choice of electrical equipment, which translates into a range of copper intensities, depends on size of the equipment required in the application as well as market offerings. Reasonable ranges for the devices are used to estimate impacts on total tonnage.

The CDA-DNV GL report is based on a generic configuration that is representative of those used to estimate the copper intensity of storage installations [1]. Storage device intensities were used to estimate unit intensities. The copper content of the associated electrical equipment was estimated based on sample configurations either planned or already in the field. The copper content of the electrical equipment was estimated based on assumptions about equipment sizes and types.

The following values represent the ranges in quantity and unit size, respectively, per installation.

- Power transformers: 1 to 11; 25 kW to 360 MW
- Breakers—1 to 20; 480V to 138 kV
- Generators—2 to 200; 100 kW to 300 MW
- Inverters—1 to 48; 5 kW to 3.5 MW
- Cooling—small window units to large rooftop systems
- Grounding—several feet to miles; #4 Cu to 4/0 Cu

4. CUMULATIVE COPPER USAGE

The demand for copper comes from not only the copper content of the energy storage units themselves but also from the electrical equipment needed to operate the energy storage with the grid. The largest near-term demand for copper is in renewable energy integration and ancillary services.

Energy storage in renewable energy integration applications is associated with strong demand for copper. Cumulative associated copper demand could range from 650 to almost 2,200 tons, depending on financial incentives. Associated incremental demand could range from 300 to more than 1,800 tons.

Energy storage for ancillary services is also associated with a strong demand for copper. The copper intensities are relatively high and so are expectations for near-term growth. Cumulative demand could range from 630 tons to almost 840 tons, depending on financial incentives. Associated incremental demand ranges from over 520 tons to over 720 tons.

Meanwhile, CES is also poised for strong growth. Although near term demand is limited, this area could have strong growth for the mid-term to long-term. Estimates range from 30 to almost 370 tons, depending on financial incentives. Associated incremental copper demand ranges from over 20 tons to over 360 tons.

Although copper intensities for transmission-related storage applications are high, these markets are limited in the near term. Estimates for cumulative copper demand are 85 tons to more than 250 tons, depending on financial incentives. The associated incremental copper demand ranges from 5 to more than 170 tons.

The market for distributed thermal storage is large but copper intensity is relatively limited. Other types of distributed energy storage have higher copper intensities, but limited market penetration. Estimates for the cumulative associated copper demand range from 100 to 160 tons, depending on financial incentives. The associated incremental copper demand ranges from over 50 tons to over 100 tons.

5. MARKET FORECAST

The demand for energy storage is strong and robust but the industry needs to continue with initiatives to reduce costs and increase experience for growth to meet expectations. The CDA-DNV GL estimated that the U.S. grid storage market could grow to between two to four gigawatts (GW) by 2017, depending on the existence of financial incentives [1]. The DOE reports 991 MW of energy storage for 314 projects (not including pumped hydropower) in the U.S. alone as of April 2015 [4].

Many new technologies are under development and some are ready for commercialization. Mature energy storage technologies currently constitute the majority of the energy storage market today. These include thermal energy storage, pumped hydropower, and CAES. In the near term, battery technologies and thermal storage are expected to have the strongest growth areas.

Second generation technologies are emerging and research is continuing to be fruitful. Venture capital investment is booming both inside and outside of the U.S.

The ability to scale quickly is seen as a challenge, as is the ability to bring down costs. Financial incentives could have a large impact on market size by defraying initial investment costs and helping to grow the market.

The U.S. markets around grid-storage applications are still evolving. Policy developments are continuing to shape the development of markets around grid applications, and demonstrations are continuing to define grid application success and inform technology value propositions.

Markets around some of the grid storage applications are expanding now, such as ancillary services, and markets for other applications that are developing now, such as peaker plant applications, will likely come to fruition after five years.

In the near future, costs are expected to come down (via improved system integration, increased production, and enhanced distribution capability), investments are expected to continue, and areas currently in the demonstration phase will likely start to commercialize.

5. CONCLUSION

Because of its potential role in supporting the integration of renewable energy, energy storage may also help bolster the copper demand associated with renewable generation.

Energy storage is one of many tools available to help address renewable intermittency. Energy storage can also assist in the integration of renewable energy by helping to address transmission constraints.

The estimated copper intensities of typical wind farms and typical centralized solar plants, according to prior CDA studies, are three to six tons per MW and two to five tons per MW, respectively.

Commissioned by the Copper Development Association Inc. (CDA), the 2012 CDA-DNV GL research report entitled "Market Evaluation for Energy Storage in the United States" evaluates the near-term market for grid energy storage in the United States (U.S.) and the copper content associated with this market. CDA is the market development, engineering, and information services arm of the copper industry, chartered to enhance and expand markets for copper and its alloys in North America. It is a member of the Copper Alliance.

While it is difficult to accurately forecast the growth in the energy storage market, let alone the copper useage in that market, the CDA-DNV GL study from 2012 shows that a robust growth in the energy storage market could result in additional demand for copper amounting to several thousand tons in the near future and perhaps much more in the long term, especially if energy storage leads to wider use of alternative energies such as wind and solar.

Acknowledgement: Portions of this paper were adapted with permission from CDA sponsored DNV GL *research paper referenced below and available online from the www.copper.org website.*

REFERENCES

[1] DNV GL (formerly KEMA), "Market Evaluation for Energy Storage in the United States," January 2012, Fairfax, Virginia, 69 pages, with 27 Figures.

[2] IEA, Technology Roadmap, Energy Storage, 2014, Box 9, page 31.

[3] U.S. Department of Energy, Sandia National Laboratories. Energy Storage for the Electricity. Grid: Benefits and market Potential Assessment Guide. February 2010. SAND2010-081

[4] The DOE Global Energy Storage Database (www.energystorageexchange.org/) is powered by Sandia Corporation (http://www.sandia.gov/) and Strategen Consulting , LLC (http://strategen.com/). Accessed April 2015.

Estimation on Autocatalytic Reaction for Ethanolamine – A Green Approach

A. C. Huang* and C. M. Shu**

*Graduate School of Engineering Science and Technology, National Yunlin University of Science and Technology (YunTech), Taiwan, ROC, d10210012@yuntech.edu.tw
**Department and Graduate School of Safety, Health, and Environmental Engineering, YunTech, Taiwan, ROC, shucm@yuntech.edu.tw

ABSTRACT

Ethanolamine, which is a derivative of ethylene oxide, is not only a major material in organic chemicals but is also the most valued product among amino alcohols. However, ethanolamine might catch fire automatically when mixed with peroxide or nitrocellulose due to its thermal instability and high sensitivity to thermal sources. Runaway reactions because of ethanolamine will lead to severe damage to equipment, people, the environment, and create a social impact. Therefore, it is necessary to provide thermal safety information regarding the decomposed mechanism and thermal hazard characteristics by using thermal analysis technology. As a green approach, our aim was to determine thermal safety parameters, such as apparent exothermic onset temperature, maximum temperature, heat of decomposition, self-heating rate, pressure rise rate, maximum pressure, by differential scanning calorimetry. The results can be used to calculate the optimal parameters for processing a safer design and establish the database of ethanolamine for proactive loss prevention protocol.

Keywords: derivative of ethylene oxide; organic chemicals; thermal safety information; decomposed mechanism; thermal analysis technology

1 Introduction

Ethanolamine is the most common product of ethylene oxide derivative which is widely used in the chemical industry, such as insecticides, intermediates of medicine, softening agent of synthetic fiber, dye stuff of dispersant and resin in recent decades. On a worldwide scale, ethanolamine is produced as approximately 1.3 million metric tons per year, and the output keeps increasing yearly [1]. Without a doubt, ethanolamine instability property and detrimental effects do still exist in usage, storage, and transporting of ethanolamine. Ethanolamine will have an obvious reaction with fire automatically when combined with peroxide [2].

In a chemical reaction, a catalyst is a substance that can keep its composition and quality but alter the rate of reaction. Autocatalysis is the product of a reaction that can increase its rate of reaction: the product of a reaction can play a catalytic role [3]. The typical graph of an autocatalysis reaction shows double bend curve. Since there is very little catalyst at the begining, the reaction carries on very slowly. As catalyst increases, the rate of reaction increases until the density of reactant decreases. The classic autocatalytic system is shown as the following:

$$\frac{du(t)}{dt} = a - u(t)v^{p}(t), \quad t > 0 \tag{1}$$

$$\frac{dv(t)}{dt} = u(t)v^{p}(t) - v(t), \ t > 0 \tag{2}$$

$$u(0) = u_0 > 0, \quad v(0) = v_0 > 0 \tag{3}$$

where a is original concentration of the reaction precursor, u is the densities of the reactant, v is autocatalyst [4].

There are two sides to every question. Despite the fact that ethanolamine was thought of as a kind of most common material to be used, the thermal hazard risk needs to be considered when it is used. Thus, to stress safety concerns, our task was to estimate the autocatalytic reaction for ethanolamine. To determine the possible potential hazard of ethanolamine, metal ion was added with ethanolamine in the experiment.

In this study, we used differential scanning calorimetry (DSC) to determine the thermokinetic model and DSC curve of ethanolamine and ethanolamine with metal ion. DSC is a highly trustworthy option as a powerful thermoanalytical technique for determining the thermokinetic data [5,6]. The primary questions that we addressed are as follows: enthalpy (ΔH_d), exothermic onset temperature (T_0), and peak temperature (T_{max}). The goal is to contribute to the decrease of hazard from the usage, storage, and transportation of ethanolamine as much as possible.

2 Experiments and methods

2.1 Material

Ethanolamine (99 mass%), which was used for testing analysis and without further purification, was purchased from Acros Organics, Thermo Fisher Scientific, New Jersey, USA. With regard to incompatibility trials, FeO_2 (98 mass%) and CuO_2 (99 mass%), that were offered from Showa Kako Corporation, Suita-shi, Osaka, Japan, were used together with ethanolamine separately. All chemicals were stored in medicine cabinets which are shady and cool [1,7]. The structure of ethanolamine is shown in Figure 1.

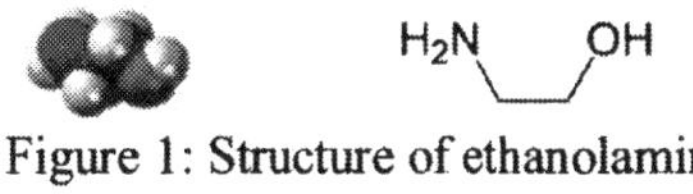

Figure 1: Structure of ethanolamine [1,7].

2.2 DSC

The thermal analysis data were measured by using a Mettler Toledo DSC-821 with nitrogen-gas atmosphere. All experiment samples were encapsulated in stainless steel aluminum pans which were placed in a test crucible that can sustain temperature up to 720.0 °C. The system of this model was tested to be capable of experiments under pressures of 100 bar [8]. Analysis of experiments was carried out by Thermal Analysis System Software to obtain thermokinetic data. The scanning rates, which were used to detect significant difference among samples, were selected for analyzing programmed ramp consisting 0.5, 1.0, 2.0, 4.0, and 8.0 °C min^{-1}. All samples, which were placed in airtight aluminum pans, were heated from 30.0 to 300.0 °C. The selected temperature range was determined by a thermal cracking test from Thermogravimctry (TG) which has the ability to investigate the thermal cracking of materials [9]. Three times of reproducibility were represented by each experiment to reconfirm the authenticity of samples.

2.3 Thermokinetic analysis

To determine the thermokinetic parameters of ethanolamine by non-isothermal kinetic model, the type of reaction must be checked and examed. Two reaction types were found regularly: Autocatalysis and *n*th order reactions. In general, an *n*th order reaction is one where a reactant reaches maximum heat flow in the initial stage or in few minutes. At this time, the exothermic output is not only the maximum but also the most dangerous moment. From another aspect, autocatalysis means the runaway reaction was suddenly brought out by the stage that the reactant received a term of calorific capacity. An *n*th order can be calculated by Arrhenius equation as shown in Eq (4).

$$k = Ae^{-Ea/RT} \tag{4}$$

where k_0 is frequency factor of reactant, E_a *is* activation energy, R is gas constant, T is absolute temperature. According **to the** Arrhenius equation, activation energy can be determined and expressed as in Eq (5).

$$k = k_0 exp\left(\frac{-E_a}{RT}\right) \tag{5}$$

When the relation between logarithm of k combine and T^{-1} is relatively a straight line, it shows that this function fits Arrhenius behavior. The slope of this line is $-E_a R^{-1}$.

$$\frac{d\alpha}{dt} = k_0 exp\left(\frac{-E_a}{RT}\right)(1-\alpha)^n \tag{6}$$

Autocatalysis reaction can be expressed in Eq (6).

$$f(\alpha) = (1-\alpha)^{n_1}(\alpha^{n_2} + z) \tag{7}$$

To acquire the E_a and k_0 of autocatalysis reaction, the autocatalysis reaction equation and Arrhenius equation are combined.

3 Results and discussion

Thermal analysis of ethanolamine with different metal oxides was carried out in DSC. Random sampling was adopted in all experiments to determine the sequence of all samples to avoid unexpected factors that may cause inaccuracy. To discriminate between the exothermic reactions of ethanolamine, ethanolamine with Fe_2O_3, and ethanolamine with CuO, experiments were designed for pure substance and substances with metal ions. The DSC tests involved ethanolamine, ethanolamine with Fe_2O_3, and ethanolamine with CuO at a heating rate of 4.0 °C min^{-1}. Of the three experiments, ethanolamine with CuO had a protruding curve compared to the exothermic reaction test of ethanolamine and ethanolamine with Fe_2O_3, as shown in Figure 2.

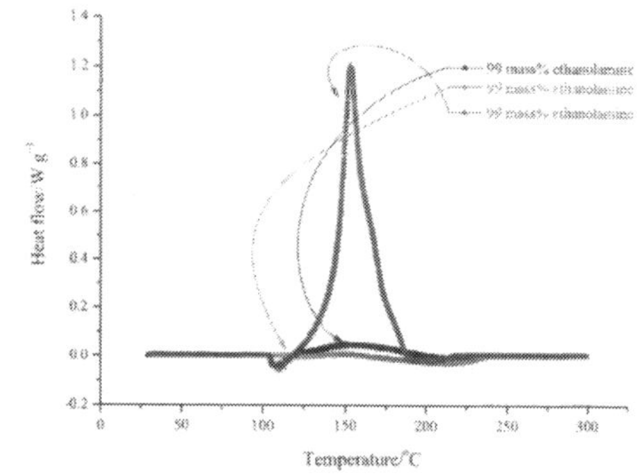

Figure 2: DSC thermal curve of temperature versus heat flow for 99 mass% ethanolamine and ethanolamine mixed with Fe_2O_3 and CuO at heating rate of 4.0 °C min^{-1}.

In this case, the reaction of ethanolamine mixed with CuO was not only presented in comparison to pure ethanolamine, but also the maximum heat flow displayed obviously. With regard to that there was no reaction in pure materials, CuO was confirmed as being an incompatible substance for ethanolamine. This may suggest that enormous heat will be generated when ethanolamine is mixed with CuO during process, usage, transportation, machining, or storage. Table 1 presents the onset temperature of ethanolamine mixed with CuO being higher than the other two groups of experiments, but the end set temperature was lower than others. The findings showed that when ethanolamine was mixed with CuO, the stage of amassing energy was accumulated in a longer period and released in a short time.

Table 1: Thermal stability data from the non-isothermal heating experiments with 99 mass% ethanolamine and ethanolamine mixed with Fe_2O_3 and CuO at a heating rate of 4.0 °C min^{-1}.

Sample	Mass/mg	T_0/°C	T_p/°C	T_f/°C	ΔH_d /J g^{-1}
Ethanolamine	3.93	134.05	150.95	229.78	31.40
Ethanolamine + Fe_2O_3	4.62	135.11	146.08	178.92	4.99

 TechConnect Briefs 2015, TechConnect.org, ISBN 978-1-4987-4728-8

Ethanolamine + CuO	3.59	143.81	153.02	165.89	395.19

Table 2: Thermal stability data from the non-isothermal heating experiments with 99 mass% ethanolamine mixed with CuO of heating rate at 0.5, 1.0, 2.0, 4.0, and 8.0 °C min^{-1}.

Heating rate /°Cmin^{-1}	T_0/°C	T_p/°C	T_f/°C	ΔH_d /J g^{-1}
0.5	132.21	132.27	132.28	178.43
1.0	123.55	136.39	146.37	157.42
2.0	137.60	144.98	153.62	379.00
4.0	143.81	153.02	165.89	395.19
8.0	158.73	170.02	187.57	285.07

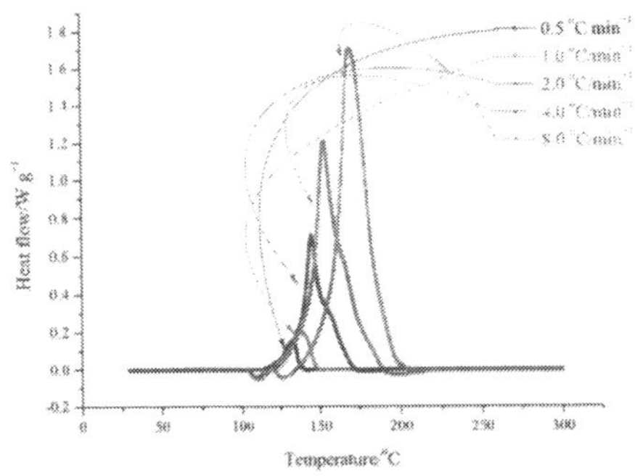

Figure 3: DSC thermal curve of temperature versus heat flow with 99 mass% ethanolamine and ethanolamine mixed with CuO at heating rates of 0.5, 1.0, 2.0, 4.0, and 8.0 °C min^{-1}.

Considering the different situations in reality, ethanolamine mixed with CuO was investigated at a heating rate of 0.5, 1.0, 2.0, 4.0, and 8.0 °C min^{-1}. With higher heating rate, the maximum temperature and end set temperature of reaction within ethanolamine mixed with CuO are also higher, as listed in Table 2. Figure 3 shows that heat flow is getting higher and exothermic onset temperature is getting later with higher heating rate on DSC. Most importantly, the maximum heat flow of 0.5 and 8.0 °C min^{-1} is 1.71 W g^{-1} and 0.16 W g^{-1} respectively. That is, there is an approximately ten-fold gap between the heat flow reasonably at heating rates of 0.5 and 8.0 °C min^{-1}.

As mentioned above, E_a was calculated by taking test methods from the American Society for Testing and Materials E698-5 in this study [10]. This conversional rate equation is expressed in Eq (8).

$$\frac{d\alpha}{dt} = Ae^{-Ea/RT}f(\alpha) \qquad (8)$$

To confirm the credibility of the thermokinetic model that we simulated, linear regression was adopted to provide R-square, as shown in Figure 4. The findings of Figure 4 show that there is an excellent relationship within the experiments. E_a and A are determined by the slope and intercept. E_a is 98.695 kJ mol^{-1}, and R-square is high to 0.933.

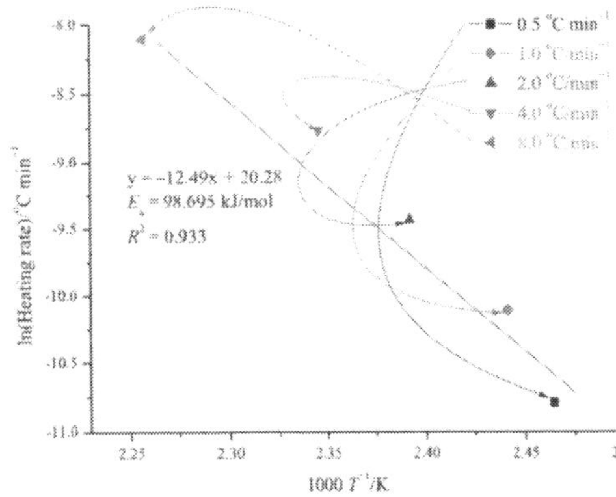

Figure 4: Apparent activation energy calculation with 99 mass% ethanolamine mixed with CuO.

In this study, we used differential isoconversional methods to assess thermokinetic parameters of reaction [10]. Equation 9 shows a differential isoconversional method, which is used to calculate E_a and A, is often seen.

$$ln\frac{d\alpha}{dt} = ln[A(\alpha)f(\alpha)] - \left(\frac{E(\alpha)}{RT(t)}\right) \qquad (9)$$

The relationship between reaction progress versus $E(\alpha)$ and reaction progress versus $ln[A(\alpha)f(\alpha)]$ is shown in Figure 5.

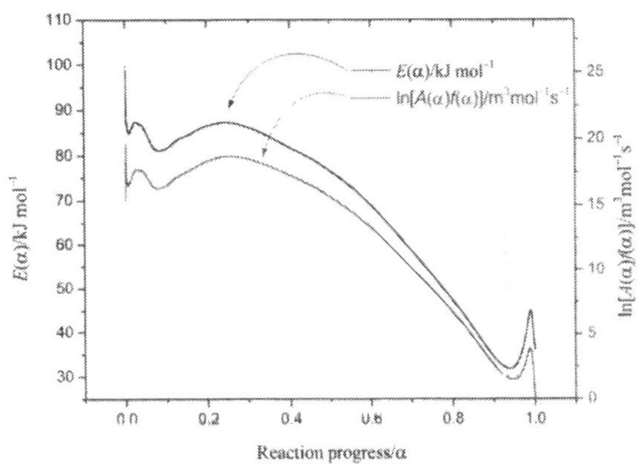

Figure 5: Apparent activation energy with degree of conversion and pre-exponential factor with degree of conversion with reaction progress from 99 mass% ethanolamine mixed with CuO.

For the reaction stage in this case, the active range of α presents decreasing curves mainly. It shows that autocatalytic reaction happening with $E(\alpha)$ is presenting a decreasing curve. According to observations, the curves of reaction progress versus with $E(\alpha)$ and reaction progress versus $ln[A(\alpha)f(\alpha)]$ are almost perfectly coinciding.

Moreover, *SADT* was investigated in this study in different mass. As observed, the curve of SADT and mass of ethanolamine mixed with CuO presents an inverse ratio, as shown in Figure 6. That is, the thermal stability of a lower mass, such as 50.0 kg, is better than the higher one.

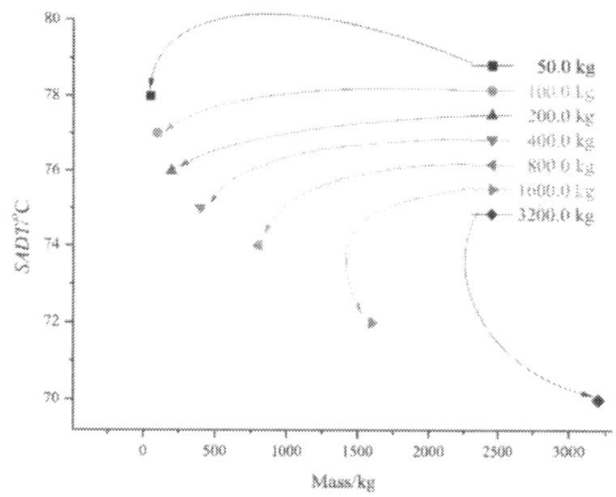

Figure 6: Simulation results on mass versus *SADT* of 99% mass% ethanolamine mixed with CuO.

4 Conclusions

This study showed that CuO is an incompatible metal material for ethanolamine. More specifically, ethanolamine is more likely to cause hazard when mixed with CuO which normally exists in industrial places. Therefore, it is necessary to choose a container or pipe with proper material of chemical property to avoid being rusty. To practice the process under upset scenarios, we can apply the thermokinetic data, which are ΔH_d, T_0, T_{max}, and SADT, that were obtained by thermal calorimetric technology and thermokinetic technology. Estimation of autocatalytic reaction by using the thermal analysis technology is fast, safe, exact, and a highly trustworthy method for the thermal stability of substance. Hence, the findings of this study can provide a better way for managing safety procedures.

5 Acknowledgments

Financial support for this study was graciously provided by the Minister of Science and Technology, Taiwan, ROC (103-2221-E-224-001-MY2).

6 REFERENCES

[1] H. Tsuneki, and A. Moriya, "Analysis and modeling of adiabatic reactor for monoethanolamine selective production," Chemical Engineering Journal, 149, 363–369, 2009.

[2] M.A. Ak, A. Ulas, B. Sümer, B. Yazıcı, C. Yıldırım, L.O. Gönc, and F.E. Orhan, "An experimental study on the hypergolic ignition of hydrogen peroxide and ethanolamine," Fuel, 90, 395–398, 2011.

[3] S.H. Liu, H.Y. Hou, and C.M. Shu, "Thermal hazard evaluation of the autocatalytic reaction of benzoyl peroxide using DSC and TAM III," Thermochimica Acta, 605, 68–76, 2015.

[4] G. Guo, B. Li, and X. Lin, "Hopf bifurcation in spatially homogeneous and inhomogeneous autocatalysis models," Computers & Mathematics with Applications, 67, 151–163, 2014.

[5] S.H. Liu, H.Y. Hou, J.W. Chen, S.Y. Weng, Y.C. Lin, and C.M. Shu, "Effects of thermal runaway hazard for three organic peroxides conducted by acids and alkalines with DSC, VSP2, and TAM III," Thermochimica Acta, 566, 226–232, 2013.

[6] P. Lang, T. Wojcik, E. Povoden-Karadeniz, A. Falahati, and E. Kozeschnik, "Thermo-kinetic prediction of metastable and stable phase precipitation in Al–Zn–Mg series aluminium alloys during non-isothermal DSC analysis," Journal of Alloys and Compounds, 609, 129–136, 2014.

[7] M.S. Masoud, A.E. Ali, H.M. Ahmed, and E.A. Mohamed, "Spectral studies and thermal analysis of new vanadium complexes of ethanolamine and related compounds," Journal of Molecular Structure, 1050, 43–52, 2013.

[8] Y.T. Tsai, M.L. You, X.M. Qian, and C.M. Shu, "Calorimetric techniques combined with various thermokinetic models to evaluate incompatible hazard of tert-butyl peroxy-2-ethyl hexanoate mixed with metal ions," Industrial & Engineering Chemistry Research, 52, 8206–8215, 2013.

[9] S.A.P. da Mota, A.A. Mancio, D.E.L. Lhamas, D.H. de Abreu, M.S. da Silva, W.G. dos Santos, D.A.R. de Castro, R.M. de Oliveira, M.E. Araújo, L.E.P. Borges, and N.T. Machado, "Production of green diesel by thermal catalytic cracking of crude palm oil (Elaeis guineensis Jacq) in a pilot plant," Journal of Analytical and Applied Pyrolysis, 110, 1–11, 2014.

[10] J.W. Tong, W.C. Chen, Y.T. Tsai, Y. Cao, J.R. Chen, and C.M. Shu, "Incompatible reaction for (3-4-epoxycyclohexane) methyl-3' -4' -epoxycyclohexyl-carboxylate (EEC) by calorimetric technology and theoretical kinetic model," Journal of Thermal Analysis and Calorimetry, 116, 1445–1452, 2014.

Wave Powered Electric Generator System, Efficient harness energy and convert into electric generator directly via mechanical

Cuong Huu Nguyen

Twin Ocean Power, 15371 Cedarwood ave, Midway City, Ca 92655, USA,

E-mail: cuong_h_nguyen@yahoo.com, Website: http://TwinOceanPower.com

ABSTRACT

Within the past decade, the boom in energy demands has driven the the research and development agendas of many alternatives for energy production. The once almighty traditional methods of burning coals and fossil fuels since the Industrial Revolution are causing irreparable damages to our planet, spewing harmful pollutants into the atmosphere, affecting many life forms in various manners[1][2]. In our search for clean and renewable energy, we turn to the largest available natural resource in existence, our oceans. Within the complex ecological system, our oceans also possess essentially unlimited energy, existing in the forms of tidal waves, marine currents and thermal energy. The primary focus hereafter will be the extraction of wave energy by harnessing the oscillating motion of the ocean waves, wherein a mechanism will capture the wave oscillating motions and converts them directly into electricity.

Keywords: wave power, wave energy, ocean power, ocean energy, clean energy, reusable energy, renewable

1 INTRODUCTION

Ocean waves represent an endless and powerful energy source. Waves are generated by winds passing over the surface of the water. When this occurs, a wind essentially transfers its kinetic energy into the water. The air pressure differences between the upwind and the lee side of a wave crest and the developing friction between the wind/water surface forces the water into a shear stress, which causes the growth of the wave. The bigger the wind, the bigger the wave.

The realized benefits from converting wave power into electricity are many. Waves give us energy for free and the giving is both relentless and limitless. Once wave energy power plants are placed into operation, the energy capture and conversion is continuous. Since the operation occurs at or near the surface of a water body and at normal atmospheric pressure and relatively normal temperature, the materials used can be easily obtained and manufactured, and the maintenance could be rather inexpensive due to ease of accessing. However, the main issue facing wave energy power plants up to this point is the inefficiency in the conversion of kinetic energy into electricity. Hence, the introduction of the Wave Power Electric Generator System (WPEGS).

Wave Powered Electric Generator System (WPEGS) device is the new wave energy technology for wave power plant. The advantages of WPEGS are various:

- Efficiently capture of ocean wave motion by converting the rotation motion directly into electricity
- Fully functional in high tide, low tide and in between
- The system floats completely above water, allowing for ease of maintenance
- Can operate anywhere in the world's oceans.
- Scalable for cost effective mass production in several size (output) ratings.

2 SYSTEM DESIGN

The WPEGS is a new technology designed to build clean, reusable energy which captures the oceans wave power to create a source of electricity more efficiently. The WPEGS has great potential in positioning itself as a leader in renewable and reusable energy.

The System is comprised of a wave powered electric generating device with three floatation devices strategically arranged to the outer floatation devices FIG1. Each floatation long body is installing parallel with the wave column. One inner floatation device is arranged to connect to the other and is securely attached by interconnecting mechanisms, referred to as side interconnectors. Each floatation device comprise a first and a second housing secured next to each other to define inwardly facing surfaces and outwardly facing surfaces with the shafts extending between the inwardly facing surfaces of said first and second housings, the torque arms being secured to the shafts at locations between said housings. The WPEGS has the capability of floating in a wave-driven body of water when the inner floatation device is placed between the outer floatation devices. A torque arm connects the pair of adjacent floatation devices. The inner floatation device includes two rotatable shafts, with each rotatable shaft being secured to one of the torque arms to convert pivotal motion of the torque arm into rotational movement of the shaft. A one-way transmission system provides one way rotation of a generator shaft in response to pivotal motion of the torque arm in both direction, and an electrical generator connected to the generator shaft generates electricity. The transmission is capable of controlling the torque, which aids in preventing damage to the system.

The interconnecting mechanism and the torque arms are configured to permit expansion and contraction of their lengths. The oscillating motion of the ocean waves create forces acting on each floatation device raises the floatation device upward, and the gravity of each floatation device pulls it downward. The torque arm captures the kinetic energy from up and down movement which is created by the ocean wave and gravity, thus rotating the driven axle in clockwise and counter clockwise motion.

The transmission connected to the driven axle converts the rotation of the axle in both directions into high speed, producing one way direction rotation and driving the electric generator to produce electricity. The System has the ability to customize amounts of electrical production by installing a certain number of inner floatation devices as needed.

3 SYSTEM PROCESS

The WPEGS is plan to stall offshore where the ocean wave is strong and high, WPEGS is installed with each floatation body is lay parallel with wave column so the driven water can apply on each floatation completely from one end to another to create the maximum kinetic energy, the kinetic and potential energy occur when the inner floatation device and outer floatation device moving up and down in different direction to each other. When the waves travel through WPEGS, the force of wave body apply on each floatation, lifts each floatation completely upward, and the floatation will stroke back down by gravity when the wave move away to another floatation. These oscillating motion of the ocean waves create kinetic energy and potential energy between inner floatation and outer floatation. The energy is then transfer to inner rotatable shaft thru torque arm to rotate the main gear, the gearing system is engaged to main gear in convert motion rotation from both direction of main gear at low rotation high torque to one way high rotation at low torque where flying wheel install in gearing system keep generator running smoother. The motion rotation is then transfers to electric generator thru mechanical to produce electricity. All process from harness energy and transfer energy to generator is using mechanical so there is no loss or minimum loss of energy when conversion.

Unlike wind power and solar power which can produce energy within curtain time, not continuously so energy storage is required to operate; WPEGS is no need storage for operating it. For more energy is needed, more devices will be installed and connect together to produce more energy or more inner floatation on the device.

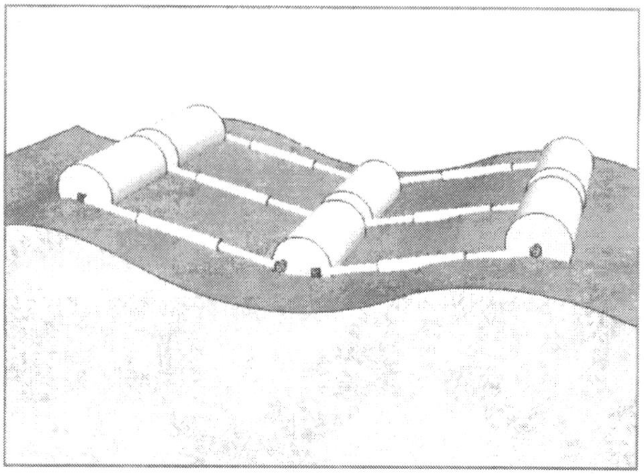

Figure 1: Wave power energy generator system device

With three floatation on each WPEGS, each ocean wave travel thru the system, the ocean wave will acting a great force on each floatation, meanwhile, the gravity off each floatation when it moving down also create the same amount of energy. The total number of wave that will go thru the system in an hour will produce the great amount of electricity.

4 CONCLUSION

WPEGS is expected to be a more competitive and economically viable solution than any other conventional and alternative energy technologies. It is definitively clean and renewable. Its design is efficiently streamlined and will not require extensive labor or maintenance to maintain the high quality of the system, lending to a longer lasting operational life. The advantages of developing and deploying such a System are endless, and can have major impacts on our environment well into the future. The primary objective is and will always be to identify areas to improve the WPEGS in energy production more efficiently. This invention is a major stake in the ground in the fight against depleting our world of resources to fulfill an insatiable appetite for energy with little or no consideration for its overall well being.

REFERENCES

1. Climate Change: Basic Information, http://www.epa.gov/climatechange/basics/

2. Health & Environmental Effects of Air Pollution, http://www.mass.gov/eea/docs/dep/air/aq/health-and-env-effects-air-pollutions.pdf

Renewable Energy Technology Gains By Interfacial Engineering With Carbon Nanomaterials

R. L. Vander Wal

The Pennsylvania State University
John and Willie Leone Family Dept. of Energy and Mineral Engineering, & The EMS Energy Institute
University Park, PA, USA, RandyVW@psu.edu

ABSTRACT

Many forms of energy utilization, conversion and storage and generation are dominated by interfacial chemistry. Therein nanomaterials as an interfacial modifier can play a critical role in these processes. Examples presented include Li ion batteries, nanolubricants, nanofluids, polymer composites, gas sensors and catalysts.

Keywords: Energy conversion, conservation, efficiency

1 INTRODUCTION

Energy engineering encompasses energy processes such as conversion, generation, storage, efficiency, conservation and control. Creating a *new* paradigm, nanomaterials, when used as interfacial modifiers have the potential to significantly alter the energy landscape, given that these processes occur predominantly at interfaces.

2 EXPERIMENTAL

Experimental details will be presented at the presentation with regards to the synthesis and utilization of the nanomaterials.

3 RESULTS AND DISCUSSION

Results illustrate the theme and support the hypothesis of using nanomaterials as interfacial modifiers to advance gains in energy engineering. Selected examples *from the author's work* include catalysis, composite materials, energy storage, sensors, thermal management, and tribology.

Catalysis using Au nanoparticles (see Fig. 1) supported on oxides such as CeO_2, TiO_2 and Fe_2O_3 offer ambient temperature oxidation of CO, volatile organic compounds (VOCs) and potentially exhaust hydrocarbons.

Within polymer composites, interfacial modification of a metal foil using carbon nanotubes directly grown upon the foil is illustrated, in Fig. 2. The purpose of the foils is to serve as a gas impermeable barrier layer [1].

Substantial gains in Li ion battery cathode and anode materials have been realized using CNTs, coating processes. Treatment of the CNT surfaces is shown to increase the Li ion capacity beyond the theoretical limit of normal graphite [2], see Fig. 3.

Exploitation of nanoscale properties of metal oxides, aided by supported noble metal catalysts led to new NO_x and O_2 sensors, operable at ambient temperature [3], shown in Fig. 4.

Nanofluids with enhanced thermal conductivity for cooling applications are demonstrated using carbon nano-onions and carbon nanotubes, (see Fig. 5), increasing water thermal conductivity by ~ 20% [4].

Carbon nanolubricants can bridge the gap between fluid and solid materials [5]. As additives with liquids or greases, synergistic properties may be realized with superior performance relative to graphite, diamond like carbon (DLC) and even Teflon, as summarized in Fig. 6.

4 CONCLUSIONS

The key concept is that the nanomaterials serve as interfacial modifiers. Since most energy related processes are dominated by interfacial reactions, nanomaterials have the potential to dramatically affect energy conversion rates and magnitudes

5 ACKNOWLEDGEMENTS

Funding through The Penn State Institutes for Energy and the Environment (PSIEE) and the Pennsylvania Keystone Innovation Starter Kit (KISK) is gratefully acknowledged.

6 REFERENCES

(1) Vander Wal, R. L., and Hall, L. J., Advanced Engineering Materials, 6, 48-52, 2004.

(2) Luo, Y., Vander Wal, R. L., and Scherson, D. A., Electrochemical and Solid State Letters, 6, A56-A58, 2003.Vander Wal, R. L., Mozes, S. D., and Pushkarev, Nanotechnology, 20, 105702-10, 2009.

(3) Hunter, G. W., Vander Wal, R. L., Liu, C. C., Xu, J. C., and Berger, G. M., Sensors and Actuators B, 138, 113-119, 2009.

(4) Vander Wal, R. L., Mozes, S. D., and Pushkarev, Nanotech, 20, 105702-10, 2009.

(5) Street, K. W., Miyoshi, K., and Vander Wal, R. L., "Application of Carbon Based Nano-particles to Aeronautics and Space Lubrication", in *Superlubricity*, Chapter 19, pp 311-340. (A. Erdemir, Ed.) Elsevier, Amsterdam (2007).

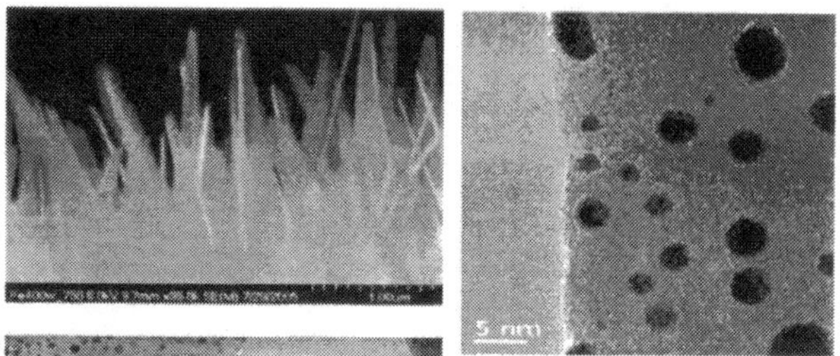

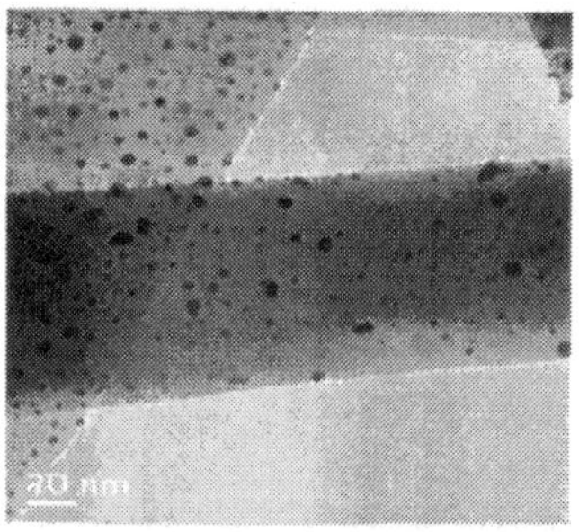

Figure 1. TEM Images Illustrating the Dispersion and Size Uniformity of Gold Nanoparticles upon Fe_2O_3 Nanoblades

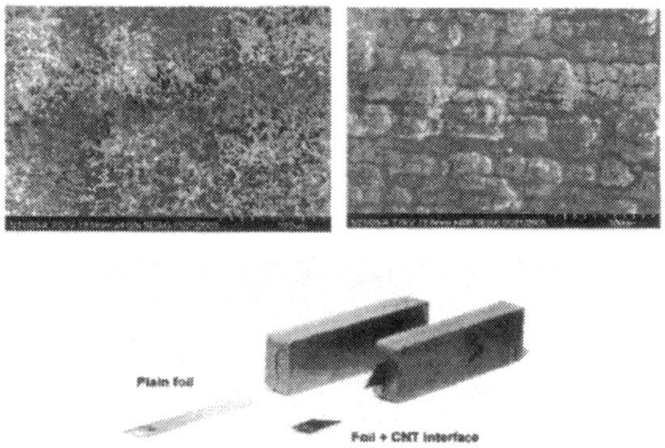

Figure 2: CNTs are used to create a three-dimensional interface Within a polymer composite. In the SEM images above, CNTs are synthesized upon the SS foil. Comparative tensile strength tests of "fuzzy" foil show that the interfacial bond between the SS foil and host polymer matrix exceeds the tensile strength of the 0.007-in. foil.

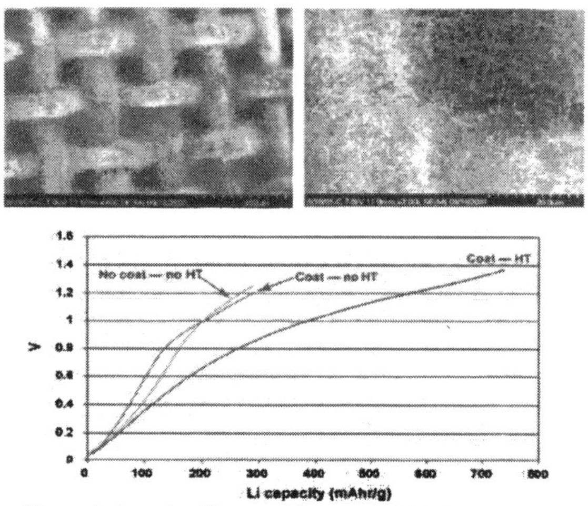

Figure 3. Scanning Electron Micrograph (SEM) Images of Carbon Nanotubes (CNTs) Synthesized Directly Upon Stainless Steel (SS) Mesh. The lower plot shows that the CNT surfaces increases the Li ion capacity beyond the theoretical limit of

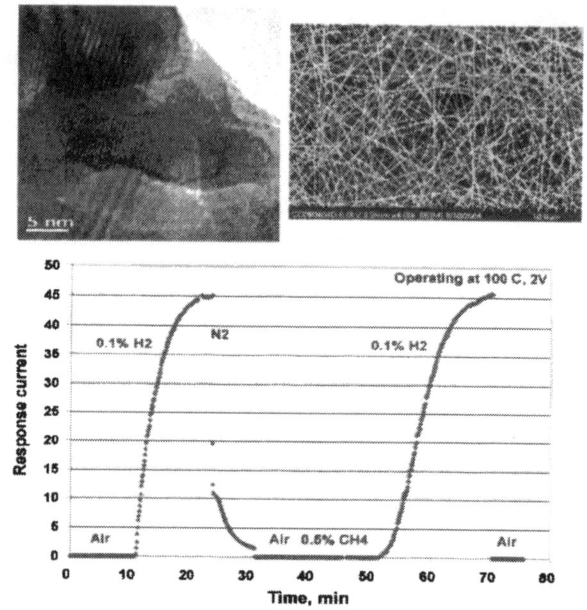

Figure 4. The above HRTEM (left) and SEM (right) images show polycrystalline nanofibers (1-d sensing elements) with demonstrated high sensitivity and selectivity to H2, at modest operating temperature with excellent cycling behavior.

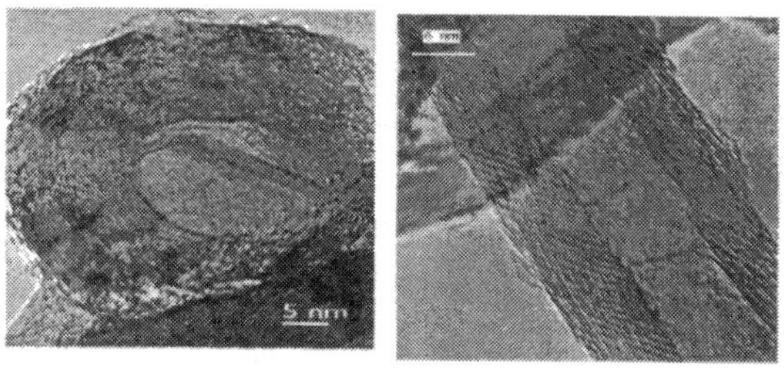

Figure 5. HRTEM images of nanocarbons that have been tested for use in nanofluids. Increases of ~ 20% in base fluid (water) thermal conductivity were measured for ~ 0.25 wt.% addition.

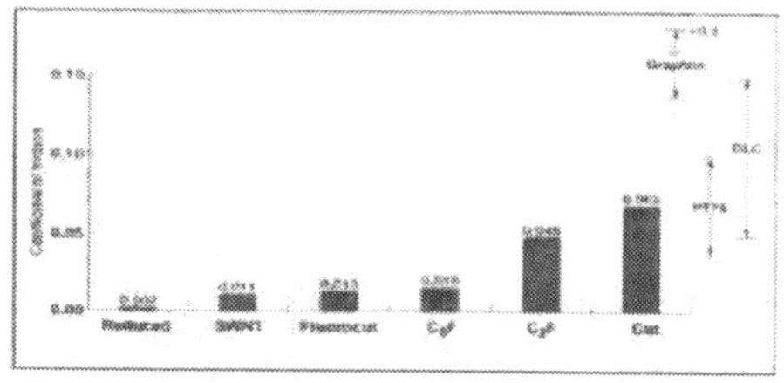

Figure 6. The bar-graph summarizes the coefficient of friction for SWNT nanomaterials with varied fluorine content. Relative to powdered graphite, DLC and even Teflon, the nanomaterials proved superior. A HRTEM image of these materials is also shown.

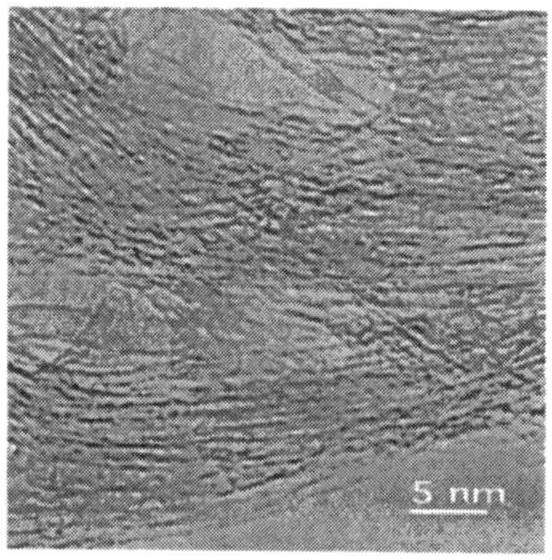

TechConnect Briefs 2015, TechConnect.org, ISBN 978-1-4987-4728-8

Index of Authors

Index of Keywords

Made in the USA
Monee, IL
13 September 2026

62976341R00181